GEOMETRY FROM EUCLID TO KNOTS

SAUL STAHL
UNIVERSITY OF KANSAS

DOVER PUBLICATIONS, INC.
Mineola, New York

Bibliographical Note

This Dover edition, first published in 2010, is an unabridged republication of the work originally published in 2003 by Pearson Education, Inc., Upper Saddle River, New Jersey. The author has provided a new Preface for this edition. This book can also be downloaded from http://www.math.ku.edu/stahl/.

Library of Congress Cataloging-in-Publication Data

Stahl, Saul
Geometry : from Euclid to knots / Saul Stahl.
p. cm.
Originally published: Upper Saddle River, N.J. : Pearson Education, 2003.
Includes bibliographical references and index.
ISBN-13: 978-0-486-47459-5
ISBN-10: 0-486-47459-3
1. Geometry. I. Title.

QA445.S74 2010
516—dc22

2009037891

Manufactured in the United States by LSC Communications
47459303 2019
www.doverpublications.com

To Aunt Chaya

Two tales, each with a moral that may have been swamped by the spirit of the times:

Egypt's king Ptolemy Soter once asked Euclid whether there was no shorter way in geometry than that of the *Elements.* "There is no royal road to geometry," replied Euclid.

A student, after learning the very first proposition in geometry, wanted to know what he would get by learning these things. Euclid then called his slave and said, "Give him a penny since he must needs make gains by what he learns."

From Isaac Todhunter's preface to *Euclid for the Use of Schools and Colleges* (published in 1874, more than two millennia after Euclid's death):

Numerous attempts have been made to find an appropriate substitute for the *Elements* of Euclid; but such attempts, fortunately, have hitherto been made in vain. The advantages attending a common standard of reference in such an important subject, can hardly be overestimated; and it is extremely improbable, if Euclid were once abandoned, that any agreement would exist as to the author who should replace him. It cannot be denied that defects and difficulties occur in the *Elements* of Euclid, and that these become more obvious as we examine the work more closely; but probably during such examination the conviction will grow deeper that these defects and difficulties are due in great measure to the nature of the subject itself, and to the place which it occupies in a course of education; and it may readily be believed that an equally minute criticism of any other work on geometry would reveal more and graver blemishes.

Contents

Preface to the Dover Edition

It gives me particular pleasure to have this book republished by Dover Publications, Inc. After all, they do list Euclid's *The Elements*, the most influential mathematical book of all time, in their catalog, and it is hard not to derive some satisfaction from this association. Especially so, as I have had so many occasions to actually consult this source in the preparation of various lectures. The mathematical world owes Dover a debt of gratitude for making this text so readily available.

I must confess that when I opened Euclid's book for the first timed during my undergraduate years, I felt rather disappointed. The questionable definitions, the lack of symbolic notation, and, most damning, the seemingly nitpicking proofs of obvious facts, all combined to dull my interest. It was not until I was assigned to teach a course in modern geometries that I returned to Euclid's opus. This time I knew more about non-Euclidean geometry, as well as mathematics in general, and so could appreciate it better. It is for this reason that the first chapter of this book is devoted to the concrete and plausible description of several alternative, non-Euclidean, geometries. Familiarity with other geometries shows the students that the obvious need not be true. This demonstrates to them the need for logical proofs. That this can be accomplished by introducing the students to the geometry of the upper halfplane, a mathematical tool that played an essential role in the recent resolutions of Fermat's and Poincar's conjectures, as well as in the classification of the finite simple groups, comes as a bonus.

The readers who know of my junior/senior level textbooks in algebra and real analysis might wonder why in writing this book I chose to depart from the historical, or evolutionary, approach to pedagogy. The reason is that since I myself found it impossible to conceive of a hyperbolic geometry without a model, I did not deem it realistic to expect my students to do so. On the other hand, perhaps it could be said that my approach in this book is also historical in the sense that it acknowledges the enormous difficulties the mathematical community experienced in trying to resolve the question of the status of Euclid's fifth postulate. It took two thousand years to do so, and not for lack of trying.

Saul Stahl
December 2009

Preface

Organization

The main purpose of this book is to provide prospective high school mathematics teachers with the geometric background they need. Its core, consisting of Chapters 2 to 5, is therefore devoted to a fairly formal (that is, axiomatic) development of Euclidean geometry. Chapters 6, 7, and 8 complement this with an exposition of transformation geometry. The first chapter, which introduces teachers-to-be to non-Euclidean geometries, provides them with a perspective meant to enhance their appreciation of axiomatic systems. The much more informal Chapters 9 through 12 are meant to give students a taste of some more recent geometric discoveries.

The development of synthetic Euclidean geometry begins by following Euclid's *Elements* very closely. This has the advantage of convincing students that they are learning "the real thing." It also happens to be an excellent organization of the subject matter. Witness the well-known fact that the first 28 propositions are all neutral. These subtleties might be lost on the typical high school student, but familiarity with Euclid's classic text must surely add to the teacher's confidence and effectiveness in the classroom. I am also in complete agreement with the sentiments Todhunter displayed in the previous excerpt: No other system of teaching geometry is better than Euclid's, provided, of course that his list of propositions is supplemented with a sufficient number of exercises. Occasionally, though, because some things have changed over two millennia, or else because of errors in the *Elements*, it was found advantageous either to expound both the modern and ancient versions in parallel or else to part ways with Euclid.

In order to convince prospective teachers of the need to prove "obvious" propositions, the axiomatic development of Euclidean geometry is preceded by the informal description of both spherical and hyperbolic geometry. The trigonometric formulas of these geometries are included in order to lend numerical substance to these alternate and unfamiliar systems. The part of the course dedicated to synthetic geometry covers the standard material about triangles, parallelism, circles, ratios, and similarity; it concludes with the classic theorems that lead to projective geometry. These lead naturally to a discussion of ideal points and lines in the extended plane. Experience indicates that the nonoptional portions of the first five chapters can be completed in about three quarters of a one semester course. During that time the typical weekly homework assignment called for about a dozen proofs.

Chapters 6 and 7 are concerned mostly with transformation geometry and symmetry. The planar rigid motions are completely and rigorously classified. This is followed by an informal discussion of frieze patterns and wallpaper designs. Inversions are developed formally and their utility for both Euclidean and hyperbolic geometry is explained.

The exposition in Chapters 8 through 12 is informal in the sense that few proofs are either offered or required. Their purpose is to acquaint students with some of the geometry that was developed in the last two centuries. Care has been taken to supply a great number of calculational exercises that will provide students with hands-on experience in these advanced topics.

The purpose of Chapter 8 is threefold. First there is an exposition of some interesting facts, such as Euler's equation and the Platonic and Archimedian solids. This is followed by a representation of the rotational symmetries of the regular solids by means of permutations, a discussion of their symmetry groups, and a visual definition of isomorphism. Both of these discussions aim to develop the prospective teacher's ability to visualize three dimensional phenomena. Finally comes the tale of Monstrous Moonshine.

Chapter 9 consists of a short introduction to the notions of homeomorphism and isotopy. Chapter 10 acquaints students with some standard topics of graph theory: traversability, colorability, and planarity. The topology of surfaces, both of the closed and the bordered varieties, is the subject of Chapter 11. Algorithms are described for identification of the topological type of any bordered surface. Two knot theoretic invariants are described in Chapter 12, including the recently discovered Jones polynomial.

Exercises

In Chapters 2 to 5 exercises are listed following every two or three propositions. This helps the professor select appropriate homework assignments and eliminates some of the guesswork for students. The great majority of these exercises call for straightforward applications of the immediately preceding propositions. On the other hand, the chapter review exercises provide problems for which the determination of the applicable propositions does require thought. In the remaining and less formal chapters the exercises appear at the end of each section. Each chapter concludes with a list of review problems. Solutions and/or hints to selected exercises are provided at the end of the book.

The exercises that are interspersed with the propositions are of four types. There are relational and constructive propositions in whose answers the students should adhere to the same format that is used in the numbered propositions. The third type of exercises has to do with the alternate spherical, hyperbolic, and taxicab geometries; in these the appropriate response usually consists of one or two English sentences. The fourth, and last, type of exercises calls for the use of some computer program, and these are marked with a (C).

In the other chapters, namely Chapter 1 and Chapters 6 through 12, exercises are listed at the end of each section. The emphasis in these is on the algorithmic aspect of geometry. They mostly require the straightforward, albeit nontrivial, application of the methods expounded in the text.

Notation and Conventions

Chapters 2 to 5 of this text present most of the content of Book I and selected topics from Books, II, III, IV, and VI of *The Elements.* In addition to the conventional labeling of propositions by *chapter, section,* and *number* these propositions are also identified by a parenthesized *roman numeral, number* that pinpoints their appearance in Euclid's book. For example, the Theorem of Pythagoras is listed as Proposition 3.3.2(I.47). Exercises are identified in a similar manner: Exercise 5.3B.6 is the sixth exercise in group B of the third section in Chapter 5. The symbol (C) is used to distinguish exercises that call for the use of computer applications.

The justifications for the various steps of the construction and proof are stated, in abbreviated form, in brackets. The abbreviations used are DFN for definition, PT for postulate, CN for common notion, and PN for proposition. The symbol $\therefore$ is used as an abbreviation for the word *therefore.* Optional sections or propositions are labeled with an asterisk. Difficult exercises are also designated by an asterisk. Propositions whose proof lies outside the scope of the text are followed by the symbol $\square$.

Acknowledgments

The author wishes to acknowledge the many valuable suggestions and corrections provided by Mark Hunacek as well as the technical help of his colleague Pawel Szeptycki. Jack Porter taught from an early draft and provided some corrections and useful suggestions. Dan Archdeacon and Marisa Debowsky helped improve portions of the manuscript. The final form of the book owes much to the suggestions of Prentice Hall's Acquisitions Editor George Lobell, Production Editor Bayani M. DeLeon, Arthur T. White, Larry W. Cusick, John Golden, and other reviewers who chose to remain anonymous. Larisa Martin converted my manuscript to LaTeX. I owe a debt of gratitude to them all.

Saul Stahl
stahl@math.ukans.edu

Credits

Illustrations

All the figures were produced by Adobe Illustrator®.

The cartoons on pp. xviii, 85, 267, and 331 are reprinted with the permission of their creator Sidney Harris.

The cartoon on p. 175 is reprinted with the gracious permission of its creator Susan H. Stahl.

Figure 6.26 is reprinted, by permission, from Owen Jones, *The Complete "Chinese Ornament,"* p. 21. Copyright © 1990 by Dover Publications, Inc.

Figure 6.36 is reprinted, by permission, from W. and G. Audsley, *Designs and Patterns from Historic Documents*, Plates 19 and 20. Copyright © 1968 by Dover Publications, Inc.

The figures in Exercises 1–34 of Section 6.7 were generated with software written at the Geometry Center, University of Minnesota.

Figures 8.1, 8.4–8.6, 8.12–8.23 were imported from Mathematica®.

Figure 8.2 is reprinted from Ernst H. P. A. Haeckel, *Report on the Scientific Results of the Voyage of H. M. S. Challenger*, Vol. XVIII, Part 3, Pl. 117. H. M. S. O., 1887.

Figure 8.3 is reprinted, by permission, from David Wells, *The Penguin Dictionary of Curious and Interesting Geometry*, pp. 6–7. Copyright © 1991 by David Wells.

Quotations

The statements of Euclid's definitions, postulates, common notions, as well as many of the propositions in Chapters 2, 3, and 4 are reprinted, by permission, from Euclid, *The Thirteen Books of the Elements* (Sir Thomas L. Heath, translator), 2nd edition. Copyright © 1956 by Dover Publications, Inc.

The recognition chart for wallpaper symmetry groups is reprinted at the end of Chapter 6, by permission, from Doris Schattschneider's article "The Plane Symmetry Groups: Their Recognition and Notation," *Amer. Math. Monthly*, **85**(1978), p. 443. Copyright © 1978 by The American Mathematical Monthly.

Appendix C is modified from G. D. Birkhoff and R. Beatley, *Basic Geometry*, 3rd edition. Copyright © 1959 by Chelsea Publishing Company.

Appendix D is reprinted from *UCSMP Geometry* by Arthur Coxford, Zalman Usiskin, and Daniel Hirschorn. © 1991 by Scott, Foresman and Company. Published by Prentice Hall, Inc. Used by permission of Pearson Education, Inc.

Appendix E is modified, by permission, from David Hilbert, *Foundations of Geometry* (L. Unger, translator). Copyright © 1971 by Open Court Publishing Co.

CHAPTER I
Call me Euclid.
S. Harris

Chapter 1

Other Geometries: A Computational Introduction

In order to provide a better perspective on Euclidean geometry, three alternative geometries are described. These are the geometry of the surface of the sphere, hyperbolic geometry, and taxicab geometry.

1.1 Spherical Geometry

Due to its relationship with geography and astronomy, spherical geometry was studied extensively by the Greeks as early as 300 B.C. Menelaus (ca. 100) wrote the book *Spherica* on spherical trigonometry, which was greatly extended by Ptolemy (100–178) in his *Almagest*. Many later mathematicians, including Leonhard Euler (1707–1783) and Carl Friedrich Gauss (1777–1855), made substantial contributions to this topic. Here it is proposed only to compare and contrast this geometry with that of the plane. Because the time to develop spherical geometry in the same manner as will be done with Euclidean geometry is not available, this discussion is necessarily informal and frequent appeals will be made to the readers' visual intuition.

Strictly speaking there are no straight lines on the surface of a sphere. Instead it is both customary and useful to focus on curves that share the "shortest distance" property with the Euclidean straight lines. The following thought experiment will prove instructive for this purpose. Imagine that two pins have been stuck in a smooth sphere in points that are not diametrically opposite and that a (frictionless) rubber band is held by the pins in a stretched state. Rotate this sphere until one of the two pins is directly

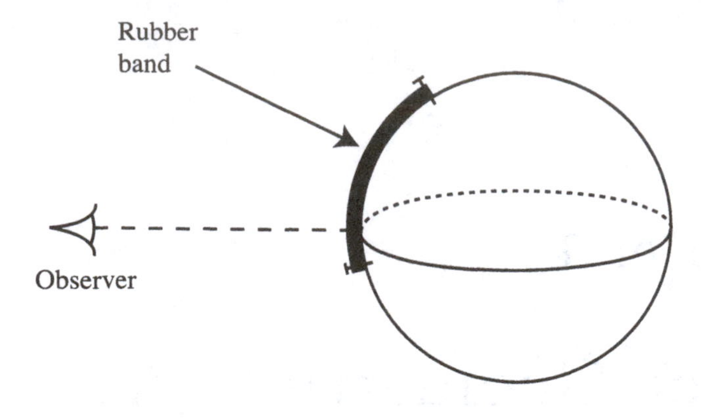

Figure 1.1. A geodesic on the sphere

above the other right in front of your mind's eye. It is then hard to avoid the conclusion that the rubber band will be stretched out along the sphere in the plane formed by the two pins and the eye—the plane of the book's page in Figure 1.1. The inherent symmetry of the sphere dictates that this plane should cut the sphere into two identical hemispheres; in other words, that this plane should pass through the center of the sphere. It is also clear that the tension of the stretched rubber band forces it to describe the shortest curve on the surface of the sphere that connects the two pins. The following may therefore be concluded.

Proposition 1.1.1 (Spherical geodesics) *If A and B are two points on a sphere that are not diametrically opposite, then the shortest curve joining A and B on the sphere is an arc of the circle that constitutes the intersection of the sphere with a plane that contains the sphere's center.* □

Such circles are called *great circles* and these arcs are called *great arcs* or *geodesic segments.* They are the spherical analogs of the Euclidean line segments.

Diametrically opposite points on the sphere present a dilemma. A stretched rubber band joining them will again lie along a great circle, but this circle is no longer uniquely determined since these points can clearly be joined by an infinite number of great semicircles. For example, assuming for the sake of argument that the earth is an exact sphere, each meridian is a great semicircle that joins the North and South Poles. Hence, the aforementioned analogy between the geodesic segments on the sphere and Euclidean line segments is not perfect. It is necessary either to exclude such meridians from the class of geodesic segments or else to accept that some points can

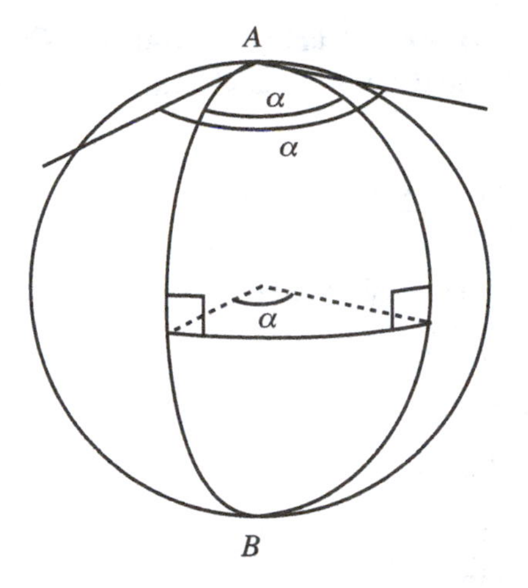

Figure 1.2. The lune α

be joined by many such segments. The first alternative is the one chosen in this text. Thus, by definition, the endpoints of geodesic segments on the sphere are never diametrically opposite.

Next, the spherical analog of the angle is defined. Any two great semicircles that join two diametrically opposite points A and B, but are not contained in the same great circle, divide the sphere into two portions each of which is called a *lune*, or a *spherical angle* (Fig. 1.2). The measure of the spherical angle is defined to be the measure of the angle between their tangent lines at A (or at B). Alternately, this equals the measure of the angle formed by the radii from the center of the sphere to the midpoints of the bounding great semicircles. For example, each meridian forms a 90° angle with the equator at their point of intersection.

In the Euclidean plane the relationships between lengths of straight line segments and measures of angles are given by well-known trigonometric identities. Some fundamental theorems of spherical trigonometry are now stated without proof.

Any three points A, B, C on the sphere, no two of which are diametrically opposite, constitute the *vertices* of a *spherical triangle* denoted by $\triangle ABC$. The three *sides* of this triangle are the geodesic segments that join each pair. The sides opposite the vertices A, B, C (and their lengths) are denoted a, b, c respectively. The *interior angle* α at the vertex A is the lune between b and c. The interior angles β and γ at B and C are defined in a similar manner. The term *spherical geometry* refers to the geometry of the sphere of radius 1.

Proposition 1.1.2 (Spherical trigonometry) *On a sphere of radius $R = 1$, let $\triangle ABC$ be a spherical triangle with sides a, b, c and interior angles α, β, γ. Then*

(i) $\cos\alpha = \dfrac{\cos a - \cos b \cos c}{\sin b \sin c}$

(i′) $\cos a = \cos b \cos c + \cos\alpha \sin b \sin c$

(ii) $\cos a = \dfrac{\cos\alpha + \cos\beta\cos\gamma}{\sin\beta\sin\gamma}$

(ii′) $\cos\alpha = \cos a \sin\beta\sin\gamma - \cos\beta\cos\gamma$

(iii) $\dfrac{\sin\alpha}{\sin a} = \dfrac{\sin\beta}{\sin b} = \dfrac{\sin\gamma}{\sin c}$ □

These are known as the first spherical law of cosines, the second spherical law of cosines, and the spherical law of sines. It should be noted that i and i′ are really the same equation as are ii and ii′, although, as will be demonstrated by the following examples, their uses are different. The *solution* of a triangle consists of the lengths of its sides and the measures of its interior angles. For any side a, radian units should be used in the evaluation of $\sin a$ and $\cos a$.

Example 1.1.3 Solve the spherical triangle with sides $a = 1$, $b = 2$, and $c = \pi/2$.

It follows from the first spherical law of cosines that

$$\cos\alpha = \frac{\cos 1 - \cos 2 \cos \pi/2}{\sin 2 \sin \pi/2} = \frac{\cos 1}{\sin 2}$$

so that

$$\alpha = \cos^{-1}\left(\frac{\cos 1}{\sin 2}\right) \approx 53.54°.$$

The angles β and γ are similarly shown to have measures 119.64° and 72.91°.

Example 1.1.4 On a sphere of radius 4000 miles, solve the triangle in which an interior angle of 50° lies between sides of lengths 7000 miles and 9000 miles, respectively.

Taking the radius as the unit, it follows that we may set

$$b = \frac{7000}{4000} = 1.75$$

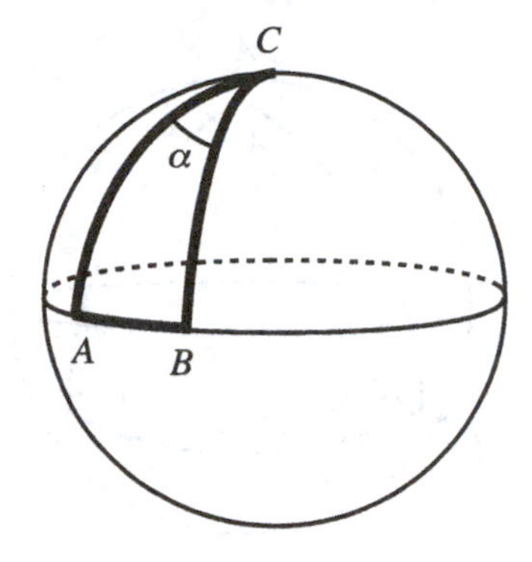

Figure 1.3. A thin spherical triangle

$$c = \frac{9000}{4000} = 2.25.$$

Hence, from the first law of cosines,

$$a = \cos^{-1}(\cos \alpha \sin b \sin c + \cos b \cos c) \approx .9221 \approx 3688 \text{ miles}.$$

Now that all three sides of the triangle are known, the method of the previous example yields

$$\beta = \cos^{-1}\left(\frac{\cos b - \cos a \cos c}{\sin a \sin c}\right) \approx 71.05^\circ$$
$$\gamma = \cos^{-1}\left(\frac{\cos c - \cos a \cos b}{\sin a \sin b}\right) \approx 131.58^\circ.$$

Note that in both of the preceding examples the sum of the angles of the spherical triangle exceeds 180°. That is in fact true for all spherical triangles.

Proposition 1.1.5 *The sum of the angles of every spherical triangle lies strictly between* 180° *and* 540°. □

A spherical triangle the sum of whose angles is close to 180° is formed by the equator together with two close meridians. Thus, the sum of the angles of the spherical $\triangle ABC$ of Figure 1.3 is $90^\circ + 90^\circ + \alpha = 180^\circ + \alpha$. A spherical triangle $A'B'C'$ with angle sum near 540° is described in Figure 1.4, where A, B, C are points that are equally spaced along a great circle. As A', B', C' approach A, B, C, respectively, the angles they form are flattened out and come arbitrarily close to 180° each. For example, since

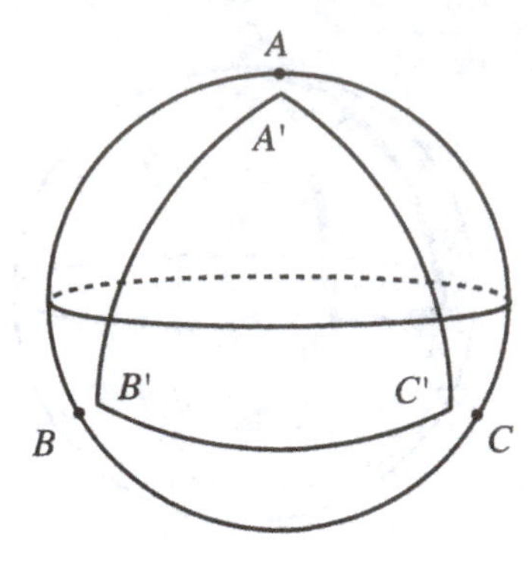

Figure 1.4. A nearly maximal spherical triangle

the spherical distance between any two of the points A, B, C is $2\pi/3$, it might be assumed that the spherical distance between any two of the points A', B', C' is $a = 2\pi/3 - 0.00001$, in which case each of the angles of $\triangle A'B'C'$ is

$$\cos^{-1}\left(\frac{\cos a - \cos a \cos a}{\sin a \sin a}\right) \approx 179.52^\circ$$

and their sum is 538.56°.

Since, by definition, each of the interior angles of the spherical triangle is less than 180°, it follows that the sum of these angles can never equal 540°. Similarly, as will be shown momentarily, the sum of these angle cannot equal the lower bound of 180° either.

The area of the spherical triangle is also of interest. An elegant proof of this formula is offered in Section 3.2.

Proposition 1.1.6 *If a triangle on a sphere of radius R has angles with radian measures α, β, γ then it has area $(\alpha + \beta + \gamma - \pi)R^2$.* □

For example, the spherical triangle formed by the equator, the Greenwich meridian and the 90° East meridian has all of its angles equal to $\pi/2$ and hence its area is

$$\left(\frac{\pi}{2} + \frac{\pi}{2} + \frac{\pi}{2} - \pi\right) R^2 = \frac{\pi R^2}{2}.$$

This answer is consistent with the fact that the said triangle constitutes one fourth of a hemisphere. Since the surface area of the sphere is $4\pi R^2$, this triangle has area

$$\frac{1}{4}\,\frac{4\pi R^2}{2} = \frac{\pi R^2}{2}$$

which agrees with the previous calculation.

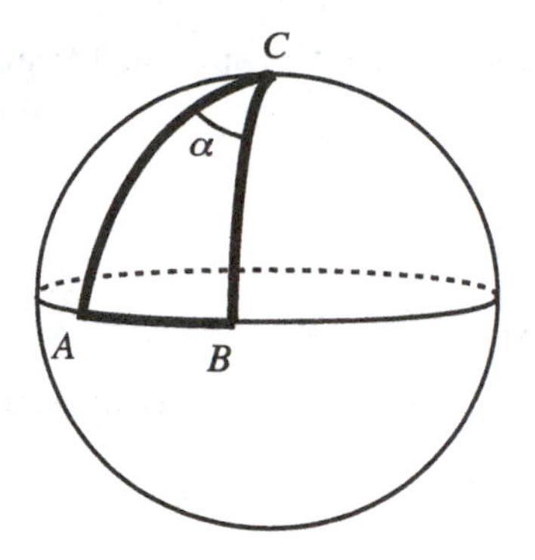

Figure 1.5. A spherical triangle

The quantity $\alpha + \beta + \gamma - \pi$ is called the *excess* of the spherical $\triangle ABC$. The preceding theorem in effect states that the area of a spherical triangle is proportional to its excess. This assertion is supported by the triangle of Figure 1.5, which has excess $\pi/2 + \pi/2 + \alpha - \pi = \alpha$ and whose area is clearly proportional to α as long as A and B vary along the equator and C remains at the North Pole.

This area formula can be used to close a gap in the preceding discussion. Since every triangle has positive area, it follows that the sum of the angles of a spherical triangle never equals π or 180°, although, as was seen previously, it can come arbitrarily close to this lower bound.

Exercises 1.1

1. Let ABC be a spherical triangle with a right angle at C. Use the formulas of spherical trigonometry to prove the following:

   (a) $\sin a = \sin\alpha \sin c$

   (b) $\tan a = \tan\alpha \sin b$

   (c) $\tan a = \cos\beta \tan c$

   (d) $\cos c = \cos a \cos b$

   (e) $\cos\alpha = \sin\beta \cos a$

   (f) $\sin b = \sin\beta \sin c$

   (g) $\tan b = \tan\beta \sin a$

   (h) $\tan b = \cos\alpha \tan c$

   (i) $\cos c = \cot\alpha \cot\beta$

   (j) $\cos\beta = \sin\alpha \cos b$

2. On a sphere of radius R, solve the spherical triangle with angles:

   (a) $60°$, $70°$, $80°$
   (b) $70°$, $70°$, $70°$
   (c) $120°$, $150°$, $170°$
   (d) θ, θ, θ where $60° < \theta < 120°$.

3. On a sphere of radius R, solve the spherical triangle with sides:

   (a) R, R, R
   (b) R, $1.5R$, $2R$
   (c) $\pi R/2$, $\pi R/2$, $\pi R/2$
   (d) d, d, d, where $0 < d < 2\pi R$
   (e) $.2R$, $.3R$, $.4R$
   (f) $.02R$, $.03R$, $.04R$
   (g) $2R$, $3R$, $4R$

4. On a sphere of radius R, solve the spherical triangle with:

   (a) $a = .5R$, $\beta = 60°$, $\gamma = 80°$
   (b) $b = R$, $\alpha = 40°$, $\gamma = 100°$
   (c) $a = 2R$, $\beta = \gamma = 10°$
   (d) $a = 2R$, $\beta = \gamma = 170°$

5. On a sphere of radius R, solve the spherical triangle with:

   (a) $a = 2R$, $b = R$, $\gamma = 100°$
   (b) $b = .5R$, $c = 1.2R$, $\alpha = 100°$
   (c) $a = 2R$, $b = R$, $\gamma = 120°$
   (d) $b = .5R$, $c = 1.2R$, $\alpha = 120°$

6. On a sphere of radius 75 cm, solve the spherical triangle with:

   (a) $a = 100$ cm, $b = 125$ cm, $c = 140$ cm
   (b) $\alpha = 100°$, $\beta = 125°$, $\gamma = 140°$
   (c) $\alpha = 100°$, $b = 125$ cm, $c = 125$ cm
   (d) $a = 100$ cm, $\beta = 125°$, $\gamma = 125°$

7. Evaluate the limits of the angles of the spherical triangles below both as $x \to 0$ and as $x \to \pi$.

   (a) $a = b = c = x$
   (b) $a = b = x$, $c = 2x$

Figure 1.6. The shrinkage that defines the hyperbolic plane

(c) $a = x$, $b = c = 2x$

8. Which of the following congruence theorems hold for spherical triangles? Justify your answer.

   (a) SSS

   (b) SAS

   (c) ASA

   (d) SAA

   (e) AAA

1.2 Hyperbolic Geometry

Imagine a two-dimensional universe, with a superimposed Cartesian coordinate system, in which the x-axis is infinitely cold. Imagine further that as the objects of this universe approach the x-axis, the drop in temperature causes them to contract (see Fig. 1.6). Thus, the inhabitants of this fictitious land will find that it takes them less time to walk along a horizontal

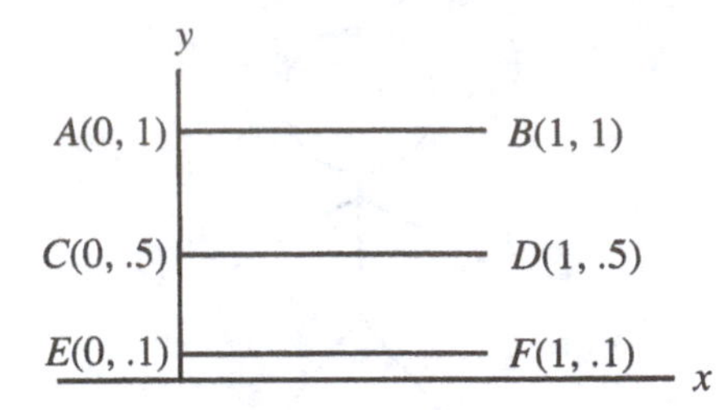

Figure 1.7. Paths of unequal hyperbolic lengths

line from $A(0,1)$ to $B(1,1)$ (Fig. 1.7) than it takes to walk along a horizontal line from $C(0,.5)$ to $D(1,.5)$. Since their rulers contract just as much as they do, this observation will not seem at all paradoxical to them. If it is assumed that the contraction is such that the outside observer sees the length of any object as being proportional to its distance from the x-axis, then the inhabitants will find that walking from $C(0,.5)$ to $D(1,.5)$ takes twice as long as walking from $A(0,1)$ to $B(1,1)$ and one-fifth of the time of walking from $E(0,.1)$ to $F(1,.1)$. To differentiate between the Euclidean length of such a segment and its length as experienced by these fictitious beings, it is customary to refer to the latter as the *hyperbolic length* of the segment. Accordingly, the hyperbolic lengths of the segments AB, CD, and EF of Figure 1.7 are 1, 2, and 10 respectively. It is customary to restrict this geometry to the *upper half-plane* [i.e., those points (x,y) for which $y > 0$]. In this half-plane, the hyperbolic length of a horizontal line segment at distance y from the x-axis is given by the formula

$$\text{hyperbolic length} = \frac{\text{Euclidean length}}{y}. \tag{1}$$

Other curves also have a hyperbolic length and a method for computing it is given in Exercise 16.

Not surprisingly, perhaps, the Euclidean straight line segment joining two points does not constitute the curve of shortest hyperbolic length between them. When setting out from $A(0,1)$ to $B(1,1)$ the inhabitants of this strange land may find that if they bear a little to the north their journey will be somewhat shorter because, unbeknownst to them, their legs are longer on this route (Fig. 1.8). However, if they stray too far north the length of the detour will offset any advantages gained by the elongation of their stride and they will find the length of the tour to be excessive. They are therefore faced with a trade-off problem. Some deviation to the north will shorten the duration of the trip from A to B, but too much will extend it. Which path, then, is it that makes the trip as short as possible?

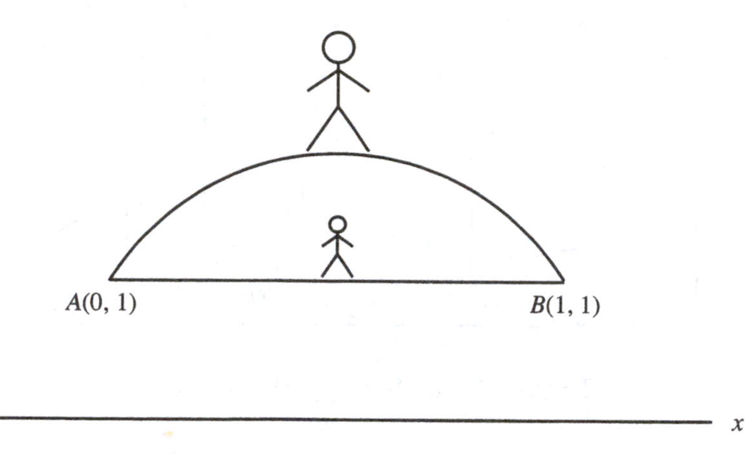

Figure 1.8. Which path has shorter hyperbolic length?

The answer to this question is surprisingly easy to describe, though not to justify. The path of shortest hyperbolic length that connects $A(0,1)$ to $B(1,1)$ is the arc of the circle that is centered at $(.5,0)$ and contains A and B (Fig. 1.9). Its hyperbolic length turns out to be $0.962\ldots$ in contrast with the hyperbolic length 1 of the horizontal segment AB. More drastically, the arc of the semicircle centered at $(50,0)$ that joins the points $A(0,1)$ and $X(100,1)$ has hyperbolic length 9.21, a mere 9% of the hyperbolic length of the segment AX.

Given any two points, their *hyperbolic distance* is the minimum of the hyperbolic lengths of all the curves joining them. As was the case for spherical geometry, the *geodesic segments* of hyperbolic geometry are those curves that realize the hyperbolic distance between their endpoints. Loosely speaking, *hyperbolic geometry*, or the *hyperbolic plane*, is the geometry that underlies the upper half-plane when distances are measured in accordance with Equa-

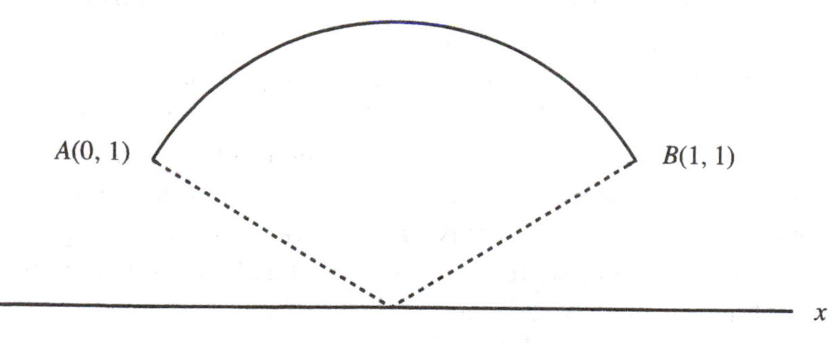

Figure 1.9. A hyperbolic geodesic

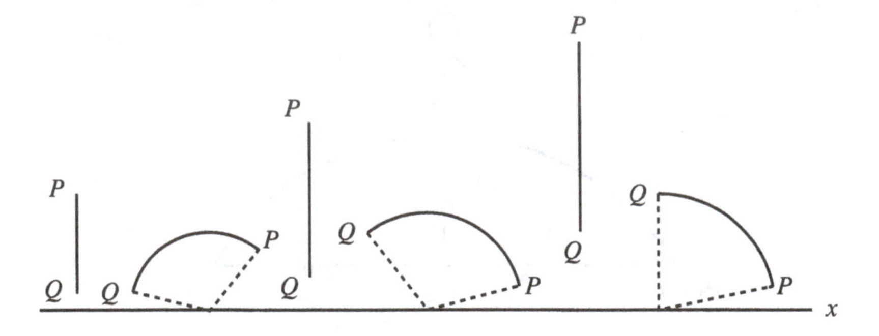

Figure 1.10. Six hyperbolic geodesics

tion (1). More precise definitions will be given in Sections 2.2 and 2.3. The upper half-plane is only one manifestation, or model, of hyperbolic geometry, and several others exist [see Greenberg, Stahl]. Nevertheless, for the purposes of this introductory exposition, readers may identify hyperbolic geometry with its representation as the distorted upper half-plane.

Proposition 1.2.1 (Hyperbolic geodesics) *The geodesic segments of the hyperbolic plane are arcs of circles centered on the x-axis and Euclidean line segments that are perpendicular to the x-axis.* □

The geodesics of the first variety are called *bowed geodesics*, whereas the vertical ones are the *straight geodesics* (see Fig. 1.10). This distinction is only meaningful to the outside observer. The inhabitants of this geometry perceive no difference between these two kinds of geodesics.

It so happens that as the inhabitants of the hyperbolic plane approach the x-axis they shrink at such a rate as to make the x-axis unattainable. Technically speaking, the hyperbolic lengths of all of the geodesic segments in Figure 1.10 diverge to infinity as the endpoints Q approach the x-axis. This claim will be given a quantitative justification at the end of this section.

A *hyperbolic angle* is the portion of the hyperbolic plane between two geodesic rays (Fig. 1.11). The measure of the angle between two geodesics is, by definition, the measure of the angle between the tangents to the geodesics at the vertex of the angle. Accordingly, two geodesics are said to form a hyperbolic right angle if and only if their tangents are perpendicular to each other as Euclidean straight lines (Fig. 1.12). Given any three points that do not lie on one hyperbolic geodesic, they constitute the vertices of a *hyperbolic triangle* formed by joining the vertices, two at a time, with hyperbolic geodesics (Fig. 1.13).

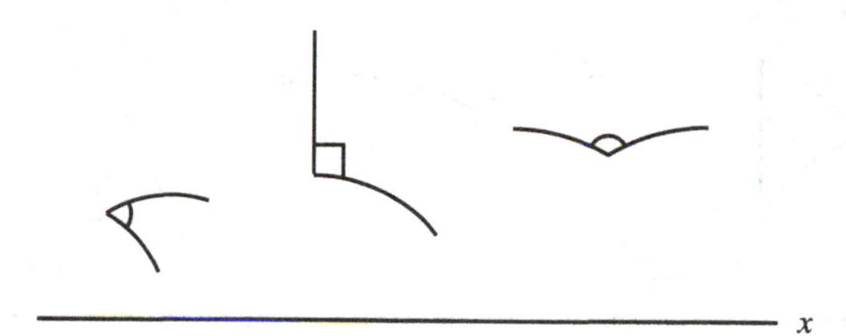

Figure 1.11. Three hyperbolic angles

The geometry of the hyperbolic plane has been studied extensively. Its trigonometric laws are surprisingly, not to say mysteriously, similar to those of spherical geometry.

Proposition 1.2.2 (Hyperbolic trigonometry) *Let $\triangle ABC$ be a hyperbolic triangle with sides a, b, c and interior angles α, β, γ. Then*

i) $\cos\alpha = \dfrac{\cosh b\cosh c - \cosh a}{\sinh b\sinh c}$

i′) $\cosh a = \cosh b\cosh c - \cos\alpha\sinh b\sinh c$

ii) $\cosh a = \dfrac{\cos\alpha + \cos\beta\cos\gamma}{\sin\beta\sin\gamma}$

ii′) $\cos\alpha = \cosh a\sin\beta\sin\gamma - \cos\beta\cos\gamma$

iii) $\dfrac{\sin\alpha}{\sinh a} = \dfrac{\sin\beta}{\sinh b} = \dfrac{\sin\gamma}{\sinh c}$. □

Example 1.2.3 Solve the hyperbolic triangle with sides $a = 1$, $b = 2$, and $c = \pi/2$.

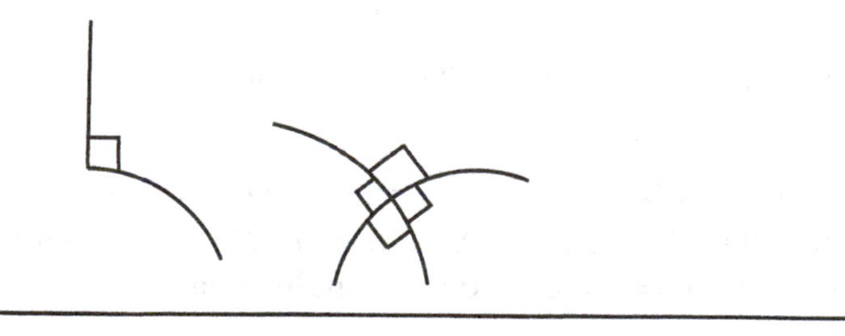

Figure 1.12. Hyperbolic right angles

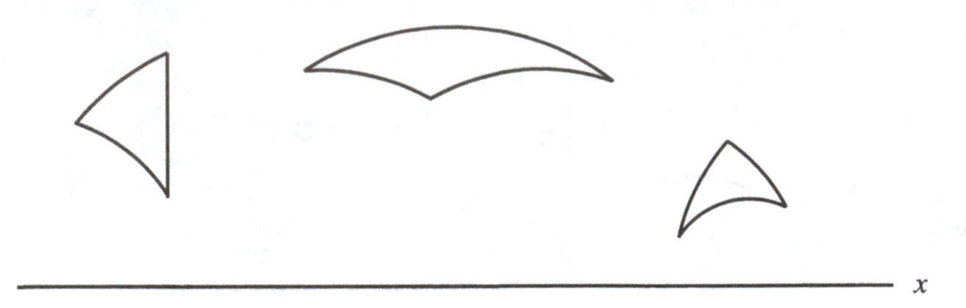

Figure 1.13. Three hyperbolic triangles

It follows from Formula i) of hyperbolic trigonometry that

$$\cos\alpha = \frac{\cosh 2\cosh\pi/2 - \cosh 1}{\sinh 2\sinh\pi/2} \approx .9461$$

and so $\alpha = \cos^{-1}(.9461) \approx 18.89°$. Similarly $\beta \approx 87.67°$ and $\gamma \approx 39.34°$.

Example 1.2.4 Solve the hyperbolic triangle with two sides of lengths 2, 3 respectively, if they are to include an angle of 30°.

Set $\alpha = 30°$, $b = 2$, $c = 3$. It follows from Formula i′) of hyperbolic trigonometry that

$$a = \cosh^{-1}(\cosh b\cosh c - \cos\alpha\sinh b\sinh c) = 2.545\ldots$$

Now that all three sides of the triangle are known, the method of the previous example yields

$$\beta = \cos^{-1}\left(\frac{\cosh a\cosh c - \cosh b}{\sinh a\sinh c}\right) \approx 16.64°$$
$$\gamma = \cos^{-1}\left(\frac{\cosh a\cosh b - \cosh c}{\sinh a\sinh b}\right) \approx 52.28°$$

As for the sum of the angles of a hyperbolic triangle, the situation is diametrically opposite to that on the sphere.

Proposition 1.2.5 *The sum of the angles of every hyperbolic triangle is less than* 180°. □

This proposition is borne out by the above two examples. Figure 1.14 demonstrates that this sum can be quite small. In fact, in the hyperbolic triangle with sides $a = b = c = 10$ each angle equals

$$\alpha = \cos^{-1}\left(\frac{\cosh 10\cosh 10 - \cosh 10}{\sinh 10\sinh 10}\right) \approx .77°.$$

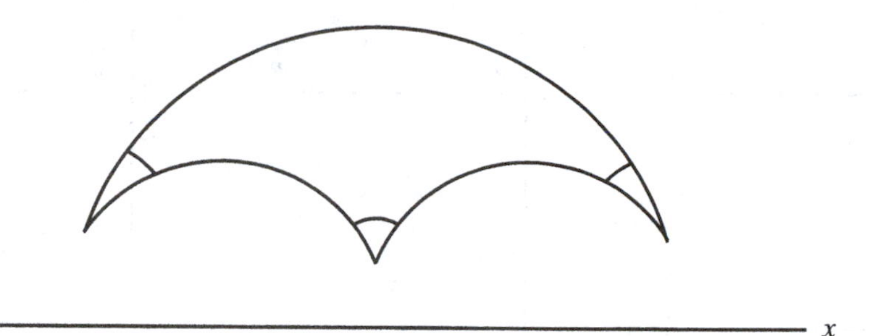

Figure 1.14. A hyperbolic triangle with three small angles

Example 1.2.6 Solve the hyperbolic triangle with $a = 2$, $\beta = \gamma = 60°$.

By Formula (ii′) of hyperbolic trigonometry,

$$\cos\alpha = \cosh 2 \sin 60° \sin 60° - \cos 60° \cos 60° \approx 2.57.$$

Since the cosine of an angle cannot exceed 1, such a hyperbolic triangle does not exist. Note that a Euclidean triangle with the same specifications does exist. Exercises 4 and 5 contain some related information.

The area of the hyperbolic triangle is, of course, of interest too. Its formula is quite surprising.

Proposition 1.2.7 *The area of the hyperbolic triangle whose angles have radian measures* α, β, γ *is* $\pi - \alpha - \beta - \gamma$. □

This formula is given some support by Figure 1.15. Note that the sum of the angles of the larger hyperbolic $\triangle ABC$ is less than the sum of the angles

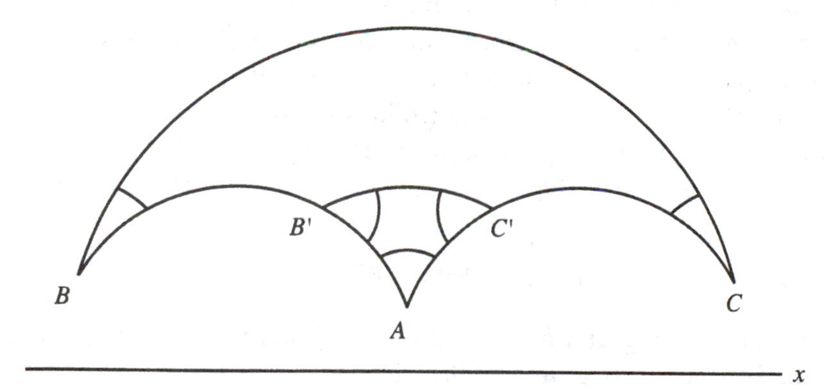

Figure 1.15. Two hyperbolic triangles with different areas and defects

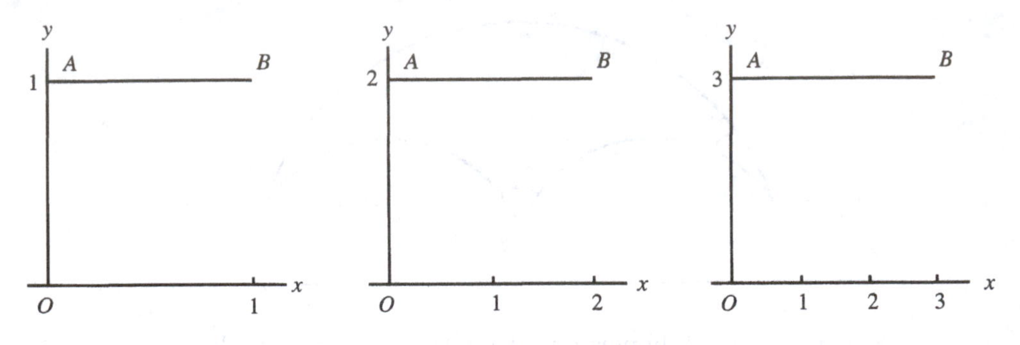

Figure 1.16. A curve with hyperbolic length 1

of the smaller hyperbolic $\triangle AB'C'$. The quantity $\pi - \alpha - \beta - \gamma$ is called, by analogy with its spherical counterpart, the *defect* of the hyperbolic triangle. Thus, the preceding theorem asserts that the area of a hyperbolic triangle is equal to its defect.

A somewhat peculiar aspect of the hyperbolic notion of length is its independence of the choice of unit of length. Regardless of what scale is chosen for the Cartesian coordinate system that is used to define the hyperbolic version of length, the hyperbolic distance between any two points remains the same. Note that the three parts of Figure 1.16 correspond to three different scales, and yet, according to Formula (1), in each of the three cases the hyperbolic length of AB is

$$\frac{1}{1} = \frac{2}{2} = \frac{3}{3} = 1.$$

It follows from this independence of scale that in hyperbolic geometry it is not necessary to specify units of length.

It might be instructive to show that this independence holds in the vertical direction as well. The hyperbolic length of a vertical segment of the hyperbolic plane is easily computed with the aid of calculus. For if dy denotes the Euclidean length of the (infinitesimally) small vertical line segment at height y above the x-axis, then its hyperbolic length is dy/y (see Fig. 1.17). Consequently the total hyperbolic length of the segment PQ is

$$\int_a^b \frac{dy}{y} = \ln b - \ln a = \ln \frac{b}{a}. \tag{2}$$

In particular, if $a = 1$ and $b = e = 2.718\ldots$, then the hyperbolic length of the y-axis between $P(0,1)$ and $Q(0,e)$ is

$$\ln \frac{e}{1} = \ln e = 1.$$

Moreover, if the scale on the axes is changed by a factor of s, then $P = (0, s)$ and $Q = (0, es)$ and again their hyperbolic distance is

$$\ln \frac{es}{s} = \ln e = 1.$$

Thus, if P and Q are points with coordinates $(0, 1)$ and $(0, e)$ relative to some unit of the Cartesian coordinate system, then the line segment joining P and Q has hyperbolic length 1 regardless of the scale that is actually used. Moreover, this Euclidean line segment PQ, which has hyperbolic length 1, is also a hyperbolic geodesic in contrast with the aforementioned segment AB which also has hyperbolic length 1 but is not a hyperbolic geodesic. Consequently, this geodesic can be taken as the *natural unit*, or *absolute unit*, of length of hyperbolic geometry.

Both spherical and hyperbolic geometry look very different from Euclidean geometry. Nevertheless, it is well known that on a large sphere, a small portion of the surface may be practically indistinguishable from a piece of a plane. This resemblance accounts for the fact that people first thought that the world was flat and small children still do so today. The same confusion could occur in the hyperbolic plane. If the portion of the hyperbolic plane that is subject to the direct experience and observation of its inhabitants is sufficiently small, their geometry would appear to them as practically indistinguishable from that of the Euclidean plane. This affinity between the hyperbolic and Euclidean planes is a topic that will be revisited many times in the subsequent discussion. The explanation of how the trigonometry of a small portion of the hyperbolic plane may be confusable with Euclidean trigonometry can be found in the references. At this point it will be demonstrated that, just like the Euclidean plane and in contrast

Figure 1.17. Hyperbolic length along the y-axis

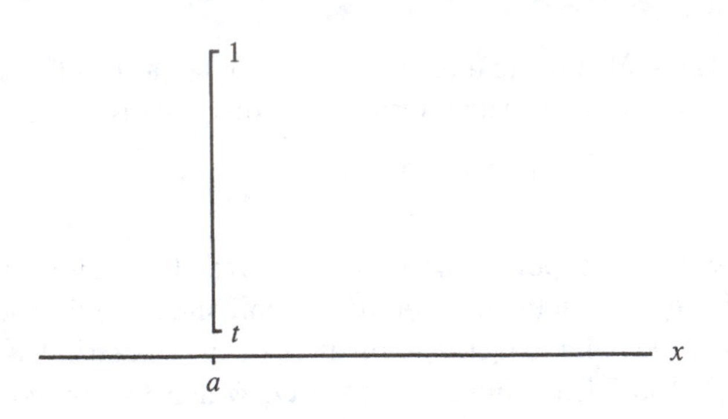

Figure 1.18. The hyperbolic plane extends indefinitely in all directions

with the sphere, hyperbolic geometry extends indefinitely in all directions. In other words, the inhabitants of the hyperbolic plane have no reason to suspect that part of their universe is missing. To see this, note that by Equation (2) the hyperbolic distance from the point $(a, 1)$ to the point (a, t) in Figure 1.18 is

$$\ln \frac{1}{t} = -\ln t.$$

Hence, if travel in the direction of the x-axis is simulated by letting t approach 0, then this quantity diverges to $-(-\infty) = \infty$. In other words, for the hyperbolic people the x-axis lies infinitely far away.

Hyperbolic Distance

The hyperbolic distance between any two points $A(x_1, y_1)$ and $B(x_2, y_2)$ can be determined by means of the following formulas:

(i) If $x_1 = x_2$ then the hyperbolic distance from A to B is

$$\left| \ln \frac{y_1}{y_2} \right|;$$

(ii) If $x_1 \neq x_2$, and $(c, 0)$ is the center of the geodesic segment that connects (x_1, y_1) and (x_2, y_2), and r is its radius, then the hyperbolic distance from A to B is

$$\left| \ln \frac{(x_1 - c - r)y_2}{(x_2 - c - r)y_1} \right|.$$

Example 1.2.8 The hyperbolic distance between the points $(5, 4)$ and $(5, 7)$ is $|\ln(4/7)| \approx 0.56$.

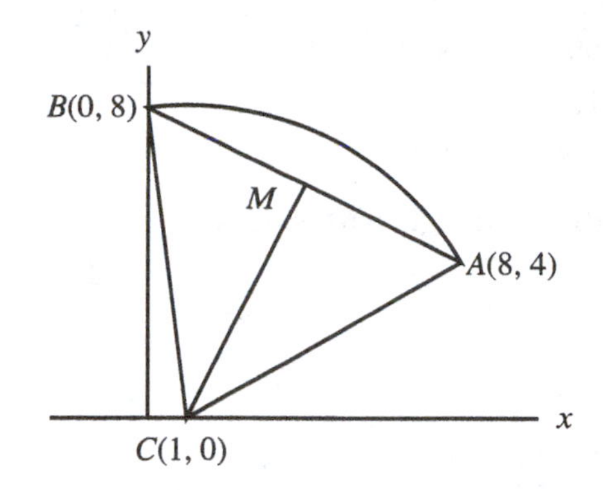

Figure 1.19. Computing the length of a hyperbolic geodesic

Example 1.2.9 To find the hyperbolic distance between $A(8,4)$ and $B(0,8)$ note that the line segment AB has slope $(8-4)/(0-8) = -1/2$ and midpoint $M(4,6)$ (see Fig. 1.19). Hence the perpendicular bisector of AB has equation $y-6 = 2(x-4)$ and is easily seen to intersect the x-axis in the point $C(1,0)$. Thus, $c = 1$ and $r = \sqrt{(0-1)^2 + (8-0)^2} = \sqrt{65}$. Hence the required distance is

$$\left| \ln \frac{(8-1-\sqrt{65})8}{(0-1-\sqrt{65})4} \right| \approx 1.45.$$

Exercises 1.2

1. Let ABC be a hyperbolic triangle with a right angle at C. Use the formulas of hyperbolic trigonometry to prove the following:

   (a) $\sinh a = \sin\alpha \sinh c$

   (b) $\tanh a = \tan\alpha \sinh b$

   (c) $\tanh a = \cos\beta \tanh c$

   (d) $\cosh c = \cosh a \cosh b$

   (e) $\cos\alpha = \sin\beta \cosh a$

   (f) $\sinh b = \sin\beta \sinh c$

   (g) $\tanh b = \tan\beta \sinh a$

   (h) $\tanh b = \cos\alpha \tanh c$

   (i) $\cosh c = \cot\alpha \cot\beta$

   (j) $\cos\beta = \sin\alpha \cosh b$

2. Solve the hyperbolic triangle with angles

   (a) $60°$, $50°$, $40°$

(b) $50°$, $50°$, $50°$

(c) $20°$, $50°$, $70°$

(d) θ, θ, θ where $0° < \theta < 60°$

3. Solve the hyperbolic triangle with sides

(a) 1, 1, 1

(b) 2, 3, 4

(c) 1/2, 1/2, 1/2

(d) d, d, d, where $0 < d$

(e) .2, .3, .4

(f) .02, .03, .04

4. Solve the hyperbolic triangle with

(a) $a = .5$, $\beta = 60°$, $\gamma = 40°$

(b) $b = .5$, $\alpha = 40°$, $\gamma = 50°$

(c) $a = 2$, $\beta = \gamma = 40°$

(d) $a = 10$, $\beta = \gamma = 40°$

(e) $a = 1$, $\beta = \gamma = 60°$

(f) $a = .1$, $\beta = \gamma = 60°$

(g) $a = .2$, $\beta = \gamma = 100°$

5. For which values of a does there exist a hyperbolic triangle with $\beta = \gamma = 60°$?

6. Solve the hyperbolic triangle with

(a) $a = 2$, $b = 1$, $\gamma = 30°$

(b) $b = .5$, $c = 1.2$, $\alpha = 120°$

(c) $a = 2$, $b = 1$, $\gamma = 45°$

(d) $b = .5$, $c = 1.2$, $\alpha = 120°$

7. Evaluate the limits of the angles of the hyperbolic triangles below both as $x \to 0$ and as $x \to \infty$.

(a) $a = b = c = x$

(b) $a = b = x$, $c = 2x$

(c) $a = x$, $b = c = 2x$

8. Does the SSS congruence theorem hold for hyperbolic triangles? Justify your answer.

9. Does the SAS congruence theorem hold for hyperbolic triangles? Justify your answer.

10. Does the ASA congruence theorem hold for hyperbolic triangles? Justify your answer.

11. Does the SAA congruence theorem hold for hyperbolic triangles? Justify your answer.

12. Explain why the AAA congruence theorem holds for hyperbolic triangles.

13. Find the hyperbolic distances between each pair of the three points $A(0,6)$, $B(10,4)$, $C(10,16)$.

14. Explain why the hyperbolic length of every Euclidean line segment that is parallel to the x-axis is independent of the unit of the underlying Cartesian coordinate system.

15. Explain why the hyperbolic length of every Euclidean line segment that is parallel to the y-axis is independent of the unit of the underlying Cartesian coordinate system.

16. Explain why

$$\int_a^b \frac{\sqrt{1+(f')^2}}{f}\,dx$$

is a reasonable formula for the hyperbolic length of a differentiable curve defined by $y = f(x)$, $a \leq x \leq b$.

17(C). Write a script that takes two distinct points as input and yields the hyperbolic geodesic joining them, as well as its hyperbolic length, as output.

18(C). Write a script that takes three distinct points as input and yields a sketch of the hyperbolic triangle they form, as well as its solution, as output. [Recall that $\cosh^{-1} x = \ln(x + \sqrt{x^2+1})$.]

1.3 Other Geometries

All school children learn about the geometry of the plane. As was seen in the previous two sections, there are other geometries that, broadly speaking, can be classified into two categories.

One way to obtain a new geometry is to distort the plane. A piece of paper can be rolled into either a cylinder or a cone. A film of soap can assume many other shapes, including that of a sphere. The best known of this type of geometries is that of the sphere, some of whose properties were presented in Section 1.1. Of course, the distortion of surfaces may or may not result in distortions of lengths of curves. Rolling a piece of paper into a cone has no such effect—the straight lines on the paper are merely twisted into spirals (or circles) but their lengths are unaffected. On the other hand, when a flat soap film waves in the air the lengths of the curves on it are continuously altered. It took mathematicians several centuries to realize, sometime around 1850, that this notion of distortion of lengths and distances could be, and should be, studied independently of the shape distortion that induced it. The geometries that are obtained by changing the way distance is measured in the plane are called *Riemannian geometries*, and hyperbolic geometry is their best known and studied instance. Riemannian geometry has found many applications in science, the most spectacular of these being the theory of relativity. Hyperbolic geometry is still the subject of much contemporary research and has had many surprising applications to other mathematical disciplines.

Every Riemannian geometry has geodesics, which are defined as the shortest curves joining two points. Such geodesics will form triangles, and these triangles will have interior angles. These angles, in turn, provide a means for quantifying the distortion, or curvature, of a geometry. If $\triangle ABC$ is a triangle of a geometry, with interior angles of radian measures α, β, γ, then the expression $\alpha + \beta + \gamma - \pi$ is called its *total curvature*. Accordingly, every triangle of the Euclidean plane has total curvature 0. It is therefore reasonable to interpret this quantity as a measure of the extent to which $\triangle ABC$ differs from a Euclidean triangle. By this definition, the total curvature of a spherical triangle is always positive and so the sphere is said to be positively curved. Hyperbolic geometry, on the other hand, is negatively curved. It was Gauss who formally defined this notion and pointed out its central role in the study of geometry.

This chapter concludes with the discussion of yet another specific geometry which, while also arising from an esoteric way of measuring distance, is not, for reasons that cannot be explained here, a Riemannian geometry. *Taxicab geometry* was first defined in 1973 but, unlike the spherical and hyperbolic geometries, has not been integrated into the mathematical mainstream. Nevertheless, is has proven useful as a pedagogical tool that sheds a light on Euclidean geometry and also provides students with an elementarily defined mathematical territory they can explore on their own. Like hyperbolic geometry, *taxicab geometry* takes the Euclidean plane as its starting point and redefines distance. The *taxicab distance* between the points $P = (x_1, y_1)$ and $Q = (x_2, y_2)$ is

$$d_t(P, Q) = |x_1 - x_2| + |y_1 - y_2|.$$

Thus, the taxicab distance between $P = (0, 0)$ and $Q = (1, 1)$ is

$$d_t(P, Q) = |0 - 1| + |0 - 1| = 2$$

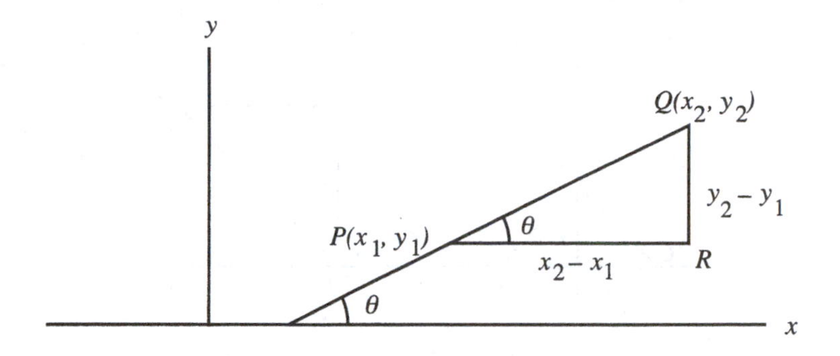

Figure 1.20. Taxicab distance

whereas the Euclidean distance $d_E(P, Q)$ between them is

$$\sqrt{(0-1)^2 + (0-1)^2} = \sqrt{2}.$$

Similarly, the taxicab distance between $(2, -3)$ and $(3, 5)$ is $1 + 8 = 9$ and their Euclidean distance is $\sqrt{1 + 64} = \sqrt{65}$. This geometry receives its name from the fact that it models the way a taxicab driver would think of distances in a city all of whose blocks are perfect squares.

It is clear that taxicab distances agree with Euclidean distances along both horizontal and vertical straight lines. If p is any other straight line, with inclination θ from the positive x-axis, then the taxicab distances along p are different from, but still proportional to, the Euclidean distances. As indicated by Figure 1.20,

$$\begin{aligned} d_t(P, Q) &= |x_2 - x_1| + |y_2 - y_1| = |\cos\theta| d_E(P, Q) + |\sin\theta| d_E(P, Q) \\ &= (|\cos\theta| + |\sin\theta|) d_E(P, Q). \end{aligned}$$

It follows that along any fixed straight line taxicab distances behave very much like Euclidean distances. In particular, the geometrical notion of betweenness can still be expressed numerically.

Proposition 1.3.1 *If the distinct points P, Q, R are collinear, then Q is between P and R if and only if*

$$d_t(P, Q) + d_t(Q, R) = d_t(P, R). \qquad \square$$

There are many other similarities between the taxicab and Euclidean geometries, and these are relegated to the exercises in the subsequent chapters. Some of these differences are qualitative. For example, note that in Figure 1.20

$$d_t(P, Q) = d_t(P, R) + d_t(R, Q).$$

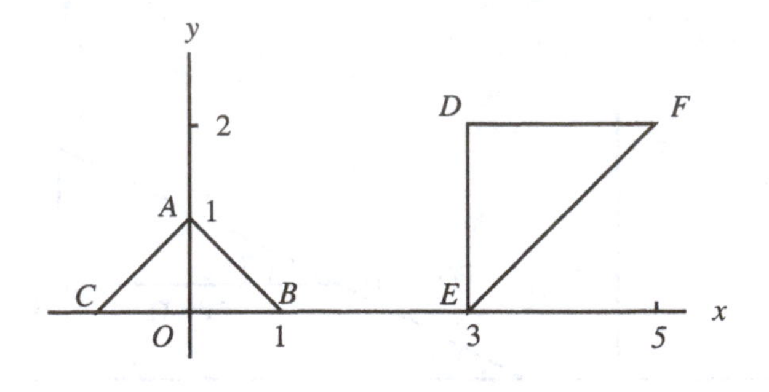

Figure 1.21. Two almost congruent triangles

In other words, the line segment PQ is not the *only* shortest path joining P and Q. In fact, if γ is any polygonal path joining P and Q all of whose segments have nonnegative slope, then the taxicab length of γ still equals $d_t(P,Q)$ (Exercise 6). Nevertheless, no path joining P and Q has taxicab length shorter than $d_t(P,Q)$ (Exercise 9) and so it is not unreasonable to agree to regard the Euclidean straight lines as the straight lines of taxicab geometry. Not surprisingly, the taxicab measure of angles agrees with their Euclidean measure. There is, however, no consensus yet on how areas should be measured in this outlandish geometry.

Only one more striking difference between the taxicab and Euclidean geometries will be mentioned here. The SAS congruence theorem does not hold for taxicab geometry. For the two triangles in Figure 1.21 we have $d_t(A,B) = d_t(D,E) = 2 = d_t(A,C) = d_t(D,F)$ and $\angle BAC = \angle EDF = 90°$ and yet $d_t(B,C) = 2 \neq 4 = d_t(E,F)$.

Maxi geometry, yet another variant of the Cartesian plane, is defined in Exercise 4. Some of its properties are the subject of the same exercise.

Exercises 1.3

1. Compute the total curvature of the triangles in Exercise 1.1.2.

2. Compute the total curvature of the triangles in Exercise 1.2.2.

3. A *metric* is a function $f(P,Q)$ of pairs of points such that for any points P, Q, R,

   (a) $f(P,Q) \geq 0$ and equality holds if and only if P and Q are identical points

   (b) $f(P,Q) = f(Q,P)$

   (c) $f(P,Q) + f(Q,R) \geq f(Q,R)$

Show that the taxicab distance is a metric.

4. The *maxi geometry* is defined on the Cartesian plane by redefining the distance between its points. The *maxi distance* between $P = (x_1, y_1)$ and $Q = (x_2, y_2)$ is

$$d_m(P, Q) = \text{Maximum of } \{|x_1 - x_2|, |y_1 - y_2|\}.$$

In other words, the maxi distance is the larger of the horizontal and vertical Euclidean distances between P and Q. (The straight lines of maxi geometry are, by definition, the Euclidean straight lines, and the taxicab measures of angles are also taken to be identical with their Euclidean measures.)

(a) Show that the maxi distance is a metric.

(b) Show that if P, Q, R are collinear, then Q is between P and R if and only if $d_m(P, Q) + d_m(Q, R) = d_m(P, R)$.

5. Determine the taxicab perimeter and curvature of the triangles with the following vertices:

(a) $(0, 0)$, $(0, 3)$, $(3, 4)$

(b) $(-1, -2)$, $(2, 3)$, $(0, 6)$

6. Suppose γ is a polygonal path joining P and Q such that all of its segments have nonnegative slopes. Show that the taxicab length of γ equals $d_t(P, Q)$. Is this also true if all of the segments have nonpositive slopes?

7. Explain why $\int_a^b (1 + |f'|)\, dx$ is a reasonable formula for the taxicab length of a differentiable curve defined by $y = f(x)$, $a \le x \le b$.

8. Let $y = f(x)$, $a \le x \le b$, be a monotone (either increasing or decreasing) differentiable function. Show that if $P = (a, f(a))$ and $Q = (b, f(b))$ then the taxicab length of the curve defined by f is $d_t(P, Q)$.

9. Explain why no path joining P and Q has taxicab length shorter than $d_t(P, Q)$.

10. Determine the maxi perimeter and total curvature of the triangles with the following vertices:

(a) $(0, 0)$, $(0, 3)$, $(3, 4)$

(b) $(-1, -2)$, $(2, 3)$, $(0, 6)$

11. Find a maxi geometry analog for

(a) Exercise 6

(b) Exercise 7

(c) Exercise 8

(d) Exercise 9

Chapter Review Exercises

1. Compute the perimeter of the triangle formed by joining the midpoints of an equilateral triangle all of whose sides have length $a = 1$ in
   (a) Euclidean geometry
   (b) spherical geometry
   (c) hyperbolic geometry
2. Repeat Exercise 1 for $a = 5$.
3. Repeat Exercise 1 for $a = 10$.
4. Repeat Exercise 1 for $a = 0.1$.
5. Compute the areas of all the triangles in
   (a) Exercise 1
   (b) Exercise 2
   (c) Exercise 4
6. Compute the total curvature of all the triangles in
   (a) Exercise 1
   (b) Exercise 2
   (c) Exercise 4
7. Are the following statements true or false? Justify your answers.
   (a) There is a spherical triangle with angles $\pi/2$, $\pi/3$, $\pi/6$.
   (b) There is a hyperbolic triangle with angles $\pi/2$, $\pi/3$, $\pi/6$.
   (c) There is a taxicab triangle with angles $\pi/2$, $\pi/3$, $\pi/6$.
   (d) If two spherical triangles have angles $\pi/2$, $\pi/2$, $\pi/2$, then they are congruent.
   (e) If two hyperbolic triangles have angles $\pi/4$, $\pi/4$, $\pi/4$, then they are congruent.
   (f) If two taxicab triangles have angles $\pi/3$, $\pi/3$, $\pi/3$, then they are congruent.
   (g) Euclidean, spherical, hyperbolic, taxicab, maxi are all the geometries there are.
   (h) Given any two points of spherical geometry, there is a unique geodesic that joins them.
   (i) Given any two points of hyperbolic geometry, there is a unique geodesic that joins them.

(j) Given any two points of taxicab geometry, there is a unique geodesic that joins them.

(k) On a sphere of radius 1 there is a triangle of area 4.

(l) In hyperbolic geometry there is a triangle of area 4.

(m) Every proposition that is valid in spherical geometry is false in hyperbolic geometry.

(n) Every proposition that is valid in hyperbolic geometry is false in spherical geometry.

Chapter 2

The Neutral Geometry of the Triangle

2.1 Introduction

Much of the mathematics that is taught nowadays in grades K through 12 was developed thousands of years ago by many cultures, the most influential of which were those of Babylonia, Egypt, China, and India. The mathematical lore of each of these civilizations usually included a collection of rules for the computation of the areas of figures and plots, as well as the volumes of solids and containers, from which geometry is descended (in Greek *geo* means "earth" and *metron* means "a measure"). Such rules were often poorly justified, if at all, and in some cases provided only an approximation to the exact value in question. Moreover, these rules were never stated in their generality. Instead, they were exemplified by means of specific computations. Thus, nowhere in the surviving writings of the Egyptian scribes is there a formula for the calculation of the area of a circle of a given radius. Instead, the Rhind Mathematical Papyrus, which dates to 1650 B.C. or earlier, contains the following computation:

> Example of a round field of a diameter 9 khet. What is its area? Take away 1/9 of the diameter, 1; the remainder is 8. Multiply 8 times 8; it makes 64. Therefore it contains 64 setat of land.

It is, of course, reasonable to second guess this papyrus's scribe and to claim that the general formula that underlies his worked out example is

$$A = \left(d - \frac{1}{9}d\right)^2 = \left(\frac{8d}{9}\right)^2$$

but the fact of the matter is that no such explicit formulas have ever been found among any of the surviving Egyptian manuscripts. They probably did not exist, and it is for this reason that this kind of geometry has been called *subconscious geometry.*

The naive geometry of these ancient cultures was transformed by the Greeks of the sixth and fifth centuries B.C. into a deductive science that is completely modern in spirit. Their contribution was twofold. They invented the abstract statements that are today called *theorems* and showed how these theorems could be *demonstrated* in a purely logical manner. The first steps in this direction seem to have been taken by Thales (ca. 640–ca. 546 B.C.) of Miletus, on the west coast of Asia Minor. While nothing is known with certainty about him, he is credited with having either stated or proved the following results:

1. *A circle is bisected by any diameter.*

2. *The base angles of an isosceles triangle are equal to each other.*

3. *The vertical angles formed by two intersecting straight lines are equal.*

4. *Two triangles are congruent if they have two angles and one side in each respectively equal.*

5. *An angle inscribed in a semicircle is a right angle.*

These results are, of course, quite elementary, and Thales's accomplishment lies in his recognition of the value of abstract statements. He is also reputed to have traveled widely, and legend has it that he amazed the Egyptian sages by computing the heights of the pyramids (probably by means of similar triangles).

Pythagoras (ca. 585–ca. 500 B.C.) is believed to have studied with Thales and then moved to southern Italy, where he founded his own school. In this school he taught a mixture of philosophy, science, and religion that also included a fair amount of mathematics. One of the tenets of the Pythagorean faith was the belief that the positive integers were the ultimate components of the universe. For example, if one taut string is double the length of another, plucking them will result in two notes that are exactly one octave apart. If the lengths of the strings have ratio 3 to 2, then the shorter one will produce a note that is recognized by musicians as being a fifth above the note of the longer string. Similarly, the visual world could be reduced to numbers by endowing the latter with shapes. Thus, 1, 3, 6, 10, ... were called *triangular numbers* because the corresponding number of dots

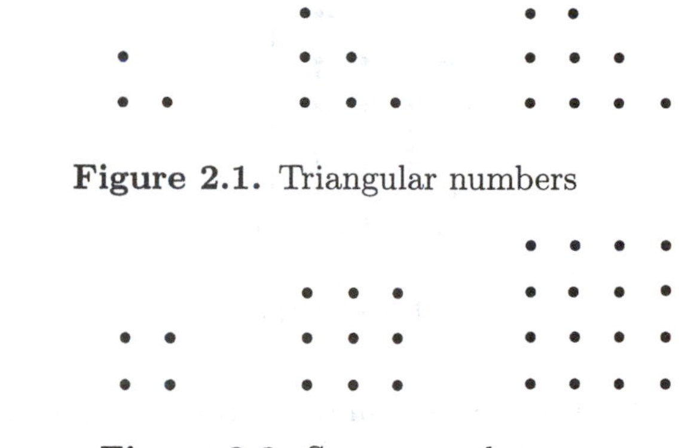

Figure 2.1. Triangular numbers

Figure 2.2. Square numbers

could be organized into triangles (Fig. 2.1). The *square numbers* were, of course, 1, 4, 9, 16, ... (Fig. 2.2). These visual representations of the integers are mathematically fruitful. Figure 2.3 makes it clear that the sum of two consecutive triangular numbers is a square number, and Figure 2.4 makes a convincing case that the sum of any sequence of consecutive odd integers that begins with 1 is in fact also a square number. Given their other discoveries, it is hard to believe that the Pythagoreans failed to notice these interesting relationships. They are supposed to have discovered all the theorems regarding triangles, polygons, and circles that are taught as part of today's high school curriculum, including the fact that the sum of the angles of a triangle is 180° and, of course, the theorem of Pythagoras. In addition, they had some knowledge of the repeating patterns and designs discussed in Sections 6.6 and 6.7, as well as the regular solids of Section 8.1.

The Pythagoreans' preoccupation with integers eventually led them to the realization that the length of the diagonal of a square of unit side is not expressible as the ratio of two integers. In other words, $\sqrt{2}$ is *irrational*. This discovery must have been disconcerting because it conflicted with the

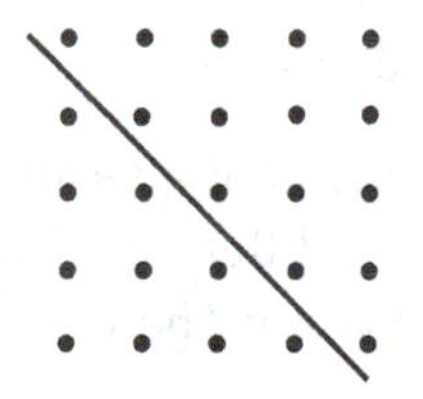

Figure 2.3. The sum of consecutive triangular numbers

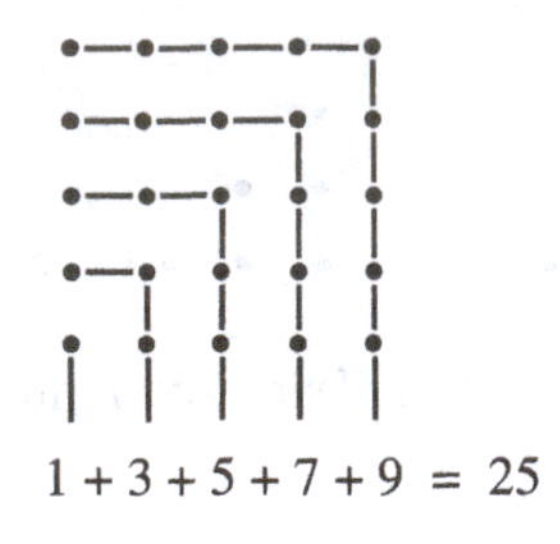

Figure 2.4.

aforementioned doctrine that the universe can be expressed in terms of integers. Its discoverer, Hippasus of Metapontum, is reported to have been drowned for his bad tidings. Eventually, the Greeks, if not the Pythagoreans themselves, resigned themselves to the existence of irrational numbers and went on to develop a theory of incommensurable numbers that made it possible to incorporate them into the general framework of mathematics.

The irrationality of $\sqrt{2}$ is, of course, an interesting and important mathematical fact. Equally significant, however, are its implications regarding the high level of rigor the Greeks of the fifth century attained. All known proofs of this irrationality are subtle; no visually immediate proof of the type displayed in Figures 2.3 and 2.4 exists. It is therefore safe to assume that by this time geometry had matured not only in the abstraction of its concepts but also in the rigor of its arguments. Anyone who can produce a convincing proof of such a surprising and counterintuitive fact as the irrationality of $\sqrt{2}$ must have logical standards at least equal to those of today's graduate mathematics students.

The sophists, a group of itinerant professional teachers, the earliest of whom studied with Pythagoras, did much to further the cause of geometry by incorporating it into their curriculum (which also included grammar, rhetoric, dialectics, eloquence, morals, astronomy, and philosophy). Since they promoted themselves as practical trainers in the art of persuasion as preparation for political and legal careers, they must have considered the abstraction and logical argumentation of geometry as a valuable pedagogical tool. As Plato (427 B.C.–327 B.C.) says in *The Republic*,

> with respect to finer reception of all studies, we surely know there is a general and complete difference between the man who has been devoted to geometry and the one who has not.

None of several geometry books that preceded Euclid's have survived. The first of those was written by Hippocrates (ca. 440 B.C.) of Chios. The most important and innovative one was, without doubt, that of Eudoxus (408–ca. 355 B.C.) who, among other accomplishments, invented the *method of exhaustion*, a calculus-like discipline.

Euclid of Alexandria wrote his famous book the *Elements* in approximately 300 B.C. This book codified some of the state of the art of geometry at the time, although much was omitted too. It actually consists of thirteen books, whose contents are as follows:

Books I–IV: The Geometry of Triangles and Circles

Books V–VI: Theory of Geometric Proportions

Books VII–IX: Theory of Numbers

Book X: Theory of Irrational Surds

Books XI–XIII: Solid Geometry

For two thousands years Euclid's *Elements* remained the standard text for geometry throughout the world. It was translated into many languages, and over a thousand printed editions have been published since 1482. After the Bible, it is Western civilization's most influential book. The first alternative exposition of geometry was offered by the French mathematician Adrien-Marie Legendre (1752–1833) in 1794. The discovery of non-Euclidean geometry in the nineteenth century forced mathematicians to re-examine the foundations of geometry from a much more critical point of view. A variety of gaps and flaws were found and this work culminated in David Hilbert's (1862–1943) *The Foundations of Geometry*, which appeared in 1899 and which set Euclidean geometry on completely rigorous grounds (see Appendix E). Unfortunately, this development, as well as those promulgated subsequently by other mathematicians, suffers from the pedagogical defect that the proofs of many "obvious" facts are quite difficult. Current high school geometry texts resolve this difficulty by listing many such facts as axioms rather than theorems. This, of course, is problematic because it opens up the possibility of simply declaring all geometrical facts to be axioms. There has been a considerable amount of experimentation with innovative ways of teaching geometry during the last fifty years but, regrettably, no consensus has been reached. It may be time to return to basics, and for this reason the author has chosen to follow Euclid's *Elements* as closely as possible in this presentation of plane geometry.

Exercises 2.1

1. Briefly describe the lives and accomplishments of the following predecessors of Euclid:
   (a) Thales
   (b) Pythagoras
   (c) Democritus
   (d) Anaxagoras
   (e) Archytas
   (f) Eudoxus
   (g) Hippocrates of Chios
2. Briefly describe the lives and accomplishments of the following successors of Euclid:
   (a) Archimedes
   (b) Apollonius
   (c) Ptolemy
   (d) Heron
   (e) Menelaus
   (f) Pappus
   (g) Proclus

2.2 Preliminaries

Appropriately, Euclid began his development with a list of definitions.

Definitions

1. *A* point *is that which has no part.*

By modern standards, at least, many of Euclid's definitions are deficient, and it has been argued that they were added by a later commentator. This particular definition is not proper because it is impossible to define a term by listing only the qualities it does not have. Instead it should be viewed as an attempt on its author's part to convey informally the notion of a point—the ultimate indecomposable particle. Definition 2 should be understood in the same way. Points will be denoted by uppercase letters A, B, C, ...

2. *A* line *is breadthless length.*

Euclid's *line* is today's *curve.*

3. *The* extremities *of a line are points.*

No new terms are defined here. Euclid is simply clarifying the relation between the points and the lines defined previously.

4. *A* straight line *is a line which lies evenly with the points on itself.*

This sentence is obscure. In the author's opinion, the *points* of this definition are the two extremities mentioned in the previous one. In other words, of all the lines having the same extremities, the straight line is that one that lies exactly in between those extremities.

It is customary to conclude from Definitions 3 and 4 that Euclid implicitly assumed a straight line to have finite length so that it necessarily has endpoints. The phrasing of Proposition 2.3.15 (among others) indicates that this view may not be entirely correct. When necessary, Euclid had no qualms about referring to the infinite line that extends indefinitely in both directions. Many of today's texts, although by no means all, use $\overline{AB}$ to denote the finite straight line segment joining A and B, AB for its length, and $\overleftrightarrow{AB}$ for the infinite straight line that contains both A and B. Like Euclid, this text will gloss over the distinction between the first two and use AB for both. When necessary, either the double arrow notation or a lowercase letter will be used for the infinite line. Similary, $\overrightarrow{AB}$ denotes the *ray* or *halfline* that consists of that half of $\overleftrightarrow{AB}$ that begins at A, extends indefinitely, and contains B.

5. *A* surface *is that which has length and breadth only.*

6. *The* extremities *of a surface are lines.*

It is implicit in this definition that every surface is necessarily of finite extent.

7. *A* plane surface *is a surface which lies evenly with the straight lines on itself.*

This should be interpreted in a manner similar to that of Definition 4. A finite plane surface is characterized by the fact that it lies exactly in between its border lines.

8. *A* plane angle *is the inclination to one another of two lines in a plane which meet one another and do not lie in a straight line.*

9. *And when the lines containing the angle are straight, the angle is called* rectilineal.

Notwithstanding the fact that Euclid seems to have displayed here an interest in curvilinear angles, he subsequently referred to such angles only once, in Proposition III.16, to make a point that is of little interest. It has been posited that the geometers of the time were dallying with such angles without coming to any serious conclusions, and that Euclid felt it necessary to acknowledge their efforts.

Since it is unclear what is meant by the *inclination* of two lines, an *angle* is defined here as either of the two portions of the plane bounded by a pair of rays that emanate from the same point. By extension, any pair of line segments with a common endpoint determines two angles. The lines forming the angle are its *sides* and the intersection of the sides is the angle's *vertex*. An angle that has vertex A and sides AB and AC will be denoted by $\angle BAC$ (or $\angle CAB$). Definition 8 excludes the possibility that AB and AC determine the same infinite straight line. That is, Euclid excluded the $0°$ and $180°$ angles from his definition, possibly because zero was not then recognized as a number. This has both advantages and disadvantages, and this text will not follow his lead in this instance. Thus, if A, B, C are consecutive points on a straight line, then $\angle ABC$ is a *straight angle* and $\angle BAC$ and $\angle BCA$ are *zero angles.*

10. *When a straight line set up on another straight line makes the adjacent angles equal to one another, each of the equal angles is* right, *and the straight line standing on the other is called a* perpendicular *to that on which it stands.*

The existence of right angles is demonstrated in Proposition 2.3.11. This is also Euclid's first mention of the notion of equality, and it is not clear whether he is referring to equality of measure or congruence. This issue is discussed in detail in the following paragraphs dealing with the common notions.

11. *An* obtuse angle *is an angle greater than a right angle.*

12. *An* acute angle *is an angle less than a right angle.*

13. *A* boundary *is that which is an extremity of anything.*

14. *A* figure *is that which is contained by any boundary or boundaries.*

15. *A* circle *is a plane figure contained by one line such that all the straight lines falling upon it from one point among those lying within the figure are equal to one another.*

16. *And that point is called the* center *of the circle.*

17. *A* diameter *of the circle is any straight line drawn through the center and terminated in both directions by the circumference of the circle, and such a straight line also bisects the circle.*

18. *A* semicircle *is the figure contained by the diameter and the circumference cut off by it. And the center of the semicircle is the same as that of the circle.*

19. Rectilineal *figures are those which are contained by straight lines,* trilateral *figures [or* triangles*] being those contained by three,* quadrilateral *those contained by four, and* multilateral *those contained by more than four straight lines.*

These rectilineal figures are today's *polygons*, and the straight line segments containing them are their *sides*. The endpoints of the sides of a polygons are its *vertices*. It is implicit in Euclid's definition that every vertex lies on only two sides and every two sides intersect only in a vertex. A straight line joining two vertices that are not the endpoints of a side is a *diagonal*. A polygon that contains all of its diagonals in its interior is said to be *convex*. The triangle with vertices A, B, C is denoted by $\triangle ABC$.

20. *Of trilateral figures, an* equilateral triangle *is that which has three sides equal, an* isosceles triangle *that which has two of its sides alone equal, and a* scalene triangle *that which has its three sides unequal.*

According to this definition equilateral triangles are *not* isosceles triangles. This unimportant distinction is contrary to modern usage, which sees an equilateral triangle as a special kind of isosceles triangle, and will be ignored here. It is customary to refer to a side of an isoceles triangle as its *base* provided the other two sides are equal to each other.

21. *Further, of trilateral figures, a* right-angled triangle *[or* right triangle*] is that which has a right angle, an* obtuse-angled triangle *that which has an obtuse angle, and an* acute-angled triangle *that which has its three angles acute.*

22. *Of quadrilateral figures, a* square *is that which is both equilateral and right-angled; an* oblong *[*rectangle*] that which is right-angled but not equilateral; a* rhombus *that which is equilateral but not right-angled; and a* rhomboid *[*parallelogram*] that which has its opposite sides and angles equal to one another but is neither equilateral nor right-angled. And let quadrilaterals other than these be called* trapezia.

Once again, Euclid's definitions do not completely agree with modern usage. To us a parallelogram is a quadrilateral whose opposite sides are parallel (see Definition 23), a rectangle is a parallelogram whose angles are **all** right angles, a rhombus is a parallelogram **all** of whose sides are equal, and a square is a figure that is both a rectangle and a rhombus. The parallelogram $ABCD$ is denoted by $\varnothing ABCD$, the rectangle $ABCD$ is denoted by $\sqsubset\!\sqsupset ABCD$, and the square $ABCD$ is denoted by $\square ABCD$.

23. Parallel *straight lines are straight lines which, being in the same plane and being produced indefinitely in both directions, do not meet one another in either direction.*

If the straight lines AB and CD are parallel, this is denoted by $AB \parallel CD$.

By the time Euclid came to write the *Elements* the Greeks had amassed thousands of theorems. As is the case today, the proofs of many of these relied on other theorems. In order to eliminate the possibility of cyclical reasoning, it was necessary to create a list of the fundamental theorems in which each proof cited only definitions, previously proved theorems, or both. It must have come as a surprise to the Greek geometers to discover that such a list was impossible. The starting point of such a list could not consist of definitions alone—some theorems must be accepted without justifications. These unproven theorems were called *axioms* or *postulates.* Just which theorems should serve as postulates is a question that must be resolved on subjective grounds. It often happens that one mathematician's postulate is another's proven theorem. It is very likely that Euclid's choice of postulates was based to a large extent on the various textbooks to which he had access. Later generations modified his choices in many ways, and some of their axiom systems appear in Appendices C, D, and E.

One informal principle that guides mathematicians in their selection of postulates is that of austerity. A postulate should be a simple, easily parsed statement. Nevertheless, as will be seen, there are important exceptions to this rule.

Euclid began with ten axioms. The first five are called postulates and the others *common notions* (the meaning of this term will be clarified soon). These were followed by 462 theorems that are called *propositions* (there is no difference between propositions and theorems). The truth of the matter is that the view of the *Elements* as a well-grounded and logically consistent ordering of theorems is to be understood only as an ideal, because Euclid's organization of geometry is flawed. There are several instances where undefined terms and unstated postulates appear in his arguments. Some of these are minor errors, but the correction of others would require considerable revision of the material. Nevertheless, because of its vision and because of its logical strength, Euclid's opus is justly regarded as one of the supreme achievements of Greek civilization in particular and of the human mind in general. Euclid's choice of postulates are now listed.

Postulates

Let the following be postulated:

1. *To draw a straight line from any point to any point.*

Taken at face value, this postulate merely states that every pair of distinct points can be joined by a straight line. However, in view of some of the arguments given in the proof of Proposition 2.3.4 and elsewhere, it is necessary to conclude that Euclid understood this statement to include the additional assumption that any two points can be joined by **exactly** one straight line.

2. *To produce a finite straight line continuously in a straight line.*

One of the implications of this postulate is that both the plane and space extend infinitely far in all directions. The reason that it is stated in terms of extendibility rather than extent is that the Greek mathematicians were aware of the logical complications inherent in the concept of infinity and therefore avoided its explicit mention whenever possible.

3. *To describe a circle with any center and distance.*

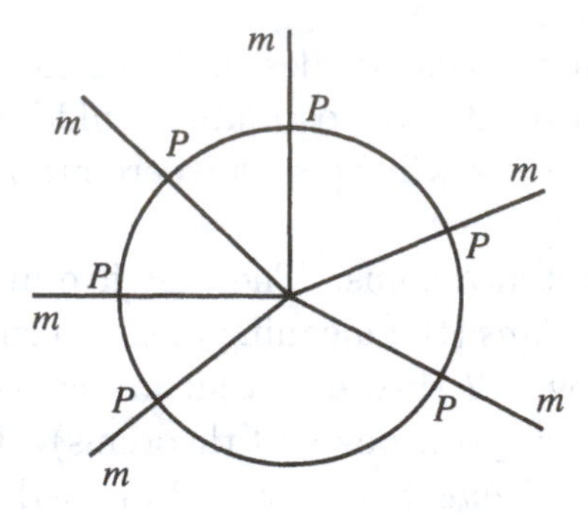

Figure 2.5. Regarding the existence of circles

In this postulate Euclid is not asserting the existence of a circle. This existence actually follows from Postulate 2, which implies that given any point A, straight line m through A, and proposed radius r, there exists a point P on m such that $AP = r$. The set of all such points P is the proposed circle with center A and radius r (Fig. 2.5). Rather, Euclid is stating here that he intends to use such circles as building blocks in the logical tower he is setting out to construct. Alternatively, we can think of Euclidean geometry as a solitaire game whose legitimate moves are these postulates.

The statement of Proposition 2.3.2 makes it clear that this postulate is to be interpreted in a very narrow sense. Namely, given a point A and a line segment AB there exists a circle with center A and radius AB. The postulate does **not** say that given a point A and a line segment BC there is a circle with center A and radius equal to BC. This latter statement, which is considerably stronger than Postulate 3, is in fact the content of Proposition 2.3.2. The distinction between Postulate 3 and Proposition 2.3.2 is rephrased by modern mathematicians by saying that their Greek predecessors used *collapsible compasses* whose legs lost their angle whenever the compass was lifted off the paper. This statement is, of course, a metaphor and should not be taken literally. The circle with center A and radius AB or r is denoted in this text by $(A; AB)$ or $(A; r)$.

4. *That all right angles are equal to one another.*

Euclid chose to use the right angle as his theoretical unit for measuring all rectilineal angles (degrees were used in practice) and this is only possible if it is known that all right angles are equal. It was therefore necessary for him to have either a theorem or a postulate that asserts this equality. It might be noted that this postulate is qualitatively different from the previous three.

Those proclaimed the legitimacy of certain constructions, whereas this one asserts the equality of some figures.

5. *That, if a straight line falling on two straight lines make the interior angles on the same side less than two right angles [in sum], the two straight lines, if produced indefinitely, meet on that side on which are the angles less than two right angles.*

Like Postulate 4, this one differs from the first three in that it makes a statement that relates the several parts of a figure. Moreover, its statement is much more complicated than any of the preceding statements. For these reasons many of Euclid's successors believed that this postulate could be proved on the basis of the others and was therefore superfluous. However, their repeated attempts, over two millennia, to substantiate this feeling were invariably unsuccessful. Finally, in the nineteenth century, it was demonstrated beyond all doubt that all such efforts must of necessity fail. It is impossible to deduce Postulate 5 from the other postulates. This issue will be discussed in greater detail in Section 2.4.

It was mentioned that Euclid's axiomatization was incomplete in the sense that his proofs make occasional use of unstated and unvalidated assumptions. This is the case in Propositions 2.3.1, 2.3.4, 2.3.8, and others. Two such assumptions are stated as additional postulates.

S (Separation). *The infinitely extended straight line, the triangle, and the circle separate the plane into two portions such that any line joining a point of one portion to a point of the other portion intersects the separating figure. In the case of the line the two portions are called the line's* sides. *In the case of a triangle or a circle the portions are called the* interior *and* exterior.

A (Application). *Given two triangles ABC and $A'B'C'$, it is possible to apply $\triangle ABC$ to $\triangle A'B'C'$ so that the vertex A falls on the vertex A', the side AB falls on the side $A'B'$, and the vertex C falls in the same side of $A'B'$ as C'.*

The common notions listed on the next page are also postulates, and in some editions they are indeed grouped together with the previous five postulates. However, these common notions do share a common theme that sets them apart. They are all concerned with *equality* and *inequality*. While Euclid did not define these notions explicitly, his sense of these terms is clarified by the way he used them. For example, Proposition 3.2.3 states that *parallelograms which are on the same base and in [between] the same parallels are equal to one another.* The conclusion that Euclid was referring to *equality of size* rather than congruence is therefore inescapable. In other words, when Euclid said that two parallelograms were equal he meant that they had the same areas, when he said that two straight lines were equal he meant that they had the same lengths, and when he said that two angles were equal he meant something of the same nature. In order for this interpretation to work it is necessary to stipulate that Euclid had an underlying, albeit unstated, assumption that all geometric objects have an aspect of *numerical size* or *magnitude.* In lieu of defining these terms, their properties were set forth in the common notions, so named because they describe the properties that are shared by (or are common to) all mensurations, regardless of whether they relate to *length*, *area*, *volume*, *weight*, or *angular size.* This explanation is supported by Euclid's failure to provide any other definition of the notions of area and volume notwithstanding his many propositions about these very concepts. When viewed in this light the common notions strikingly resemble the standard modern axiomatic definition of a *Haar measure.* Section 3.2 contains a more detailed discussion of area.

The symbol "=" is used to denote equality in magnitude, regardless of the specific aspect that is being measured. Thus,

$$\triangle ABC = \triangle DEF$$

means that the two said triangles have equal areas. Similarly,

$$AB = CD$$

means that the straight line segments joining A to B and C to D have equal lengths. Of course, in this case, they are also congruent, but that is accidental. The equality in length of line segments and the equality of angular measures of angles happens to imply their congruence. On the other hand, the equality of the areas of regions and volumes of solids does not entail the stronger relation of congruence.

Common Notions

1. *Things which are equal to the same thing are also equal to one another.*

2. *If equals be added to equals, the wholes are equal.*

3. *If equals be subtracted from equals, the remainders are equal.*

From the modern point of view the third common notion is redundant, since it seems to be already subsumed by the previous one. The Greeks, however, did not recognize the existence of negative numbers, and so it was necessary for Euclid to include both common notions 2 and 3 in his list.

4. *Things that coincide with one another are equal to one another.*

In view of the proof of Proposition 2.3.4 (see page 52) this should be understood as saying that things that can be *made to coincide* with one another have equal sizes. This can be interpreted as Euclid's first mention of congruence in the sense of a rigid transformation. The contradictory attitudes Euclid displayed in the proofs of Propositions 2.3.2 and 2.3.4 make it clear that he had ambivalent feelings about the use of such transformations. The first of these propositions could have been proved by simply moving a given line segment to a given location. Instead, Euclid produced an elaborate and ingenious proof that, quite properly, made no use of such movements. The proof of Proposition 2.3.4, on the other hand, begins with an *application* of one triangle to another. In other words, one triangle is lifted and placed on top of the other. Since none of the definitions or postulates provide for such applications, this is a clear-cut relinquishment of standards on the part of Euclid. The fact that Euclid used this device sparingly implies that he was in all likelihood aware of its impropriety. The lack of a framework for the treatment of transformations and congruence constitutes one of the more serious flaws of the *Elements.*

5. *The whole is greater than the part.*

This common notion turns out to be very useful in a variety of proofs by contradiction, that of Proposition 2.3.6 being the first instance. In addition, it also ensures that, whatever aspect it is that is being measured, some geometrical figure has a nonzero size. This is logically necessary since otherwise it could be possible to trivialize the notion of size by assigning to every figure the measure of zero. Note that this trivial zero measure does have the properties stipulated in the the first four notions and only fails to satisfy the fifth one.

Euclid's postulates are now reexamined in the context of the *surface of the sphere.* Since great semicircles are not geodesics, diametrically opposite points on the sphere cannot be joined by a geodesic and so, strictly speaking, Euclid's *Postulate 1* fails to hold on the sphere. However, the next best thing is true: *Every two points on the sphere that are not diametrically opposite can be joined by a (unique) geodesic segment.* The uniqueness is guaranteed

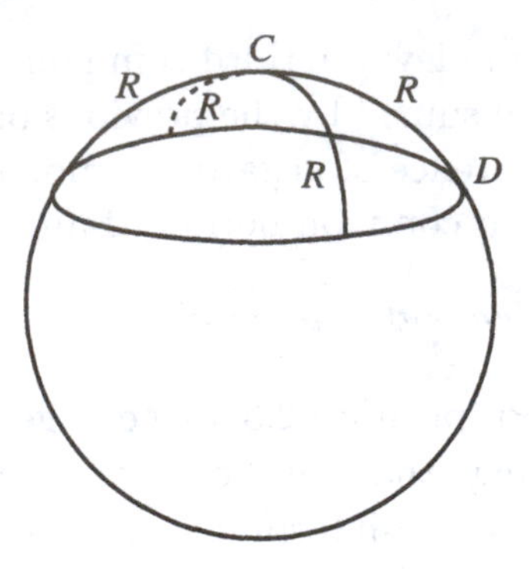

Figure 2.6. A spherical circle of spherical radius R and center C

by the fact that the plane containing the geodesic segment must also contain the center of the sphere.

Euclid's *Postulate 2* fails to hold on the sphere in a rather dramatic fashion. *No geodesic on the sphere of radius R can be extended to a length equal to or greater than* $2\pi R$. In fact, it could be argued that no spherical geodesic can be extended to a length of πR or more, since if that length were reached the geodesic would have to contain a pair of diamterically opposite points.

Euclid's *Postulate 3* does hold on the sphere. Given any point C and any geodesic segment CD, there clearly is a circle of spherical radius CD on the sphere (see Fig. 2.6).

Euclid's *Postulate 4* also holds on the sphere. A right spherical angle is formed by any two geodesic arcs whose defining planes are at right angles to each other. For example, every meridian cuts the equator in a spherical right angle. It is also true that given any two spherical right angles, there is a rotation of the sphere that transforms one into the other.

Postulate 5 holds trivially, regardless of the angles between the lines, since any two geodesic segments, if sufficiently extended, will intersect (in a pair of diametrically opposite points). The reason for this is that both segments extend to great circles each of which divides the sphere into two equal halves. If these circles did not intersect these equalities could not hold.

With the proper reinterpretation *Postulate S* can be made to hold on the sphere. Infinitely extended straight lines have to be replaced with great circles, and the exterior sides are no longer infinite. With these modifications the figures in question all separate the surface of the sphere into two parts that can be seen as sides.

Postulate A also holds on the sphere. The requisite applications are rotations of the sphere about an axis that passes through its center and reflections in planes that contain the center.

Exercises 2.2A

1. Let A and B be two distinct points on the surface of a sphere. Describe a rigid motion of the sphere that moves A to B's location.

2. Let AB and CD be two equal geodesic segments on the surface of the sphere. Describe a rigid motion that moves AB to the position of CD. (*Note:* It is permissible to describe a rigid motion as a composition of several rigid motions.)

3. Let $\angle ABC$ and $\angle DEF$ be equal spherical angles. Describe a rigid motion of the sphere that moves the first angle onto the position of the second so that BA falls along ED and BC falls along EF.

Euclid's postulates are now reexamined in the context of hyperbolic geometry. Euclid's *Postulate 1* holds here too. To verify this assertion two cases must to be considered. If the two points in question have coordinates (a, b_1) and (a, b_2), then clearly the vertical Euclidean line segment joining them is the only geodesic that joins them. If, on the other hand, they have coordinates (a_1, b_1) and (a_2, b_2) with $a_1 \neq a_2$, then there is one and only one semicircle that contains them both and has its center on the x-axis. This semicircle's center is the intersection of the x-axis with the perpendicular bisector to the line segment joining those two points.

Euclid's *Postulate 2* holds for hyperbolic geometry because the x-axis is infinitely far for all of its inhabitants (see Section 1.2).

As was noted in the first discussion of *Postulate 3*, the existence of the circle $(A; AB)$ follows from Postulate 2, which is already known to hold in the hyperbolic plane. Consequently, all such circles exist in hyperbolic geometry as well. It follows that these circles are legitimate building blocks in this context too.

Since two geodesics form a hyperbolic right angle if and only if their tangents form a Euclidean right angle, the equality of hyperbolic right angles follows from the equality of Euclidean right angles. Thus, *Postulate 4* holds in the hyperbolic plane.

Rather surprisingly, *Postulate A* does hold for the hyperbolic plane. This is difficult to explain at this point because not enough tools have been developed to describe the rigid motions of the hyperbolic plane. However, this statement is borne out by Exercises 9 through 11 in Section 7.3.

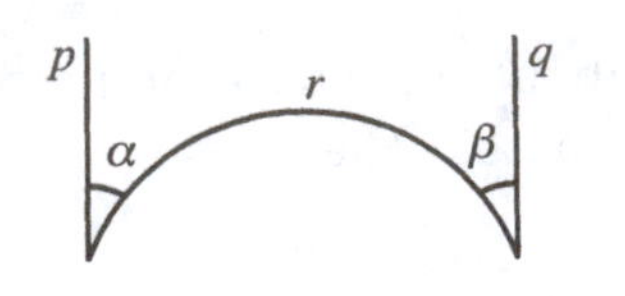

Figure 2.7. A hyperbolic counterexample to Euclid's Postulate 5

Postulate S also holds in the hyperbolic plane. This is intuitively plausible since, in fact, every closed curve C that does not intersect itself has a well defined inside and outside, and every curve that joins the inside to the outside must of necessity cross C at some point.

Hyperbolic geometry can now be defined somewhat more precisely as the geometry in which Euclids Postulates 1, 2, 3, 4, A, S hold but Postulate 5 does not. Since Postulates 1, 2, 3, 4, A, S hold in both Euclidean and hyperbolic geometry it follows that all the propositions that can be proved on the basis of these postulates alone also hold for both geometries. This common set of propositions is known as *neutral geometry* or *absolute geometry*. It includes Propositions 1 through 28 of Book I of Euclid as well as a variety of others that are listed in the neutral sections of this text. *Postulate 5*, on the other hand, does not hold in the hyperbolic plane. In Figure 2.7 the geodesics p and q make interior angles α and β with the geodesic r, where the sum of α and β is quite clearly less than 180°, and yet the geodesics p and q, no matter how far extended, do not meet. Postulate 5 is therefore not a part of neutral geometry.

Exercises 2.2B

1. Comment on the following postulates in the context of taxicab geometry:
   (a) 1 (b) 2 (c) 3 (d) 4 (e) 5 (f) S (g) A
2. Comment on the following postulates in the context of maxi geometry:
   (a) 1 (b) 2 (c) 3 (d) 4 (e) 5 (f) S (g) A
3. Draw a taxicab circle of radius 1.
4. Draw a maxi circle of radius 1.

2.3 Propositions 1 through 28

Having laid a foundation in the form of definitions, postulates, and common notions, Euclid proceeded to list several hundred logical conclusions each of which is called a *proposition*. There are two kinds of propositions. One kind asserts a *relationship* between some geometrical objects. Such, for example is Proposition 2.3.5, which states that the angles at the base of an isosceles triangle are equal to each other. The second kind of proposition asserts the *constructibility* of a certain geometrical object. Thus, Proposition 2.3.9 states that angles can be bisected, or, in other words, that angle bisectors can be constructed. Since Euclid's postulates only mention the existence of points, straight lines, and circles, it has become customary to say that the only tools that are allowed in Euclidean constructions are rulers and compasses. This too is to be understood only as a metaphor. Surely the idea of using a physical ruler to draw a straight line is just as objectionable as that of using the top of a can to draw a circle. It would be more accurate to say that Euclid constrained himself to a discussion of logical constructs definable by means of points, straight lines, and circles alone.

Ideally, the justification of both kinds of propositions must rely only on definitions, postulates, common notions, and previously justified propositions. In fact, both Euclid and the author allow for some deviation from this strict standard.

Euclid's conventions also allowed him to only use and/or make relational propositions about geometrical figures whose constructibility had been demonstrated in a previous constructive proposition. Thus, Proposition 2.3.20, whose proof requires the midpoint of the side of a triangle, must be preceded by Proposition 2.3.10, which asserts the constructibility of the midpoint of a straight line segment.

The format that Euclid used for stating and validating his propositions is still in use today. The verbal description of the proposition is followed by a (preferably symbolic) description of the context (GIVEN), and the relational or constructive asseveration (TO PROVE or TO CONSTRUCT). In the case of a relational proposition this is followed by a proof (PROOF) that validates the proposition. In the justifications of the steps of a proof the terms *definition*, *common notion*, *postulate*, and *proposition* will be respectively abbreviated as DFN, CN, PT, and PN.

Euclid invariably signed off on his proofs with the phrase "what it was required to prove," which is rendered as "quod erat demonstrandum" in Latin. The initials of this phrase, Q.E.D., served as the traditional end of proof symbol for hundreds of years until it became fashionable in the last thirty or so years to replace it with some variant of a small rectangle. In this text the traditional "Q.E.D." denotes the end of a proof. In the case of a constructive proposition, the aforementioned asseveration is followed by a description of the construction (CONSTRUCTION). This, in turn, should be followed by a proof that the constructed figure does indeed possess the

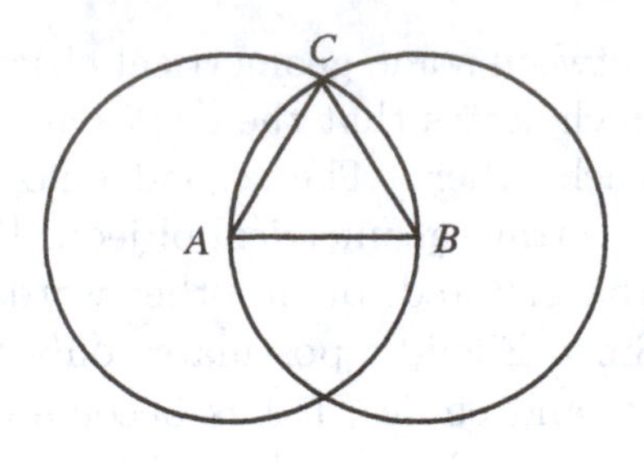

Figure 2.8.

required properties. However, the proof of the validity of a construction is often so straightforward that it should, and will be, omitted.

As part of his exposition Euclid incorporated many propositions that constitute geometrical analogs of algebraic rules. Because of the lack of algebraic symbols in his times, this resulted in many obvious propositions with tedious proofs. In order to avoid these inconveniences, basic algebraic manipulations will be permitted in this text's proofs.

For the most part, the statements of the propositions and proofs in this text are essentially the same as those that appear in Sir Thomas L. Heath's century old translation of the *Elements*. For the sake of clarity the proofs are presented with modern notation and some of the statements of the propositions are paraphrased. Some of Euclid's more cumbersome or erroneous proofs are replaced by improved ones. All such instances are explicitly noted. The reader is reminded that the paranthesized numbering of the propositions refers to their position within the *Elements*.

Propositions

Proposition 2.3.1 (I.1) *On a given finite straight line to construct an equilateral triangle.*

GIVEN: Line segment AB (Fig. 2.8).

TO CONSTRUCT: $\triangle ABC$ such that $AB = BC = CA$.

CONSTRUCTION: Draw the circles $(A; AB)$ and $(B; AB)$ [PT 3]. Let C be one of their intersections [PT S]. Then $\triangle ABC$ is the required triangle.

PROOF:

$$AB = AC \quad \text{[DFN 15]}$$
$$BA = BC \quad \text{[DFN 15]}$$
$$\therefore \quad AC = BC \quad \text{[CN 1]}$$

Q.E.D.

This proposition demonstrates both some of Euclid's strengths and some of his weaknesses. On the positive side he was a careful thinker and expositor who was unwilling to accept as obvious the existence of a triangle which most people take for granted. Unfortunately, he was not careful enough. Specifically, he implicitly accepted that the two auxiliary circles drawn in this proof necessarily intersect (there is no version of Postulate S in the *Elements*). Now, as physical objects, these figures must clearly intersect, but as abstract entities, whose properties must be reducible to Euclid's definitions, postulates, and common notions, this claim calls for verification. This is not a minor point. The fact is that Euclid failed to provide a framework within which the interiors and exteriors of configurations can be discussed and this is one of the major defects of his logical edifice. The need for such a framework is underscored by the paradoxical nature of Exercise 3.3A.10. In this text, of course, it is Postulate S that provides the rationale for the existence of the intersection point C in the preceding proof.

There was no compelling reason for Euclid to choose the construction of equilateral triangles as his first proposition. Other texts employ different starting points. It is noteworthy that the last few propositions of the last of the thirteen books that comprise Euclid's *Elements* deal with the construction of the five regular solids (see Section 8.1) that are the three-dimensional analogs of the equilateral triangle in particular and of the regular polygons in general. In fact, the faces of three of these five solids are themselves equilateral triangles. Thus, Euclid may have chosen his starting point and ending point on esthetic grounds. They gave his work an artistic form.

It is convenient to denote the intersection of two lines p and q by $p \cap q$. In the case where p or q is a circle, this will be used to denote only one of the intersection points.

Proposition 2.3.2 (I.2) *To place at a given point (as an extremity) a straight line equal to a given straight line.*

GIVEN: Point A, line segment BC (Fig. 2.9).

TO CONSTRUCT: A point D such that $AD = BC$.

CONSTRUCTION: Let $\triangle ABE$ be equilateral [PN 2.3.1], let $F = EB \cap (B; BC)$ [PT 3], and let $D = EA \cap (E; EF)$ [PT 3]. Then D is the required point.

PROOF:

$ED = EF$ [Radii of the same circle]

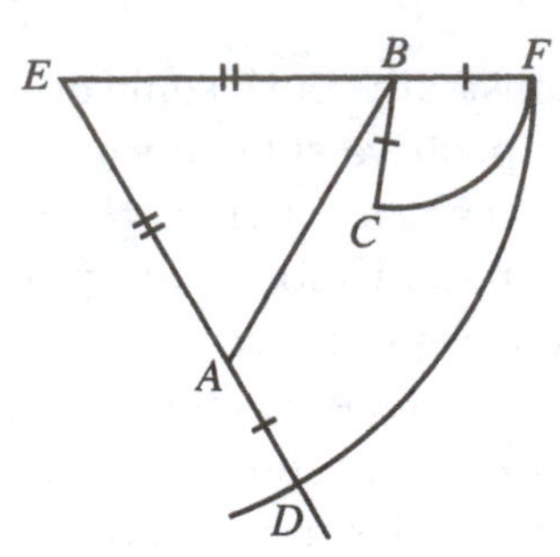

Figure 2.9.

	$EA = EB$	[Sides of an equilateral triangle]
$\therefore$	$AD = BF$	[CN 3]
But	$BF = BC$	[Radii of the same circle]
$\therefore$	$AD = BC$	[CN 1]

Q.E.D.

Informally speaking, this proposition asserts that any construction that can be carried out with a non-collapsible compass, i.e., a compass that does not lose its angle when lifted off the plane, can also be accomplished with a collapsible one, albeit by a process that requires several more steps. More succinctly, *the constructive power of the collapsible compass equals that of the rigid compass.*

This proposition and its relation to Postulate 3 clarify one of the principles that underlie Euclid's strategy in choosing his postulates. Postulates should have as little content as possible. After all, Euclid could have chosen Proposition 2.3.2 as Postulate 3 with no resulting loss of any subsequent propositions or complications in any subsequent proofs. That he chose not to do this indicates that he enjoyed flexing his mental muscles just for the sake of using them.

Proposition 2.3.3 (I.3) *Given two unequal straight lines, to cut off from the greater a straight line segment equal to the less.*

GIVEN: Line segments $AB > CD$ (Fig. 2.10).

TO CONSTRUCT: A point E on AB such that $AE = CD$.

CONSTRUCTION: Let F be a point such that $AF = CD$ [PN 2.3.2] and let $E = AB \cap (A; AF)$ with AB. Then E is the required point.

PROOF: By construction, $AE = AF = CD$. Q.E.D.

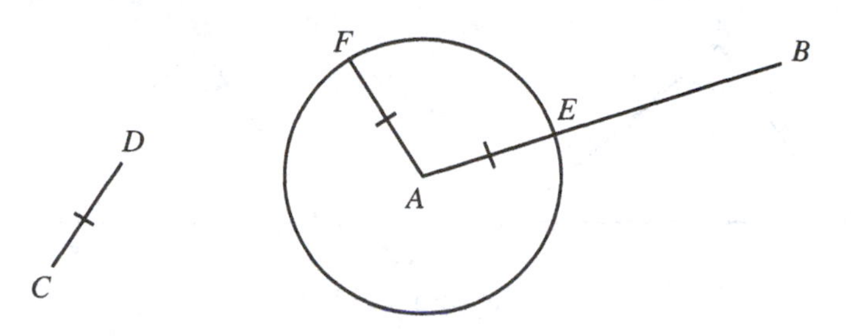

Figure 2.10.

Exercises 2.3A

1. Construct a line segment whose length is double that of a given line segment.
2. At a given point P, construct a line segment whose length is double that of a given line segment AB.
3. Construct a line segment whose length is the sum of the lengths of two given line segments AB and CD.
4. At a given point P, construct a line segment whose length equals the sum of the lengths of two given line segments AB and CD.
5. At a given point P, construct a line segment whose length equals the difference of the lengths of two given unequal line segments AB and CD.
6. At a given point P, construct a line segment whose length is triple that of a given line segment AB.
7. Comment on Proposition 2.3.1 in the context of the following geometries:
   (a) spherical (b) hyperbolic (c) taxicab (d) maxi
8. Comment on Proposition 2.3.2 in the context of the following geometries:
   (a) spherical (b) hyperbolic (c) taxicab (d) maxi
9. Comment on Proposition 2.3.3 in the context of the following geometries:
   (a) spherical (b) hyperbolic (c) taxicab (d) maxi

10(C). Perform the construction of Proposition 2.3.1 using a computer application.

11(C). Perform the construction of Proposition 2.3.2 using a computer application.

12(C). Perform the construction of Proposition 2.3.3 using a computer application.

The next proposition is the well-known SAS (side-angle-side) congruence theorem.

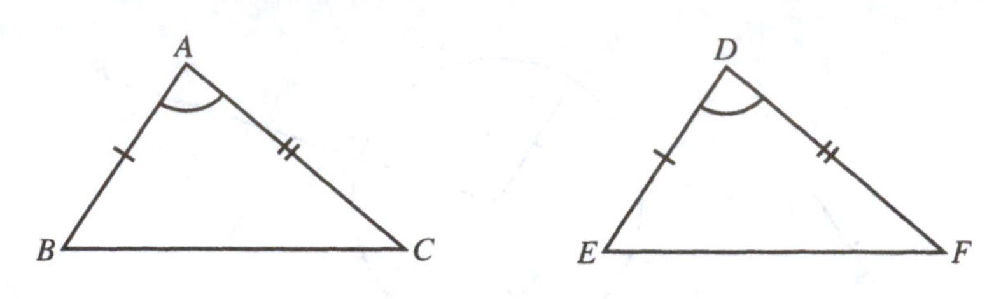

Figure 2.11.

Proposition 2.3.4 (I.4) *If two triangles have two sides equal to two sides respectively, and have the angles contained by the equal straight lines equal, they will also have the base equal to the base, the triangle will be equal to the triangle, and the remaining angles will be equal to the remaining angles, respectively, namely those which equal sides subtend.*

GIVEN: $\triangle ABC$, $\triangle DEF$, $AB = DE$, $AC = DF$, $\angle BAC = \angle EDF$ (Fig. 2.11).

TO PROVE: $BC = EF$, $\triangle ABC = \triangle DEF$, $\angle ABC = \angle DEF$, $\angle ACB = \angle DFE$.

PROOF: Apply $\triangle ABC$ to $\triangle DEF$ so that A falls on D, AB falls along DE [PT A], and C falls on the same side of DE as F. Then

	B falls on E	[AB = DE, given]
	AC falls along DF	[$\angle BAC = \angle EDF$, given]
	C falls on F	[$AC = DF$, given]
	BC falls on EF	[PT 1]
$\therefore$	$BC = EF$, $\triangle ABC = \triangle DEF$,	
	$\angle ABC = \angle DEF$, $\angle ACB = \angle DFE$.	[CN 4]

Q.E.D.

This is the first of several instances wherein Euclid uses the undefined notion of *application*. Postulate A can be used to justify this step.

The triangles ABC and DEF of Proposition 2.3.4 are such that

$$AB = DE, \quad BC = EF, \quad AC = DF,$$
$$\angle BCA = \angle EFD, \quad \angle CAB = \angle FDE, \quad \angle ABC = \angle DEF$$

(i.e., their respective sides and their respective angles are equal). Such triangles are today called *congruent* since it is intuitively clear that given any two corporeal congruent triangles, one of them can be picked up and placed on top of the other so that the corresponding sides and angles will coincide. It is important to note, though, that this term does not appear in the *Elements*. The congruence of these triangles is denoted today as $\triangle ABC \cong \triangle DEF$. It is both customary and helpful to list the vertices of these triangles in an order that is consistent with the equality of their parts. In other words, if $\triangle PQR \cong \triangle XYZ$, then it is implicitly understood that $\angle QRP = \angle YZX$, $PR = XZ$, and so on. Similarly, two polygons are said to be *congruent* provided their respective sides and angles are all equal. Here too it is taken for granted that given any two corporeal congruent polygons, one can be applied to the other so that the corresponding sides and angles will coincide.

Exercises 2.3B

1. Use the notion of an application to create a simpler "construction" for Proposition 2.
2. Use the notion of an application to provide a "proof" of Postulate 4.
3. Use the notion of an application to prove the ASA congruence theorem [PN 2.3.29] (i.e., that if two triangles have two angles of one equal, respectively, to two angles of the other, and if the sides joining these angles are also equal, then the triangles are congruent).
4. Comment on Proposition 2.3.4 in the context of the following geometries: (a) spherical (b) hyperbolic
5. Let $A = (0,0)$, $B = (1,1)$, $C = (-1,1)$, $A' = (2,0)$, $B' = (4,0)$, $C' = (2,-2)$. Show that $\triangle ABC$ and $\triangle A'B'C'$ satisfy the hypothesis of Proposition 2.3.4 in taxicab geometry but are not congruent in it.
6. Let $\triangle ABC$ be given. Show that there exists a $\triangle DEF$ such that $\triangle ABC \cong \triangle DEF$ in the Euclidean sense but not in the taxicab sense.
7. Comment on Proposition 2.3.4 in the context of maxi geometry.

Proposition 2.3.5 (I.5) *In isosceles triangles the angles at the base are equal to one another; and, if the equal straight lines be produced further, the angles under the base will be equal to one another.*

GIVEN: $\triangle ABC$, $AB = AC$ (Fig. 2.12).

TO PROVE: $\angle 1 = \angle 2$, $\angle 3 = \angle 4$.

PROOF: $\triangle ABC \cong \triangle ACB$ by SAS because

$$AB = AC \qquad \text{[Given]}$$

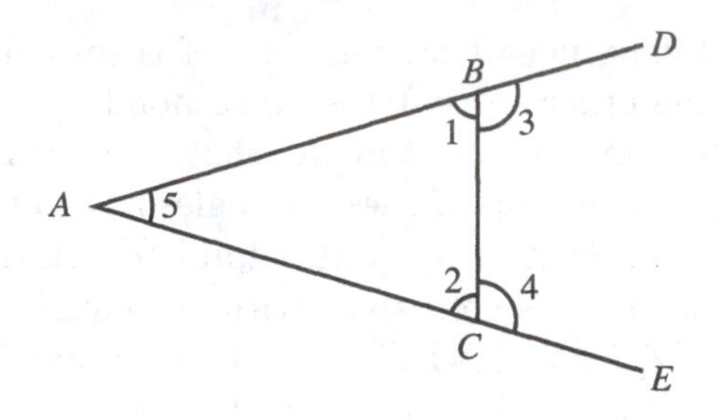

Figure 2.12.

	$AC = AB$	[Given]
	$\angle 5 = \angle 5$	
$\therefore$	$\angle 1 = \angle 2.$	
But	$\angle ABD = \angle ACE$	[Both are straight angles]
$\therefore$	$\angle 3 = \angle 4.$	[CN 3]

Q.E.D.

Euclid's proof of Proposition 2.3.5 is surprisingly long and intricate. This earned it the nickname *pons asinorum* or *ass's bridge* in the days when every schoolboy studied Euclid. The logician and philosopher Charles S. Pierce (1834–1914) commented that it "made so many boys conclude they have no capacity for geometry because this proof, the first one of any difficulty in Euclid, leaves the proposition to their minds less evident than they found it." This is the attitude that is evident in the following modern-day limerick:

In Greek mathematical forum
Your Euclid was present to bore 'em.
He spent all his time
Drawing circles sublime
And crossing the Pons Asinorum.

The proof given here is due to Pappus (ca. 300), also an Alexandrine, who wrote a now lost commentary on Euclid's *Elements*.

Exercises 2.3C

1. Prove that the three angles of an equilateral triangle are all equal to each other. What adjustments would your proof require if you accepted Euclid's definition of an isosceles triangle?

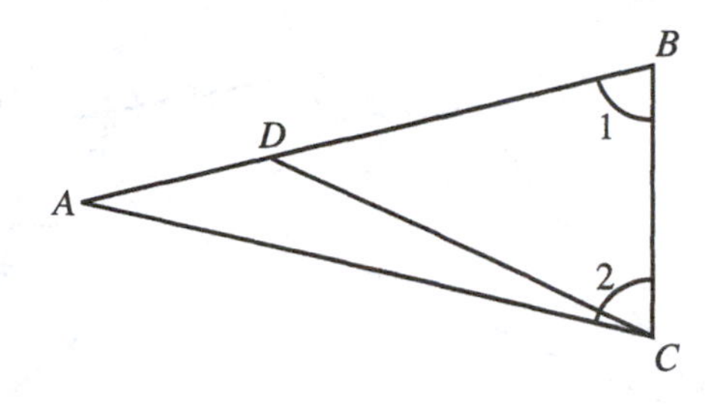

Figure 2.13.

2. In $\triangle ABC$, K and L are points on the equal sides AB and AC, respectively, such that $AK = AL$. Prove that $CK = BL$.
3. In $\triangle ABC$, $AB = AC$ and D and E are points on the side BC such that $BD = CE$. Prove that $AD = AE$.
4. Suppose $AC = AD$ and $BC = BD$, where C and D are points in the opposite sides of the straight line AB. Prove that $\triangle ABC \cong \triangle ABD$.
5. Comment on Proposition 2.3.5 in the context of the following geometries: (a) spherical (b) hyperbolic (c) taxicab (d) maxi.

Proposition 2.3.6 (I.6) *If in a triangle two angles be equal to one another, the sides which subtend the equal angles will also be equal to one another.*

GIVEN: $\triangle ABC$, $\angle 1 = \angle 2$ (Fig. 2.13).

TO PROVE: $AC = AB$.

PROOF: By contradiction. Suppose AB and AC are unequal and assume without loss of generality that $AB > AC$. Let D be that point in the interior of side AB such that $BD = AC$. Then $\triangle BCD \cong \triangle CBA$ by SAS because

$$\begin{aligned} & DB = AC && \text{[Construction]} \\ & BC = CB && \\ & \angle 1 = \angle 2 && \text{[Given]} \\ \therefore \quad & \triangle BCD = \triangle CBA. && \end{aligned}$$

This, however, contradicts common notion 5 since $\triangle DCB$ is clearly only a part of $\triangle ACB$. Hence, the original supposition was false and so $AB = AC$. Q.E.D.

The *converse* of a statement of the form *if p then q* is the statement *if q then p*. Thus, the converse of the statement *if* $a = b$ *then* $a^2 = b^2$ is the

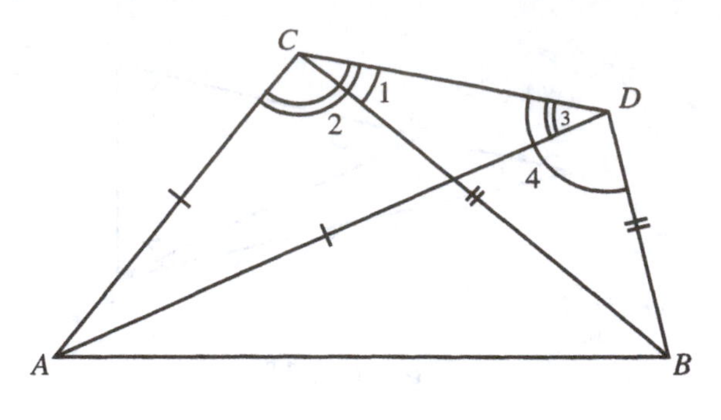

Figure 2.14.

statement *if* $a^2 = b^2$ *then* $a = b$. Similarly, each of Propositions 2.3.5 and 2.3.6 is the other's converse. As these examples attest, there is no apparent necessary relation between the logical validity of a statement and that of its converse. In the first instance the original statement *if* $a = b$ *then* $a^2 = b^2$ is valid whereas its converse *if* $a^2 = b^2$ *then* $a = b$ is demonstrably false. Similarly, the false statement *if* $a^2 = b^2$ *then* $a = b$ has the valid converse *if* $a = b$ *then* $a^2 = b^2$. Finally, the converse Propostions 2.3.5 and 2.3.6 are both valid.

Exercises 2.3D

1. Prove that if all three angles of a triangle are equal to each other then the triangle is equilateral.
2. Comment on Proposition 2.3.6 in the context of the following geometries:
   (a) spherical (b) hyperbolic (c) taxicab (d) maxi
3. Give an example of a proposition and its converse that are both false.

Proposition 2.3.7 (I.7) *Given two straight lines constructed on a straight line (from its extremities) and meeting in a point, there cannot be constructed on the same straight line (from its extremities) and on the same side of it, two other straight lines meeting in another point and equal to the former two respectively, namely, each to that which has the same extremity.*

GIVEN: Line segment AB, points C, D on the same side of AB, $CA = DA$, $CB = DB$ (Fig. 2.14).

TO PROVE: The points C and D coincide.

PROOF: By contradiction. Suppose C and D are distinct and draw the line segment CD. Since $\triangle ACD$ is isosceles, it follows from Proposition 2.3.5

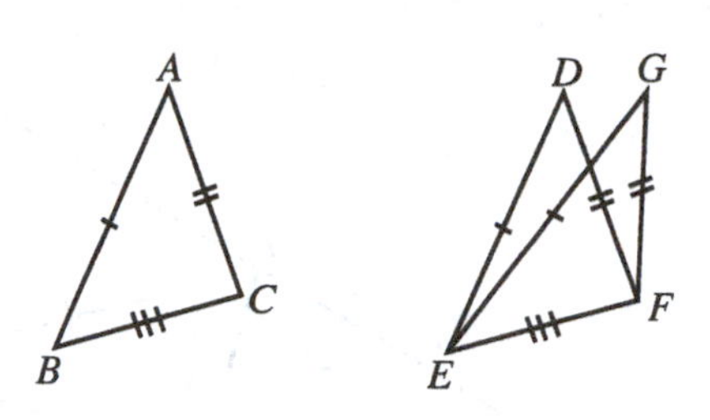

Figure 2.15.

that $\angle 2 = \angle 3$. Consequently,

$$\angle 1 < \angle 2 = \angle 3 < \angle 4.$$

This, however, contradicts the fact that $\angle 1 = \angle 4$ [because $\triangle BDC$ is also isosceles].

Hence the points C and D cannot be distinct. Q.E.D.

Exercises 2.3E

1. The proof of Proposition 2.3.7 is incomplete because it depends on the position of D relative to $\triangle ABC$. Complete this proof by considering three additional cases: one where D is inside $\triangle ABC$, one where it is in the interior of one of the sides, and one where it coincides with one of the vertices.
2. Comment on Proposition 2.3.7 in the context of the following geometries:
   (a) spherical (b) hyperbolic (c) taxicab (d) maxi

Proposition 2.3.8 (I.8) *If two triangles have their respective sides equal, then they are congruent.*

GIVEN: $\triangle ABC$, $\triangle DEF$, $AB = DE$, $BC = EF$, $AC = DF$ (Fig. 2.15).

TO PROVE: $\triangle ABC \cong \triangle DEF$.

PROOF: Apply $\triangle ABC$ to $\triangle DEF$ so that B, C fall on E, F respectively and so that the vertex A falls in position G on that side of EF that contains the point D [PT A]. Then

$$GE = AB = DE \quad \text{and} \quad GF = AC = DF$$

and hence, by Proposition 2.3.7, it follows that the points G and D coincide. Consequently, $\angle ABC = \angle GEF = \angle DEF$ and so, $\triangle ABC \cong \triangle DEF$ [SAS]. Q.E.D.

Proposition 2.3.8 is called the SSS congruence theorem.

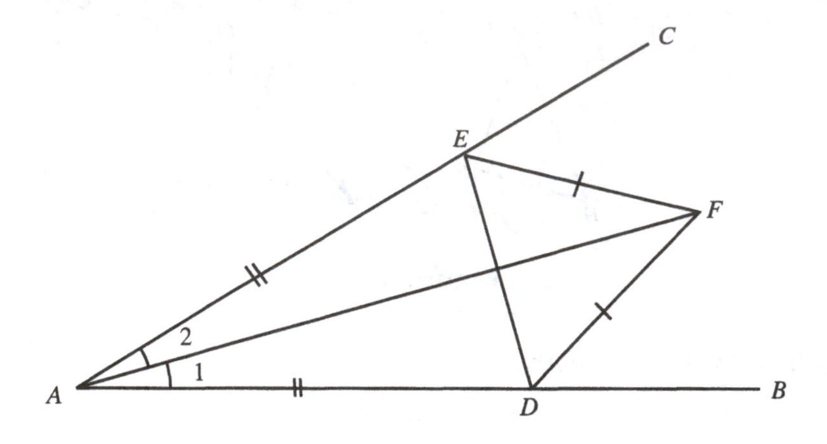

Figure 2.16.

Exercises 2.3F

1. Use Exercise 2.3C.4 to provide an alternate proof of Proposition 2.3.8 that does not rely on Proposition 2.3.7.
2. Comment on Proposition 2.3.8 in the context of the following geometries:
   (a) spherical (b) hyperbolic (c) taxicab (d) maxi

Proposition 2.3.9 (I.9) *To bisect a given rectilineal angle.*

GIVEN: $\angle BAC$ (Fig. 2.16).

TO CONSTRUCT: A line AF such that $\angle BAF = \angle FAC$.

CONSTRUCTION: Let D be any point on AB and let E be a point on AC such that $AE = AD$ [PN 2.3.3]. Let DEF be an equilateral triangle [PN 2.3.1]. Then $\angle 1 = \angle 2$.

PROOF: $\triangle DAF \cong \triangle EAF$ by SSS because

$$\begin{aligned} AD &= AE && \text{[Construction]} \\ DF &= EF && \text{[Construction]} \\ AF &= AF && \\ \therefore \quad \angle 1 &= \angle 2. && \end{aligned}$$

Q.E.D.

The line AF of this proposition is called the *angle bisector* of $\angle BAC$. In a $\triangle ABC$ it is common to refer to the portion of the angle bisector of

$\angle BAC$ that lies between the vertex A and the side BC also as one of the triangle's angle bisectors. The method used by Euclid to construct angle bisectors is different from the one commonly taught in high schools wherein the equilateral $\triangle DEF$ is replaced by two arcs of equal radii that are centered at D and E, respectively, and intersect at F. This latter construction is properly speaking not Euclidean since it assumes the compass to be rigid. Of course, this method could be justified by Proposition 2.3.2, but Euclid chose to work with Proposition 2.3.1 instead.

Proposition 2.3.9 begs the question of how to divide an arbitrary angle into any number of equal parts. There is of course no difficulty in dividing an arbitrary angle into 4, 8, or 2^n equal parts, where n is any positive integer. On the other hand, dividing an angle into three equal parts, otherwise known as the *angle trisection* problem turns out to be impossible. This problem was first formulated by the Greeks and continued to draw the interest of both professional and amateur mathematicians for over 2000 years. In 1837 the impossibility of this construction was finally demonstrated by Pierre Laurent Wantzel (1814–1848). Unfortunately, the proof is not easy and lies beyond the grasp of many amateurs, who still continue to search for a ruler and compass angle trisection. Sometimes they "succeed" and produce a construction that either misuses rulers and/or compasses or else simply produces an approximation. An example of such a popular misconstruction appears in Exercise 3.1C.32. The reader is referred to Section 4.5 for more information about this topic.

Exercises 2.3G

1. Find an alternate method of bisecting a given angle using Euclid's "collapsible" compass.
2. Prove that the angle bisectors of an isosceles triangle divide the equal sides into respectively equal line segments.
3. Prove that the bisector of the angle opposite to the base of an isosceles triangle also bisects the base and is perpendicular to it.
4. Comment on Proposition 2.3.9 in the context of the following geometries:
   (a) spherical (b) hyperbolic (c) taxicab (d) maxi
5. Divide a given angle into four equal parts.
6. Given positive integers m and n and an angle of measure α, prove that it is possible to construct an angle of measure $m\alpha/2^n$.

7(C). Perform the construction of Proposition 2.3.9 using a computer application.

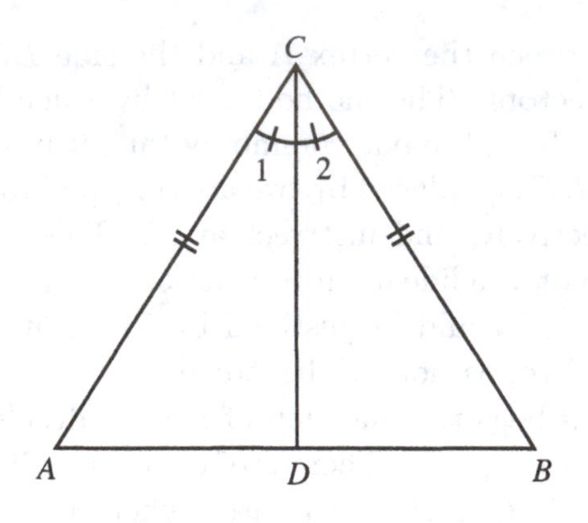

Figure 2.17.

Proposition 2.3.10 (I.10) *To bisect a given finite straight line.*

GIVEN: Line segment AB (Fig. 2.17).

TO CONSTRUCT: A point D on AB such that $AD = DB$.

CONSTRUCTION: Construct an equilateral $\triangle ABC$ [PN 2.3.1] and let D be the intersection of the bisector of $\angle ACB$ with AB. Then $AD = DB$.

PROOF: $\triangle DAC \cong \triangle DBC$ by SAS because

$$AC = BC \qquad \text{[Construction]}$$
$$\angle 1 = \angle 2 \qquad \text{[Construction]}$$
$$CD = CD$$
$$\therefore \quad AD = DB.$$

Q.E.D.

That point M of the line segment AB such that $AM = MB$ is called the *midpoint* of AB. The line segment joining a vertex of a triangle to the midpoint of the opposite side is called a *median*. The median joining vertex A of $\triangle ABC$ to the midpoint of the side BC is denoted by either m_a or m_{BC}.

Exercises 2.3H

1. Prove that the triangle formed by joining the midpoints of the three sides of an isosceles triangle is also isosceles.

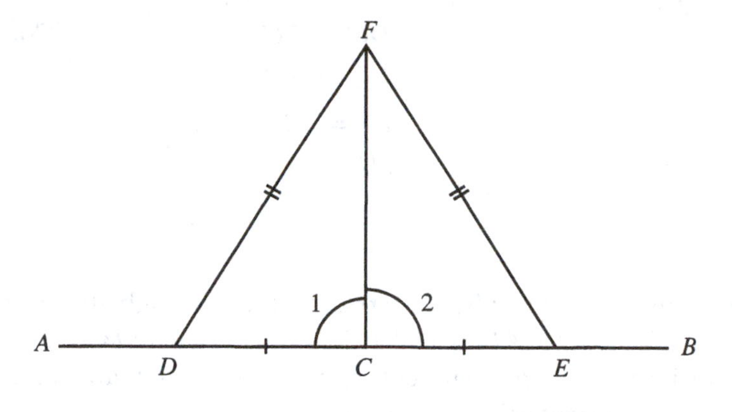

Figure 2.18.

2. Prove that the triangle formed by joining the midpoints of the three sides of an equilateral triangle is also equilateral.
3. Prove that the medians to the equal sides of an isosceles triangle are equal to each other.
4. Prove that the medians to the equal sides of an isosceles triangle divide each other into respectively equal segments.
5. Prove that the median to the base of an isosceles triangle is perpendicular to the base and bisects the opposite angle.
6. Divide a given line segment into four equal parts.
7. Comment on Proposition 2.3.10 in the context of the following geometries: (a) spherical (b) hyperbolic (c) taxicab (d) maxi

8(C). Perform the construction of Proposition 2.3.10 using a computer application.

Proposition 2.3.11 (I.11) *To draw a straight line at right angles to a given straight line.*

GIVEN: Point C on straight line AB (Fig. 2.18).

TO CONSTRUCT: Straight line $CF \perp AB$.

CONSTRUCTION: Let D be any point on AC and let E be point on BC such that $CD = CE$ [PN 2.3.3]. Construct equilateral $\triangle DEF$. Then $CF \perp AB$.

PROOF: $\triangle DCF \cong \triangle ECF$ by SSS because

$$DF = EF \qquad \text{[Construction]}$$

$$DC = EC \qquad \text{[Construction]}$$
$$FC = FC$$
$$\therefore \quad \angle 1 = \angle 2$$
$$\therefore \quad CF \perp AB. \qquad \text{[DFN 10]}$$

Q.E.D.

Given a line segment AB, the straight line through its midpoint that is also perpendicular to AB is called its *perpendicular bisector*. The next two propositions about perpendicular bisectors do not appear in Euclid's *Elements* but are nevertheless quite useful.

Proposition 2.3.12 *Every point on the perpendicular bisector of a line segment is equidistant from the segment's endpoints.*

See Exercise 1.

Proposition 2.3.13 *Every point that is equidistant from the endpoints of a line segment is on its perpendicular bisector.*

See Exercise 2.

The geometrical word for the notion of *set* or *collection* is *locus*. For example, given a point C and a real number r, the locus of all the points at distance r from C is the circle $(C; r)$. This term is frequently used to describe collections of points that have geometrically interesting properties. Thus, it is customary to combine Propositions 2.3.12 and 2.3.13 into the following one.

Proposition 2.3.14 *The locus of all the points that are equidistant from two distinct points is the perpendicular bisector to the line segment determined by these points.*

As this proposition does not say anything that isn't already contained in either Proposition 2.3.12 or Proposition 2.3.13, no proof is required.

Proposition 2.3.15 (I.12) *To a given infinite straight line, from a given point which is not on it, to draw a perpendicular straight line.*

GIVEN: Straight line $\overleftrightarrow{AB}$; point C not on $\overleftrightarrow{AB}$ (Fig. 2.19).

TO CONSTRUCT: A straight line CH such that $CH \perp \overleftrightarrow{AB}$.

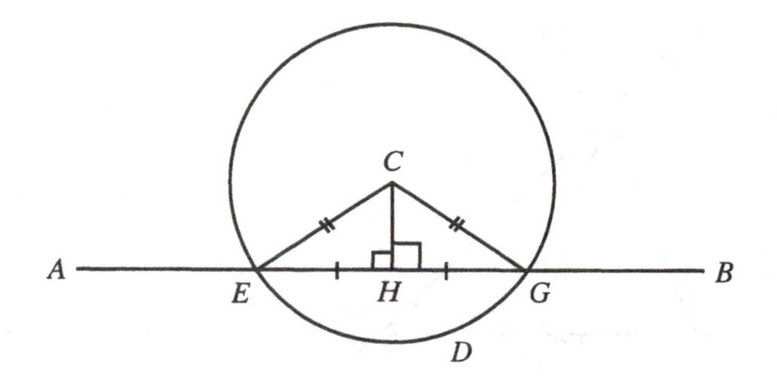

Figure 2.19.

CONSTRUCTION: Let D be any point on the side of $\overleftrightarrow{AB}$ that does not contain C. Let the circle $(C; CD)$ [PT 3] intersect the straight line in the points E and G [PT S]. Let H be the midpoint of EG [PN 2.3.10]. Then $CH \perp \overleftrightarrow{AB}$.

PROOF: Draw CG and CE. Then $\triangle CGH \cong \triangle CEH$ by SSS because

$$CG = CE \qquad \text{[Radii of the same circle]}$$
$$GH = EH \qquad \text{[Construction]}$$
$$CH = CH$$
$$\therefore \quad \angle GHC = \angle EHC$$
$$\therefore \quad CH \perp \overleftrightarrow{AB}. \qquad \text{[DFN 10]}$$

Q.E.D.

The straight line segment that joins a vertex of a triangle to a point on the opposite side and is perpendicular to that side is called an *altitude* of the triangle. The altitudes that contain the vertices A, B, C are denoted by h_a, h_b, h_c, respectively, or else h_{BC}, h_{AC}, h_{AB}.

Example 2.3.16 *Construct $\triangle ABC$ given the data β, a, m_c.* By this is meant that a $\triangle ABC$ is to be constructed in which the magnitudes of $\angle ABC$, the side BC, and the median from C to AB are prespecified. Thus, following the standard format for construction problems,

GIVEN: Angle β, line segments a, m_c (Fig. 2.20).

TO CONSTRUCT: $\triangle ABC$ such that $\angle ABC = \beta$, $BC = a$, $m_{BC} = m_a$.

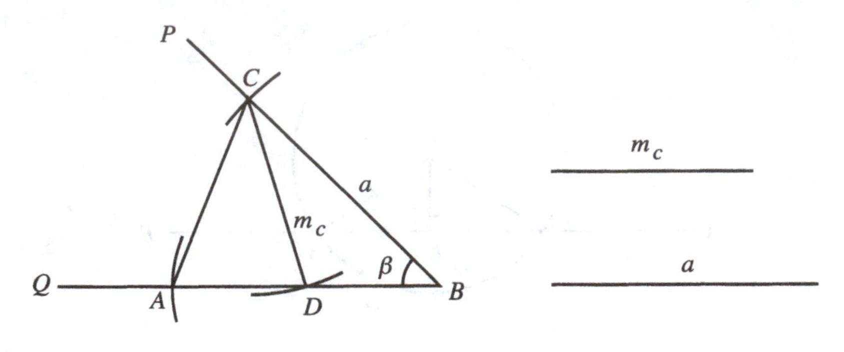

Figure 2.20.

CONSTRUCTION: Label the vertex of β with B, and its two sides with $\overleftrightarrow{BP}$ and $\overleftrightarrow{BQ}$.

Set $C = \overleftrightarrow{BP} \cap (B; a)$ [PN 2.3.3]

Set $D = \overleftrightarrow{BQ} \cap (C; m_c)$ [PN 2.3.3]

Set $A = \overleftrightarrow{BQ} \cap (D; DB)$ [PT 2.3.3]

PROOF: Self-evident. Q.E.D.

COMMENTS: If m_c is too short, there will be no solution. If m_c is long enough, there may be two solutions, depending on whether the given angle β is acute, right, or obtuse.

Exercises 2.3I

1. Prove Proposition 2.3.12.
2. Prove Proposition 2.3.13.
3. Prove that if one of the altitudes of a triangle is also a median, then the triangle is isosceles.
4. Construct an isosceles triangle in which a and h_a are given (b and c being the equal sides).
5. The point D lies either inside or outside $\angle BAC$. Construct a straight line that contains D and cuts off equal segments on the sides of $\angle BAC$.
6. Construct $\triangle ABC$ given the data b, c, h_a.
7. Construct $\triangle ABC$ given the data a, b, h_b.
8. Construct $\triangle ABC$ given the data a, m_a, h_b.

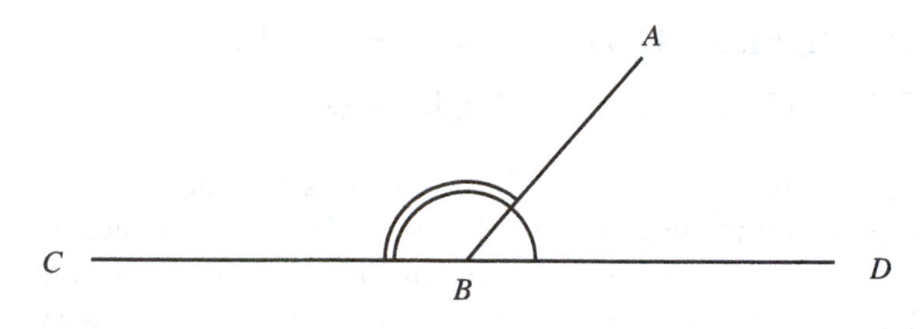

Figure 2.21.

9. Construct $\triangle ABC$ given the data a, m_b, h_b.

10. Construct $\triangle ABC$ given the data c, a, m_a.

11. Construct $\triangle ABC$ given the data a, m_a, β.

12. Construct $\triangle ABC$ given the data a, $b + c$, β.

13. Construct $\triangle ABC$ given the data a, $b - c$, γ.

14. Construct $\triangle ABC$ given the data a, $b - c$, β.

15. Comment on Proposition 2.3.11 in the context of the following geometries:
(a) spherical (b) hyperbolic (c) taxicab (d) maxi

16. Comment on Proposition 2.3.12 in the context of the following geometries:
(a) spherical (b) hyperbolic (c) taxicab (d) maxi

17. Comment on Proposition 2.3.13 in the context of the following geometries:
(a) spherical (b) hyperbolic (c) taxicab (d) maxi

18. Comment on Proposition 2.3.14 in the context of the following geometries:
(a) spherical (b) hyperbolic (c) taxicab (d) maxi

19. Comment on Proposition 2.3.15 in the context of the following geometries:
(a) spherical (b) hyperbolic (c) taxicab (d) maxi

20(C). Perform the construction of Proposition 2.3.11 using a computer application.

If A is a point not on the straight line $\overleftrightarrow{CD}$ and B is a point on it between C and D, then $\angle ABC$ and $\angle ABD$ are said to be *supplementary* (Fig. 2.21).

Proposition 2.3.17 (I.13) *The sum of two supplementary angles is equal to two right angles.*

GIVEN: Straight lines AB and CD with B on CD.

TO PROVE: $\angle CBA + \angle ABD = 2$ right angles.

This proposition will not be proved in this text as the sum of supplementary angles quite obviously equals a straight angle. Euclid, however, was somewhat hobbled by his arbitrary exclusion of the straight angle from the realm of angles (see DFN 8). Consequently, he found it necessary to replace this obvious equality by the statement that the sum of supplementary angles equals that of two right angles. Moreover, he gave a proof that is, in the author's opinion, unnecessary. The same attitude is adopted regarding the next proposition, which is the converse of the previous one.

Proposition 2.3.18 (I.14) *If with any straight line, and at a point on it, two straight lines not lying on the same side make the adjacent angles equal to two right angles, the two straight lines will be in a straight line with one another.*

GIVEN: $\angle ABC$ and $\angle ABD$ with C and D lying in distinct sides of AB. $\angle ABC + \angle ABD = 2$ right angles.

TO PROVE: BC and BD form a single straight line.

Exercises 2.3J

1. Prove that the bisectors of $\angle ABC$ and $\angle ABD$ of Proposition 2.3.17 are perpendicular to each other.
2. Comment on Proposition 2.3.17 in the context the following geometries:
   (a) spherical (b) hyperbolic (c) taxicab (d) maxi
3. Comment on Proposition 2.3.18 in the context the following geometries:
   (a) spherical (b) hyperbolic (c) taxicab (d) maxi

Proposition 2.3.19 (I.15) *If two straight lines cut one another, they make the vertical angles equal to one another.*

GIVEN: Straight lines AB and CD intersecting at E (Fig. 2.22).

TO PROVE: $\angle 1 = \angle 2$.

PROOF:

$$\angle 1 + \angle 3 = 2 \text{ right angles} \qquad [\text{PN } 2.3.17]$$

$$\angle 2 + \angle 3 = 2 \text{ right angles} \qquad [\text{PN } 2.3.17]$$

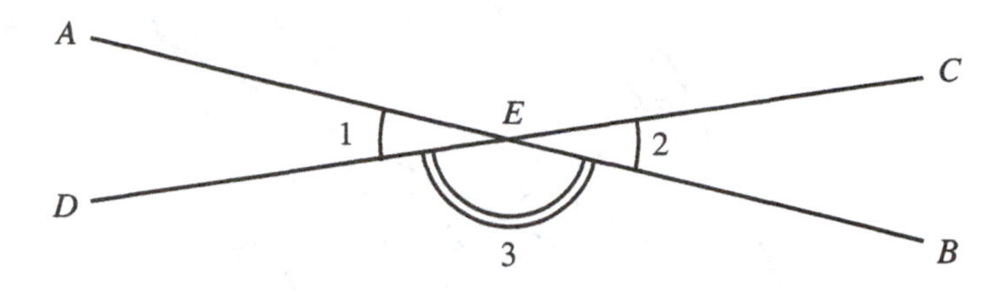

Figure 2.22.

$\therefore \quad \angle 1 = \angle 2.$ [By subtraction of $\angle 3$ & CN 3]

Q.E.D.

The angles whose equality is asserted by Proposition 2.3.19 are said to be *vertically opposite.*

Exercises 2.3K

1. Prove that if an angle's bisector is extended into the vertically opposite angle, then it bisects the latter too.
2. Prove that the bisectors of vertically opposite angles lie in the same straight line.
3. Prove that if one of the triangle's medians is also an angle bisector, then the triangle is isosceles.
4. Prove that if a quadrilateral's diagonals bisect each other, then the quadrilateral's opposite sides are equal to each other and so are its opposite angles.
5. Comment on Proposition 2.3.19 in the context of the following geometries:
   (a) spherical (b) hyperbolic (c) taxicab (d) maxi

The angle formed by one side of a polygon with the extension of an adjacent side is called an *exterior angle.*

Proposition 2.3.20 (I.16) *In any triangle, if one of the sides be produced, the exterior angle is greater than either of the interior and opposite angles.*

GIVEN: $\triangle ABC$, side BC produced to BD (Fig. 2.23).

TO PROVE: $\angle 5 > \angle 3$.

PROOF: Let E be the midpoint of AC [PN 2.3.10] and extend BE to F so that $BE = EF$ [PT 2, PN 2.3.3]. Then $\triangle AEB \cong \triangle CEF$ by SAS because

$$AE = EC \qquad \text{[Construction]}$$

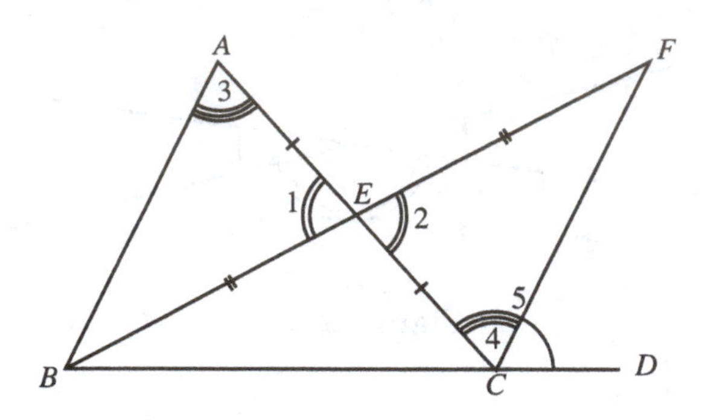

Figure 2.23.

	$\angle 1 = \angle 2$	[PN 2.3.19]
	$BE = FE$	[Construction]
$\therefore$	$\angle 3 = \angle 4.$	
But	$\angle 4 < \angle 5$	[CN 5]
$\therefore$	$\angle 3 < \angle 5.$	

Q.E.D.

Exercises 2.3L

1. Let D be a point in the interior of $\triangle ABC$. Prove that $\angle ADC > \angle ABC$.
2. Points are said to be *collinear* if they lie on one straight line. Prove that three collinear points cannot all be equidistant from the same point.
3. Prove that through a given point P there is only one straight line perpendicular to a given straight line AB.
4. Comment on Proposition 2.3.20 in the context of the following geometries: (a) spherical (b) hyperbolic (c) taxicab (d) maxi

Proposition 2.3.21 (I.17) *In any triangle two angles taken together in any manner are less than two right angles.*

GIVEN: $\triangle ABC$ (Fig. 2.24).

TO PROVE: $\angle 1 + \angle 2 < 2$ right angles.

PROOF: Extend BC to D [PT 2]. Then

	$\angle 3 + \angle 2 = 2$ right angles	[PN 2.3.17]
	$\angle 3 > \angle 1$	[PN 2.3.20]
$\therefore$	$\angle 1 + \angle 2 < 2$ right angles.	

Q.E.D.

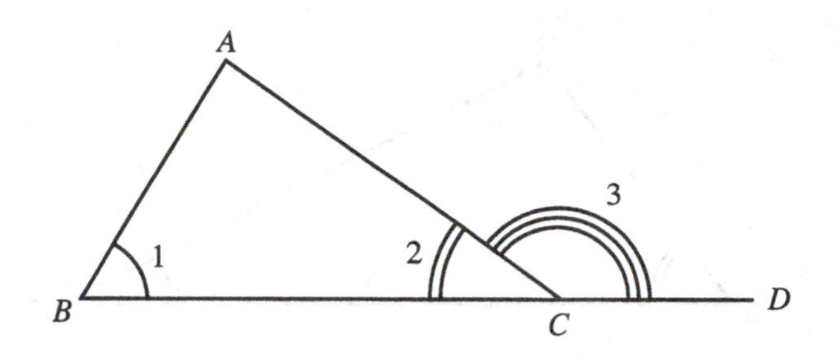

Figure 2.24.

Exercises 2.3M

1. One of the angles of $\triangle ABC$ is obtuse. Prove that the other two are acute.
2. Prove that if one of a triangle's angles is right, then the other two are acute.
3. Prove that the angles at the base of an isosceles triangle are acute.
4. Prove that every triangle has at least one altitude that is interior to it.
5. One side of a triangle is extended in both directions. Prove that the sum of the two exterior angles so formed is greater than two right angles.
6. Comment on Proposition 2.3.21 in the context of the following geometries:
   (a) spherical (b) hyperbolic (c) taxicab (d) maxi

Proposition 2.3.22 (I.18) *In any triangle the greater side subtends the greater angle.*

GIVEN: $\triangle ABC$, $AC > AB$ (Fig. 2.25).

TO PROVE: $\angle 1 > \angle 4$.

PROOF: Let D be that point inside AC such that $AD = AB$ [PN 2.3.3], and draw BD. Then

$$\angle 1 > \angle 2 \qquad \text{[CN 5]}$$

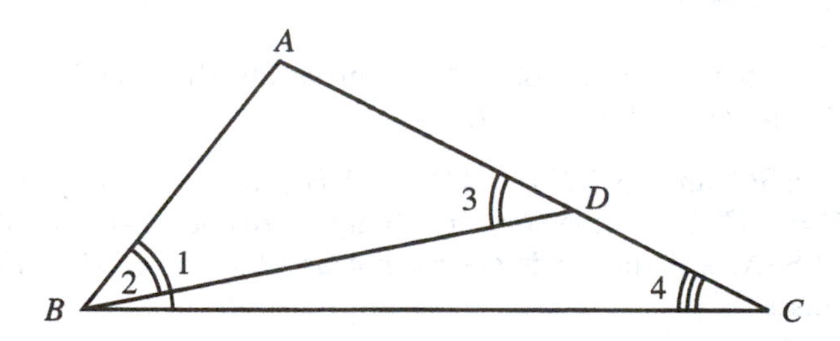

Figure 2.25.

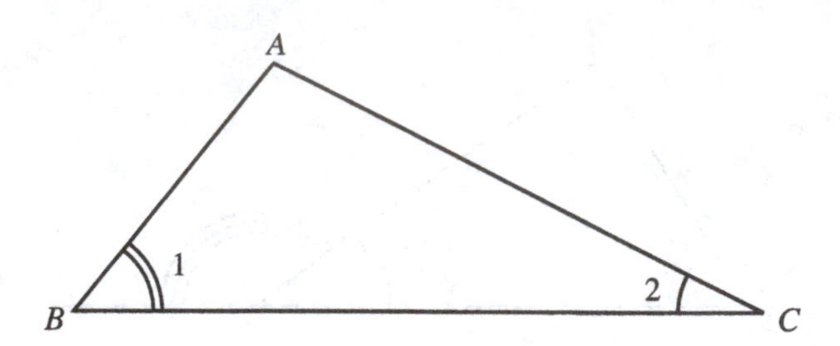

Figure 2.26.

	$\angle 2 = \angle 3$	[$\triangle ABD$ is isosceles, PN 2.3.5]
	$\angle 3 > \angle 4$	[Exterior versus interior, PN 20]
$\therefore$	$\angle 1 > \angle 4$.	

Q.E.D.

Proposition 2.3.23 (I.19) *In any triangle the greater angle is subtended by the greater side.*

GIVEN: $\triangle ABC$, $\angle 1 > \angle 2$ (Fig. 2.26).

TO PROVE: $AC > AB$.

PROOF: By contradiction. Suppose that $AC \not> AB$. Then either $AC = AB$ or $AC < AB$. However,

if $AC = AB$ then $\angle 1 = \angle 2$	[PN 2.3.5]
if $AC < AB$ then $\angle 1 < \angle 2$.	[PN 2.3.22]

Since both of the preceding conclusions contradict the given $\angle 1 > \angle 2$, it follows that $AC > AB$. Q.E.D.

Exercises 2.3N

1. If the bisector of $\angle BAC$ of $\triangle ABC$ intersects the side BC in the point D, then $AB > BD$ and $AC > CD$.
2. If in $\triangle ABC$ and $\triangle A'B'C'$ $AB = A'B'$, $BC = B'C'$, $AB < BC$, and $\angle BAC = \angle B'A'C'$, then the two triangles are congruent. (This is sometimes dubbed SSA, although it is clearly not as powerful as the other congruence theorems.)
3. Prove that in a right triangle the *hypotenuse* (the side opposite the right angle) is greater than either of the *legs* (the other two sides).

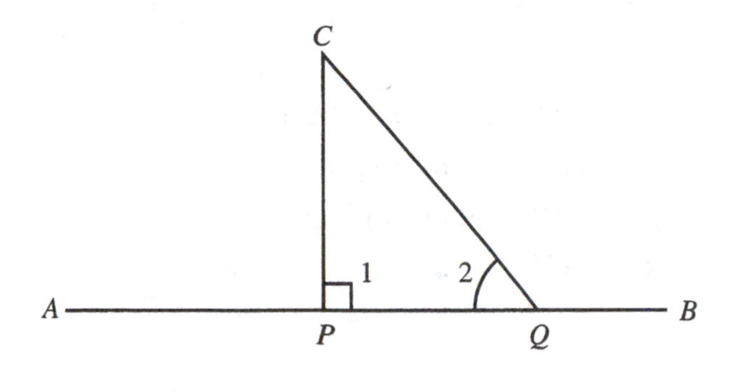

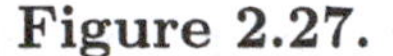
Figure 2.27.

4. Prove that if the leg and hypotenuse of one right triangle are equal to the leg and hypotenuse of another, respectively, then the two triangles are congruent.

5. In $\triangle ABC$, $AB < AC$ and N is the intersection of the bisectors of the angles at B and C. Prove that $NB < NC$.

6. Prove that the line segment joining any point in the interior of the base of an isosceles triangle to the opposite vertex is shorter than the two other sides.

7. In $\triangle ABC$, $AB < BC$ and E is the midpoint of AC. Prove that $\angle CBE < \angle ABE$.

8. In $\triangle ABC$, $AB < BC$ and the bisector of $\angle ABC$ intersects the side AC in the point D. Prove that $AD < CD$.

9. One of the angle bisectors of a triangle is also a median. Prove that the triangle is isosceles.

10. Comment on Proposition 2.3.22 in the context of the following geometries:
(a) spherical (b) hyperbolic (c) taxicab (d) maxi

11. Comment on Proposition 2.3.23 in the context of the following geometries:
(a) spherical (b) hyperbolic (c) taxicab (d) maxi

The following useful proposition does not appear explicitly in the *Elements*, but it is implicit in Definition 5 of Book III.

Proposition 2.3.24 *Of all the line segments joining a point to a straight line, the shortest is the one that is perpendicular to the given straight line.*

GIVEN: Straight line $\overleftrightarrow{AB}$, point C not on AB, points $P \neq Q$ on $\overleftrightarrow{AB}$, $CP \perp \overleftrightarrow{AB}$ (Fig. 2.27).

TO PROVE: $CP < CQ$.

PROOF:

	$\angle 1 + \angle 2 < 2$ right angles	[PN 2.3.21]
	$\angle 1 =$ right angle	[PC ⊥ AB]
$\therefore$	$\angle 2 <$ right angle $= \angle 1$	
$\therefore$	$CP < CQ$.	[PN 2.3.23]

Q.E.D.

The *distance* of the line $\overleftrightarrow{AB}$ from a point C not on it is the shortest of the segments CP, where P is any point on $\overleftrightarrow{AB}$. It follows from Proposition 2.3.24 that this distance is realized by that point P such that $CP \perp \overleftrightarrow{AB}$.

Proposition 2.3.25 (I.20) *In any triangle two sides taken together in any manner are greater than the remaining one.*

GIVEN: $\triangle ABC$ (Fig. 2.28).

TO PROVE: $AB + AC > BC$.

PROOF: Extend BA to a point D such that $AD = AC$ [PT 2, PN 2.3.3] and draw CD. Then

	$\angle 3 > \angle 2$	[CN 5]
	$\angle 2 = \angle 1$	[$\triangle ACD$ is isosceles, PN 2.3.5]
$\therefore$	$\angle 3 > \angle 1$	
$\therefore$	$BD > BC$	[PN 2.3.23 in $\triangle BCD$]
$\therefore$	$AB + AC = BA + AD = BD > BC$.	

Q.E.D.

The *perimeter* of a triangle is the sum of the lengths of its sides.

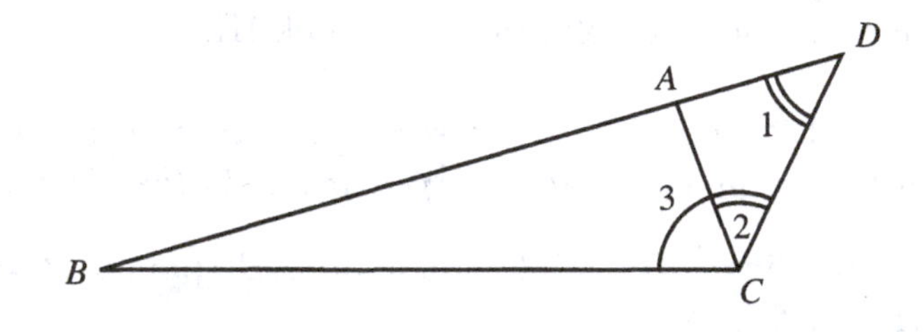

Figure 2.28.

Exercises 2.3O

1. Prove that the difference of the lengths of two sides of a triangle is less than the third side.

2. Prove that each of the triangle's sides is less than half of the triangle's perimeter.

3. Prove that half of the triangle's perimeter is greater than any line segment that joins a vertex to a point in the interior of the opposite side.

4. Prove that in a triangle one side's median is less than half the sum of the other two sides.

5. Prove that the sum of the lengths of the three line segments that join any point to the three vertices of a triangle is greater than half of that triangle's perimeter. (The point may be inside the triangle, outside it, or on one of its sides.)

6. Prove that the sum of the lengths of the four line segments that join a point to the vertices of a quadrilateral is greater than the sum of the lengths of the quadrilateral's diagonals, unless that point is the intersection of the diagonals.

7. Prove that the sum of the lengths of a quadrilateral's diagonals is less than its perimeter.

8. Prove that the sum of the lengths of the diagonals of a convex quadrilateral is greater than half its perimeter.

9. Prove that if AD is the bisector of the exterior $\angle CAE$ of $\triangle ABC$, then $BD + DC > BA + AC$.

10*. Let A, B be two points on the same side of the straight line m, and let P be an arbitrary point on m. Prove that $AP + PB$ is least when m forms equal angles with AP and BP.

11. Comment on Proposition 2.3.24 in the context of the following geometries:
(a) spherical (b) hyperbolic (c) taxicab (d) maxi

12. Comment on Proposition 2.3.25 in the context of the following geometries:
(a) spherical (b) hyperbolic (c) taxicab (d) maxi

Proposition 2.3.26 (I.21) *If on one of the sides of a triangle, from its extremities, there be constructed two straight lines meeting within the triangle, the straight lines so constructed will be less than the remaining two sides of the triangle but will contain a greater angle.*

See Exercise 1.

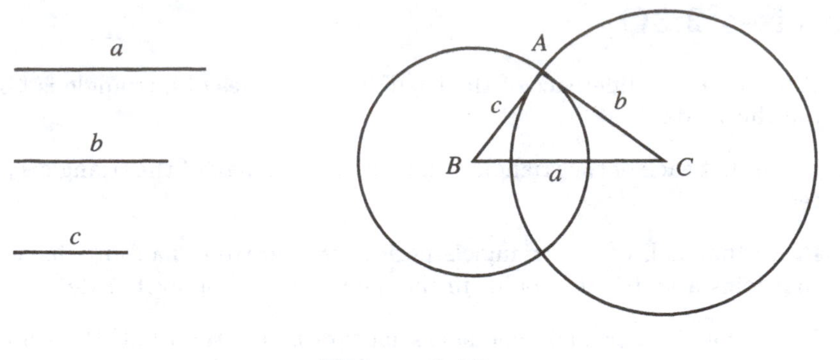

Figure 2.29.

Proposition 2.3.27 (I.22) *Out of three straight lines, which are equal to three given straight lines, to construct a triangle: thus it is necessary that two of the straight lines taken together in any manner should be greater than the remaining one.*

GIVEN: Line segments a, b, c, $a + b > c$, $b + c > a$, $c + a > b$ (Fig. 2.29).

TO CONSTRUCT: $\triangle ABC$ such that $BC = a$, $CA = b$, $AB = c$.

CONSTRUCTION: Because of the symmetry of the given inequalities, it may be assumed, without loss of generality, that $a \geq b \geq c$. Let BC be any line segment of length a. If $A = (B; c) \cap (C; b)$, then $\triangle ABC$ is the required triangle.

PROOF: It follows from the inequality $b + c > a = BC$ that the two drawn circles intersect in a point A not on BC. The choice of radii makes it clear that $AC = b$ and $AB = c$. Q.E.D.

Euclid's proof of the preceding proposition is unsatisfactory because he did not prove that his circles intersect. This was fixed in this text by assuming that $a \geq b \geq c$.

Proposition 2.3.28 (I.23) *On a given straight line and at a point on it to construct a rectilineal angle equal to a given rectilineal angle.*

GIVEN: Straight line AB and $\angle DCE$ (Fig. 2.30).

TO CONSTRUCT: $\angle FAB = \angle DCE$.

CONSTRUCTION: Draw DE. Construct $\triangle AFG$ so that G is on AB, $AG = CE$, $AF = CD$, and $FG = DE$ [PN 2.3.27]. Then $\angle FAG$ is the required angle.

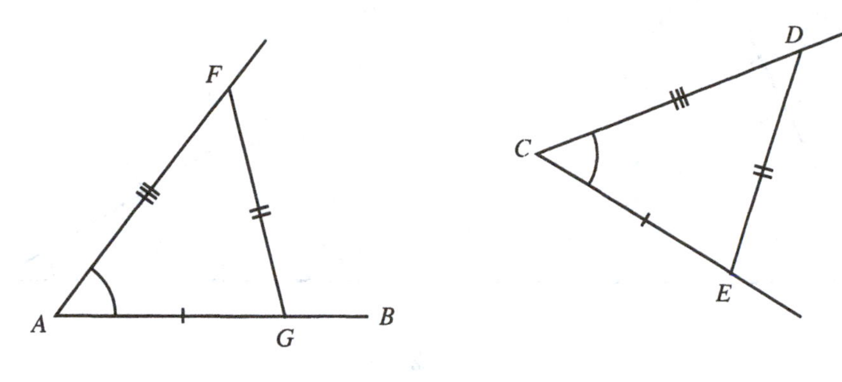

Figure 2.30.

PROOF: $\triangle AFG \cong \triangle CDE$ by SSS and so $\angle FAG = \angle DCE$. Q.E.D.

It would have been much easier to accomplish this construction by applying the given angle to the given line. That Euclid chose not to do so indicates that he was uncomfortable with this device.

Exercises 2.3P

1. Prove Proposition 2.3.26.
2. For which of the following triples of numbers does there exist a triangle whose sides have those lengths?
   (a) 6, 7, 8
   (b) 4, 6, 8
   (c) 3, 5, 8
   (d) 103, 104, 105
3. Find all numbers x such that there is a triangle the lengths of whose sides are 6, 8, x.
4. Let a and b be two positive numbers. Find all numbers x such that there is a triangle the lengths of whose sides are a, b, x.
5. Let a and b be two positive numbers such that $a > b > a/2$. Prove that there exists a triangle whose sides have lengths a^2, b^2, and $2ab$ units.
6. Construct an angle that equals the sum of two given angles.
7. Construct an angle that equals the difference of two given angles.
8. Construct $\triangle ABC$ given the data a, β, γ.
9. Construct $\triangle ABC$ given the data b, h_a, α.

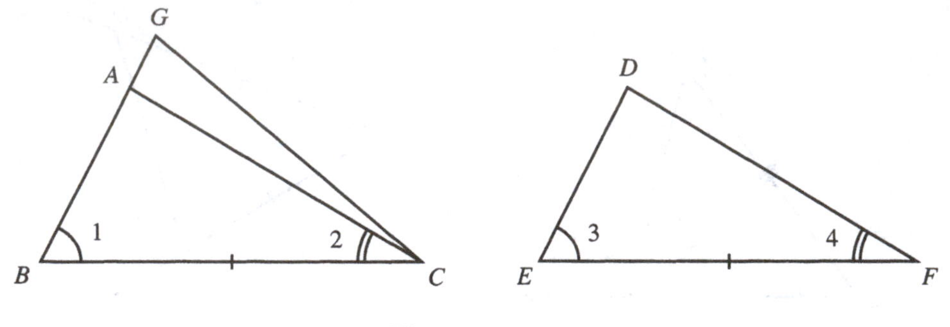

Figure 2.31.

10. Comment on Proposition 2.3.26 in the context of the following geometries:
    (a) spherical (b) hyperbolic (c) taxicab (d) maxi

11. Comment on Proposition 2.3.27 in the context of the following geometries:
    (a) spherical (b) hyperbolic (c) taxicab (d) maxi

12. Comment on Proposition 2.3.28 in the context of the following geometries:
    (a) spherical (b) hyperbolic (c) taxicab (d) maxi

13(C). Perform the construction of Proposition 2.3.28 using a computer application.

As Propositions I.24 and I.25 are not important for the development in this text, their statements and proofs have been relegated to Exercises 5 and 6, respectively.

Euclid's Proposition I.26 consists of two congruence theorems that are different enough to require separate proofs. These two are stated here separately. The following proposition is known as the ASA congruence theorem. Proposition 2.3.30, on the other hand, could be called the AAS congruence theorem. An alternative proof for ASA that makes use of Postulate A was mentioned in Exercise 2.3B.3.

Proposition 2.3.29 (I.26) *If two triangles have the two angles equal to the two angles, respectively, and the sides joining the angles are also equal, then they are congruent.*

GIVEN: $\triangle ABC$, $\triangle DEF$, $BC = EF$, $\angle 1 = \angle 3$, $\angle 2 = \angle 4$ (Fig. 2.31).

TO PROVE: $\triangle ABC \cong \triangle DEF$.

PROOF: It is first proven, by contradiction, that $AB = DE$. Suppose not, then it may be assumed without loss of generality that $AB < DE$. Let G

be a point in the extension of the segment AB such that $GB = DE$. Then $\triangle GBC \cong \triangle DEF$ by SAS because

$$\begin{array}{lll} & GB = DE & \text{[Construction]} \\ & \angle 1 = \angle 3 & \text{[Given]} \\ & BC = EF & \text{[Given]} \\ \therefore & \angle GCB = \angle 4 & \\ \therefore & \angle GCB = \angle 2. & [\angle 4 = \angle 2, \text{ given}] \end{array}$$

This last conclusion, however, contradicts CN 5, and hence it may be concluded that $AB = DE$. That $\triangle ABC \cong \triangle DEF$ now follows by SAS. Q.E.D.

Exercises 2.3Q

1. Prove that if one of the triangle's altitudes is also an angle bisector, then the triangle is isosceles.
2. Prove that if a triangle has two equal angles, then their bisectors are equal.
3. In quadrilateral $ABCD$ the diagonal AC bisects the interior angles at A and C. Prove that the interior angles at B and D are equal.
4. In $\triangle ABC$, $AB = AC$ and D, E are points in the interiors of sides AB and AC, respectively, such that $AD = AE$. Prove that if CD and BE intersect in the point O, then AO bisects $\angle BAC$.

5*. In $\triangle ABC$ and $\triangle DEF$, $AB = DE$, $AC = DF$, and $\angle BAC > \angle EDF$. Prove that $BC > EF$. (This is Proposition I.24.)

6. In $\triangle ABC$ and $\triangle DEF$, $AB = DE$, $AC = DF$, and $BC > EF$. Prove that $\angle BAC > \angle EDF$. (This is Proposition I.25.)
7. Comment on Proposition 2.3.29 in the context of the following geometries: (a) spherical (b) hyperbolic (c) taxicab (d) maxi

Proposition 2.3.30 (I.26) *If two triangles have two angles equal to two angles, respectively, and the side opposite one of the angles equal to the side opposite the corresponding angle, then the two triangles are congruent.*

See Exercise 1.

There is an easy nonneutral proof of Proposition 2.3.30 that relies on the fact that the sum of the angles of every triangle is 180°. Nevertheless, it will prove convenient to possess a neutral proof. The same is true for the following restricted, but very useful, congruence theorem, which does not appear in the *Elements*. Its intricate neutral proof is outlined in Exercises 2.3N.2–4.

Proposition 2.3.31 *If the leg and the hypothenuse of one right triangle are respectively equal to the leg and hypothenuse of another right triangle, then the triangles are congruent.*

The following two propositions about angle bisectors do not appear in the *Elements*. They are, however, standard fare in most elementary geometry texts.

Proposition 2.3.32 *Every point on an angle's bisector is equidistant from that angle's sides.*

See Exercise 5.

Proposition 2.3.33 *Every point inside an angle that is equidistant from its sides lies on its bisector.*

See Exercise 6.

The next proposition merely combines the previous two. It requires no new proof.

Proposition 2.3.34 *The locus of all the points that are equidistant from two intersecting lines are the two straight lines that bisect the four angles formed by them.*

Exercises 2.3R

1. Prove Proposition 2.3.30.
2. Prove that the altitudes to the equal sides of an isosceles triangle are also equal.
3. Prove that the altitudes to the equal sides of an isosceles triangle divide each other into segments that are respectively equal.
4. Prove that the altitude to the base of an isosceles triangle bisects both the base and the angle opposite to it.
5. Prove Proposition 2.3.32.
6. Prove Proposition 2.3.33.

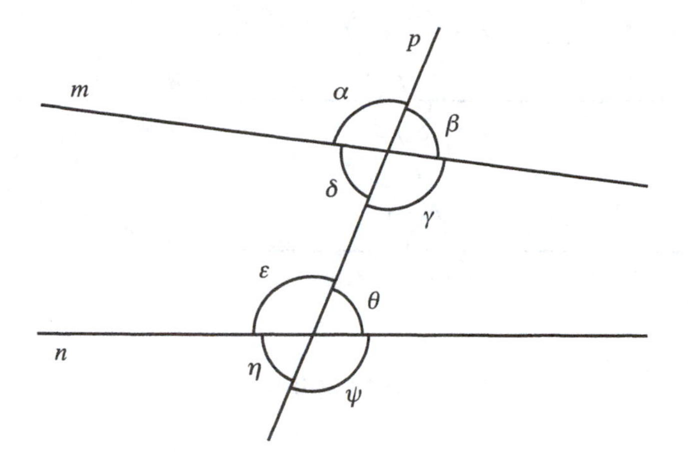

Figure 2.32.

7. Comment on Proposition 2.3.30 in the context of the following geometries:
   (a) spherical (b) hyperbolic (c) taxicab (d) maxi
8. Comment on Proposition 2.3.31 in the context of the following geometries:
   (a) spherical (b) hyperbolic (c) taxicab (d) maxi
9. Comment on Proposition 2.3.32 in the context of the following geometries:
   (a) spherical (b) hyperbolic (c) taxicab (d) maxi
10. Comment on Proposition 2.3.33 in the context of the following geometries:
   (a) spherical (b) hyperbolic (c) taxicab (d) maxi

If a straight line p intersects two other straight lines m and n, then it is convenient to attach the following nomenclature to the listed angles and pairs of angles (Fig. 2.32).

Alternate angles:	$\{\varepsilon, \gamma\}, \{\delta, \theta\}, \{\alpha, \psi\}, \{\beta, \eta\}$
Corresponding angles:	$\{\alpha, \varepsilon\}, \{\beta, \theta\}, \{\delta, \eta\}, \{\gamma, \psi\}$
Interior angles:	$\gamma, \delta, \varepsilon, \theta$
Exterior angles:	$\alpha, \beta, \eta, \psi$

Proposition 2.3.35 (I.27) *If a straight line falling on two straight lines make the alternate angles equal to one another, the straight lines will be parallel to one another.*

GIVEN: Straight line EF intersects the straight lines AB and CD in the points E and F, respectively, $\angle 1 = \angle 2$ (Fig. 2.33).

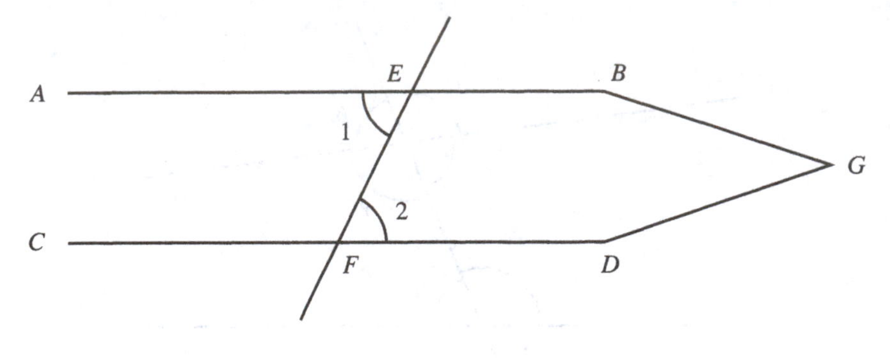

Figure 2.33.

TO PROVE: $AB \parallel CD$.

PROOF: By contradiction. Suppose AB and CD are not parallel. Then it may be assumed without loss of generality that $\overleftrightarrow{AB}$ and $\overleftrightarrow{CD}$ intersect at some point G that lies on the same side of EF as B and D. Then, however, $\angle 2$ is interior to $\triangle EFG$ whereas $\angle 1$ is exterior to it. By Proposition 2.3.20, $\angle 2 < \angle 1$, contradicting the given equality. Hence $AB \parallel CD$. Q.E.D.

Proposition 2.3.36 (I.28) *If a straight line falling on two straight lines make the exterior angle equal to the interior and opposite angle on the same side, or the interior angles on the same side equal to two right angles, the straight lines will be parallel to one another.*

See Exercise 3.

These two propositions provide a method for constructing parallel straight lines and imply the following corollary.

Proposition 2.3.37 *Through a point not on a given straight line there exists a straight line that is parallel to the given straight line.*

See Exercise 6.

Exercises 2.3S

1. If both pairs of the opposite sides of a quadrilateral are equal to each other, then they are also parallel to each other.
2. If the diagonals of a quadrilateral bisect each other, then the quadrilateral's opposite sides are parallel to each other.

3. Prove Proposition 2.3.36.
4. Comment on Proposition 2.3.35 in the context of the following geometries:
   (a) spherical (b) hyperbolic (c) taxicab (d) maxi
5. Comment on Proposition 2.3.36 in the context of the following geometries:
   (a) spherical (b) hyperbolic (c) taxicab (d) maxi
6. Given a straight line m and a point P not on m, construct a straight line n that contains P and is parallel to m.

2.4 Postulate 5 Revisited

So far no use whatsoever has been made of Postulate 5. It is somewhat puzzling that Euclid should shy away from the use of this postulate for so long, especially as this deferral results in the need for an unnatural, though ingenious, proof for the apparently incomplete Proposition 2.3.20(I.16). The reason this proposition might be viewed as incomplete is that, as will be seen in Proposition 3.1.6(I.32), the exterior angle actually equals the sum of the two interior and opposite angles—a considerably more elegant and satisfying observation. Some scholars read into this ordering of the propositions a conscious reluctance on Euclid's part to make use of the fifth postulate. They believe that Euclid felt that this postulate was unnecessary and that, with enough effort, it could be made superfluous by proving that it is a consequence of the other postulates and common notions. By 450 A.D. the search for such a proof of the Postulate 5 became an acknowledged goal of geometry. Over the centuries, the existence of a proof of this postulate turned into a holy grail for mathematicians. Many geometers occupied themselves with this project and many purported proofs were produced. All of these proofs were subsequently demonstrated to be faulty in that they turned out to rely on yet other unstated assumptions that were in fact logically equivalent to Euclid's Postulate 5. By the early 1800s the mathematical community began acknowledging the possibility that such a proof might not exist. Gauss, János Bolyai (1802–1862), Nikolay Ivanovich Lobachevsky (1792–1856), and Ernst Minding (1806–1885) found strong evidence supporting this view and in 1868 Eugenio Beltrami (1835–1900) proved conclusively that

Postulate 5 cannot be proved on the basis of the other postulates and common notions alone.

Beltrami accomplished this feat by pointing out that with the exception of Postulate 5, all of Euclid's postulates and common notions were valid in the hyperbolic geometry that had been created by Joseph Liouville (1809–1882) two decades earlier in a completely different context and for completely different purposes and that was described in Section 1.2. Postulate 5, on the other hand, does not hold in this geometry as is illustrated by Figure 2.7 where $\alpha + \beta < 180°$ and yet the geodesics p and q do not intersect. This difference between Postulate 5 and the other postulates and common notions demonstrates that this postulate cannot be a logical consequence of them alone.

Chapter Review Exercises

1. Let E be the midpoint of the median AD of the equilateral $\triangle ABC$. Prove that $AE < BE$.

2. The nonoverlapping isosceles triangles CAD and CBD have the common base CD. Prove that AB bisects both $\angle CAD$ and $\angle CBD$.

3. Let P be a point in the interior of $\triangle ABC$ in which $\angle BAC$ is obtuse. Prove that if $D = \overleftrightarrow{BP} \cap \overleftrightarrow{AC}$ and $E = \overleftrightarrow{CP} \cap \overleftrightarrow{AB}$, then $BD + CE > BE + ED + DC$.

4. Let P be a point in the interior of $\triangle ABC$ and set $D = \overleftrightarrow{AP} \cap \overleftrightarrow{BC}$ and $E = \overleftrightarrow{BP} \cap \overleftrightarrow{AC}$. Prove that if $PA = PB$ and $PD = PE$, then $\triangle ABC$ is isosceles.

5. Prove that the sum of the three altitudes of a triangle is less than the sum of its three sides.

6. Given $\triangle ABC$ and point P in its interior, let $D = \overleftrightarrow{AP} \cap \overleftrightarrow{BC}$ and $E = \overleftrightarrow{BP} \cap \overleftrightarrow{AC}$. Prove that if $\angle PBC = \angle PAC$ and $\angle PBA = \angle PAB$, then $\triangle ABC$ is isosceles.

7. Show that if a quadrilateral is convex, then the sum of the diagonals is greater than the sum of each pair of opposite sides.

8. In quadrilateral $ABCD$, $AB = AD$ and $\angle ABC = \angle ADC$. Prove that AC is the perpendicular bisector of BD.

9. Let D be the midpoint of the base BC of the isosceles $\triangle ABC$, and let E be any point on AC. Prove that the difference between DB and DE is smaller than the difference between AB and AE.

10. Let C be the midpoint of AB and let D and E be points on the same side of AB such that $AD = BE$ and $CD = CE$. Prove that $AE = BD$.

11. Let BD, CE, AF be equal segments on the respective sides BC, CA, AB of the equilateral $\triangle ABC$. Prove that if the points $M = AD \cap BE$, $N = BE \cap CF$, $P = CF \cap AD$ are distinct, then they form an equilateral triangle.

12. Given points A and B on the same side of the straight line m, determine on m a point P such that the difference $AM - MB$ is as large as possible.

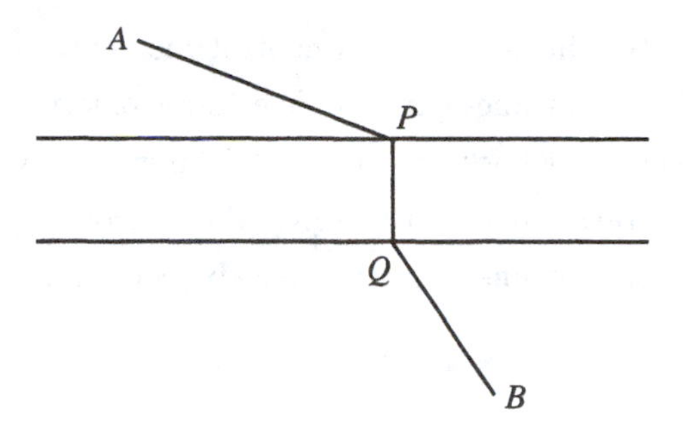

Figure 2.34.

13. It is planned to place a bridge across a straight river so that each of two given locations on the river's opposite sides are at equal distances from the nearest entry to the bridge. Where should the bridge be placed, assuming that it runs at right angles to the banks? (See Fig. 2.34.)

14**. Prove that if two of the angle bisectors of a triangle are equal, then the triangle is isosceles.

15. Explain the difference between Euclid's axioms and common notions.

16. Explain the difference between collapsible and rigid compasses. What is the significance of this difference in Euclidean geometry?

17. Are the following statements true or false? Justify your answers.

    (a) Euclid's development of geometry is error free.
    (b) Euclid gave an axiomatic definition of area.
    (c) Every proposition of Euclidean geometry is also valid in neutral geometry.
    (d) Every proposition of neutral geometry is also valid in Euclidean geometry.
    (e) Every proposition of spherical geometry is also valid in neutral geometry.
    (f) Every proposition of neutral geometry is also valid in spherical geometry.
    (g) Every proposition of hyperbolic geometry is also valid in neutral geometry.
    (h) Every proposition of neutral geometry is also valid in hyperbolic geometry.
    (i) Every proposition of taxicab geometry is also valid in neutral geometry.
    (j) Every proposition of neutral geometry is also valid in taxicab geometry.

(k) Parallel straight lines exist in neutral geometry.

(l) Parallel straight lines exist in spherical geometry.

(m) Parallel straight lines exist in Euclidean geometry.

(n) Parallel straight lines exist in hyperbolic geometry.

(o) Parallel straight lines exist in taxicab geometry.

"YOU WANT PROOF? I'LL GIVE YOU PROOF!"

Chapter 3

Nonneutral Euclidean Geometry

This chapter's propositions differ from those of the previous one in that they depend on Postulate 5 for their validity. Their proofs are therefore not valid in the context of hyperbolic geometry.

3.1 Parallelism

The first of the nonneutral propositions is the converse of Proposition 2.3.36, the last proposition of the previous chapter.

Proposition 3.1.1 (I.29) *A straight line falling on parallel straight lines makes the alternate angles equal to one another, the corresponding angles equal to one another, and the interior angles on the same side equal to two right angles.*

GIVEN: Straight lines $AB \parallel CD$, straight line EF intersecting AB and CD at G and H, respectively (Fig. 3.1).

TO PROVE: $\angle 1 = \angle 2$, $\angle 3 = \angle 4$, $\angle 2 + \angle 3 = 2$ right angles.

PROOF: By contradiction. Suppose $\angle 1$ and $\angle 2$ are unequal. Then it may be assumed without loss of generality that

$$\angle 1 > \angle 2$$

$$\therefore \quad \angle 1 + \angle 3 > \angle 2 + \angle 3$$

$$\text{But} \quad \angle 1 + \angle 3 = 2 \text{ right angles} \qquad \text{[PN 2.3.17]}$$

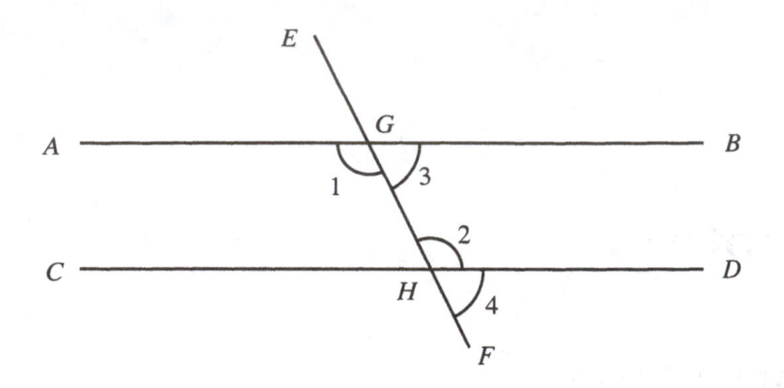

Figure 3.1.

$\therefore$ 2 right angles $> \angle 2 + \angle 3$

$\therefore$ AB and CD intersect [PT 5]

This however, contradicts the fact that $AB \parallel CD$ and so $\angle 1 = \angle 2$. Moreover,

$$\angle 1 + \angle 3 = \angle 2 + \angle 4 = 2 \text{ right angles} \quad \text{[PN 2.3.17]}$$

$$\therefore \quad \angle 3 = \angle 4 \quad \text{[CN 3]}$$

$$\text{and also} \quad \angle 2 + \angle 3 = 2 \text{ right angles.}$$

Q.E.D.

The following proposition provides an alternative and intuitively appealing characterization of parallel lines.

Proposition 3.1.2 *The locus of all points on one side of a straight line that are equidistant from it is a straight line.*

See Exercise 11.

Exercises 3.1A

1. Prove that if two parallel straight lines are cut by a third line then the two bisectors of a pair of alternate interior angles are parallel to each other.
2. Prove that if a straight line is perpendicular to one of two parallel straight lines then it is also perpendicular to the other one.
3. Suppose $AB \perp KL$ and $CD \perp MN$ are all straight lines such that $KL \parallel MN$. Prove that $AB \parallel CD$.

4. Suppose $AB \perp KL$ and $CD \perp MN$ are all straight lines such that $KL \nparallel MN$. Prove that $AB \nparallel CD$.

5. Prove that two angles whose sides are respectively parallel are either equal or supplementary.

6. Prove that two angles whose sides are respectively perpendicular are either equal or supplementary.

7. In $\triangle ABC$, AD is the bisector of $\angle BAC$ and E is a point on AC such that $DE \parallel AB$. Prove that $AE = DE$.

8. For a given $\triangle ABC$, $AD \parallel BC$ and $AD = AB$. Prove that BD bisects either the interior angle or the exterior angle at B.

9. Prove that if the points A, B are on the same side of the straight line m and at the same distance from m, then $AB \parallel m$.

10. Prove that if the points A, B are such that $AB \parallel m$, then they are at the same distance from m.

11. Use the above two exercises to prove Proposition 3.1.2.

12. Prove that the internal bisectors of each pair of angles of a triangle intersect.

13. Given two distinct parallel lines, construct a straight line that is parallel to both and also equidistant from both.

14. Comment on Proposition 3.1.1 in the context of the following geometries:
    (a) spherical; (b) hyperbolic; (c) taxicab; (d) maxi.

15. Comment on Proposition 3.1.2 in the context of the following geometries:
    (a) spherical; (b) hyperbolic; (c) taxicab; (d) maxi.

Proposition 3.1.3 (I.30) *(Distinct) Straight lines parallel to the same straight line are also parallel to one another.*

GIVEN: Distinct straight lines $AB \parallel EF$, $CD \parallel EF$ (Fig. 3.2).

TO PROVE: $AB \parallel CD$.

PROOF: By contradiction. Suppose $\overleftrightarrow{AB}$ and $\overleftrightarrow{CD}$ intersect in some point I. Join I to any point J of EF. Then

$$\angle 1 = \angle 3 \qquad [\text{PN 3.1.1, } AB \parallel EF]$$

$$\angle 2 = \angle 3 \qquad [\text{PN 3.1.1, } CD \parallel EF]$$

$$\therefore \quad \angle 1 = \angle 2 \qquad [\text{CN 1}]$$

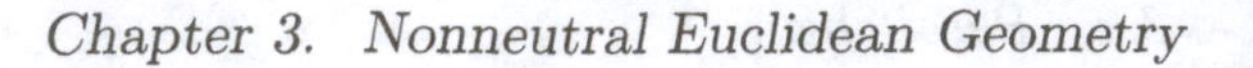

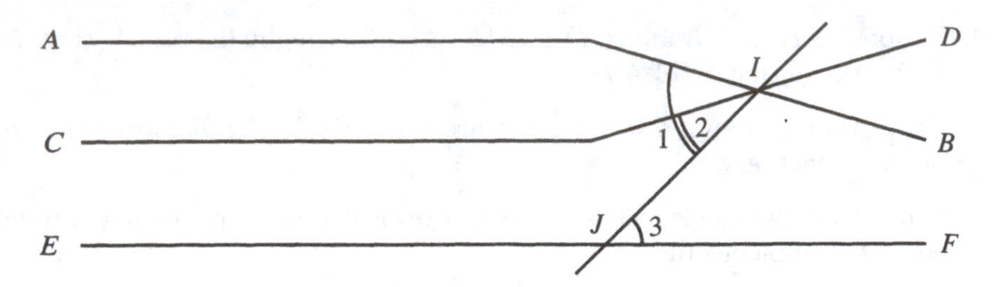

Figure 3.2.

but this is impossible since the straight lines $\overleftrightarrow{AB}$ and $\overleftrightarrow{CD}$ are distinct. Hence $\overleftrightarrow{AB} \parallel \overleftrightarrow{CD}$. Q.E.D.

Euclid began his proof of this proposition by drawing a straight line PQ that intersects all the three given lines. While intuitively plausible, the existence of such a line calls for a justification and Euclid's proof is therefore incomplete. The need for such a justification is demonstrated by Figure 3.3, which exhibits three pairwise parallel hyperbolic geodesics such that no single geodesic intersects all three.

Exercises 3.1B

1. Prove that if a straight line intersects one of two parallel straight lines (in only one point), then it also intersects the other one.
2. Comment on Proposition 3.1.3 in the context of the following geometries:
   (a) spherical (b) hyperbolic (c) taxicab (d) maxi

Proposition 3.1.4 (I.31) *Through a given point to draw a straight line parallel to a given straight line.*

GIVEN: Straight line BC, point A not on $\overleftrightarrow{BC}$ (Fig. 3.4).

TO CONSTRUCT: A straight line AE such that $AE \parallel BC$.

Figure 3.3. Three hyperbolic parallel straight lines that are *not* intersected by the same hyperbolic straight line

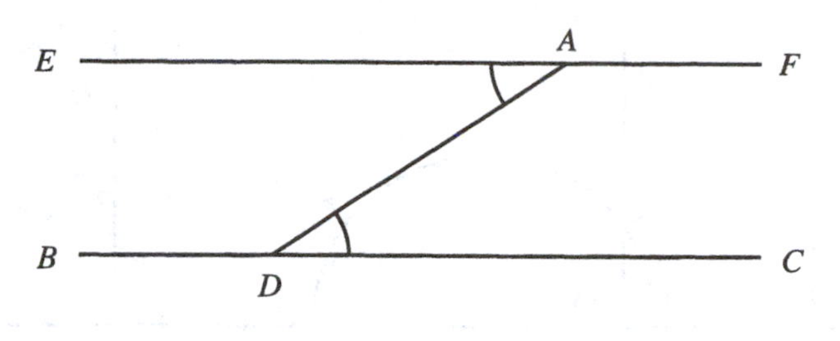

Figure 3.4.

CONSTRUCTION: Let D be any point on BC and draw AD. Construct $\angle DAE = \angle ADC$ [PN 2.3.28]. Then AE is the required straight line.

PROOF:

$$\angle EAD = \angle CDA \qquad \text{[Construction]}$$

$$\therefore \quad AE \parallel BC \qquad \text{[PN 2.3.35]}$$

Q.E.D.

The following proposition has supplanted Euclid's Postulate 5 in many texts where it is known as *Playfair's Postulate*. Although this will not be demonstrated here, the two are in fact logically equivalent.

Proposition 3.1.5 (Playfair's Postulate) *Through a point not on a given straight line there exists exactly one straight line that is parallel to the given line.*

See Exercises 1 and 2.

Just like Postulate 5, Playfair's postulate does not hold in hyperbolic geometry. Figure 3.5 exhibits three distinct geodesics p, q, r, all of which contain the same point P and all of which are parallel to the same geodesic m.

It is now possible to give a more precise definition of hyperbolic geometry. This calls for negating Playfair's postulate, which is equivalent to Euclid's Postulate 5. In view of Proposition 2.3.37, the following postulate is the proper negation of Playfair's postulate.

H (Hyperbolic). *There exists a straight line that is parallel to two distinct intersecting straight lines.*

Hyperbolic geometry is the geometry based on Euclid's Postulates 1, 2, 3, 4, A, S and Postulate H.

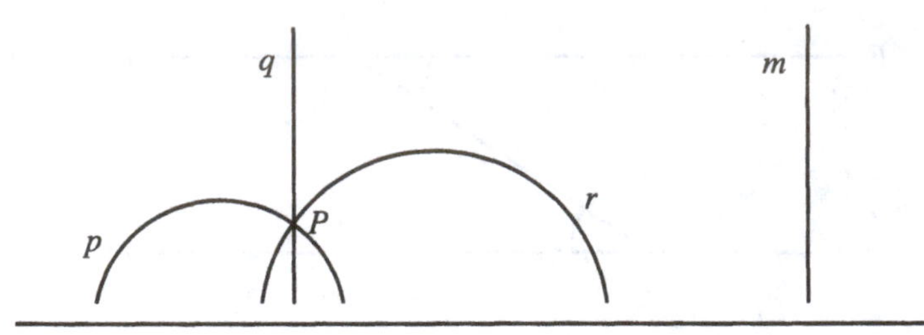

Figure 3.5. A hyperbolic counterexample to Playfair's postulate

Proposition 3.1.6 (I.32) *In any triangle, if one of the sides be produced, the exterior angle is equal to the two interior and opposite angles, and the three interior angles of the triangle are equal to two right angles.*

GIVEN: $\triangle ABC$, side BC extended to D (Fig. 3.6).

TO PROVE: $\angle ACD = \angle ABC + \angle CAB$,

$$\angle ABC + \angle BCA + \angle CAB = 2 \text{ right angles.}$$

PROOF: Draw $CE \parallel AB$ [PN 3.1.4]. Then

	$\angle 5 = \angle 2$	[PN 3.1.1]
	$\angle 6 = \angle 1$	[PN 3.1.1]
$\therefore$	$\angle 4 = \angle 2 + \angle 1$	CN 2
$\therefore$	$\angle 4 + \angle 3 = \angle 1 + \angle 2 + \angle 3$	[CN 2]
$\therefore$	2 right angles $= \angle 1 + \angle 2 + \angle 3$.	[PN 2.3.17]

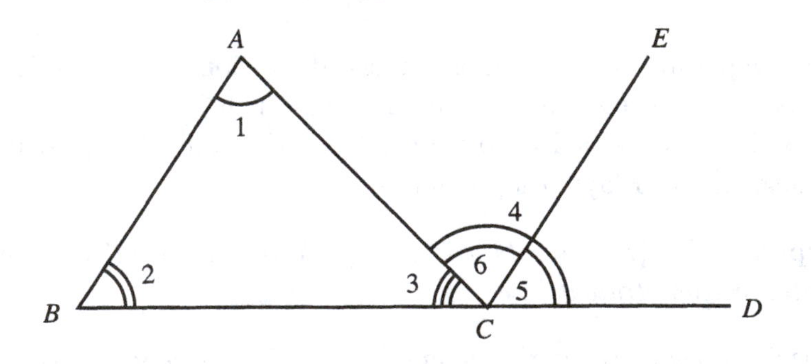

Figure 3.6.

The neutral analog of Proposition 3.1.6 is Proposition 2.3.20, which asserts that the exterior angle is greater than either of the two interior and opposite angles. Recall also that according to Chapter 1, the sum of the angles of every spherical triangle is greater than 180° [PN 1.1.5] whereas the sum of the angles of every hyperbolic triangle is less than 180° [PN 1.2.6].

Exercises 3.1C

1. Prove Proposition 3.1.5.
2. Prove that in the context of neutral geometry Playfair's postulate implies Euclid's Postulate 5.
3. Prove that in $\triangle ABC$ the bisector of the exterior angle at A is parallel to BC if and only if $AB = AC$.
4. Prove that a straight line that is parallel to one side of an isosceles triangle cuts off another isosceles triangle. (*Note:* There are two distinct cases to be considered here.)
5. A straight line cuts off an isosceles triangle from a given isosceles triangle. Prove that the straight line is parallel to one of the sides of the given isosceles triangle.
6. In an isosceles $\triangle ABC$, a line perpendicular to the base BC intersects $\overleftrightarrow{AB}$ and $\overleftrightarrow{AC}$ in the points D and E, respectively. Prove that $\triangle ADE$ is also isosceles.
7. In $\triangle ABC$, $\angle BAC = 90°$ and $\angle ACB = 30°$. Prove that $BC = 2AB$.
8. In $\triangle ABC$, $\angle ABC = 60°$ and $BC = 2AB$. Prove that $\triangle ABC$ is a right triangle.
9. Prove that in a right triangle the angle between the altitude to the hypotenuse and one of the legs equals the angle opposite that leg.
10. Let D be that point on side BC of $\triangle ABC$ such that AD is the bisector of $\angle BAC$. Prove that $\angle ADC$ is half the sum of the interior angle at B and the exterior angle at C.
11. Prove that in $\triangle ABC$ the bisectors of the interior angle at B and the exterior angle at A form an angle that is half the interior angle at C.
12. Prove that in $\triangle ABC$ the angle bisector and the altitude at A form an angle that is half the difference between the interior angles at B and C.
13. Prove that in a right $\triangle ABC$ the bisector of $\angle ABC$, the altitude to the hypotenuse BC, and the side AC form an isosceles triangle.
14. The point D on the hypotenuse BC of the right isosceles $\triangle ABC$ is such that $BD = AB$. Prove that $\angle BAD = 67.5°$.

15. Prove that if the diagonals of quadrilateral $ABCD$ are equal and the sides $AB = CD$, then $AD \parallel BC$.

16. Prove that the sum of the interior angles of a quadrilateral is 360°.

17. Prove that in quadrilateral $ABCD$ the bisectors of the interior angles at A and B form an angle that is half the sum of the interior angles at C and D, and, if the bisectors of the interior angles at A and C intersect, they form an angle that is half the difference between the angles at B and D.

A polygon is said to be convex if all of its diagonals fall in its interior.

17. Prove that the sum of the interior angles of a convex n-sided polygon is $(n-2)180°$.

18. Prove that the sum of the interior angles of an arbitrary n-sided polygon is $(n-2)180°$.

19. Prove that the number of acute interior angles of a convex polygon cannot exceed 3.

20. Prove that the sum of the exterior angles of a convex polygon is 360°. (Try to prove this without making use of Exercise 17.)

21. Construct $\triangle ABC$ given the data a, α, β.

22. Construct an isosceles triangle given one of its angles and one of its sides.

23. Construct $\triangle ABC$ given the data a, $b+c$, α.

24. Construct $\triangle ABC$ given the data $b+c$, α, β.

25. Comment on Proposition 3.1.4 in the context of the following geometries:
    (a) spherical (b) hyperbolic (c) taxicab (d) maxi

26. Comment on Proposition 3.1.5 in the context of the following geometries:
    (a) spherical (b) hyperbolic (c) taxicab (d) maxi

27. Comment on Proposition 3.1.6 in the context of the following geometries:
    (a) spherical (b) hyperbolic (c) taxicab (d) maxi

28. Explain why there are no rectangles in spherical geometry.

29. Explain why there are no rectangles in hyperbolic geometry.

30. Are there rectangles in taxicab geometry?

31. Are there rectangles in maxi geometry?

32. The following method for trisecting an arbitrary angle is credited to Archimedes. If that attribution is correct, he must have been aware of its shortcomings as a construction in the sense of Euclid.

    Let α be a given angle with vertex A (Fig. 3.7). Draw a circle centered at A of radius $AB = AC$. On a ruler mark two points D and E such that

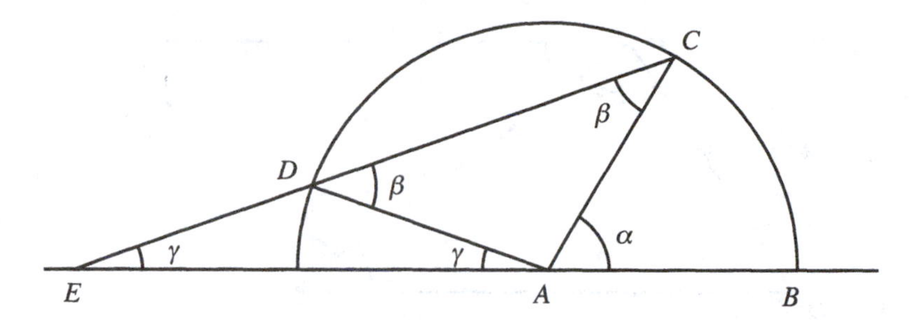

Figure 3.7. An angle "trisection"

$DE = AB = AC$ and place the ruler on the page so that the point E falls on the extension of AB, the point D falls on the circle $(A; AB)$ and the ruler also passes through the point C. Prove the following assertions:

(a) $\angle ADC = \angle ACD\ (= \beta)$

(b) $\angle AED = \angle EAD\ (= \gamma)$

(c) $\alpha = 3\gamma$, or $\gamma = \alpha/3$

Explain why this "trisection" of α does not meet Euclid's standards for a construction.

33. Criticize the following "neutral proof" of Playfair's postulate, offered by Proclus (410–485): "I say that if any straight line cuts one of two parallels, it will cut the other also. For let AB, CD be parallel and let EFG cut AB [at F, with G between AB and CD]; I say that it will cut CD also. For, since BF, FG are two straight lines from one point F, they have, when produced indefinitely, a distance greater than any magnitude, so that it will be greater than the interval between the parallels. Whenever, therefore, they are at a distance from one another greater than the distance between the parallels, FG will cut CD."

34. Criticize the following "proof" of the fact that the sum of the interior angles of a triangle is 180°. Let ABC be a given triangle let d be a line segment that lies on the straight line AB with its center at A. Slide d along AB until its center falls on B and then rotate it through the exterior of the triangle, about B as a pivot, until it falls along side BC. Next slide d along BC until its center reaches C and rotate it about C as a pivot through the exterior of the triangle until it falls along CA. Finally, slide d along CA until its center reaches A and rotate it about A as pivot so that it comes into its initial position. If the triangle's interior angles are α, β, γ, then the segment d has been rotated successively by the angles $180° - \beta$, $180° - \gamma$, and $180° - \alpha$ before it returns to its original position. Consequently, $(180° - \beta) + (180° - \gamma) + (180° - \alpha) = 360°$, from which it follows that $\alpha + \beta + \gamma = 180°$.

35. Construct $\triangle ABC$ given the data α, β, h_c.

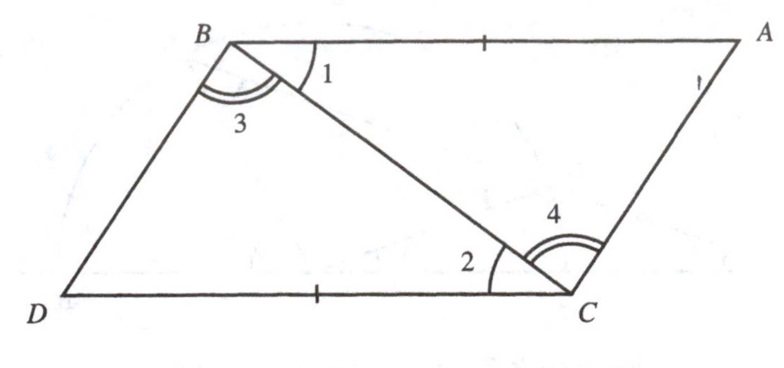

Figure 3.8.

36. Construct $\triangle ABC$ given the data α, β, h_a.

37(C). Perform the construction of Proposition 3.1.4 using a computer application.

38(C). Use a computer application to verify Proposition 3.1.6.

Euclid's statement of the following proposition is awkward, and so it appears here in a paraphrased form.

Proposition 3.1.7 (I.33) *A quadrilateral in which two opposite sides are both equal and parallel to each other is a parallelogram.*

GIVEN: Quadrilateral $ABCD$, $AB \parallel CD$, $AB = CD$ (Fig. 3.8).

TO PROVE: $AC \parallel BD$, $AC = BD$.

PROOF: Draw BC. Then $\triangle ABC \cong \triangle DCB$ by SAS because

$$AB = DC \qquad \text{[Given]}$$
$$\angle 1 = \angle 2 \qquad \text{[Alternating angles, } AB \parallel CD\text{, PN 3.1.1]}$$
$$BC = CB$$
$$\therefore \quad AC = DB \quad \text{and} \quad \angle 4 = \angle 3$$
$$\therefore \quad AC \parallel BD. \qquad \text{[Equal alternating angles, PN 2.3.35]}$$

Q.E.D.

Proposition 3.1.8 (I.34) *If both pairs of opposite sides of a quadrilateral are parallel to one another, then they as well as the opposite angles are equal to one another, and the diameter bisects the area.*

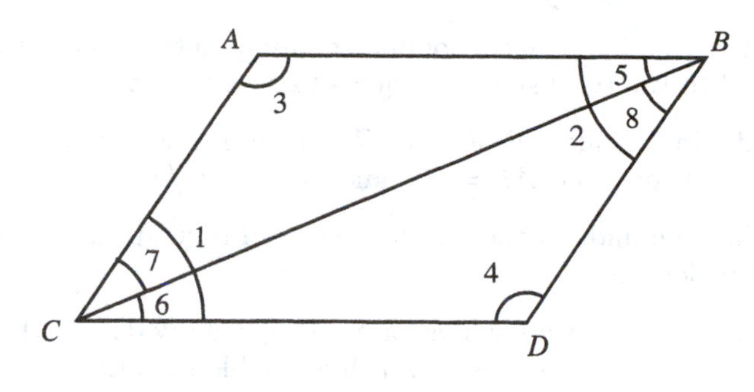

Figure 3.9.

GIVEN: Quadrilateral $ACDB$, $AB \parallel CD$, $AC \parallel BD$ (Fig. 3.9).

TO PROVE: $AB = CD$, $AC = BD$, $\angle CAB = \angle BDC$, $\angle ABD = \angle DCA$, $\triangle ABC = \triangle DCB = (1/2)\square ABDC$.

PROOF: $\triangle ABC \cong \triangle DCB$ by ASA because

$\angle 5 = \angle 6$ [Alternating angles, $AB \parallel CD$, PN 3.1.1]

$BC = CB$

$\angle 7 = \angle 8$ [Alternating angles, $AC \parallel DB$, PN 3.1.1]

$\therefore$ $AB = CD, \quad AC = BD, \quad \angle 3 = \angle 4$

$\triangle ABC = \triangle DCB = (1/2)\square ABDC$

Also $\angle 1 = \angle 2.$ [CN 2]

Q.E.D.

Exercises 3.1D

1. Both pairs of opposite sides of a quadrilateral are equal to each other. Prove that the quadrilateral is a parallelogram.
2. Both pairs of opposite angles of a quadrilateral are equal to each other. Prove that the quadrilateral is a parallelogram.
3. Prove that the diagonals of a parallelogram bisect each other.
4. Prove that if the diagonals of a quadrilateral bisect each other, then it is a parallelogram.
5. Prove that a parallelogram is a rectangle if and only if its diagonals are equal to each other.
6. Prove that a parallelogram is a rhombus if and only if its diagonals are perpendicular to each other.

7. Prove that the line segment joining the midpoints of two sides of a triangle is parallel to the third side and equals half its length.

8. The midpoint of side AB of $\triangle ABC$ is D and E is a point of AC such that $DE \parallel BC$. Prove that $AE = EC$ and $DE = BC/2$.

9. Prove that the midpoints of the four sides of a quadrilateral are the vertices of a parallelogram.

10. Prove that each of two medians of a triangle is divided by their intersection point into two segments one of which is double the other.

11. Prove that the three medians of a triangle all pass through one point.

12*. Point E in the interior of square $ABCD$ is such that $\angle ABE = \angle BAE = 15°$. Prove that $\triangle CDE$ is equilateral.

13. In $\triangle ABC$, $AB = AC$, and D, E, F are points on the interiors of sides BC, AB, AC, respectively, such that $DE \perp AB$ and $DF \perp AC$. Prove that the value of $DE + DF$ is independent of the location of D.

14. Prove that the three segments joining the midpoints of the three sides of a triangle divide it into four congruent triangles.

15. Prove that three parallel straight lines that cut off equal line segments on one straight line also cut off equal line segments on every straight line that intersects them.

16. A *trapezoid* is a quadrilateral two of whose sides are parallel. Prove that the line segment joining the midpoints of the nonparallel sides of a trapezoid is parallel to the other two sides and equals half the sum of their lengths.

17. Construct angles of the following magnitudes:

    (a) 60° (b) 30° (c) 120° (d) 75°

18. Through a given point construct a straight line such that its portion between two given parallel straight lines is equal to a given line segment.

19. Let A be a point in the interior of an angle. Construct a straight line whose segment between the sides of the angle has A as its midpoint.

20. A pair of parallel straight lines is intersected by another pair of parallel straight lines. Through a given point construct another straight line on which the two given pairs cut off equal line segments.

21. Given an angle, determine the locus of all the points the sums of whose distances from the sides of the angle equals a given magnitude.

22. In a given $\triangle ABC$, construct points M on AB and N on BC such that $AM + NC = MN$ and $MN \parallel BC$.

23. Construct $\triangle ABC$ given the data

(a) a, h_a, β

(b) a, h_a, h_b

(c) a, h_b, α

(d) h_b, h_c, α

(e) h_b, m_c, α

(f) α, h_c, $b+c$

(g) $a+b+c$, β, γ

(h) $a+b+c$, β, h_a

24. Construct a parallelogram given

    (a) two adjacent sides and the included angle

    (b) two adjacent sides and a diagonal

    (c) two adjacent sides and the distance between two opposite sides

    (d) a side and the two diagonals

    (e) the diagonals and the angles between them

25. Construct a rectangle given one side and the diagonal.

26. Construct a rhombus given

    (a) its side and one of its angles

    (b) its side and one diagonal

    (c) both diagonals

27. Construct a square given
    (a) its side (b) its diagonals

28. Construct $\triangle ABC$ given the data
    (a) a, c, m_b (b) a, h_a, m_b

29. Comment on Proposition 3.1.7 in the context of the following geometries:
    (a) spherical (b) hyperbolic (c) taxicab (d) maxi

30. Comment on Proposition 3.1.8 in the context of the following geometries:
    (a) spherical (b) hyperbolic (c) taxicab (d) maxi

3.2 Area

Euclid defined the concept of area by means of axioms that he called *common notions*. This axiomatic approach is customary today as well, although the specific axioms are different from those used by Euclid. The modern approach to area stipulates that a certain unit of length, called *unit*, has

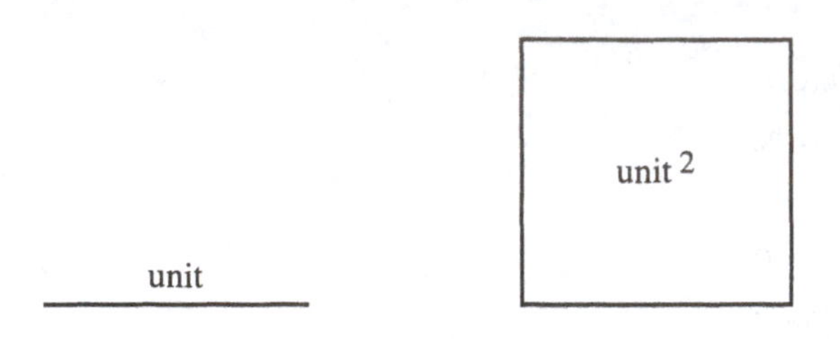

Figure 3.10.

been chosen. The square the length of whose side is 1 unit (Fig. 3.10) is denoted by *unit square* or *unit*^{2} and serves as the unit for measuring areas. It is then assumed that area is a measurement of figures that satisfies the following three properties (or axioms):

UNIT: *The unit square has area* 1 unit^2.

ADDITIVITY: *If a figure is divided by a line into two subfigures, then the area of the figure equals the sum of the areas of the subfigures.*

INVARIANCE: *Congruent polygons have equal areas.*

The additivity and invariance axioms were stated by Euclid as common notions 2 and 4, respectively. The unit axiom, however, has no analog in Euclid's system. As a consequence, Euclid's *Elements* contains no proposition that computes areas *explicitly.* Instead, Euclid made comparative statements such as

> *parallelograms on equal bases and between the same parallels are equal*

and

> *if a parallelogram have the same base with a triangle and be in the same parallels, the parallelogram is double the triangle.*

This has the theoretical advantage of dispensing with units and the practical disadvantage of not answering the reasonable question of *what is the area of a rectangle of dimensions 3 and 5 units?* Greek mathematicians did of course make use of units and could resolve such questions with ease. It is just that Euclid, for reasons that can only be guessed at (and that in the author's opinion were probably esthetic), decided to develop his geometry without any units whatsoever.

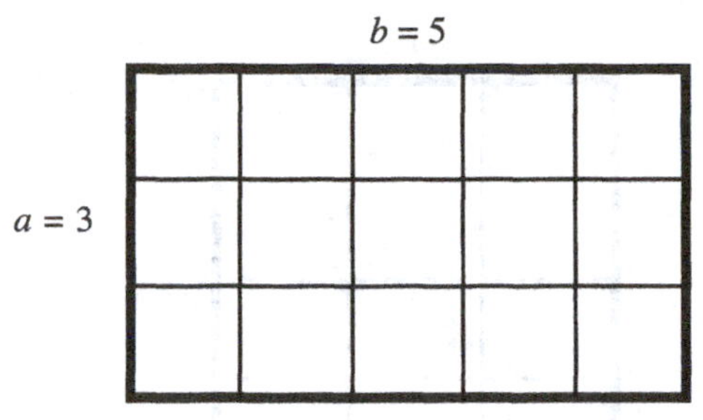

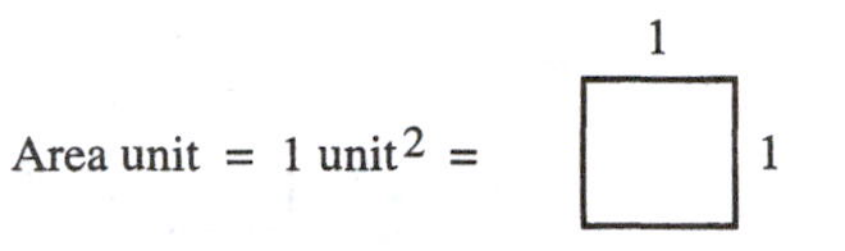

Figure 3.11.

Propositions 3.2.1 and 3.2.2 are the explicit modern-day analogs of Euclid's I.35 (PN 3.2.3). They give explicit formulas for the areas of rectangles and parallelograms. Their complete proofs unfortunately contain elements that are beyond the scope of this text. Specifically, we run into difficulties inherent in proving propositions regarding line segments with irrational (nonfractional) lengths. These difficulties were first encountered by the Greeks in the sixth century B.C. and eventually surmounted by Eudoxus two hundred years later. Euclid's book did incorporate Eudoxus's treatment of irrational numbers, but it would be impractical to expound this theory here. Instead, a mere supporting argument is offered for the fact that the area of a rectangle is given by the product of the lengths of its sides. It is customary in today's high school geometry textbooks to circumvent these difficulties by stating this formula as yet another axiom, the *rectangle axiom*. In the author's opinion this is a misguided solution to a pedagogical problem since it opens up the possibility of stating many other interesting and nontrivial geometrical facts as axioms, even when elementary and convincing, albeit logically incomplete, validating arguments are available.

Proposition 3.2.1 *If a rectangle has dimensions a units and b units, then it has area ab unit^2.*

GIVEN: $\square ABCD$ with sides a, b units.

TO PROVE: Area of $\square ABCD = ab$ unit^2.

SUPPORTING ARGUMENT: If a and b are positive integers, then a rectangle of dimensions a and b can be divided into ab unit squares by means of straight lines that are parallel to its sides (Fig. 3.11). Consequently, by the additivity property, the given rectangle has area ab unit^2.

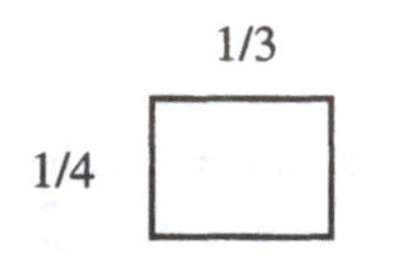

Area of rectangle = 1/12 unit^2

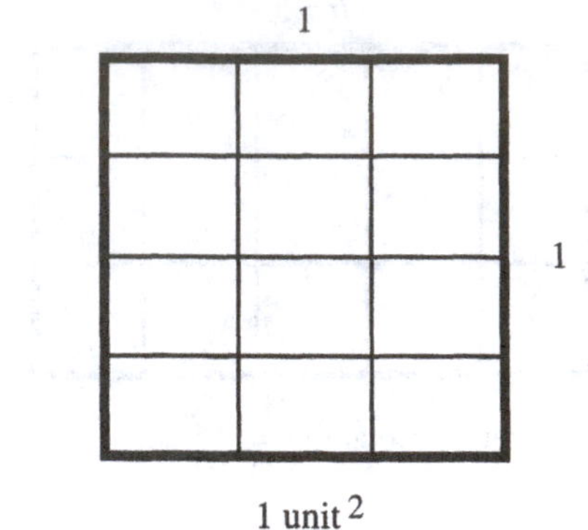

Figure 3.12.

Similarly, if a rectangle has dimensions $a = 1/m$ and $b = 1/n$ for some positive integers m and n, then the unit square can be divided into mn copies of the given rectangle all of which, by the invariance property, have the same area (Fig. 3.12). Hence the given rectangle has area

$$\frac{1 \text{ unit}^2}{mn} = \frac{1}{m}\frac{1}{n} \text{ unit}^2 = ab \text{ unit}^2.$$

Next, if a rectangle has dimensions $a = m/n$ and $b = p/q$, where m, n, p, q are all positive integers (Fig. 3.13), then it can be decomposed into mp rectangles each of which has dimensions $1/n$ and $1/q$. Since each of these latter rectangles is now known to have area $1/nq$ unit^2, it follows from the

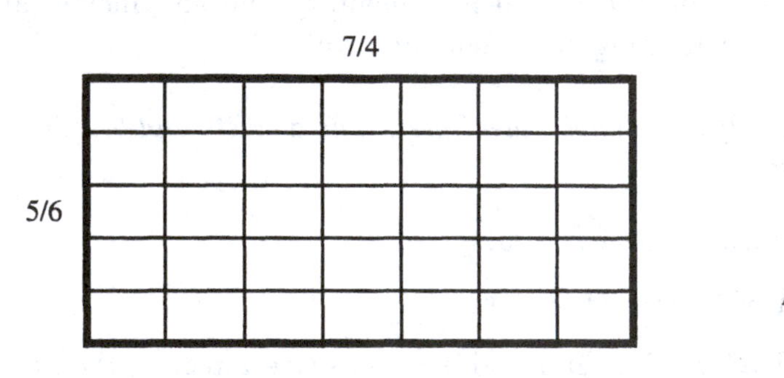

Area of rectangle = 35/24 unit^2

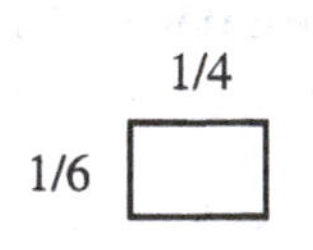

Area = 1/24 unit^2

Figure 3.13.

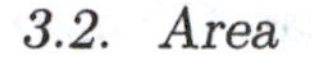

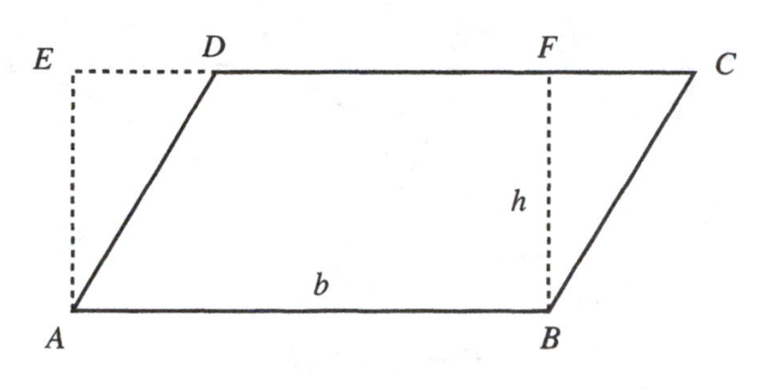

Figure 3.14.

additivity property that the given rectangle has area

$$mp\left(\frac{1}{nq}\right)\text{ unit}^2 = \frac{m}{n}\frac{p}{q}\text{ unit}^2 = ab\text{ unit}^2.$$

This verifies the proposition for all rectangles with fractional dimensions. As was mentioned previously, the extension of this formula to rectangles with arbitrary real dimensions lies beyond the scope of this text. Q.E.D.

An *altitude* of a parallelogram is any line segment cut off by two opposite sides from a straight line that is perpendicular to both of them. It follows from Proposition 3.1.2 that all the altitudes joining the same pair of opposite sides of a parallelogram have equal length.

Proposition 3.2.2 *The area of a parallelogram with base b units and altitude h units is bh unit*^{2}.

GIVEN: $\square ABCD$ with base b and altitude h units (Fig. 3.14).

TO PROVE: Area of $\square ABCD = bh$ unit^2.

PROOF: In the given parallelogram draw $AE \perp \overleftrightarrow{CD}$ and $BF \perp \overleftrightarrow{CD}$ with E, F on $\overleftrightarrow{CD}$. Thus, $ABFE$ is a rectangle with area bh unit^2. Since $\triangle ADE \cong \triangle BCF$, it follows that they have the same area, and hence, by the additivity property,

$$\square ABCD = \square ABFE = bh\text{ unit}^2.$$

Q.E.D.

Euclid's version of Propositions 3.2.1 and 3.2.2 is now stated together with his proof, as well as another proof that is more consistent with modern pedagogy.

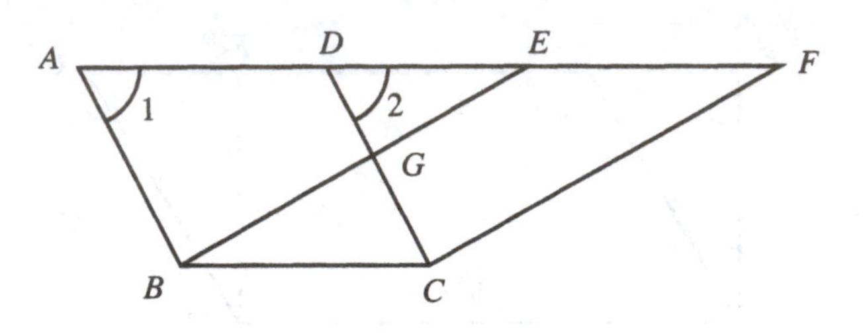

Figure 3.15.

Proposition 3.2.3 (I.35) *Parallelograms which are on the same base and in the same parallels are equal to one another.*

GIVEN: $\square ABCD$ and $\square EBCF$ such that A, D, E, F are collinear (Fig. 3.15).

TO PROVE: $\square ABCD = \square EBCF$.

PROOF (Euclid): Since $AD = BC = EF$ [PN 34], it follows from CN 2 that $AE = DF$. Then $\triangle BAE \cong \triangle CDF$ by SAS because

	$AE = DF$	[See above]
	$\angle 1 = \angle 2$	[Corresp. angles, $AB \parallel DC$, PN 3.1.1]
	$AB = DC$	[Parallelogram $ABCD$, PN 3.1.8]
$\therefore$	$\triangle EAB = \triangle FDC$	
$\therefore$	$ABGD = EGCF$	[Subtract $\triangle DGE$, CN 3]
$\therefore$	$\square ABCD = \square EBCF$.	[Add $\triangle GBC$, CN 2]

Q.E.D.

Euclid's proof of Proposition 3.2.3 is incomplete (albeit easily fixed) because it depends on the relative position of the points A, D, E, F on their common line (Exercise 12).

PROOF (modern): Draw HJ perpendicular to AD and BC (Fig. 3.16). It then follows from PN 3.2.2 that $\square ABCD = BC \cdot HJ = \square EBCF$.

Q.E.D.

Proposition 3.2.4 (I.36) *Parallelograms which are on equal bases and in the same parallels are equal to one another.*

See Exercise 1.

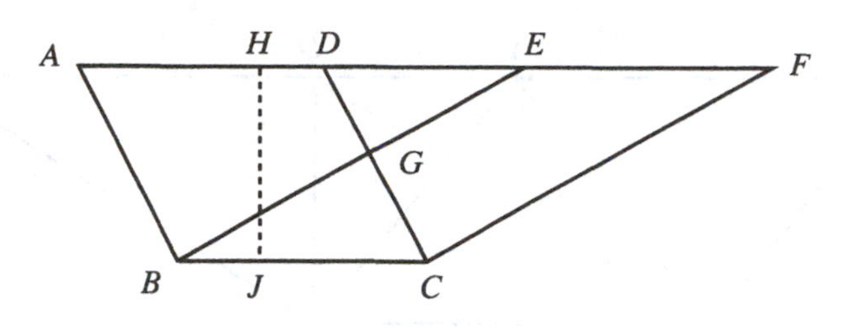

Figure 3.16.

The area of the triangle will be given the same dual treatment as that of the rectangle. First the modern formula is offered.

Proposition 3.2.5 *The area of a triangle with base b units and altitude h units is bh/2 unit*^{2}.

PROOF: Through the vertices B and C of $\triangle ABC$ draw straight lines parallel to AC and AB, respectively, and let their intersection be D (Fig. 3.17). It is clear that $ACDB$ is a parallelogram and hence, by Proposition 3.1.8,

$$\triangle ABC = \frac{1}{2} \square ACDB = \frac{bh}{2} \text{ unit}^2.$$

Q.E.D.

Next comes Euclid's version.

Proposition 3.2.6 (I.37) *Triangles which are on the same base and in the same parallels are equal to one another.*

GIVEN: $\triangle ABC$, $\triangle DBC$, $AD \parallel BC$ (Fig. 3.18).

TO PROVE: $\triangle ABC = \triangle DBC$.

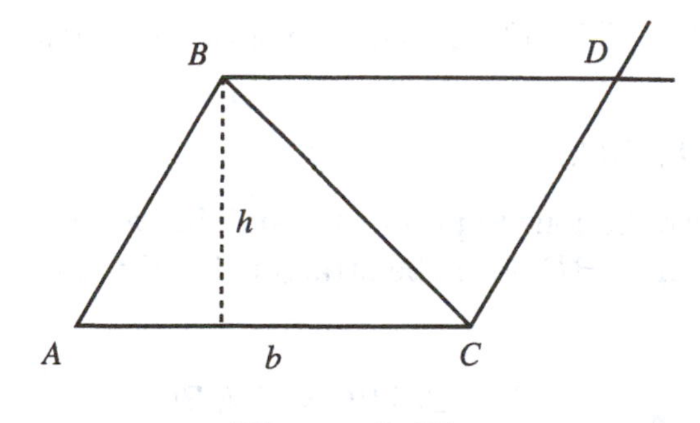

Figure 3.17.

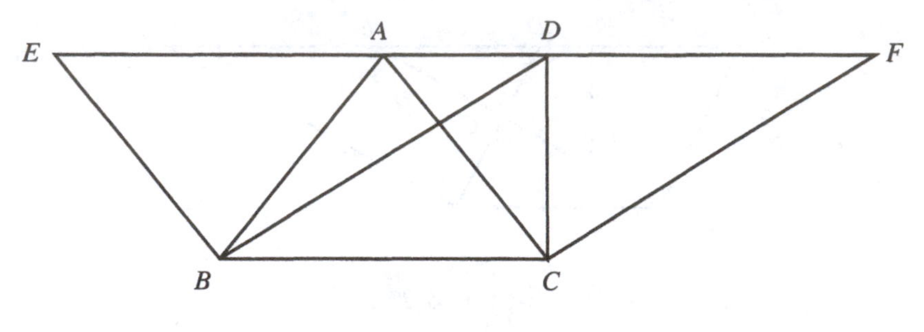

Figure 3.18.

PROOF: Let E be the intersection of $\overleftrightarrow{AD}$ with the straight line through B parallel to AC and let F be the intersection of $\overleftrightarrow{AD}$ with the straight line through C parallel to BD [PN 3.1.4]. Then

$$\square AEBC = \square DBCF \qquad \text{[PN 3.2.2]}$$

$$\triangle ABC = \frac{1}{2}\square AEBC \qquad \text{[PN 3.1.8]}$$

$$\triangle DBC = \frac{1}{2}\square DBCF \qquad \text{[PN 3.1.8]}$$

$$\therefore \quad \triangle ABC = \triangle DBC.$$

Q.E.D.

Proposition 3.2.7 (I.38) *Triangles which are on equal bases and in the same parallels are equal to one another.*

See Exercise 2.

Proposition 3.2.8 (I.39) *Equal triangles which are on the same base and on the same side are also in the same parallels.*

GIVEN: $\triangle ABC = \triangle DBC$, A and D are on the same side of BC (Fig. 3.19).

TO PROVE: $AD \parallel BC$.

PROOF: By contradiction. Suppose AD and BC are not parallel and let E be the intersection of BD with the straight line through A parallel to BC. Then

$$\triangle ABC = \triangle EBC \qquad \text{[PN 3.2.6]}$$

$$\triangle EBC < \triangle DBC \qquad \text{[CN 5]}$$

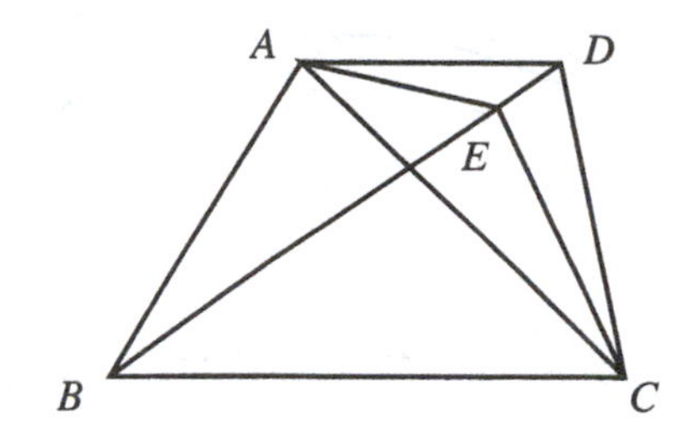

Figure 3.19.

$$\therefore \qquad \triangle ABC < \triangle DBC.$$

This, however, contradicts the given equality of the two triangles. Hence $AD \parallel BC$. Q.E.D.

Proposition 3.2.9 (I.40) *Equal triangles which are on equal bases and on the same side are also in the same parallels.*

According to Heath (Euclid's translator), Proposition 3.2.9 is an interpolation into the *Elements* by a later geometer (Exercise 2).

Proposition 3.2.10 (I.41) *If a parallelogram have the same base with a triangle and be in the same parallels, the parallelogram is double of the triangle.*

See Exercise 2.

There is no analog of Proposition 3.2.5 for the area of a general quadrilateral. In practice, any such quadrilateral can be divided into triangles by means of a diagonal and then the area of each of the parts can be evaluated by means of Proposition 3.2.5. A similar procedure can be used to dissect any polygon, regardless of the number of its sides, into triangles.

Neither Euclid's nor the modern approach to areas is applicable to spherical geometry. Both of these approaches rely heavily on the notion of parallelism, and the sphere has no parallel geodesics. Thus, another approach is required in order to develop a theory of spherical areas. As spherical polygons can also be dissected into spherical triangles, it suffices to provide a formula for the latter.

It is clear that the area of a lune with fixed vertices and variable angle is proportional to the angle. Since the lune of angle 2π radians has area $4\pi R^2$ (the sphere's total surface area), the following lemma is obtained.

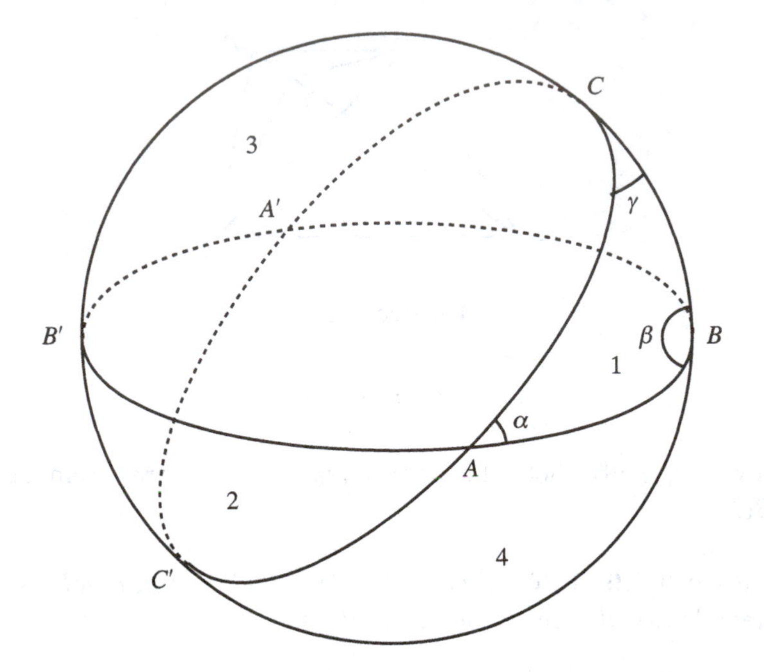

Figure 3.20.

Lemma 3.2.11 *On a sphere of radius R the area of a lune of angle α radians is $2\alpha R^2$ unit^2.*

The following theorem was first discovered by the Flemish mathematician Albert Girard (1595–1632). The proof presented here is due to Euler.

Proposition 3.2.12 *On a sphere of radius R, the area of the spherical triangle ABC with angles of radian measures α, β, γ is $(\alpha + \beta + \gamma - \pi)R^2$ unit^2.*

GIVEN: Spherical $\triangle ABC$ with interior angles α, β, γ (measured in radians).

TO PROVE: $\triangle ABC = (\alpha + \beta + \gamma - \pi)R^2$ unit^2.

PROOF: Let A', B', C' be the respective antipodes of A, B, C (Fig. 3.20). Draw the great circles that contain the geodesic segments AB, BC, and CA. The hemisphere in front of the great circle $BCB'C'$ is thereby divided into four spherical triangles ABC, $AB'C'$, $AB'C$, ABC', whose areas are denoted, respectively, by T_1, T_2, T_3, T_4.

From the construction it follows that the spherical $\triangle A'BC$ is congruent to the spherical $\triangle AB'C'$ of area T_2. Hence,

$$T_1 + T_2 = \text{lune } \alpha.$$

Similarly,

$$T_1 + T_3 = \text{lune } \beta$$

and

$$T_1 + T_4 = \text{lune } \gamma.$$

Consequently,

$$\begin{aligned} 2T_1 &= \text{lune } \alpha + \text{lune } \beta + \text{lune } \gamma - (T_1 + T_2 + T_3 + T_4) \\ &= (2\alpha + 2\beta + 2\gamma - 2\pi)R^2 \text{ unit}^2 \end{aligned}$$

and the statement of the theorem now follows immediately. Q.E.D.

Exercises 3.2A

1. Prove Proposition 3.2.4.
2. Use Proposition 3.2.5 to prove
   (a) Proposition 3.2.6
   (b) Proposition 3.2.7
   (c) Proposition 3.2.8
   (d) Proposition 3.2.9
   (e) Proposition 3.2.10.
3. One of the triangle's sides is divided into n equal segments and the division points are joined to the opposite vertex. Prove that the triangle is divided into n equal parts.
4. Prove that the area of the trapezoid equals the product of half the sum of its parallel sides with the distance between them.
5. Prove that the diagonals of a parallelogram divide it into four equal triangles.
6. Prove that the line segment joining the midpoints of two sides of a triangle cuts off a triangle that is equal to one fourth of the original triangle.
7. Prove that the parallelogram formed by the midpoints of the sides of a quadrilateral equals one-half of that quadrilateral.
8. Prove that the triangle's medians divide it into six equal triangles.
9. The diagonals of a quadrilateral divide it into four equal triangles. Prove that the quadrilateral is a parallelogram.
10. Prove that if the point P lies in the interior of $\square ABCD$, then the parallelogram equals twice the sum of $\triangle ABP$ and $\triangle CDP$.

11. Each of the sides AB, BC, CA of an equilateral triangle is extended by their common length to points D, E, F, respectively, all in the same sense. Prove that $\triangle DEF = 7\triangle ABC$.
12. Complete Euclid's proof of Proposition 3.2.3.

Both the taxicab and maxi areas of a figure are defined to equal its Euclidean area.

13. Comment on Propositions 3.2.1, 3.2.2, and 3.2.5 in the context of taxicab geometry.
14. Comment on Proposition 3.2.3, 3.2.4, and 3.2.6–10 in the context of taxicab geometry.
15. Comment on Propositions 3.2.1, 3.2.2, and 3.2.5 in the context of maxi geometry.
16. Comment on Proposition 3.2.3, 3.2.4, and 3.2.6–3.2.10 in the context of maxi geometry.

Euclid's Propositions I.42–I.45 are of limited interest. They are included here only in order to facilitate the later discussion of the golden ratio (Proposition 3.4.1).

Proposition 3.2.13 (I.42) *To construct, in a given rectilineal angle, a parallelogram equal to a given triangle.*

See Exercise 1.

The preceding proposition is an example of a *conversion*, which consists of the construction of a polygon Π' , of some prespecified nature, that is equal to a given polygon Π.

Proposition 3.2.14 (I.43) *In any parallelogram the complements of the parallelograms about the diagonal are equal to one another.*

GIVEN: $\square ABCD$, K is a point on the diagonal AC, $\square BGKE$, $\square KFDH$ (Fig. 3.21).

TO PROVE: $\square BGKE = \square KFDH$.

PROOF: See Exercise 2.

Proposition 3.2.15 (I.44) *To a given straight line to apply, in a given rectilineal angle, a parallelogram equal to a given triangle.*

See Exercise 3.

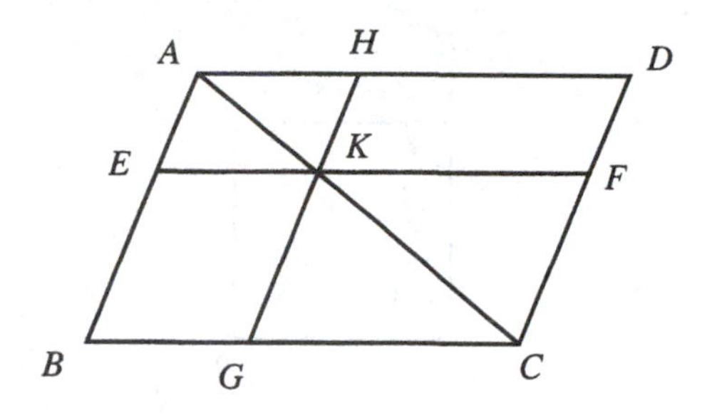

Figure 3.21.

Proposition 3.2.16 (I.45) *To construct, in a given rectilineal angle, a parallelogram equal to a given rectilineal figure.*

See Exercise 4.

Exercises 3.2B

1. Prove Proposition 3.2.13.
2. Prove Proposition 3.2.14.
3. Prove Proposition 3.2.15.
4. Prove Proposition 3.2.16.
5. Convert a given parallelogram into a rectangle with the same base.
6. Convert a given parallelogram into a rhombus with the same base.
7. Convert a given parallelogram into another parallelogram with the same base and a given angle.
8. Convert a given parallelogram into a triangle with the same base and a given angle.
9. Convert a given triangle into a right triangle with the same base.
10. Convert a given triangle into an isosceles triangle with the same base.
11. Convert a given triangle into another triangle with the same base and a given angle.
12. Bisect the area of a parallelogram by means of a straight line that is parallel to a given straight line.
13. Given $\triangle ABC$, construct a point O in its interior such that the triangles AOB, BOC, COA all have equal areas.

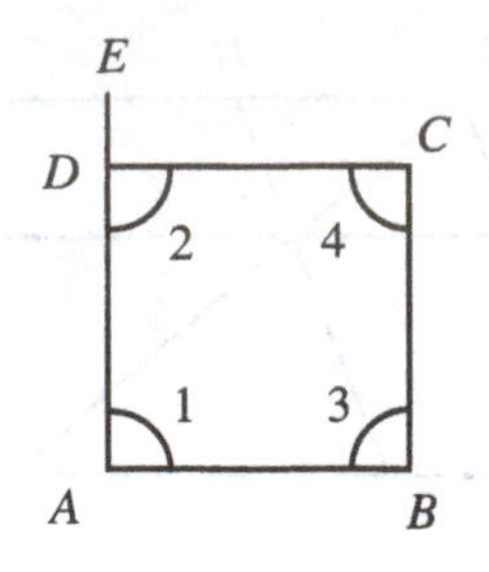

Figure 3.22.

3.3 The Theorem of Pythagoras

The theorem of Pythagoras was discovered independently by several cultures and has been given more different proofs than any other theorem. It is considered by many mathematicians to be the most important of all theorems, and has the dubious distinctions of being misquoted in the classic movie *The Wizard of Oz* and of being the subject of popular jokes. It will be presented following an easy lemma.

Proposition 3.3.1 (I.46) *On a given straight line to describe a square.*

GIVEN: Line segment AB (Fig. 3.22).

TO CONSTRUCT: $\square ABCD$.

CONSTRUCTION: Draw $EA \perp AB$ [PN 2.3.11] and let D on AE be such that $AD = AB$. Let C be the intersection of the straight lines through B and D that are parallel to AD and AB, respectively [PN 3.1.4]. Then quadrilateral $ABCD$ is the required square.

PROOF: By construction, $ABCD$ is a parallelogram. Since $AB = AD$ it follows that $AB = AD = DC = CB$. It remains to show that all of the angles of $ABCD$ are right angles. However,

$$\angle 1 + \angle 2 = 2 \text{ right angles} \qquad \text{[PN 3.1.1]}$$
$$\therefore \quad \angle 2 = \text{right angle} \qquad [\angle 1 \text{ is a right angle}]$$
$$\therefore \quad \angle 3 = \angle 4 = \text{right angle.} \qquad \text{[PN 3.1.8]}$$

Q.E.D.

Proposition 3.3.2 (I.47, The Theorem of Pythagoras) *In right-angled triangles the square on the side subtending the right angle is equal to the squares on the sides containing the right angle.*

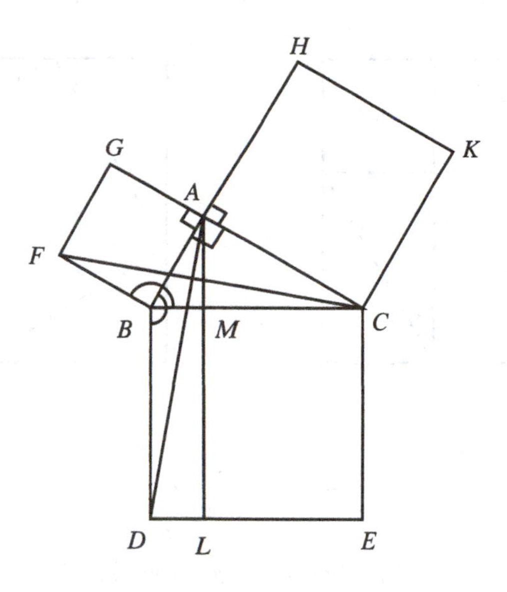

Figure 3.23.

GIVEN: $\triangle ABC$, $\angle BAC =$ right angle, $\square ABFG$, $\square ACKH$, $\square BCED$ (Fig. 3.23).

TO PROVE: $\square BCED = \square ABFG + \square ACKH$.

PROOF: Let L, M be the respective intersections of the straight lines DE and BC with the straight line through A parallel to BD and CE [PN 3.1.4]. Note that

the points G, A, C are collinear
[$\angle GAB + \angle BAC = 2$ right angles, PN 2.3.18]

the points B, A, H are collinear
[$\angle HAC + \angle BAC = 2$ right angles, PN 2.3.18]

$\triangle ABD \cong \triangle FBC$ by SAS because

$BD = BC$ [Sides of the same square]

$\angle ABD = \angle FBC$ [Both equal $\angle ABC +$ right angle]

$AB = FB$ [Sides of the same square]

$\therefore$ $\triangle ABD = \triangle FBC$

$\therefore$ $\square BDLM = \square ABFG$. [Doubles of equal triangles, PN 3.2.10]

A similar argument yields the equation $\square CELM = \square ACKH$ and hence

$$\square BCED = \square BDLM + \square CELM = \square ABFG + \square ACKH. \qquad \text{Q.E.D.}$$

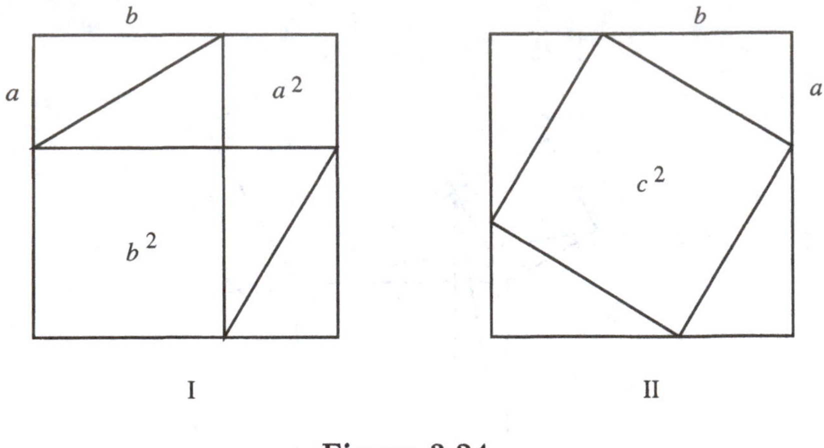

Figure 3.24.

Two other proofs of this theorem are now sketched.

If a and b are the legs and c is the hypotenuse of a right triangle, then the square of side $a+b$ can be dissected in the two ways depicted in Figure 3.24. The dissection of I calls for no explication. That of II requires a proof that the interior quadrilateral labeled as c^2 is indeed a square (see Exercise 8). However, once these dissections are granted, it is clear from Figure 3.24 that $a^2+b^2=c^2$. This proof is attributed to the Chinese mathematician Chou-pei Suan-ching, who lived circa 250 B.C.

The next proof is due to the Indian mathematician Bhaskara (1114–1185). The square of side c can be dissected in the manner depicted in Figure 3.25. It then follows that

$$c^2 = 4\frac{ab}{2} + (a-b)^2 = 2ab + a^2 - 2ab + b^2 = a^2 + b^2.$$

Yet another proof of the theorem of Pythagoras in indicated in Exercise 17. This one is due to former U.S. president James Garfield (1831–1881).

The theorem of Pythagoras has a converse.

Proposition 3.3.3 (I.48) *If in a triangle the square on one of the sides be equal to the squares on the remaining two sides of the triangle, the angle contained by the remaining two sides is right.*

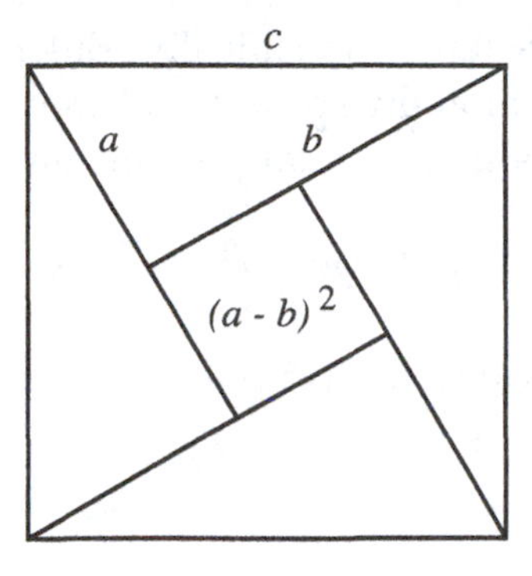

Figure 3.25.

See Exercise 9.

Since $3^2 + 4^2 = 5^2$, it follows that any triangle whose sides have lengths 3 units, 4 units, and 5 units is necessarily a right triangle. So is the triangle whose sides have lengths 5, 12, 13. Triples of integers a, b, c such that

$$a^2 + b^2 = c^2$$

are known as *Pythagorean triples*, but the interest in such triples precedes Pythagoras by over a thousand years. The Babylonian tablet PLIMPTON 322, dated between 1900 and 1600 B.C. contains fifteen Pythagorean triples, the largest of which consists of 12709, 13500, and 18541. Although it is highly unlikely that the Babylonians found these numbers by trial and error, it is not known what method they used to generate these triples. Not surprisingly, the earliest method for generating Pythagorean triples appears in Euclid's *Elements.* Lemma 1 to Proposition 29 of Book X states that if $m > n$ are any positive integers, then

$$(2mn)^2 + (m^2 - n^2)^2 = (m^2 + n^2)^2$$

(Exercise 14) so that $2mn$, $m^2 - n^2$, $m^2 + n^2$ form a Pythagorean triple. For example, $m = 5$ and $n = 4$ yield the triple

$$(2 \cdot 5 \cdot 4)^2 + (5^2 - 4^2)^2 = (5^2 + 4^2)^2$$

or

$$40^2 + 9^2 = 41^2.$$

Pierre Fermat (1601?–1665) took it for granted that Euclid's method can be used to generate *all* the Pythagorean triples, and this fact was proven by Euler a hundred years later. Specifically, Euler proved that if a, b, c are numbers whose only common divisor is 1 and that constitute a Pythagorean triple, then there exists a pair of relatively prime integers m, n such that

$$\{a, b, c\} = \{2mn, m^2 - n^2, m^2 + n^2\}.$$

All other Pythagorean triples are proportional to these.

Exercises 3.3A

1. Decide which of the following triples of numbers are the lengths of the sides of a right triangle:

   (a) 7, 10, 15

   (b) 5, 12, 13

   (c) 203750, 364056, 417194

   (d) 57302, 491714, 650463

2. Show that an equilateral triangle of side a has area $a^2\sqrt{3}/4$.

3. An isosceles right triangle has a hypotenuse of length c. Compute its other sides and its area.

4. A right triangle has an angle of 30° and a hypotenuse of length 1. Compute its other sides and its area.

5. Compute the area of a rhombus whose sides equal 13 and one of whose diagonals has length 10.

6. Compute the area of a parallelogram whose sides have lengths 11 and 8 and one of whose angles is 45°.

7. The diagonals and one side of a parallelogram have lengths 30, 16, 17, respectively. Prove that it is a rhombus and compute its area.

8. Show that the interior quadrilateral in Dissection II of Figure 3.24 is indeed a square of area c^2.

9. Prove Proposition 3.3.3.

10. Find the error in the following "proof" of the "proposition" that every triangle is isosceles:

    GIVEN: $\triangle ABC$.

    TO PROVE: $AB = AC$.

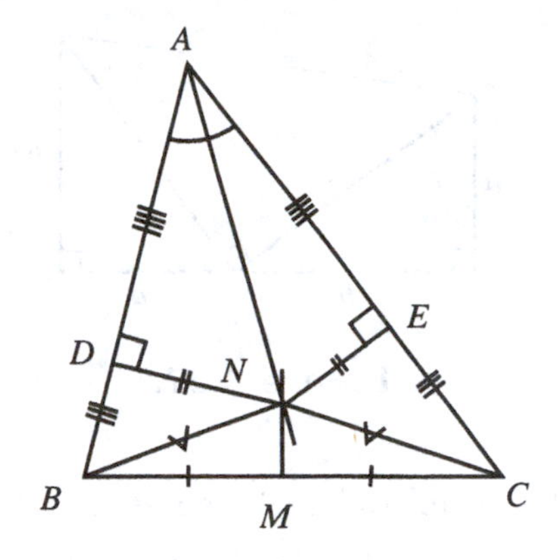

Figure 3.26.

PROOF: Let N be the intersection of the bisector of $\angle BAC$ and the perpendicular bisector of side BC, M the midpoint of BC, and $ND \perp AB$, $NE \perp AC$ (Fig. 3.26). Then

	$ND = NE$	[PN 2.3.34]
$\therefore$	$AD = AE$	[Pythagoras]
Also	$BN = CN$	[PN 2.3.14]
$\therefore$	$BD = CE$	[Pythagoras]
$\therefore$	$AB = AC.$	[CN 2]

Q.E.D.

11. Given a square of side a, construct a square of double its area.
12. Given a square of side a and a positive integer n, construct a square whose area equals n times that of the given square.
13. Construct a square whose area equals the sum of three given squares.
14. Use algebra to prove that Euclid's method does indeed generate Pythagorean triples.
15. Assume that a line segment of length 1 inch is given. Prove that line segments of the following lengths can be constructed.

    (a) $\sqrt{2}$ inch
    (b) $\sqrt{3}$ inch
    (c) $\sqrt{5}$ inch
    (d) $\sqrt{n}$ inch, where n is any positive integer
    (e) $1/\sqrt{2}$ inch

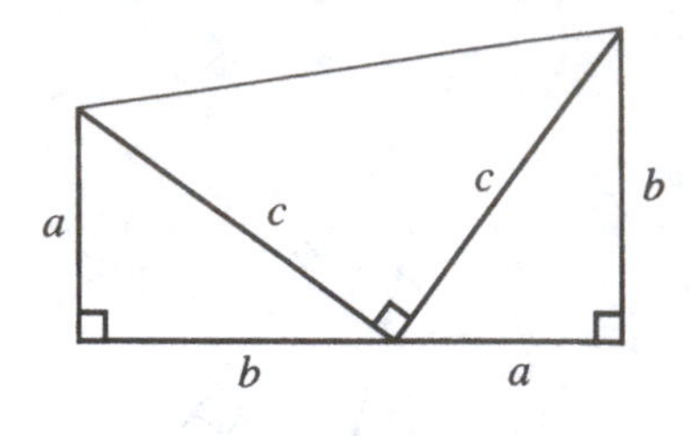

Figure 3.27.

16. Find all the Pythagorean triples whose least member is no greater than 10.
17. Prove the theorem of Pythagoras by applying Exercise 3.2A.4 to Figure 3.27.

18(C). Perform the construction of Proposition 3.3.1 using a computer application.

19(C). Use a computer application to verify the theorem of Pythagoras.

Both spherical and hyperbolic geometry have their own versions of the theorem of Pythagoras. Their proofs follow directly from the appropriate trigonometries (Exercises 1, 3).

Proposition 3.3.4 (The spherical theorem of Pythagoras)
If the spherical $\triangle ABC$ has a right angle at C, then

$$\cos c = \cos a \cos b.$$

Proposition 3.3.5 (The hyperbolic theorem of Pythagoras)
If the hyperbolic $\triangle ABC$ has a right angle at C, then

$$\cosh c = \cosh a \cosh b.$$

Exercises 3.3B

1. Derive the spherical theorem of Pythagoras from Proposition 1.1.2.
2. Find the length of the hypotenuses of the three spherical isosceles right triangles whose legs have lengths 1, .1, .01, respectively. Compare the answers to the lengths of the hypotenuses of the three Euclidean isosceles right triangles both of whose legs have lengths 1, .1, .01, respectively.
3. Derive the hyperbolic theorem of Pythagoras from Proposition 1.2.2.

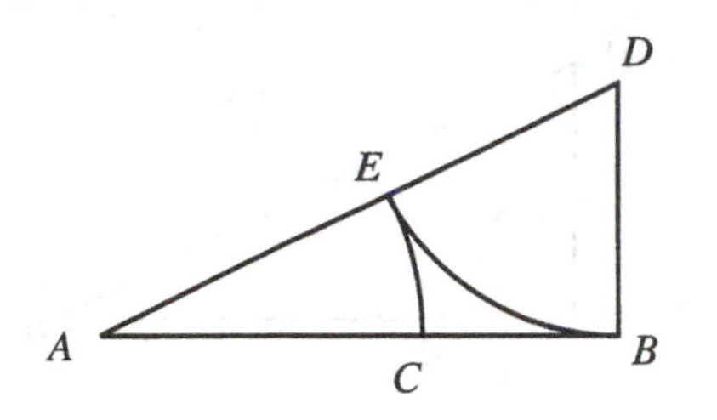

Figure 3.28.

4. Find the length of the hypotenuses of the three hyperbolic right triangles whose legs have lengths 1, .1, .01, respectively. Compare the answers to the lengths of the hypotenuses of the three Euclidean triangles both of whose legs have lengths 1, .1, .01, respectively.
5. Is there a taxicab version of the theorem of Pythagoras?
6. Is there a maxi version of the theorem of Pythagoras?

3.4 Consequences of the Theorem of Pythagoras (optional)

Book II of Euclid's *Elements* contains a variety of consequences of the theorem of Pythagoras of which only a sample are presented here. The first of these is tantamount to a construction of the *golden ratio*. This proposition will be used later in the construction of the regular pentagon.

Proposition 3.4.1 (II.6) *To cut a given line segment so that the rectangle contained by the whole and one of the segments is equal to the square on the remaining segment.*

GIVEN: Line segment AB (Fig. 3.28).

TO CONSTRUCT: A point C on AB such that $AB \cdot BC = AC^2$.

CONSTRUCTION: At B construct $BD \perp AB$ and $BD = (1/2)AB$. Join AD, let E be the point on AD such that $DE = DB$, and let C be the point on AB such that $AC = AE$. Then C is the required point.

PROOF: Set $AB = 2a$. Then $DE = BD = a$ and

$$AC = AE = AD - DE = \sqrt{(2a)^2 + a^2} - a = (\sqrt{5} - 1)a$$

so that

$$AC^2 = (\sqrt{5} - 1)^2 a^2 = (5 - 2\sqrt{5} + 1)a^2 = (6 - 2\sqrt{5})a^2$$

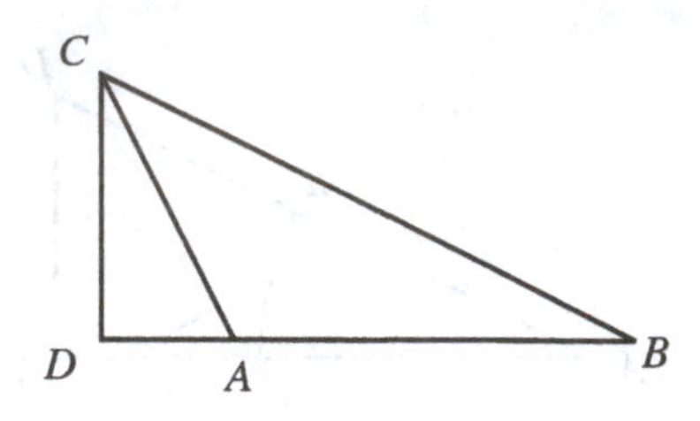

Figure 3.29.

and

$$\begin{aligned} AB \cdot BC &= AB(AB - AC) \\ &= 2a[2a - (\sqrt{5}a - a)] = 2(3 - \sqrt{5})a^2 = AC^2. \end{aligned}$$ Q.E.D.

In the context of the above proposition, the common value τ of the ratios

$$\frac{AC}{BC} = \frac{AB}{AC} = \frac{2a}{(\sqrt{5}-1)a} = \frac{2}{\sqrt{5}-1} = \frac{\sqrt{5}+1}{2} = 1.618\ldots$$

is called the *golden ratio*. While of demonstrated mathematical interest, this quantity has also been the subject of much nonsensical speculation. Typical of this latter variety is an article that reports that the average ratio of the height of a man's navel off the ground to his height equals the golden ratio.

The following two propositions constitute Euclid's analog of the modern day law of cosines. Their proofs are relegated to the Exercises.

Proposition 3.4.2 (II.12) *In obtuse-angled triangles the square on the side subtending the obtuse angle is greater than the squares on the sides containing the obtuse angle by twice the rectangle contained by one of the sides about the obtuse angle, namely that on which the perpendicular falls, and the straight line cut off outside by the perpendicular towards the obtuse angle.*

GIVEN: $\triangle ABC$, $\angle BAC >$ right angle, $CD \perp AB$ (Fig. 3.29).

TO PROVE: $BC^2 = AB^2 + AC^2 + 2AB \cdot AD$.

PROOF: See Exercise 2.

Proposition 3.4.3 (II.13) *In acute-angled triangles the square on the side subtending the obtuse angle is less than the squares on the sides containing the obtuse angle by twice the rectangle contained by one of the sides about the*

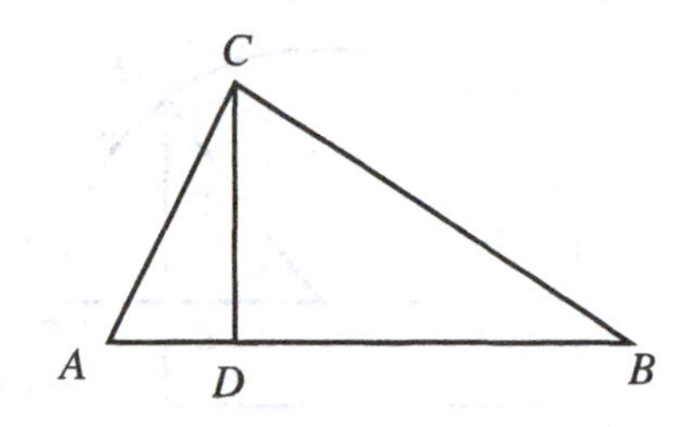

Figure 3.30.

obtuse angle, namely that on which the perpendicular falls, and the straight line cut off within by the perpendicular towards the acute angle.

GIVEN: $\triangle ABC$, $\angle BAC <$ right angle, $CD \perp AB$ (Fig. 3.30).

TO PROVE: $BC^2 = AB^2 + AC^2 - 2AB \cdot AD$.

PROOF: See Exercise 3.

Exercises 3.4A

1. Explain the relation of Proposition 3.4.2 to the law of cosines of the trigonometry of the Euclidean plane.
2. Prove Proposition 3.4.2.
3. Explain the relation of Proposition 3.4.3 to the law of cosines of the trigonometry of the Euclidean plane.
4. Prove Proposition 3.4.3.

Assuming a unit length and a segment of length r units, the next proposition deals with the construction of a line segment of length $\sqrt{r}$ units. The statement that appears here is weaker than that of Euclid's, but it is sufficient for this text's purposes and obviates the need for Proposition 3.2.6 (I.45). Euclid's version appears in Exercise 2.

Proposition 3.4.4 (II.14) *To construct a square equal to a given rectangle.*

GIVEN: $\square BCDE$ (Fig. 3.31).

TO CONSTRUCT: Line segment EH such that $EH^2 = \square BCDE$.

CONSTRUCTION: If $BE = ED$, then BE is the required line segment. Otherwise, it may be assumed without loss of generality that $BE > ED$.

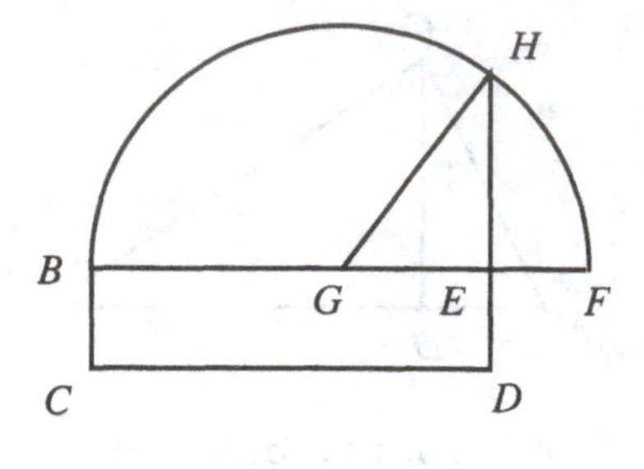

Figure 3.31.

Extend BE to F so that $EF = DE$ and let G be the midpoint of BF. Let H be an intersection of the straight line through E perpendicular to BF with the circle $(G; BG)$. Then EH is the required line segment.

PROOF: By the theorem of Pythagoras,

$$EH^2 = GH^2 - GE^2 = (GH + GE)(GH - GE) = (BG + GE)(GF - GE)$$
$$= BE \cdot EF = BE \cdot ED = \Box BCDE. \qquad \text{Q.E.D.}$$

Exercises 3.4B

1. Assume that a line segment of length 1 inch is given. Construct a line segment whose length is
   (a) $\sqrt{6}$ inch
   (b) $\sqrt{11}$ inch
   (c) $\sqrt[4]{2}$ inch
   (d) $\sqrt[4]{6}$ inch
   (e) $\sqrt[4]{11}$ inch
2. Prove Euclid's Proposition II.14: To construct a square equal to a given polygon.

3.5 Proportion and Similarity

Euclid's definition of proportion is too intricate for the context and purpose of this text. Instead, a shortcut provided by the real number system is used. Recall Euclid's tacit assumption that all geometrical figures have an aspect of size or magnitude. Thus, lines, regions, solids, and angles have lengths, areas, volumes, and angular measure, respectively. If two figures of the same types have sizes a and b (relative to some unit), it is said that their *ratio* is the real number a/b. The numbers $a, b, c, \ldots$ are said to be *proportional* to the numbers $a', b', c', \ldots$ provided that

$$\frac{a}{a'} = \frac{b}{b'} = \frac{c}{c'} = \cdots.$$

The next five propositions set up some algebraic preliminaries. The first of these is, of course, none other than the distributive law and so requires no proof. The rest are basic observations regarding proportions.

Proposition 3.5.1 (V.1) *If m, a, b, c, are any numbers, then*

$$ma + mb + mc + \cdots = m(a + b + c + \cdots).$$

Proposition 3.5.2 (V.12) *If $a, b, c, \ldots$ are proportional to $a', b', c', \ldots$, then*

$$\frac{a}{a'} = \frac{b}{b'} = \frac{c}{c'} = \cdots = \frac{a + b + c + \cdots}{a' + b' + c' + \cdots}.$$

GIVEN: $\frac{a}{a'} = \frac{b}{b'} = \frac{c}{c'} = \cdots$.

TO PROVE: $\frac{a}{a'} = \frac{b}{b'} = \frac{c}{c'} = \cdots = \frac{a + b + c + \cdots}{a' + b' + c' + \cdots}$.

PROOF: Let $k = \frac{a}{a'} = \frac{b}{b'} = \frac{c}{c'} = \cdots$. Then

$$a = ka', \quad b = kb', \quad c = kc', \ldots$$

$$\therefore \quad a + b + c + \cdots = ka' + kb' + kc' + \cdots = k(a' + b' + c' + \cdots)$$

$$\therefore \quad \frac{a + b + c + \cdots}{a' + b' + c' + \cdots} = k = \frac{a}{a'} = \frac{b}{b'} = \frac{c}{c'} = \cdots.$$

Q.E.D.

Proposition 3.5.3 (V.16) $\frac{a}{a'} = \frac{b}{b'}$ *if and only if* $\frac{a}{b} = \frac{a'}{b'}$.

See Exercise 8.

Proposition 3.5.4 (V.17) *If* $\frac{a}{a'} = \frac{b}{b'}$, *then* $\frac{a - a'}{a'} = \frac{b - b'}{b'}$.

See Exercise 9.

Proposition 3.5.5 (V.18) . *If* $\frac{a}{a'} = \frac{b}{b'}$, *then* $\frac{a + a'}{a'} = \frac{b + b'}{b'}$.

See Exercise 10.

Exercise 3.5A

In Exercises 1–7 prove the stated equalities on the basis of the assumption that a, b, c, d are proportional to a', b', c', d'.

1. $\dfrac{a}{a'} = \dfrac{4a - 3b + 2c + 7d}{4a' - 3b' + 2c' + 7d'}$
2. $\dfrac{a^2}{a'^2} = \dfrac{4a^2 - 3b^2 + 2c^2 + 7d^2}{4a'^2 - 3b'^2 + 2c'^2 + 7d'^2}$
3. $\dfrac{a}{a'} = \dfrac{\sqrt{a^2 + 2b^2 - 3c^2 + 5d^2}}{\sqrt{a'^2 + 2b'^2 - 3c'^2 + 5d'^2}}$
4. $\dfrac{a^3}{a'^3} = \dfrac{a^3 - b^3 + c^3 - d^3}{a'^3 - b'^3 + c'^3 - d'^3}$
5. $\dfrac{a + 3b}{a - 3b} = \dfrac{a' + 3b'}{a' - 3b'}$
6. $\dfrac{a - 3c}{a + 3c} = \dfrac{a' - 3c'}{a' + 3c'}$
7. $\dfrac{a^2 + 3d^2}{a^2 - 3d^2} = \dfrac{a'^2 + 3d'^2}{a'^2 - 3d'^2}$
8. Prove Proposition 3.5.3.
9. Prove Proposition 3.5.4.
10. Prove Proposition 3.5.5.

Proposition 3.5.4 (VI.2) *If a straight line meets two sides of a triangle, then it is parallel to the third side if and only if it cuts them into proportional segments.*

GIVEN: $\triangle ABC$, points D and E on $\overleftrightarrow{AB}$ and $\overleftrightarrow{CD}$ respectively (Fig. 3.32).

TO PROVE: 1. If $DE \parallel BC$ then $\dfrac{AD}{DB} = \dfrac{AE}{EC}$

2. If $\dfrac{AD}{DB} = \dfrac{AE}{EC}$ then $DE \parallel BC$.

PROOF of 1: Join CD and BE. Then, since $DE \parallel BC$

$$\triangle BDE = \triangle CDE \qquad \text{[PN 3.2.6]}$$
$$\triangle ADE = \triangle ADE$$
$$\therefore \quad \triangle ABE = \triangle ACD \qquad \text{[CN 2]}$$

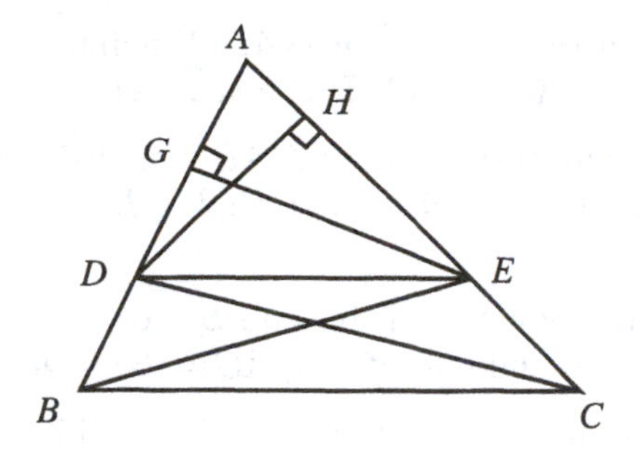

Figure 3.32.

$$\therefore \quad \frac{\triangle ABE}{\triangle ADE} = \frac{\triangle ACD}{\triangle ADE}.$$

Let EG and DH be altitudes of $\triangle ADE$. Then it follows from the foregoing that

$$\frac{AB \cdot GE}{AD \cdot GE} = \frac{AC \cdot DH}{AE \cdot DH} \qquad \text{[PN 3.2.5]}$$

$$\therefore \quad \frac{AB}{AD} = \frac{AC}{AE}$$

$$\therefore \quad \frac{DB}{AD} = \frac{EC}{AE} \qquad \text{[PN 3.5.4]}$$

$$\therefore \quad \frac{AD}{DB} = \frac{AE}{EC}.$$

PROOF of 2: Using the same construction as before, it is only necessary to reverse the order of the steps of the preceding argument (Exercise 1).Q.E.D.

The proof of the above theorem depends superficially on the additional assumption that the point D lies between A and B. Exercises 15–17 rectify this minor flaw.

Exercise 3.5B

1. Complete the proof of Proposition 3.5.6.
2. Prove that the straight line that bisects one side of a triangle and is parallel to a second side also bisects the third side.
3. Use Proposition 3.5.6 (twice) to prove that the line segment joining the midpoints of two sides of the triangle is parallel to the third side and equals half its length.

4. The point K is on the side AB of $\triangle ABC$, points L, M are on side AC, so that $KL \parallel BM$ and $KM \parallel BC$. Prove that $AL/AM = AM/AC$.

5. Point O is not on any of the sides of $\triangle ABC$ or their extensions, and K, L, M are on OA, OB, OC, respectively, so that $KL \parallel AB$ and $LM \parallel BC$. Prove that $KM \parallel AC$.

6. Prove that if D is any point on the side BC of $\triangle ABC$, then AD bisects the interior angle at A if and only if $AB/AC = BD/DC$.

7. Prove that if E is any point on the extension of side BC of $\triangle ABC$, then AE bisects the exterior angle at A if and only if $AB/AC = BE/EC$.

8. Prove that if the straight lines $m_1, m_2, \ldots, m_n$ are all parallel to one side of a triangle and they cut off equal segments on a second side, then they also cut off equal segments on the third side.

9. Divide a given line segment into three equal parts.

10. Let n be a given positive integer. Divide a given line segment into n equal parts.

11. Let m and n be given positive integers. Divide a given line segment into two parts whose ratio is m/n (VI.9).

12. Let a, b, c be three given line segments. Divide a into two parts whose ratio equals b/c (VI.10).

13. Let a, b, c be three given line segments. Construct a line segment x such that $a/b = c/x$ (VI.12).

14. Let a, b be two given line segments. Construct a line segment x such that $a/x = b/a$.

15. Show that the proof of Proposition 3.5.6 still holds, with minor modifications, when A lies in between B and D.

16. Show that the proof of Proposition 3.5.6 still holds, with minor modifications, when B lies in between A and D.

17. Does the proof of Proposition 3.5.6 require any other corrections?

18. Comment on Proposition 3.5.6 in the context of the following geometries:
    (a) spherical (b) hyperbolic (c) taxicab (d) maxi

19. Use the spherical trigonometry formulas to experiment with comparing the length of the line joining the midpoints of two sides of a spherical triangle with that of the third side (see Exercise 3). Form a conjecture regarding the relative sizes of these two geodesic segments.

20. Use the hyperbolic trigonometry formulas to experiment with comparing the length of the line joining the midpoints of two sides of a hyperbolic triangle with that of the third side (see Exercise 3). Form a conjecture regarding the relative sizes of these two geodesic segments.

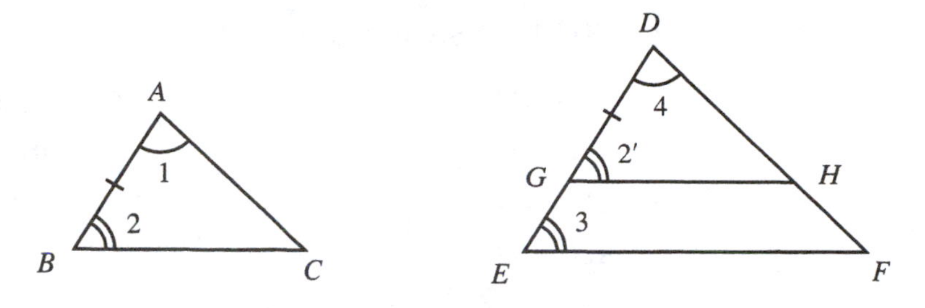

Figure 3.33.

21. Comment on Exercise 19 in the context of taxicab geometry.
22. Comment on Exercise 19 in the context of maxi geometry.
23. Construct $\triangle ABC$ given the following data:
    (a) a, m_b, m_c (b) m_a, m_b, m_c

24(C). Use a computer application to verify part 1 of Proposition 3.5.6.

Similar polygons are those whose corresponding angles are equal and whose corresponding sides are proportional. Congruent triangles are similar (Exercise 2) and the relation of similarity is transitive (Exercise 3). If $\triangle ABC$ and $\triangle DEF$ are similar, this is denoted by $\triangle ABC \sim \triangle DEF$, where it is implicit that A, B, C correspond to D, E, F, respectively, as was the case for congruent triangles. The next proposition is known as the AAA similarity theorem.

Proposition 3.5.7 (VI.4) *Equiangular triangles are similar.*

GIVEN: $\triangle ABC$, $\triangle DEF$, $\angle ABC = \angle DEF$, $\angle ACB = \angle DFE$, $\angle BAC = \angle EDF$ (Fig. 3.33).

TO PROVE: $\dfrac{AB}{DE} = \dfrac{BC}{EF} = \dfrac{AC}{DF}$.

PROOF: If any side of $\triangle ABC$ equals the corresponding side of $\triangle DEF$ then the two triangles are congruent [ASA] . Hence it may be assumed without loss of generality that $AB < DE$. Let G be a point in the interior of DE such that $DG = AB$. Let H be the intersection of the straight line parallel to EF through G with DF. Then

$$\frac{GE}{DG} = \frac{HF}{DH} \qquad \text{[PN 3.5.6]}$$

$$\therefore \quad \frac{DE}{DG} = \frac{DF}{DH}. \qquad \text{[PN 3.5.5]}$$

However, $\triangle ABC \cong \triangle DGH$ by ASA because

$$\begin{aligned} &\angle 2 = \angle 2' && \text{[Both equal } \angle 3\text{]} \\ &AB = DG && \text{[Construction]} \\ &\angle 1 = \angle 4 && \text{[Given]} \\ \therefore\quad &AC = DH && \\ \therefore\quad &\frac{AB}{DE} = \frac{DG}{DE} = \frac{DH}{DF} = \frac{AC}{DF}. && \end{aligned}$$

A similar argument can be used to prove that $\dfrac{BC}{EF} = \dfrac{AC}{DF}$. Q.E.D.

The following similarity theorems are known as the SSS and SAS similarity theorems, not to be confused with the SSS and SAS congruence theorems, and their proofs are relegated to Exercises 5 and 6, respectively.

Proposition 3.5.8 (VI.5) *If two triangles have their sides proportional then the triangles are similar.*

Proposition 3.5.9 (VI.6) *If two triangles have one angle equal to one angle and the sides about those angles are proportional, then the triangles are similar.*

In view of the fact that the sum of the angles of every Euclidean triangle is 180°, it follows that the conclusion of Proposition 3.5.7 holds even when only *two* of the angles of one triangle are known to be equal to the corresponding angles of the other triangle.

Exercise 3.5C

1. Complete the proof of Proposition 3.5.7.
2. Prove that congruent triangles are similar.
3. Prove that if $\triangle ABC$ is similar to $\triangle A'B'C'$ and $\triangle A'B'C'$ is similar to $\triangle A''B''C''$, then $\triangle ABC$ is similar to $\triangle A''B''C''$.
4. Prove Proposition 3.5.8.
5. Prove Proposition 3.5.9.
6. Prove that in similar triangles corresponding altitudes are proportional to corresponding sides.
7. Prove that in similar triangles corresponding medians are proportional to corresponding sides.

8. Prove that in similar triangles the corresponding angle bisectors are proportional to the corresponding sides.

9. Prove that the areas of similar triangles are proportional to the squares of their corresponding sides.

10. Prove that the areas of similar polygons are proportional to the squares of their corresponding sides (VI.19).

11. Three parallel straight lines cut the straight lines m and n in the points A, B, C, and K, L, M, respectively. Prove that $AB/BC = KL/LM$.

12. In $\square ABCD$ the straight line $BKLM$ cuts the diagonal AC in the point K and the (possibly extended) sides AD and CD in the points L and M, respectively. Prove that $BK/KL = KM/BK$.

13. In $\square ABCD$ the straight line BKN cuts the (possibly extended) sides CD and AD in the points K and N, respectively. Prove that $AD/DN = CK/KD$.

14. Prove that the intersection point of two of a triangle's altitudes divides them so that the product of each altitude's segments equals the product of the other's segments.

15. Prove Euclid's Proposition VI.8: If in a right triangle a perpendicular be drawn from the right angle to the opposite side, then the triangles so formed are similar to each other and to the given triangle.

16. Prove Euclid's Proposition VI.13: The square of the altitude to the hypotenuse of a right triangle equals the product of the segments it determines on the hypotenuse.

17. Prove Euclid's Proposition VI.31: If similar polygons are constructed on the sides of a right triangle, then the polygon on the hypotenuse equals the sum of the polygons on the other two sides. (*Hint:* Use Exercise 10.)

18. Describe the locus of all the points whose distances from the sides of a given angle have a ratio equal to that of two given line segments.

19. Given an angle and a point A inside it, construct through A a straight line whose portion between the sides of the angle is divided by A into segments whose ratio equals that of two given line segments.

20. Construct $\triangle ABC$ given the data:

    (a) β, γ, h_a (b) β, γ, m_a

21. Prove that the perimeters of similar triangles are proportional to their corresponding sides.

22. Prove that the perimeters of similar polygons are proportional to their corresponding sides.

23. A straight line through the intersection of the diagonals of a trapezoid is parallel to its parallel sides. Prove that the segment between the nonparallel sides is bisected by the intersection of the diagonals.
24. Prove that in a trapezoid the line joining the midpoint of one of the parallel sides to the intersection of the diagonals bisects both the parallel sides.
25. Prove that in a trapezoid which is not a parallelogram the straight line joining the intersection of the diagonals to the intersection of the nonparallel sides bisects both the parallel sides.
26. Comment on Proposition 3.5.7 in the context of the following geometries:
    (a) spherical (b) hyperbolic (c) taxicab (d) maxi

Chapter Review Exercises

1. Let P be a point in the interior of the equilateral $\triangle ABC$. Prove that the sum of the perpendicular segments from P to the sides of the triangle is constant.
2. Prove that the diagonals of a trapezoid cut each other into segments that are proportional to the parallel sides of the trapezoid.
3. Suppose the nonparallel sides of a trapezoid are equal. Prove the following:
    (a) The two angles adjacent to the same base are equal.
    (b) The diagonals are equal.
    (c) The diagonals divide each other into respectively equal segments.
    (d) The midpoints of the four sides form a rhombus.
4. Find a simple expression for the angle between two of a triangle's angle bisectors.
5. A straight line through the vertex of a triangle divides it into two triangles that are similar to each other and to the original triangle. Prove that the given triangle is a right triangle.
6. In $\square ABCD$ a straight line parallel to AB intersects AD, AC, and BC in the points P, Q, R, respectively. Prove that $\triangle APR = \triangle AQD$.
7. In $\triangle ABC$, $m_a = BC/2$. Prove that $\angle CAB$ is a right angle.
8. In $\triangle ABC$, $\angle BAC = 2\angle ABC$. Prove that $a^2 = b(b + c)$.
9. In $\square ABCD$, M and N are the midpoints of the opposite sides AB and CD. Prove that the straight lines DM and BN divide the diagonal AC into three equal segments.
10. Suppose that $\angle ACB$ of $\triangle ABC$ is obtuse and the perpendicular bisectors to AC and BC intersect AB in the points D and E, respectively. Prove that $\angle DCE = 2(\angle ACB - 90°)$.

11. From a point on the base of an isosceles triangle straight lines parallel to the triangle's other sides are drawn. Prove that the perimeter of the parallelogram thus formed is independent of the position of the point on the base.

12. Prove that the bisectors of the two angles formed by the opposite pairs of sides of a convex quadrilateral intersect in an angle that equals half the sum of two opposite angles of the quadrilateral.

13. Prove that the median and the altitude to the hypotenuse of a right triangle form an angle that equals the difference of the triangle's two other angles.

14*. In $\triangle ABC$, $AB = AC$, E and D are on the sides AB and AC, respectively, and $\angle ABD = 20°$, $\angle CBD = 60°$, $\angle BCE = 50°$, and $\angle ACE = 30°$. Find $\angle EDB$.

15. Are the following statements true or false? Justify your answers.
    (a) Playfair's postulate is valid in neutral geometry.
    (b) Playfair's postulate is valid in Euclidean geometry.
    (c) Playfair's postulate is valid in spherical geometry.
    (d) Playfair's postulate is valid in hyperbolic geometry.
    (e) Playfair's postulate is valid in taxicab geometry.
    (f) If in a quadrilateral one pair of opposite sides are equal, as are one pair of opposite angles, then the quadrilateral is a parallelogram.
    (g) There is a Euclidan right triangle with sides 287, 816, 865.
    (h) There is a neutral right triangle with sides 287, 816, 865.
    (i) There is a spherical right triangle with sides 287, 816, 865.
    (j) There is a hyperbolic right triangle with sides 287, 816, 865.
    (k) There is a taxicab right triangle with sides 287, 816, 865.
    (l) If the corresponding sides of two triangles are proportional, then so are their corresponding angles.
    (m) Equiangular triangles are similar.
    (n) The corresponding sides of equiangular quadrilaterals are proportional.
    (o) If the corresponding angles of two quadrilaterals on the surface of a sphere are equal, then so are their areas.
    (p) Equiangular quadrilaterals are similar.

16. Find all the Pythagorean triplets whose largest member is no greater than 50.

Chapter 4

Circles and Regular Polygons

Circles and regular polygons are the subject of Books III and IV of the *Elements*. Euclid's abstract exposition of the interrelation of chords, arcs, and tangents lines is augmented with the computation of the circle's circumference and area.

4.1 The Neutral Geometry of the Circle

Equal circles are circles that have equal radii. A *chord* of a circle is a line segment that joins two of its points. A *diameter* is a chord that contains the center of the circle. An *arc* of a circle is a portion of the circle that joins two of its points. Every chord determines two arcs of the circle. Consequently, it takes at least three points to denote an arc unambiguously, and the two arcs of the circle of Figure 4.1 with endpoints A and B should be denoted, properly speaking, by $A\overset{\frown}{E}B$ and $A\overset{\frown}{F}B$. Nevertheless, it is customary to label both of these arcs $\overset{\frown}{AB}$ and to rely on the context for clarification. A *segment* of a circle is the portion between a chord and either of its arcs. A *sector* of a circle is the portion between two radii. The arcs determined by a diameter are each called a *semicircle*. That the two semicircles determined by a diameter are equal (in length) is a proposition that Euclid mentions in Definition 17 (Chapter 2). This observation is proved as part of Proposition 4.1.1 in this chapter.

A *central angle* of a circle is one both of whose sides are radii. Every arc subtends a central angle that is either greater or less than 180° according as the arc is greater or less than a semicircle. Every chord subtends a central angle that is at most 180°.

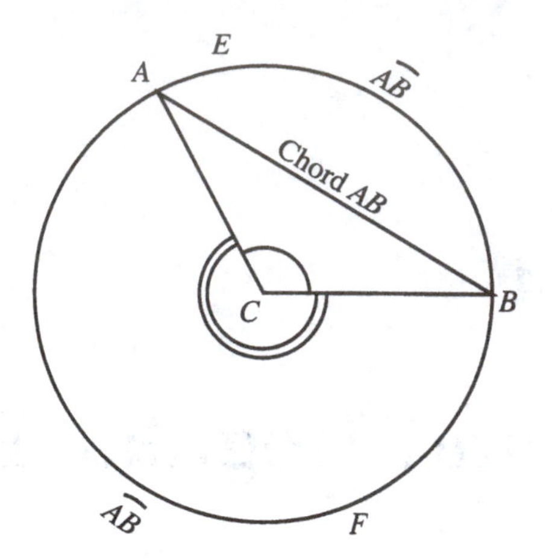

Figure 4.1.

The following four propositions of Euclid's are established here with a single unified proof.

Proposition 4.1.1 (III.26) *In equal circles equal central angles stand on equal arcs.*

Proposition 4.1.1 (III.27) *In equal circles central angles standing on equal arcs are equal to one another.*

Proposition 4.1.1 (III.28) *In equal circles equal chords cut off equal arcs, the greater equal to the greater and the less to the less.*

Proposition 4.1.1 (III.29) *In equal circles, equal arcs are subtended by equal chords.*

GIVEN: Equal circles centered at E and E', respectively. Points A, B on the first circle and points A', B' on the second (Fig. 4.2).

TO PROVE: The following are equivalent:

1. $\overset{\frown}{AB}=\overset{\frown}{A'B'}$
2. $AB = A'B'$
3. $\angle AEB = \angle A'E'B'$

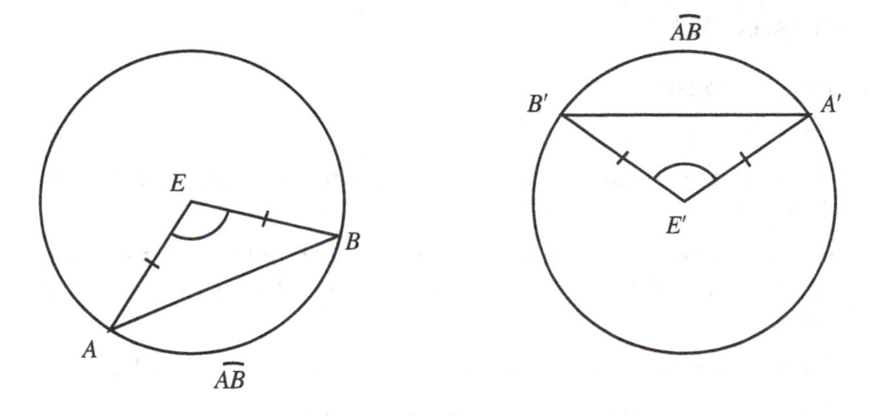

Figure 4.2.

PROOF: 1 ⇒ 2: Assume $\overset{\frown}{AB}=\overset{\frown}{A'B'}$. Since the given circles are equal, it is possible to apply the circle centered at E to that centered at E' so that E and A fall on E' and A', respectively, and $\overset{\frown}{AB}$ falls along $\overset{\frown}{A'B'}$. The two arcs having equal lengths, B falls on B'. It follows from PT 1 that the chord AB falls on the chord $A'B'$ and hence, by CN 4, $AB = A'B'$.

2 ⇒ 3: Assume $AB = A'B'$. $\triangle AEB \cong \triangle A'E'B'$ by SSS because

$$\begin{aligned} AB &= A'B' && \text{[Given]}\\ AE &= A'E' && \text{[Given]}\\ BE &= B'E' && \text{[Given]}\\ \therefore\quad \angle AEB &= \angle A'E'B' \end{aligned}$$

3 ⇒ 1: Assume $\angle AEB = \angle A'E'B'$. Then it is possible to apply the first circle to the second so that E falls on E' and these angles coincide. Since the circles have equal radii, it follows that $\overset{\frown}{AB}$ falls on $\overset{\frown}{A'B'}$. Consequently, these arcs have equal lengths. Q.E.D.

Corollary 4.1.2 *In a circle all the semicircles are equal to each other.*

See Exercise 1.

Proposition 4.1.3 (III.3) *In a circle, a radius bisects a chord not through the center if and only if the radius and the chord are perpendicular to each other.*

See Exercise 2.

Exercises 4.1A

1. Prove Corollary 4.1.2.
2. Prove Proposition 4.1.3.
3. Prove that in a circle, a diameter is greater than any chord that is not a diameter.
4. Prove that two chords of a circle are equal if and only if they are at equal distances from its center. (Exercise 2.3N.2 can be used to produce a neutral proof.)
5. Prove that a circle cannot contain three collinear points (III.2).
6. Prove that in a circle, the radius perpendicular to a chord bisects that chord's central angle and arc.
7. Prove that in a circle two equal intersecting chords cut each other into, respectively, equal segments.
8. Prove that of two unequal chords in a circle, the greater one is closer to the center. (This is Proposition III.15. It can be easily proved on the basis of the theorem of Pythagoras, but such a proof would not be neutral. Euclid's neutral proof is based on Proposition I.24 [Exercise 2.3Q.5].)
9. Construct the midpoint of a given arc on a given circle.
10. Given an arc of a circle, construct the center of the circle.
11. Given points A, B, C, D, construct a circle through A and B whose center is equidistant from C and D.
12. Given a point A inside a circle, construct a chord that is bisected by A. Prove that this chord is the shortest of all the chords through A.
13. Given an angle α and a line segment a, construct a circle whose center is on one side of α and which cuts a segment equal to a on the other side.
14. Comment on Proposition 4.1.1 in the context of the following geometries:
    (a) spherical (b) hyperbolic (c) taxicab (d) maxi
15. Comment on Proposition 4.1.3 in the context of the following geometries:
    (a) spherical (b) hyperbolic (c) taxicab (d) maxi

An infinitely extended straight line is said to be *tangent* to a circle if they have exactly one point in common, and that point is called their *point of contact.*

Proposition 4.1.4 (III.16, 18) *If a straight line intersects a circle, then they are tangent if and only if the straight line is perpendicular to the radius through the point of contact.*

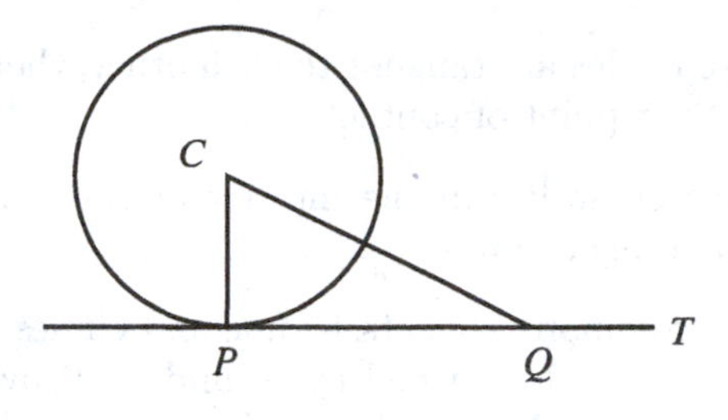

Figure 4.3.

GIVEN: Circle $(C; CP)$, straight line $\overleftrightarrow{PT}$ (Fig 4.3).

TO PROVE: $\overleftrightarrow{PT}$ is tangent to $(C; CP)$ if and only if $CP \perp \overleftrightarrow{PT}$.

PROOF: Suppose first that $\overleftrightarrow{PT}$ is tangent to $(C; CP)$. Hence, if Q is any point of $\overleftrightarrow{PT}$ that is distinct from P, it must lie outside the circle so that $CP < CQ$. Consequently, $CP \perp PT$ [PN 2.3.24].

Conversely, suppose that $CP \perp \overleftrightarrow{PT}$. Then, by Proposition 2.3.24, for any point Q of $\overleftrightarrow{PT}$ that is distinct from P, $CP < CQ$. Consequently no such point Q can lie on the circle $(C; CP)$ and hence $\overleftrightarrow{PT}$ is tangent to $(C; CP)$. Q.E.D.

Exercises 4.1B

1. Suppose S and T are the contact points of the tangents to a circle from a point P outside it. Prove that $PS = PT$.
2. Prove that in a circle the tangents at the endpoints of a diameter are parallel.
3. Prove that the straight line that joins the center of a circle to the intersection of two of its tangents bisects the angle between these tangents.
4. Prove that for each side of the triangle there is a circle that is tangent to that side at one of its interior points and tangent to the other two sides at points on their extensions. Construct these circles.

Two circles are said to be tangent if they intersect in exactly one point. If one circle lies inside the other the tangency is said to be internal; otherwise it is external.

5. Prove that if two circles are externally tangent, then the line segment joining their centers contains the point of contact.
6. Prove that if two circles are internally tangent, then the line joining their centers contains the point of contact.

7. Prove that if two circles are tangent to each other, then they have a common tangent line at their point of contact.

8. Prove that if one circle lies in the interior of another, then the two circles have no common tangent lines.

9. Let m and n be common tangents to unequal circles such that both circles lie inside one of the angles formed by m and n. Prove that the line joining the centers of the circles bisects this angle.

10. Let m and n be common tangents to unequal circles such that the circles lie in vertically opposite angles formed by m and n. Prove that the line joining the centers of the circles bisects these angles.

11. Given two circles with the same center and unequal radii, prove that all the chords of the larger circle that are tangent to the smaller circle have the same length.

12. Construct a circle with a given radius tangent to a given line.

13. Construct a circle containing a given point and tangent to a given straight line at a given point on the line.

14. Construct a circle that is tangent to two given intersecting straight lines.

15. Construct a circle that is tangent to two given parallel straight lines as well as a third straight line that intersects them.

16. Construct a circle that is tangent to two parallel straight lines.

17. Given a circle p and a point A, construct a straight line containing A such that its segment inside p has a given length.

18. Construct a point such that the lengths of the tangents from it to two given circles are given.

19. Comment on Proposition 4.1.4 in the context of the following geometries:
(a) spherical (b) hyperbolic (c) taxicab (d) maxi

The following proposition was proved by Euclid in its entirety. The proof offered in this text is incomplete in two ways. In the first place, the argument is restricted to *rational* values of the ratios in question. Moreover, given an angle $\angle ABC$ and a positive integer m, this argument makes use of the angle $(\angle ABC)/n$ even though it has not been demonstrated that such an angle can be constructed within Euclid's system.

Proposition 4.1.5 (VI.33) *In equal circles, central angles are proportional to the arcs on which they stand.*

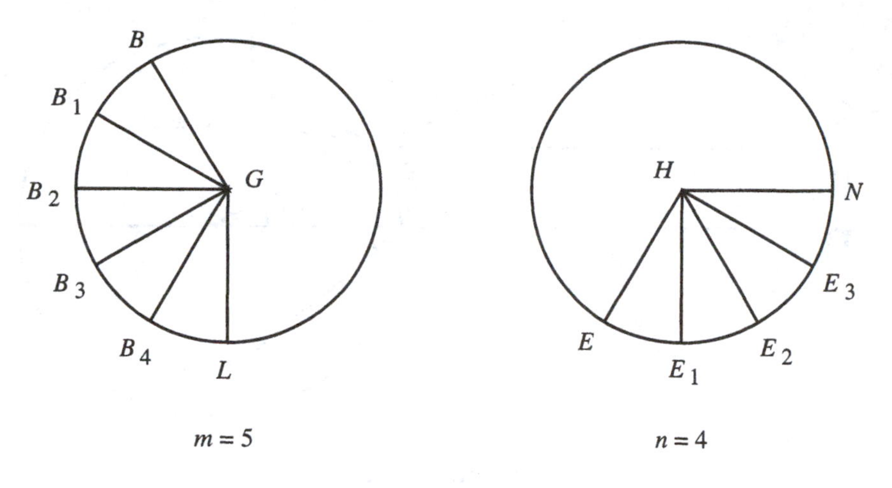

Figure 4.4.

GIVEN: Equal circles with centers G and H, respectively (Fig. 4.4).

TO PROVE: $\dfrac{\angle BGL}{\angle EHN} = \dfrac{\overset{\frown}{BL}}{\overset{\frown}{EN}}$

SUPPORTING ARGUMENT: The argument is limited to the case where the given ratios are rational. In other words, it is assumed that there exist positive integers m and n such that

$$\frac{\angle BGL}{\angle EHN} = \frac{m}{n} \quad \text{i.e.,} \quad \frac{\angle BGL}{m} = \frac{\angle EHN}{n}.$$

Let α be an angle such that

$$\alpha = \frac{\angle BGL}{m} = \frac{\angle EHN}{n}.$$

It follows that there exist points $B_1, B_2, \ldots, B_{m-1}$ on $\overset{\frown}{BL}$ and points $E_1, E_2, \ldots, E_{n-1}$ on $\overset{\frown}{EN}$ such that

$$\begin{aligned} \angle BGB_1 &= \angle B_1GB_2 = \cdots = \angle B_{m-1}GL \\ &= \angle EHE_1 = \angle E_1HE_2 = \cdots = \angle E_{n-1}HN = \alpha. \end{aligned}$$

Hence, by Proposition 4.1.1,

$$\overset{\frown}{BB_1} = \overset{\frown}{B_1B_2} = \cdots = \overset{\frown}{B_{m-1}L} = \overset{\frown}{EE_1} = \overset{\frown}{E_1E_2} = \cdots = \overset{\frown}{E_{m-1}L}.$$

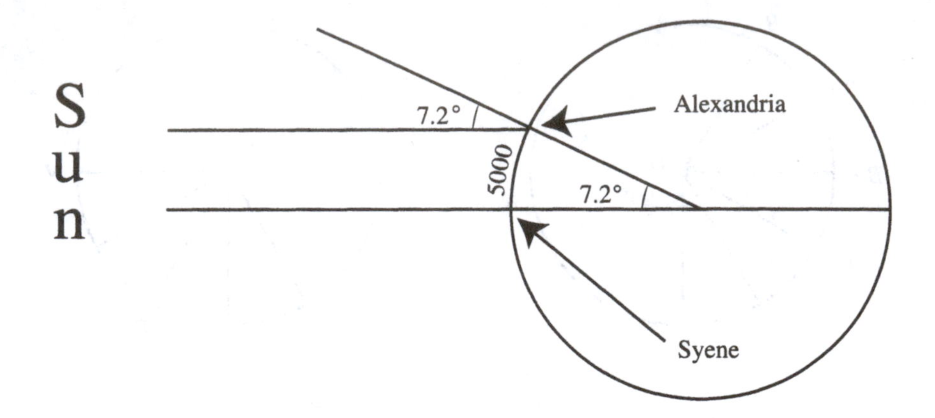

Figure 4.5.

If the common length of these arcs is denoted by β, then

$$\frac{\widehat{BL}}{\widehat{EN}} = \frac{m\beta}{n\beta} = \frac{m}{n} = \frac{\angle BGL}{\angle EHN}.$$

Q.E.D.

Proposition 4.1.5 was used by Eratosthenes (ca. 275–194 B.C.), director of the Alexandrian library, to obtain a remarkably accurate estimate of the circumference of the earth. He knew that on the summer solstice the sun shone down at midday directly into a well in the city of Syene whereas in Alexandria, 5000 stadia to the north, the shadows indicated that the sun's rays formed an angle of $1/50$ of $360°$ ($7.2°$) with the vertical. Assuming that the sun is so far away that its rays can be considered to be parallel when they reach the earth (Fig. 4.5), he then used Proposition 4.1.5 to obtain the equation

$$\frac{\text{Circumference of the Earth}}{\text{Distance from Alexandria to Syene}} = \frac{360°}{7.2°} = 50$$

from which he concluded that the circumference is $50 \cdot 5000 = 250{,}000$ stadia. In order to make his answer divisible by 60 (probably because of the influence of the Babylonian sexagesimal number system), he adjusted this result to 252,000 stadia. The standard stade of the time had a length of 178.6 meters, which converts this rounded estimate to 45,007 km, an overestimate of 12.3%, since the circumference of the earth is actually 40,075 km.

Exercises 4.1C

1. A circle has circumference 10 ft. Find the lengths of the arcs that subtend the following angles at the center of the circle:
(a) 10° (b) 30° (c) 90° (d) 110° (e) 120° (f) 180°
2. A location on earth has latitude 25° N. Find its distance from the equator and from the North Pole.
3. A location on earth has latitude 70° N. Find its distance from the equator and from the North Pole.
4. A location on earth has latitude 30° S. Find its distance from the equator and from the North Pole.
5. A location on earth has latitude 80° S. Find its distance from the equator and from the North Pole.
6. A location on earth lies 2000 km north of the equator. Find its latitude.
7. A location on earth lies 1234 km north of the equator. Find its latitude.
8. A location on earth lies 1000 km south of the equator. Find its latitude.
9. A location on earth lies 617 km south of the equator. Find its latitude.
10. Comment on Proposition 4.1.5 in the context of the following geometries:
(a) spherical (b) hyperbolic (c) taxicab (d) maxi

4.2 The Nonneutral Euclidean Geometry of the Circle

The next proposition is one of the most surprising in the *Elements*. Unlike those appearing the previous section, its implications are quite unexpected.

Proposition 4.2.1 (III.20) *In a circle, the angle at the center is double of the angle at the circumference, when the angles have the same arc as base.*

GIVEN: Points A, B, C on the circumference of a circle centered at E (Fig. 4.6).

TO PROVE: $\angle BEC = 2\angle BAC$.

PROOF: It is necessary to distinguish three cases.

Case 1: One of the sides of $\angle BAC$ contains the center E.

$$\angle BEC = \angle BAC + \angle ECA \qquad \text{[Exterior, PN 3.1.6]}$$
$$\angle BAC = \angle ECA \qquad [AE = EC\text{, PN 2.3.5]}$$

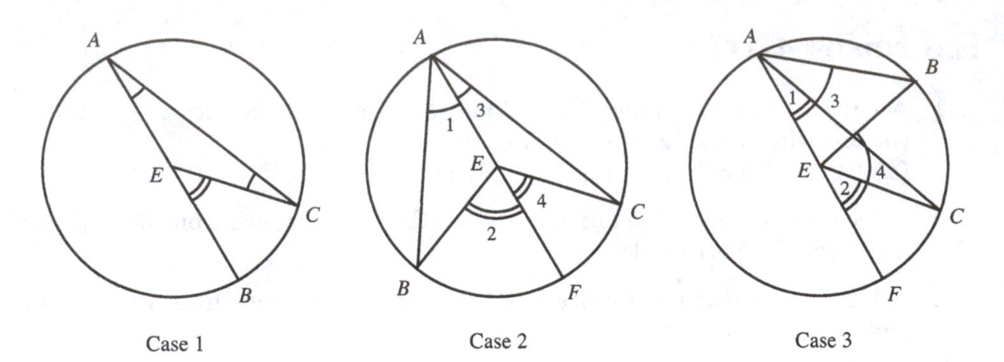

Figure 4.6.

$$\therefore \qquad \angle BEC = 2\angle BAC$$

Case 2: The center E lies in the interior of $\angle BAC$. Let F be the other intersection of $\overleftrightarrow{AE}$ with the circumference of the given circle. Then

$$\angle 2 = 2\angle 1 \qquad \text{[Case 1]}$$
$$\angle 4 = 2\angle 3 \qquad \text{[Case 1]}$$
$$\therefore \qquad \angle BEC = 2\angle BAC. \qquad \text{[CN 2]}$$

Case 3: The center E lies outside of $\angle BAC$. Let F be the other intersection of $\overleftrightarrow{AE}$ with the circumference of the given circle. Then

$$\angle 2 = 2\angle 1 \qquad \text{[Case 1]}$$
$$\angle 4 = 2\angle 3 \qquad \text{[Case 1]}$$
$$\therefore \qquad \angle BEC = 2\angle BAC. \qquad \text{[CN 3]}$$

Q.E.D.

Proposition 4.2.1 has several corollaries whose proofs are relegated to the exercises.

Proposition 4.2.2 (III.21) *In a circle, the angles in the same segment are equal to one another.*

GIVEN: Points A, B, C, D on the circumference of a circle such that A and D lie on the same side of BC (Fig. 4.7).

TO PROVE: $\angle BAC = \angle BDC$.

PROOF: See Exercise 1.

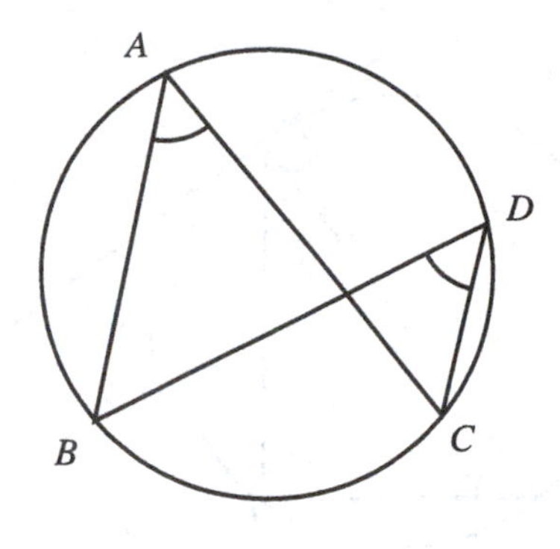

Figure 4.7.

This proposition is somewhat counterintuitive. Suppose the points A and B in Figure 4.8 are fixed whereas P slides clockwise around the circle occupying positions $P_1, P_2, \ldots, P_5$ successively. Proposition 4.2.2 implies that as long as the point P remains in the interior of the upper (or longer) $\overset{\frown}{AB}$ the angle APB retains a constant (acute) value. When P passes through A or B, APB is no longer an angle. Finally, when P is in the interior of the shorter (or lower), $\overset{\frown}{AB}$ the angle APB assumes a different (obtuse) value. In other words, even though the point P moves in a *continuous* manner, $\angle APB$ varies as a *discontinuous* function of the position of P.

Proposition 4.2.3 (III.31) *In a circle, the angle subtended by a diameter from any point on the circumference is a right angle.*

See Exercise 2.

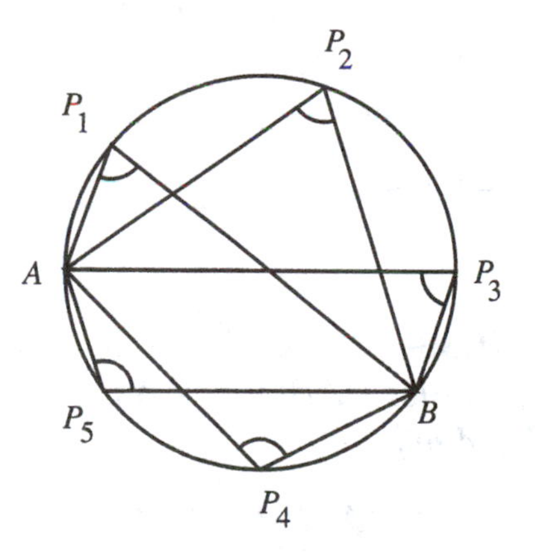

Figure 4.8. A discontinuous function.

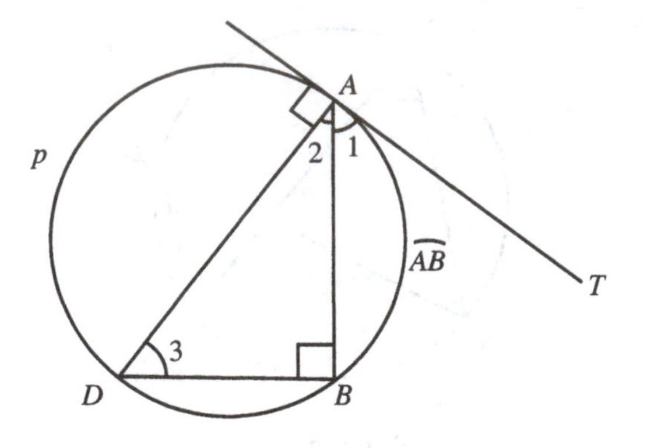

Figure 4.9.

Proposition 4.2.4 (III.32) *Let AB be a chord of a circle and let AT be any straight line at A. Then the line $\overleftrightarrow{AT}$ is tangent to the circle if and only if $\angle TAB$ is equal to the angle at the circumference subtended by the intercepted arc.*

GIVEN: Circle p with chord AB, straight line $\overleftrightarrow{AT}$, $\widehat{AB}$ (Fig. 4.9).

TO PROVE: $\overleftrightarrow{AT}$ is tangent to p if and only if $\angle TAB$ equals the angle at the circumference of p subtended by $\widehat{AB}$.

PROOF: Let AD be the diameter of the circle containing A, and join BD.

By Proposition 4.2.3 $\angle ABD = 90°$. Hence the following statements are all equivalent to each other:

$$\overleftrightarrow{AT} \text{ is tangent to the circle } p$$
$$\angle DAT = 90°$$
$$\angle 1 = 90° - \angle 2$$
$$\angle 1 = \angle 3.$$

Q.E.D.

Proposition 4.2.5 (III.36–37) *Let P be a point outside a given circle p let T be a point on p, and let PAB be a secant line with chord AB. Then $\overleftrightarrow{PT}$ is tangent to p if and only if*

$$PA \cdot PB = PT^2.$$

GIVEN: Point P outside circle p, straight lines $\overleftrightarrow{PT}$ and $\overleftrightarrow{PAB}$ that intersect p in T, A, B (Fig. 4.10).

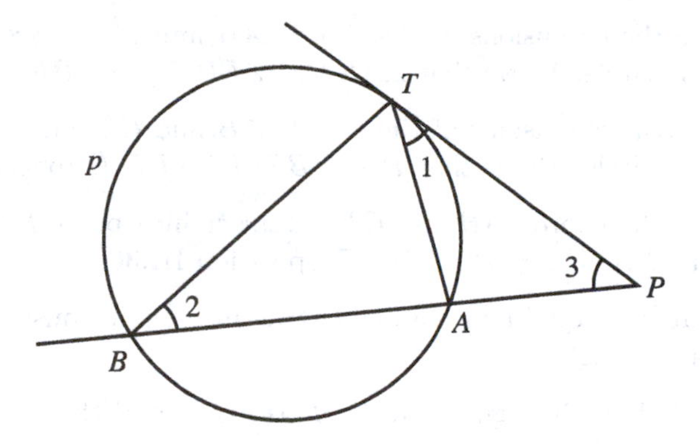

Figure 4.10.

TO PROVE: $\overleftrightarrow{PT}$ is tangent to p if and only if $PA \cdot PB = PT^2$.

PROOF: The following statements are all equivalent to each other:

The line $\overleftrightarrow{PT}$ is tangent to p

$\angle 1 = \angle 2$; [PN 4.2.4]

$\triangle TPA$ and $\triangle BPT$ are similar to each other [PN 3.5.7]

$$\frac{PA}{PT} = \frac{PT}{PB}$$

$PA \cdot PB = PT^2$.

Q.E.D.

A polygon is said to be cyclic if all of its vertices lie on a circle.

Proposition 4.2.6 (III.22) *The opposite angles of a cyclic quadrilateral are equal to two right angles.*

See Exercise 3.

Exercises 4.2A

1. Prove Proposition 4.2.2.
2. Prove Proposition 4.2.3.
3. Prove Proposition 4.2.6.
4. Prove that in a circle parallel chords enclose equal arcs.
5. Prove that if the quadrilateral $ABCD$ is cyclic, then the exterior angle at A equals the interior angle at C.

6. In a circle the extensions of the chords AB and KL intersect in a point P outside the circle. Prove that $\angle AKP = \angle LBP$ and $\angle BKP = \angle LAP$.

7. In a circle the extensions of the chords AB and CD intersect in a point P outside the circle. Prove that $PA \cdot PB = PC \cdot PD$ (Proposition III.35).

8. In a circle the chords AB and CD intersect in a point P inside the circle. Prove that $PA \cdot PB = PC \cdot PD$ (Proposition III.36).

9. Prove that two equal and parallel chords in a circle constitute the opposite sides of a rectangle.

10. Prove that if the hexagon $ABCDEF$ is cyclic and the interior angles at A and D are equal, then $BC \parallel EF$.

11. In the cyclic quadrilateral $ABCD$, $AD = BC$. Prove that the interior angles at A and B are equal to each other (as are those at C and D).

12. Prove that the sum of the interior angles at A, C and E in the cyclic hexagon $ABCDEF$ is four right angles.

13*. Prove that if the perpendicular chords AB and CD of a circle intersect at the point M (inside the circle) then the straight line through M that is perpendicular to AD bisects the chord BC.

14. Prove that every cyclic rhombus is a square.

15. Prove that if A and B are two distinct points and D is any other point on AB then the locus of all the points P in the plane such that $AP/PB = AD/DB$ is a circle. (This is the *circle of Apollonius.*)

16. State and prove the converse of Proposition 4.2.6.

17. Given a line segment AB, construct the circle which consists of all the points from which AB subtends an angle of $90°$.

18. Given a line segment AB, construct the arc which consists of all the points from which AB subtends an angle of $60°$.

19. Given a line segment AB, construct the arc which consists of all the points from which AB subtends an angle of $120°$.

20. Given a line segment AB, construct the arc which consists of all the points from which AB subtends an angle equal to a given angle α.

21. Construct a triangle given the data:

    (a) a, h_b, h_c

    (b) a, h_a, α

    (c) a, m_a, α

    (d) $a + b + c$, h_a, α.

22. Construct a parallelogram given its two diagonals and one of its angles.

23. Given line segment AB and CD and angles α and β, construct a point P such that $\angle APB = \alpha$ and $\angle CPD = \beta$.

24. In a given $\triangle ABC$ construct a point P such that $\angle APB = \angle BPC = \angle CPA$.

25. Comment on Proposition 4.2.1 in the context of the following geometries: (a) spherical; (b) hyperbolic; (c) taxicab; (d) maxi.

26. Comment on Proposition 4.2.2 in the context of the following geometries: (a) spherical; (b) hyperbolic; (c) taxicab; (d) maxi.

27. Comment on Proposition 4.2.3 in the context of the following geometries: (a) spherical; (b) hyperbolic; (c) taxicab; (d) maxi.

28. Comment on Proposition 4.2.6 in the context of the following geometries: (a) spherical; (b) hyperbolic; (c) taxicab; (d) maxi.

29(C). Use a computer application to verify the following propositions: (a) 4.2.1 (b) 4.2.2 (c) 4.2.3b (d) 4.2.4.

30. Given an angle and a point A inside it, construct a point on one side of the angle that is equidistant from A and the angle's other side.

31. Given an angle and a point A inside it, construct a circle that is tangent to the angle's sides and contains the point A.

32. Let a, b be two given line segments. Construct a line segment x such that $a/x = x/b$ (VI.13).

Three (or more) straight lines are said to be *concurrent* if they all contain the same point. The following is the first of several concurrence theorems.

Proposition 4.2.7 *The three perpendicular bisectors of the sides of a triangle are concurrent.*

GIVEN: $\triangle ABC$; DD', EE', FF' are the perpendicular bisectors of AB, AC, and BC respectively (Fig. 4.11).

TO PROVE: DD', EE', FF' are concurrent.

PROOF: Exercise 3.1A.4 guarantees that DD' and EE' intersect in some point M. Draw AM, BM, CM. Then

$$AM = BM \qquad \text{[PN 2.3.12]}$$
$$AM = CM \qquad \text{[PN 2.3.12]}$$

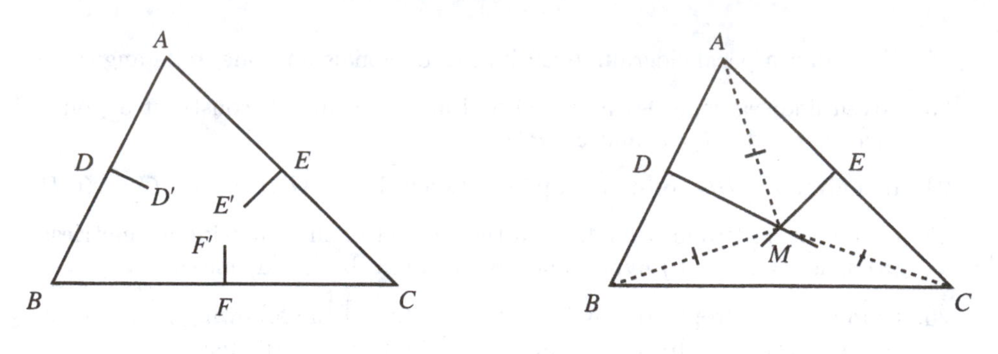

Figure 4.11.

$\therefore$ $BM = CM$ [CN 1]

$\therefore$ M is on the perpendicular bisector to BC. [PN 2.3.13]

Q.E.D.

A circle is said to *circumscribe* a triangle if all of the triangle's vertices are on the circle. Its center and radius are, respectively, the triangle's *circumcenter* and *circumradius*.

Proposition 4.2.8 (IV.5) *About a given triangle to circumscribe a circle.*

See Exercise 1.

Proposition 4.2.9 *The bisectors of the three interior angles of a triangle are concurrent.*

GIVEN: $\triangle ABC$, AA', BB', CC' are the bisectors of $\angle BAC$, $\angle ACB$, $\angle ABC$, respectively (Fig. 4.12).

TO PROVE: AA', BB', CC' are concurrent.

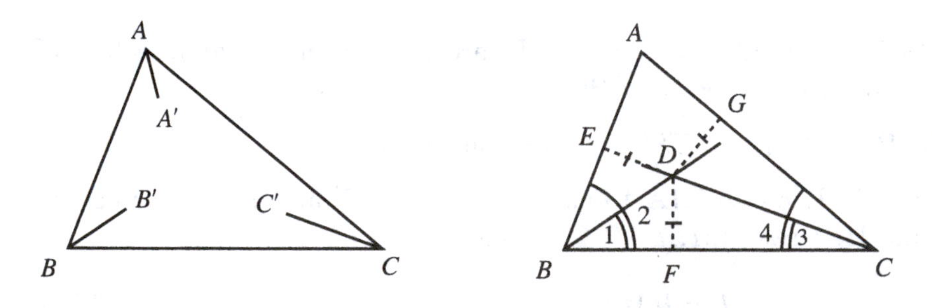

Figure 4.12.

PROOF: Since

$$\angle 1 + \angle 3 = \frac{1}{2}(\angle 2 + \angle 4) < \frac{1}{2}180° \qquad \text{[PN2.3.21]}$$

it follows from Postulate 5 that BB' and CC' intersect in some point D. Let E, F and G be those points on AB, BC, CA, respectively, such that $DE \perp AB$, $DF \perp BC$, and $DG \perp AC$. Then

$$DE = DF = DG \qquad \text{[PN 2.3.34]}$$

$$\therefore \quad DE = DG \qquad \text{[CN 1]}$$

$$\therefore \quad DA \text{ bisects } \angle BAC. \qquad \text{[PN 2.3.34]}$$

Q.E.D.

A circle that lies in the interior of a triangle and is tangent to all of its sides is said to be *inscribed* in the triangle. Its center and radius are, respectively, the triangle's *incenter* and *inradius*.

Proposition 4.2.10 (IV.4) *In a given triangle to inscribe a circle.*

See Exercise 3.

Exercises 4.2B

1. Prove Proposition 4.2.8.
2. Prove that similar triangles have circumradii that are proportional to their sides.
3. Prove Proposition 4.2.10.
4. Prove that similar triangles have inradii that are proportional to their sides.
5. Prove that the circumcenter of a right triangle is the midpoint of its hypotenuse.
6. Prove that if the circumcenter of $\triangle ABC$ lies inside the triangle, then the triangle is acute, if the center is on a side the triangle is right, and if the center is outside the triangle, then the triangle is obtuse.
7. Prove that the circumcenter of an acute triangle lies inside the triangle.
8. Prove that the circumcenter of an obtuse triangle lies outside it.
9. Prove that the area of the triangle equals the product of half its perimeter with the inradius.
10. In a given circle inscribe a triangle similar to a given triangle.

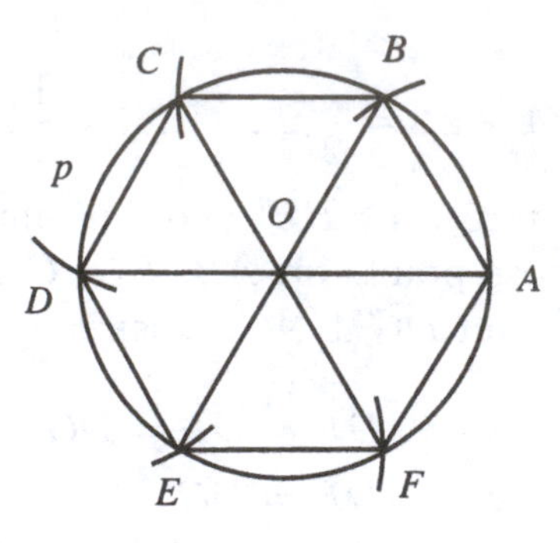

Figure 4.13.

11. Prove that the altitudes of the triangle are concurrent.

12. Comment on Proposition 4.2.7 in the context of the following geometries:
    (a) spherical (b) hyperbolic (c) taxicab (d) maxi

13. Comment on Proposition 4.2.8 in the context of the following geometries:
    (a) spherical (b) hyperbolic (c) taxicab (d) maxi

14. Comment on Proposition 4.2.9 in the context of the following geometries:
    (a) spherical (b) hyperbolic (c) taxicab (d) maxi

15. Comment on Proposition 4.2.10 in the context of the following geometries:
    (a) spherical (b) hyperbolic (c) taxicab (d) maxi

16(C). Use a computer application to verify the following propositions:
    (a) 4.2.7 (b) 4.2.9

4.3 Regular Polygons

A polygon is *regular* if all of its sides and all of its interior angles are equal. The equilateral triangles are the regular triangles and they are the subject matter of Proposition 1 of Book I. Squares are the regular quadrilaterals and they are constructed in Proposition 3.3.1. Book IV of the *Elements* is mostly concerned with the constructibility of other regular polygons and their inscription in circles. Regular hexagons are also easily constructed.

Proposition 4.3.1 (IV.15) *In a given circle to inscribe a regular hexagon.*

GIVEN: Circle $p = (O; r)$ (Fig. 4.13).

TO CONSTRUCT: Points A, B, C, D, E, F on p such that $ABCDEF$ is a regular hexagon.

Figure 4.14.

CONSTRUCTION: Let A be an arbitrary point on the circle p. Let B be the intersection of an arc of radius r and center A with p. Let C be the intersection of an arc of radius r and center B with p, and let D, E, F be constructed in a similar manner. Then $ABCDEF$ is a regular hexagon.

PROOF: By construction, $\triangle AOB$, $\triangle BOC$, $\triangle COD$, $\triangle DOE$, and $\triangle EOF$ are all equilateral so that $\angle AOB = \angle BOC = \angle COD = \angle DOE = \angle EOF = 60°$. It follows that $\angle FOA = 360° - 5 \cdot 60° = 60°$ and hence the isosceles $\triangle FOA$ is also equilateral. Thus, $FA = OA$ and so each of the sides of $ABCDEF$ has length r. It also follows that each of the interior angles of $ABCDEF$ equals $120°$. Thus, $ABCDEF$ is a regular hexagon. Q.E.D.

A slight variation on the construction of the regular hexagon yields the flower-like configuration of Figure 4.14.

The construction of the regular pentagon is a considerably more difficult matter. Some of the technically demanding details are isolated in the following lemma. Others appeared as propositions previously.

Proposition 4.3.2 (IV.10) *To construct an isosceles triangle having each of the angles at the base equal to double of the remaining one.*

TO CONSTRUCT: $\triangle ABC$ such that $\angle ABC = \angle ACB = 2\angle BAC$ (Fig. 4.15).

CONSTRUCTION: Let AB be an arbitrary line segment and let D be a point such that $AB \cdot BD = AD^2$ [PN 3.4.1]. Then the required $\triangle ABC$ is that triangle such that $AC = AB$ and $BC = AD$ [PN 2.3.27].

PROOF: By construction, $BC^2 = AD^2 = AB \cdot BD$. It therefore follows that from Proposition 4.2.5 that BC is tangent to the circle p that circumscribes $\triangle ACD$. Hence,

$$\angle 1 = \angle 2 \qquad \text{[PN 4.2.4]}$$

$$\therefore \quad \angle 3 = \angle 1 + \angle 4 = \angle 2 + \angle 4 = \angle 5 = \angle 6$$

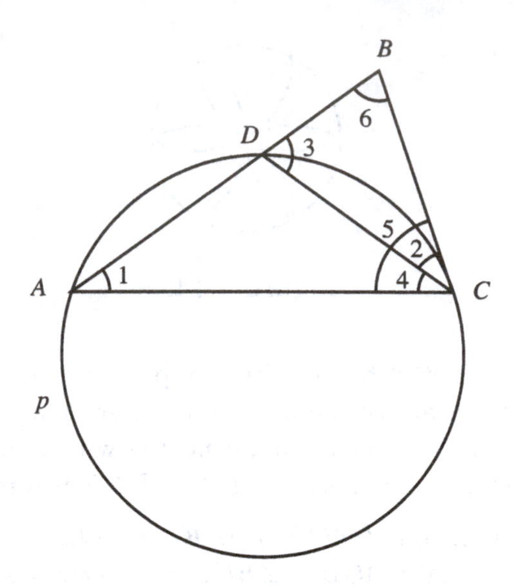

Figure 4.15.

$$\therefore \quad DC = BC = AD$$

$$\therefore \quad \angle 1 = \angle 4 = \frac{1}{2}\angle 3 = \frac{1}{2}\angle 6 = \frac{1}{2}\angle 5.$$

Q.E.D.

If the smallest of the angles of the triangle of Proposition 4.3.2 is denoted by x, then the other two angles are each $2x$ and so, by Proposition 3.1.6,

$$180^\circ = x + 2x + 2x = 5x$$

from which it follows that $x = 36^\circ$. Hence the following corollary holds.

Proposition 4.3.3 *To construct angles of* 36° *and* 72°*.*

We are now ready to construct the regular pentagon.

Proposition 4.3.4 (IV.11) *In a given circle to inscribe a regular pentagon.*

GIVEN: Circle $p = (O; r)$ (Fig. 4.16).

TO CONSTRUCT: Points A, B, C, D, E on p such that $ABCDE$ is a regular pentagon.

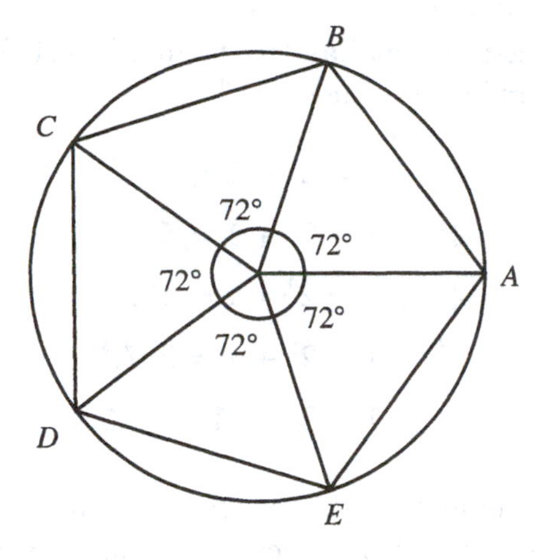

Figure 4.16.

CONSTRUCTION: At the center O of the circle construct five nonoverlapping central angles of 72° [PN 4.3.3]. Label the successive intersections of their sides with the circle A, B, C, D, E. Then $ABCDE$ is the required pentagon.

PROOF: The five constructed isosceles triangles are all congruent by SAS. It follows that the five sides AB, BC, CD, DE, and EA are all equal. Moreover, the base angles of these triangles are $(1/2)(180° - 72°) = 54°$ each and hence all of the pentagon's interior angles are equal (to 108° each).
Q.E.D.

Euclid took the trouble to prove that the regular 15-sided polygon is constructible (Exercise 3). It is reasonable to suppose that this was his way of pointing out that there is an interesting question to be pondered here. Namely, for which integers n can the regular n-sided polygon be constructed? It has already been shown above that this is possible for $n = 3$, 4, 5, 6. Some more such n can be easily produced by simply doubling the number of sides of any constructible regular polygon (see Exercises 1, 2, 4). However, this does not answer the question for such numbers as 7, 9, 11, 13, 14, 17,.... The surprising intricacy of the construction of the regular pentagon indicates that such polygons might pose an even greater challenge. This problem continued to excite the interest of mathematicians after Euclid, but no progress was made for over 2000 years until the young Gauss demonstrated the constructibility of the regular 17-sided polygon in 1796. Actually, he did much more. Using the newly emergent theory of complex numbers Gauss proved that a regular p-sided polygon can be constructed for every prime integer p that has the form $2^{2^n} + 1$ for some nonnegative integer n. These include the values

$$2^{2^0} + 1 = 2^1 + 1 = 3,$$
$$2^{2^1} + 1 = 2^2 + 1 = 5,$$
$$2^{2^2} + 1 = 2^4 + 1 = 17,$$
$$2^{2^3} + 1 = 2^8 + 1 = 257,$$
$$2^{2^4} + 1 = 2^{16} + 1 = 65{,}537.$$

Curiously, the next number in this sequence, namely $2^{2^5} + 1 = 2^{32} + 1 =$ 4,294,967,297, fails to be a prime since it can be factored as $641 \cdot 6{,}700{,}417$, a fact that had already been noted by Euler over fifty years earlier. The same is true for all the numbers of this form for $n = 6, 7, \ldots, 16$ and several other values including $n = 1945$. In fact, it is not known whether there are any more primes p that can be expressed in this form above and beyond the five listed previously.

Gauss completely resolved the issue of the constructibility of regular polygons as follows.

Proposition 4.3.5 *It is possible to construct (in the sense defined by Euclid) a regular g-gon ($g \geq 3$) if and only if the factorization of g into primes has the form*

$$g = 2^k p_1 p_2 \cdots p_m$$

where $m \geq 0$ and $p_1, p_2, \ldots, p_m$ are distinct primes each of which has the form $2^{2^n} + 1$. □

Thus, the regular 2040-gon is constructible because $2040 = 2^3 \cdot 3 \cdot 5 \cdot 17$ whereas the regular 28-sided and 100-sided polygons are not constructible because $28 = 2^2 \cdot 7$ and $100 = 2^2 \cdot 5^2$.

Exercises 4.3

1. Prove that the regular octagon is constructible.
2. Prove that the regular decagon is constructible.
3. Prove that the regular 15-sided polygon is constructible (Proposition IV.16).
4. Let g be a positive integer. Prove that if the regular g-sided polygon is constructible, so is the regular $2g$-sided polygon.

5. Let p and q be two prime integers. Prove that if the regular p-sided and q-sided polygons are constructible so is the regular pq-sided polygon.
6. Let g and h be relatively prime integers such that the regular g-sided and h-sided polygons can be constructed. Prove that the regular gh-sided polygon can be constructed.
7. Let $g, h > 1$ be integers such that h is an integer multiple of g. Prove that if the regular h-sided polygon is constructible so is the regular g-sided polygon.
8. Use a computer to prove that $2^{2^6} + 1$ is not a prime integer.
9. For which $n = 3, 4, \ldots, 100$ is the regular n-gon constructible?
10. For which $n = 101, 102, \ldots, 200$ is the regular n-gon constructible?
11. Comment on Proposition 4.3.1 in the context of the following geometries: (a) spherical; (b) hyperbolic; (c) taxicab; (d) maxi.
12. Comment on Proposition 4.3.2 in the context of the following geometries: (a) spherical; (b) hyperbolic; (c) taxicab; (d) maxi.
13. Comment on Proposition 4.3.4 in the context of the following geometries: (a) spherical; (b) hyperbolic; (c) taxicab; (d) maxi.
14. Show that in taxicab geometry equilateral triangles are not necessarily equiangular. Can all three of the angles of an equilateral taxicab triangle be distinct?

15(C). Perform the construction of Proposition 4.3.1 using a computer application.

4.4 Circle Circumference and Area

The fundamental observation that the circumference of a circle is proportional to its diameter (and hence also its radius) goes back several millennia. Surprisingly, Euclid said nothing on this topic in *The Elements*.

Proposition 4.4.1 *Circumferences of circles are proportional to their radii.*

GIVEN: Circles $p_1 = (O_1; r_1)$ and $p_2 = (O_2; r_2)$ of circumferences c_1 and c_2 respectively (Fig. 4.17).

TO PROVE: $\dfrac{c_1}{c_2} = \dfrac{r_1}{r_2}$.

SUPPORTING ARGUMENT: It follows from Proposition 4.3.1 that it is possible to inscribe a regular hexagon in each of the given circles. By repeatedly bisecting the central angles subtended by the sides of the polygons

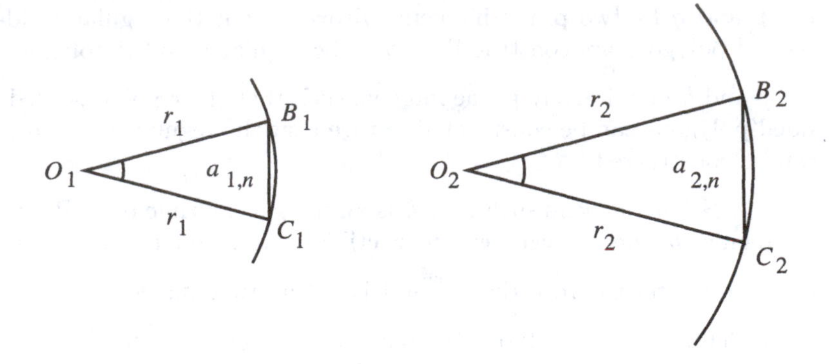

Figure 4.17.

it is possible to inscribe in each circle p_i, $i = 1, 2$, a regular n-sided polygon of side, say, $a_{i,n}$ where n is an integer of the form $3 \cdot 2^m$. It is clear that $\angle B_1O_1C_1 = \angle B_2O_2C_2 = 360°/n$. Since $\triangle B_1O_1C_1$ and $\triangle B_2O_2C_2$ are isosceles they must be equiangular [PN 3.1.6] and hence they are similar [PN 3.5.7]. In particular, the sides of the inscribed polygons are proportional to the radii. Making the reasonable assumption that for large n the difference between the circumferences of each circle and that of its inscribed polygon is negligible, it follows that

$$\frac{c_1}{c_2} = \frac{na_{1,n}}{na_{2,n}} = \frac{a_{1,n}}{a_{1,n}} = \frac{r_1}{r_2}.$$

Q.E.D.

An alternate supporting argument that makes use of calculus is described in Exercise 1.

It follows from the above proposition that if c and r denote the circumference and radius of an arbitrary circle, then the ratio

$$\frac{c}{r}$$

has a constant value, say, α. This constant number can be used to restate the above proposition in the following form.

Proposition 4.4.2 *There is a number α such that if c and r are, respectively, the circumference and radius of any circle, then $c = \alpha r$.*

The numerical value of α is, of course, of interest and will be estimated at the end of this section. Next, the area of the circle is examined. The following proposition was proved by Euclid using the *method of exhaustion* which was the Greeks' version of the integral calculus. This method was developed by Euclid's predecessor Eudoxus, whom Archimedes credited with this and other similar propositions.

Proposition 4.4.3 (XII.2) *The areas of circles are proportional to the squares of their radii.*

GIVEN: Circles $p_1 = (O_1; r_1)$ and $p_2 = (O_2; r_2)$ of areas A_1 and A_2, respectively (Fig 4.17).

TO PROVE: $\dfrac{A_1}{A_2} = \dfrac{r_1^2}{r_2^2}$.

SUPPORTING ARGUMENT: It follows from Proposition VI.19 (Exercise 3.5E.10) that the areas of $\triangle O_1B_1C_1$ and $\triangle O_2B_2C_2$ are proportional to the squares of the radii r_1 and r_2. Making the reasonable assumption that for large n the difference between the areas of each circle and that of its inscribed polygon is negligible, it follows that

$$\frac{A_1}{A_2} = \frac{n(\triangle O_1B_1C_1)}{n(\triangle O_2B_2C_2)} = \frac{\triangle O_1B_1C_1}{\triangle O_2B_2C_2} = \frac{r_1^2}{r_2^2}.$$

Q.E.D.

An alternate supporting argument makes use of calculus (see Exercise 2).

It follows from the foregoing proposition that if A and r denote the area and radius of an arbitrary circle, then the ratio

$$\frac{A}{r^2}$$

has a constant value, say π. This number can be used to restate the above proposition in the following form.

Proposition 4.4.4 *There is a number π such that if A and r are, respectively, the area and radius of any circle, then $A = \pi r^2$.*

The numerical value of π is, of course, of interest, but the relationship between α and π needs to be addressed first. The discovery of this relationship was attributed by Proclus to Archimedes.

Proposition 4.4.5 *The proportionality constants of the circumference and area of a circle are related by the equation $\alpha = 2\pi$.*

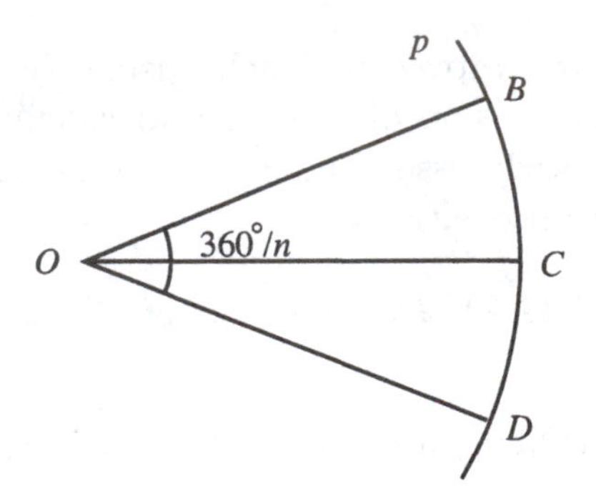

Figure 4.18.

SUPPORTING ARGUMENT: Suppose the circle p is divided into n equal sectors each of which has a central angle of $360°/n$, and let OBD be a typical sector (Fig. 4.18). If n is large it may be assumed that OBD is a triangle with altitude $OC = r$. Applying Proposition 3.2.5, it follows that this triangle has area $(r/2)\ \overset{\frown}{BD}$. Hence the circle p has area

$$A = n(\overset{\frown}{BD})\frac{r}{2} = c\frac{r}{2} = (\alpha r)\frac{r}{2} = \frac{\alpha}{2}r^2.$$

It now follows from Proposition 4.4.4 that $\pi = \alpha/2$ or $\alpha = 2\pi$. Q.E.D.

Corollary 4.4.6 *The circumference of a circle of radius r is $2\pi r$.*

An alternate supporting argument for Proposition 4.4.5 can be based on Figure 4.19. Imagine that the circle of radius r on the left is filled with circular strands. Cut the circle along the vertical dashed radius and straighten out all the strands as indicated until they form an isosceles triangle (that the sides of this triangle are straight follows from Proposition 4.4.2). It follows from Proposition 3.2.5 that the area of this triangle, and hence also the area of the circle, is

$$A = (\text{circumference} \cdot r)/2 = cr/2 = \alpha r \cdot r/2 = (\alpha/2)r^2,$$

and the rest of the argument proceeds as before. Appealing as this argument is, it is fraught with logical perils, which are discussed in Exercise 21.

Yet another alternative argument in support Proposition 4.4.5 calls for slicing the circle into an even number of equal sectors and rearranging these to form the near-parallelogram at the top of Figure 4.20. As the number of sectors increases to infinity, the near-parallelogram converges to the rectangle below it, whose area is clearly

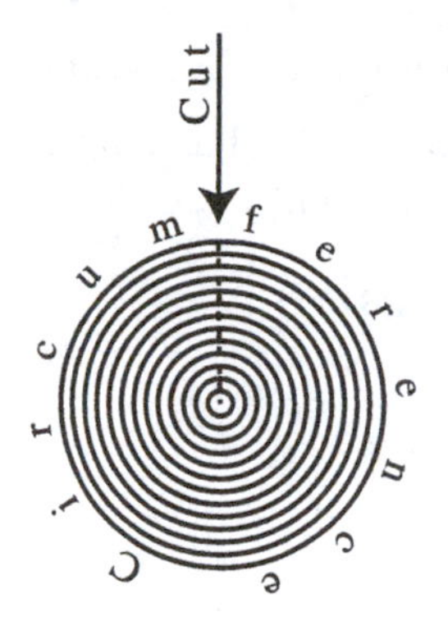

Figure 4.19.

$$A = (\text{half circumference})r = cr/2 = \alpha r \cdot r/2 = (\alpha/2)r^2.$$

The procedure of successive approximations used in Proposition 4.4.1 also yields a method for obtaining numerical estimates of the constant of proportion π. This was first carried out by Archimedes and constitutes the first of a long (and still ongoing) series of scientific estimations of π. For each positive integer $n \geq 3$, let a_n denote the length of the chord AB, which

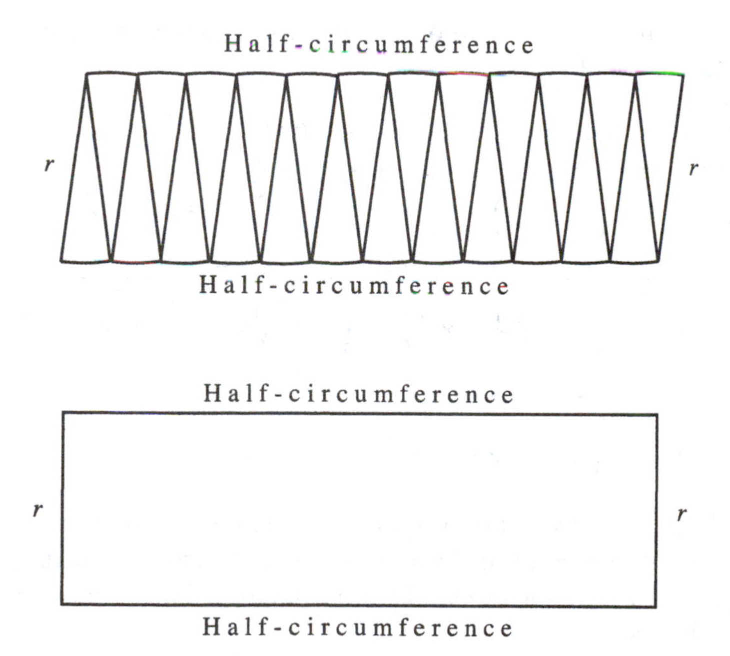

Figure 4.20.

is one side of the regular n-gon inscribed in a circle of radius 1 (Fig. 4.21). If C is the midpoint of the corresponding $\overset{\frown}{AB}$, then the chord AC has length a_{2n}. Two applications of the Theorem of Pythagoras then yield

$$\begin{aligned} a_{2n}^2 = AC^2 = AD^2 + DC^2 &= \left(\frac{a_n}{2}\right)^2 + (OC - OD)^2 \\ &= \left(\frac{a_n}{2}\right)^2 + \left(1 - \sqrt{OA^2 - AD^2}\right)^2 \\ &= \left(\frac{a_n}{2}\right)^2 + \left(1 - \sqrt{1 - \left(\frac{a_n}{2}\right)^2}\right)^2 \\ &= \left(\frac{a_n}{2}\right)^2 + \left(1 - 2\sqrt{1 - \left(\frac{a_n}{2}\right)^2} + 1 - \left(\frac{a_n}{2}\right)^2\right) \\ &= 2 - \sqrt{4 - a_n^2}. \end{aligned}$$

Hence,

Proposition 4.4.7 *If a_n denotes the length of the side of the regular polygon with n sides that is inscribed in a circle of radius 1, then*

$$a_{2n} = \sqrt{2 - \sqrt{4 - a_n^2}}.$$

Since the length of the side of the inscribed regular hexagon equals the radius, it follows that $a_6 = 1$ and so

$$a_{12} = \sqrt{2 - \sqrt{4 - 1}} = \sqrt{2 - \sqrt{3}} = .5176380902\ldots$$

$$a_{24} = \sqrt{2 - \sqrt{4 - \left(2 - \sqrt{3}\right)}} = \sqrt{2 - \sqrt{2 + \sqrt{3}}} = .2610523844\ldots$$

$$a_{48} = \sqrt{2 - \sqrt{4 - \left(2 - \sqrt{2 + \sqrt{3}}\right)}} = \sqrt{2 - \sqrt{2 + \sqrt{2 + \sqrt{3}}}} = .1308062585\ldots$$

$$a_{96} = \sqrt{2 - \sqrt{2 + \sqrt{2 + \sqrt{2 + \sqrt{3}}}}} = .0654381656\ldots.$$

Since the length of any arc exceeds that of its chord (another one of those reasonable assumptions), the circumference of the circle exceeds that of any inscribed polygon. As the circle of radius 1 has circumference 2π it follows that for each positive integer n,

$$\pi > \frac{na_n}{2}$$

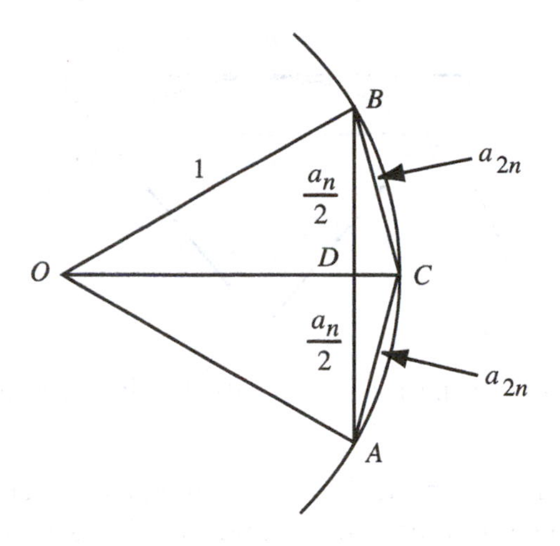

Figure 4.21.

and, in particular,

$$\pi > 48a_{96} = 3.14103195089\cdots.$$

To obtain upper bounds for the value of π, Archimedes examined regular n-gons whose sides were tangent to the circle. Suppose a regular n-gon is inscribed in a circle of radius 1 and at each of the vertices a straight line tangent to the circle is constructed. It is easily verified that the resulting polygon surrounding the circle is also a regular n-gon of side, say, b_n (Fig. 4.22). The side b_n can be estimated by showing that (see Exercise 8)

$$b_6 = \frac{2}{\sqrt{3}} \quad \text{and} \quad b_{2n} = \frac{2\left(\sqrt{4+b_n^2}-2\right)}{b_n}. \tag{1}$$

Alternately, it can be shown that (Exercise 22)

$$b_n = \frac{2a_n}{\sqrt{4-a_n^2}}. \tag{2}$$

This gives us $b_{96} = .0654732208\cdots$ and hence

$$\pi < \frac{96b_{96}}{2} = 3.1427145996\cdots.$$

Archimedes did not have the decimal number system at his disposal and

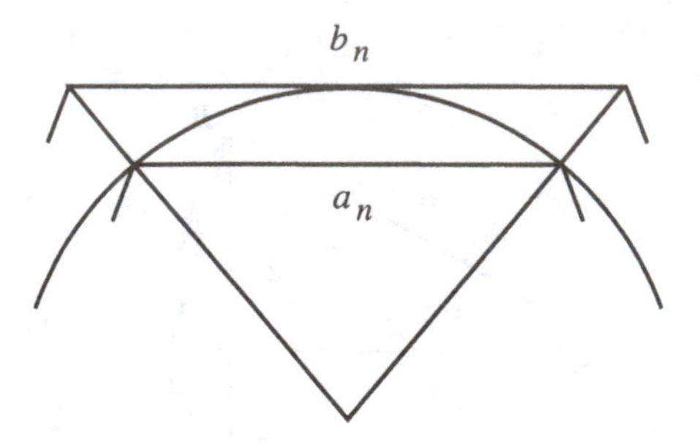

Figure 4.22. Comparing a circumscribed polygon with an inscribed one

he had to work within the much more cumbersome systems that were then current. Using some complicated methods for the estimation of square roots by means of fractions, he showed that

$$48a_{96} > \frac{6336}{2017\frac{1}{4}} > 3\frac{10}{71}(= 3.1408\ldots)$$

and

$$48b_{96} < \frac{14688}{4673\frac{1}{2}} < 3\frac{1}{7}(= 3.1428\ldots).$$

Thus, Archimedes proved that

Proposition 4.4.8 $3\frac{10}{71} < \pi < 3\frac{1}{7}$.

This section and chapter conclude with a discussion of some paradoxes and problems regarding the areas of circles. Consider Figure 4.23, where $\triangle ABC$ is both right and isosceles with legs of length a and hypotenuse of length $a\sqrt{2}$. The outside arc is a semicircle with diameter AB and radius $a\sqrt{2}/2$ whereas the inside arc is a quarter-circle centered at C with radius a. Regions bounded between two such arcs are called *lunes* (but these are different from the spherical lunes). Note that the area of this lune is the difference between the entire figure and the quarter-circle centered at C. Thus,

$$\begin{aligned}\text{area of lune} &= \text{area of semicircle} + \text{area of } \triangle ABC - \text{area of quarter-circle}\\ &= \frac{1}{2}\pi\left(\frac{a\sqrt{2}}{2}\right)^2 + \frac{a^2}{2} - \frac{\pi a^2}{4} = \frac{\pi a^2}{4} + \frac{a^2}{2} - \frac{\pi a^2}{4} = \frac{a^2}{2}\\ &= \text{area of } \triangle ABC.\end{aligned}$$

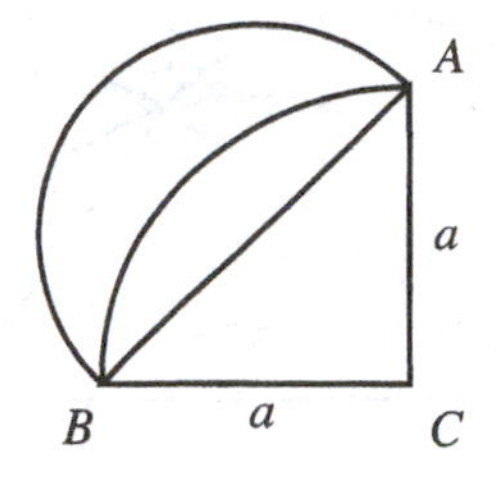

Figure 4.23.

This surprising equation is due to Hippocrates of Chios. A similar equation appears in Exercise 10. There are two unexpected aspects to this observation. First, the area of the lune, whose boundary consists of circular arcs, turns out to have an expression that is free of π. Second, this curvilinear figure has the same area as a triangle. This leads naturally to the question of whether it is possible to construct a triangle, or a square, for that matter, whose area equals that of a given circle (the simplest of all the curvilinear regions). The operative word here is *construct*. It is clear that any circle of radius r has the same area as a square of side $r\sqrt{\pi}$. The difficulty lies in constructing $r\sqrt{\pi}$ within the framework of Euclidean geometry. This problem drew the attention of many mathematicians and nonmathematicians, both in classical times and during the subsequent two and a half millennia. Although many individuals dedicated their lives to the solution of this problem, and some even deluded themselves into believing they had discovered the construction, all their efforts were in fact misdirected. In 1882 the German mathematician C. L. Ferdinand von Lindemann (1852–1939) proved a theorem that had the following corollary amongst many others:

> *Given a line segment a it is impossible to construct (in the sense of the* Elements*) a square whose area equals that of the circle of radius a.*

Exercises 4.4

1. The length of the graph of the function $y = f(x)$ for $a \leq x \leq b$ is

$$\int_a^b \sqrt{1 + f'^2}dx.$$

   Use this formula to prove Proposition 4.4.1.

2. Use calculus to prove Proposition 4.4.4.

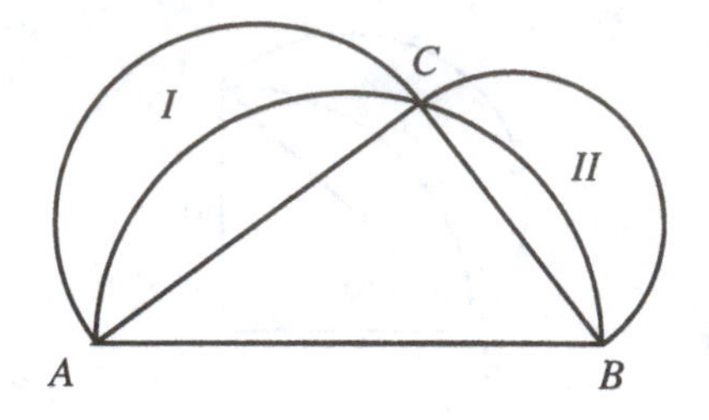

Figure 4.24.

3. Using 3.14 for the value of π, in a circle of radius 10″ find the length of the arc and the area of a sector determined by a central angle of
(a) 60° (b) 20° (c) 90° (d) 100° (e) 180° (f) 230°

4. Compute the radical expressions for a_{192} and b_{192} and use them (and a calculator) to obtain decimal bounds of the value of π.

5. Compute the radical expressions for a_{384} and b_{384} and use them (and a calculator) to obtain decimal bounds of the value of π.

6. Compute the radical expressions for a_{768} and b_{768} and use them (and a calculator) to obtain decimal bounds of the value of π.

7. Compute the radical expressions for a_{1536} and b_{1536} and use them (and a calculator) to obtain decimal bounds of the value of π.

8. Prove Equations (1).

9. Prove that of two circular arcs joining two given points, the one with the longer radius has shorter length.

10. Semicircles are constructed on the sides of a right $\triangle ABC$ (Fig. 4.24). Prove that the sum of the areas of the two lunes I and II equals the the area of $\triangle ABC$.

11. Show that the circumference of a circle of spherical radius r on a sphere of radius R is $2\pi R\sin(r/R)$.

12. Show that the area of a spherical circle of spherical radius r on a sphere of radius R is $2\pi R^2[1-\cos(r/R)]$.

13. Comment on Proposition 4.4.1 in the context of the following geometries:
(a) spherical (b) hyperbolic (c) taxicab (d) maxi

14. Comment on Proposition 4.4.2 in the context of the following geometries:
(a) spherical (b) hyperbolic (c) taxicab (d) maxi

15. Comment on Proposition 4.4.3 in the context of the following geometries:
(a) spherical (b) hyperbolic (c) taxicab (d) maxi

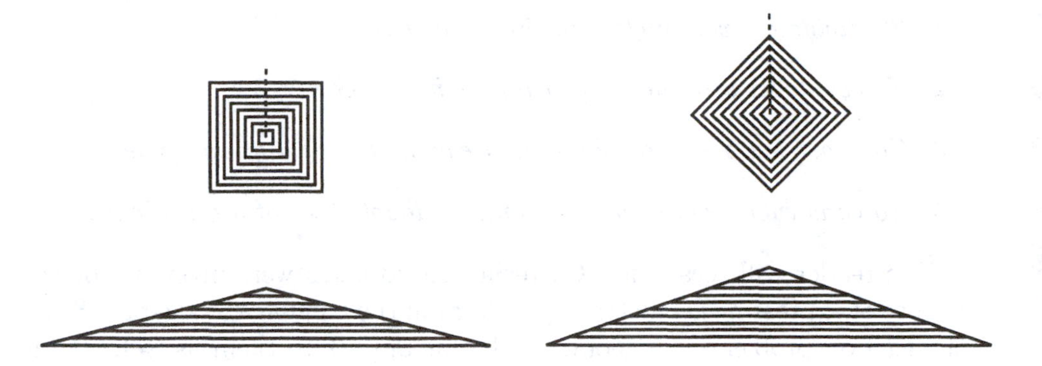

Figure 4.25.

16. Comment on Proposition 4.4.4 in the context of the following geometries:
(a) spherical (b) hyperbolic (c) taxicab (d) maxi
17. Comment on Proposition 4.4.5 in the context of the following geometries:
(a) spherical (b) hyperbolic (c) taxicab (d) maxi
18. Comment on Proposition 4.4.6 in the context of the following geometries:
(a) spherical (b) hyperbolic (c) taxicab (d) maxi
19. Comment on Proposition 4.4.7 in the context of the following geometries:
(a) spherical (b) hyperbolic (c) taxicab d) maxi
20. Comment on Proposition 4.4.8 in the context of the following geometries:
(a) spherical (b) hyperbolic (c) taxicab (d) maxi
21. Explain the following paradox. Suppose the method that was used to convert a circle into a triangle (see paragraph following Corollary 4.4.6) is applied to the same square in two different manners—first by slicing from a corner to the center and second by slicing from the middle of a side to the center (Fig. 4.25). The two triangles so obtained have their bases equal to the perimeter of the square but their altitudes are clearly different. Why are two triangles of different areas obtained?
22. Prove Equation (2).

4.5 Impossible Constructions

Part of the legacy that the Greek mathematicians passed on to their successors was a collection of construction problems they could not resolve by ruler and compass alone. While most of these problems have already been discussed (Sections 2.3 and 4.3), it might be a good idea to reexamine this topic here in order to provide a better perspective on its outcome. We begin by listing the specific construction problems in question.

1. *To divide a given angle into three equal parts*

2. *To construct a regular n-gon for each integer $n \geq 3$*

3. *To construct a square whose area equals that of a given circle*

4. *To construct a cube whose volume is double that of a given cube*

The reader will recall that Cartesian coordinates were invented for the purpose of expressing geometrical problems in the language of algebra. Since construction problems are geometrical, this applies to them as well. Some of this relation between geometry and algebra has already been pointed out. If a and b are the lengths of two given line segments, then it is possible to construct line segments of lengths $a + b$ (Exercise 2.3A.3) and $a - b$ (Proposition 2.3.3). Assuming a to be a unit length, Exercise 4.2A.32 shows how, given segments of lengths b and c, it is possible to construct a segment of length bc. Assuming c to be a unit length, the same exercise can be used to construct, for any given segments of lengths a and b, a segment of length a/b. Finally, assuming b to have unit length, Exercise 4.2A.32 can be used to construct, for any given segment of length a, a line segment of length $\sqrt{a}$. Thus, the four arithmetic operations, as well as the taking of square roots, can be mimicked by ruler and compass constructions.

The Cartesian coordinate system can be used to argue that the power of ruler and compass constructions cannot be extended beyond these five algebraic operations. To do this, it is first necessary to formalize some notions. A *configuration* is a set of points, straight lines, and circles. An *elementary ruler and compass construction* is any of the following five operations:

1. *Draw the line joining two given points.*

2. *Draw a circle with a given center and radius.*

3. *Find the intersection of two given straight lines.*

4. *Find the intersection of a given circle and a given straight line.*

5. *Find the intersection of two given circles.*

A configuration T is said to be constructible from configuration S provided every element of T can be obtained from the elements of S by a succession of elementary ruler and compass constructions. In particular, note that every construction problem stipulates a given configuration and aims at the derivation of a desired configuration.

Assume now that a Cartesian coordinate system has been chosen to which all the configurations below are referred. The *numerical aspects* of the point (x, y) are x and y. The *numerical aspects* of the straight line with equation $ax + by + c = 0$ are a, b, and, c. The *numerical aspects* of the circle with equation $x^2 + y^2 + ax + by + c = 0$ are a, b, and c.

A real number r is said to be a *Hippasian function* of the set S provided it is obtainable from the elements of S and the rational numbers by rational operations and extractions of real square roots (this terminology honors Hippasus of Metapontum, the discoverer of the irrationality of $\sqrt{2}$). For example, the following numbers are all Hippasian functions of the set $S = \{\pi, \sqrt[3]{2}, e\}$:

$$1, \quad \frac{3}{5}, \quad \frac{\sqrt[3]{2}}{e}, \quad \frac{\pi + 3e}{2 - \sqrt[3]{2}}, \quad \frac{3\sqrt{\pi} - 4\sqrt[3]{e}}{\sqrt{10 + \sqrt[3]{2}}}.$$

The *Hippasian numbers* are those that are obtainable from the rational numbers alone by the rational operations and the extractions of real square roots. In fact, this is tantamount to saying that they are obtainable from the number 1 by the said operations. The following numbers are all Hippasian numbers:

$$1, \quad \frac{3}{5}, \quad -\sqrt{2}, \quad \sqrt{2 + \sqrt{2 + \sqrt{3}}}, \quad \frac{\sqrt{1 + \sqrt{2}}}{5 + \sqrt{35 - \sqrt{13}}}.$$

The following theorem formalizes the intuitively plausible connection between constructibility and Hippasian functions.

Theorem 4.5.1 *If configuration T is constructible from configuration S, then the numerical aspects of T are Hippasian functions of the numerical aspects of S.*

OUTLINE OF PROOF: Suppose configuration T is obtained from configuration S by the elementary construction i, where $i = 1, 2, 3, 4, 5$. If $i = 5$, for example, let the two given circles have equations

$$x^2 + y^2 + ax + by + c = 0$$

and

$$x^2 + y^2 + a'x + b'y + c' = 0$$

By Exercise 5, the intersection point of these two circles, if it exists, has coordinates that are Hippasian functions of a, b, c, a', b', c'. The proof of the cases $i = 1, 2, 3, 4$ is similar (see Exercises 1–4), and we conclude that if T is constructible from S by any ruler and compass operations, then the numerical aspects of T are obtainable from those of S in the desired manner. Q.E.D.

We note in passing that the converse of Theorem 4.5.1 is also valid, albeit somewhat harder to prove. As this converse is not needed for the proof of the following impossibilities, it is relegated to Exercise 6.

The strategy for proving the nonfeasibility of a ruler and compass construction calls for demonstrating that the numerical aspects of the desired configuration are not Hippasian functions of those of the given configuration. Matters can, and will be, simplified by setting things up so that the numerical aspects of the given data are either integers or Hippasian numbers, and hence it will suffice to show that the desired configuration has a nonHippasian number as one of its numerical aspects

The following proposition, whose proof is omitted, provides an easily applied criterion for recognizing nonHippasian numbers. It is found, in a more general form, in many undergraduate modern algebra texts.

Proposition 4.5.2 *Let x be a real solution of the equation*

$$ax^3 + bx^2 + cx + d = 0, \tag{1}$$

where a, b, c, d are integers. Then x is Hippasian if and only if this equation has a rational solution. □

Unlike the previous proposition, the next one is found in many precalculus texts and is easily proven (Exercise 7).

Proposition 4.5.3 (The Rational Zeros Theorem) *Let*

$$P(x) = a_n x^n + a_{n-1} x^{n-1} + \cdots + a_1 x + a_0,$$

where all the a_i's are integers. If p/q is a rational number in lowest terms such that

$$P(p/q) = 0,$$

then p is a factor of a_0 and q is a factor of a_n.

The four construction problems of antiquity are now reexamined one at a time. The simplest of these turns out to be the doubling of the cube.

DOUBLING THE CUBE: If it were possible to double any given cube by a ruler and compass construction, then it would certainly be possible to construct a cube of volume 2. The length of the side of such a cube would be $\sqrt[3]{2}$ and it would have to be a Hippasian number. However, $\sqrt[3]{2}$ is clearly a solution of the equation

$$x^3 - 2 = 0 \tag{2}$$

and hence, by Proposition 4.5.2, this equation would have to have a rational solution, say p/q in lowest terms. By Proposition 4.5.3, p must be a factor of 2 and q a factor of 1. Hence p/q must be one of the numbers $\pm 1/1$, $\pm 2/1$, none of which, by Exercise 9, is a solution of Equation (2). Thus, the supposed feasibility of a ruler and compass doubling of the cube has lead to a contradiction, and we conclude that

The cube cannot be doubled by ruler and compass alone.

ANGLE TRISECTION: We next argue that there is no method for trisecting angles by ruler and compass alone. Suppose, to the contrary, that such a method exists and is used to trisect the $60°$ angle of an equilateral triangle whose side has unit length. Here it may be supposed that the given configuration consists of the three points $O(0,0)$, $A(0,1)$, $B(1/2, \sqrt{3}/2)$ (Fig. 4.26) all of whose numerical aspects are Hippasian numbers. The hypothetical construction yields an angle of $20°$ that may be placed at the origin with one side on the x-axis (Fig. 4.26). The (constructible) intersection P of this angle's other side with the circle $(O; 1)$ has coordinates $(\cos 20°, \sin 20°)$, and hence it follows from Theorem 4.5.1 that $\cos 20°$ is a Hippasian function of Hippasian numbers. Consequently, $\cos 20°$ is a Hippasian number. We now go on to obtain an analog of Equation (2). If $x = \cos 20°$, then, by Exercise 8,

$$1/2 = \cos 60° = 4\cos^3 20° - 3\cos 20° = 4x^3 - 3x$$

and hence

$$8x^3 - 6x - 1 = 0. \tag{3}$$

By Proposition 4.5.2, this equation has a rational solution, say p/q, where p and q are integers. By Proposition 4.5.3, p/q must be one of the fractions

$$\pm 1, \quad \pm 1/2, \quad \pm 1/4, \quad \pm 1/8$$

none of which, by Exercise 10, is a root of Equation (3). Thus, the assumption of the feasibility of a ruler and compass construction for trisecting angles has resulted in a contradiction, and hence

There is no method for trisecting angles by ruler and compass alone.

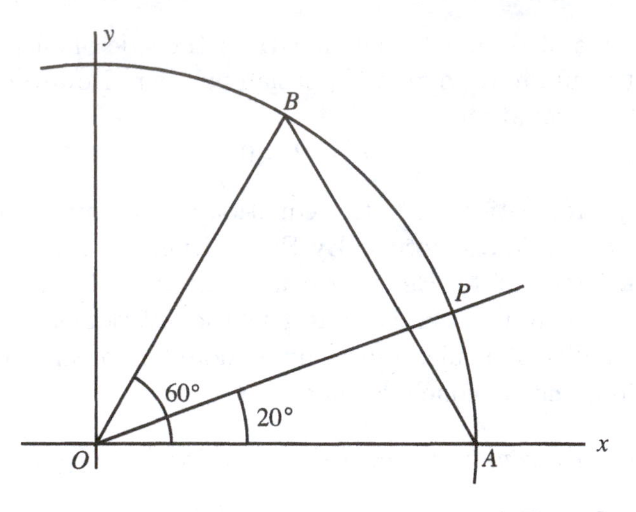

Figure 4.26.

REGULAR n-GONS: Whether or not a regular n-gon is constructible by ruler and compass turns out to depend on the value of n. In general, such a polygon can be constructed if and only if an angle of $360°/n$ can be constructed. When this angle is placed at the origin with one side on the x-axis, the other side intersects the circle $(O;1)$ in the point $(\cos 360°/n, \sin 360°/n)$. By Theorem 4.5.1 and Exercise 18, this general ruler and compass construction is feasible if and only if $\cos 360°/n$ is a Hippasian number. We now show that there is no ruler and compass construction for the regular 7-gon. Set $A = 360°/7$ and $x = \cos 360°/7$. If such a method existed, then x would be a Hippasian number. However, by Exercises 8 and 12,

$$\cos 3A = 4\cos^3 A - 3\cos A = 4x^3 - 3x$$
$$\cos 4A = 8\cos^4 A - 8\cos^2 A + 1 = 8x^4 - 8x^2 + 1.$$

Since $3A + 4A = 360°$ it follows that $\cos 3A = \cos 4A$ and hence

$$4x^3 - 3x = 8x^4 - 8x^2 + 1$$
$$8x^4 - 4x^3 - 8x^2 + 3x + 1 = 0$$
$$(x-1)(8x^3 + 4x^2 - 4x - 1) = 0.$$

However, $\cos 360°/7 \neq 1$ and hence

$$8x^3 + 4x^2 - 4x - 1 = 0.$$

It follows from Proposition 4.5.3 that the only possible rational solutions of this equation are again ± 1, $\pm 1/2$, $\pm 1/4$, $\pm 1/8$. Since, by Exercise 10, none of these is a solution, it follows from Proposition 4.5.2 that $\cos 360°/7$ is not a Hippasian number and hence no such method for constructing regular 7-gons can exist.

According to Exercise 11, $\cos 360°/5$ is a Hippasian number and so the regular pentagon is indeed constructible, as demonstrated in Proposition 4.3.4. As was mentioned previously, the regular 17-gon is also constructible and hence $\cos 360°/17$ must also be a Hippasian number. In fact, it is known to equal

$$\begin{aligned} &-1/16 + \sqrt{17}/16 + (1/16)\sqrt{34 - 2\sqrt{17}} \\ &\quad - (1/8)\sqrt{17 + 3\sqrt{17} - \sqrt{34 - 2\sqrt{17}} - 2\sqrt{34 + 2\sqrt{17}}}. \end{aligned}$$

SQUARING THE CIRCLE: Squaring a circle of unit radius is tantamount to constructing a line segment of length $a = \sqrt{\pi}$. As it happens, neither π nor $\sqrt{\pi}$ are the solutions of any equation of form (1). In fact, it was proven by Lindemann that there exists no polynomial $P(x)$ with integer coefficients such that either π or $\sqrt{\pi}$ are solutions of $P(x) = 0$. Since every Hippasian number is known to be the solution of such an equation, it follows that $\sqrt{\pi}$ is not a Hippasian number and hence

It is impossible to square a circle by ruler and compass alone.

Exercises 4.5

1. Show that the straight line joining the points (a, b) and (a', b') has equation
$$(b' - b)x + (a - a')y + (a'b - ab') = 0.$$

2. Show that the circle with center (a, b) and radius r has equation
$$x^2 + y^2 + (-2a)x + (-2b)y + (a^2 + b^2 - r^2) = 0.$$

3. Show that if the two lines with equations $ax+by+c = 0$ and $a'x+b'y+c' = 0$ intersect, then their point of intersection has coordinates $((bc' - b'c)/(ab' - a'b), (a'c - ac')/(ab' - a'b))$.

4. Show that if the line with equation $ax+by+c = 0$ and the circle with equation $x^2 + y^2 + a'x + b'y + c' = 0$ intersect, then their points of intersection have coordinates $x = (-B \pm \sqrt{B^2 - 4AC})/(2A)$ and $y = (-ax - c)/b$, where $A = a^2 + b^2$, $B = 2ac + a'b^2 - abb'$, and $C = c^2 - bb'c + c'b^2$.

5. Show that if the two circles with equations $x^2 + y^2 + ax + by + c = 0$ and $x^2 + y^2 + a'x + b'y + c' = 0$ intersect, then their points of intersection have coordinates $x = (-B \pm \sqrt{B^2 - 4AC})/(2A)$ and $y = (-a''x - c'')/b''$ where $A = a''^2 + b''^2$, $B = 2a''c'' + a''b''^2 - a''b''b'$, $C = c''^2 - b''b'c'' + c'b''^2$ and $a'' = a - a'$, $b'' = b - b'$, $c'' = c - c'$.

*6. State and prove the converse of Theorem 4.5.1.

7. Prove Proposition 4.5.3.

8. Prove that $\cos 3A = 4\cos^3 A - 3\cos A$.

9. Verify that none of the numbers $\pm 1/1$, $\pm 2/1$, is a solution of the equation $x^3 - 2 = 0$.

10. Verify that none of the numbers $\pm 1/1$, $\pm 1/2$, $\pm 1/4$, $\pm 1/8$ is a solution of the equation $8x^3 - 6x - 1 = 0$.

11. Prove that if $x = \cos 72°$, then $4x^2 + 2x - 1 = 0$.

12. Prove that $\cos 4A = 8\cos^4 A - 8\cos^2 A + 1$.

13. Prove that it is impossible to construct a regular 9-gon by ruler and compass alone.

14. Prove that it is impossible to triple a cube by ruler and compass alone.

15. Prove that it is impossible to halve a cube by ruler and compass alone.

16. Prove that the following numbers are not Hippasian:
    (a) $\sqrt[3]{5}$ (b) $2 + \sqrt[3]{5}$ (c) $1/(2 - \sqrt[3]{7})$

17. Is it possible to construct an angle of 1°?

18. Let A be any number. Explain why $\sin A$ is a Hippasian number if and only if $\cos A$ is a Hippasian number.

Chapter Review Exercises

1. Circle p intersects two concentric circles. Prove that the arcs of p cut off by the two circles are equal.

2. Prove that if each of the sides of a square is extended in both directions by the length of the radius of the circle that circumscribes the square, we obtain the vertices of a regular octagon.

3. Two circles, centered at C and D, respectively, intersect at a point A. Prove that if PAQ is the "double chord" that is parallel to CD, then $PQ = 2CD$.

4. The area of the annular region bounded by two concentric circles equals that of a circle whose diameter is a chord of the greater circle that is tangent to the smaller one.

5. In a circle, a diameter bisects the angle formed by two intersecting chords. Prove that the chords are equal.
6. Prove that every equiangular polygon all of whose sides are tangent to the same circle is regular.
7. Through the center of a circle passes a second circle of greater radius and their common tangents are drawn. Prove that the chord joining the contact points of the greater circle is tangent to the smaller circle.
8. Prove that every cyclic equilateral pentagon is regular.

*9. Three circles through the point O and of radius r intersect pairwise in the additional points A, B, C. Prove that the circle circumscribed about $\triangle ABC$ also has radius r.

10. Prove that in the regular hexagon $ABCDEF$ the diagonals AC and AE cut the diagonal BF into three equal segments.
11. Each of the sides of a cyclic quadrilateral is the chord of a new circle. Prove that the other four intersection points of these new circles also form a cyclic quadrilateral.
12. A circle of radius r is inscribed in $\triangle ABC$ in which $\angle ACB$ is a right angle. Prove that $a + b = c + 2r$.
13. The chord AB of a circle of radius 1 has the property that if the circle is folded along AB so as to bring AB's arc into the circle, then the arc passes through the center of the circle. Compute the lengths of the chord AB and its arc.
14. In a given $\triangle ABC$ construct a point whose distances from the sides of the triangle are proportional to three given line segments.

*15. Through the midpoint M of a chord PQ of a circle, any other chords AB and CD are drawn; chords AD and BC meet PQ at points X and Y. Prove that M is the midpoint of XY. (This is the notorious butterfly problem.)

*16. The points of intersection of the adjacent trisectors of any triangle are the vertices of an equilateral triangle. (This is known as Morley's theorem.)

17. Let A be a given point and p a circle centered at C. If the point P moves along the circle p, prove that the midpoint of AP describes a circle centered at the midpoint of CA.
18. Given a circle p and a point A outside it, construct a straight line through A which cuts the circle so that the section of m from A to the circle equals the section of m inside the circle.
19. Are the following statements true or false? Justify your answers.

    (a) The Greeks believed that the world is flat.

(b) The area of a plane Euclidean figure whose perimeter is composed of circular arcs must involve π in its expression.

(c) The Greeks knew that $\pi = 3.14$.

(d) If two sectors have equal angles, then their arcs are proportional to their radii.

(e) Given a line segment a, it is impossible to construct (in the sense of the *Elements*) a square whose area equals that of the circle of radius a.

(f) Given a line segment a, it is impossible to construct (in the sense of the *Elements*) an equilateral triangle whose area equals that of the circle of diameter a.

(g) Of two equal chords in unequal circles, the one in the larger circle lies further from the center.

(h) The diameter is the circle's longest chord.

(i) It is possible to construct a regular 340-sided polygon (in the sense of the *Elements*).

(j) It is possible to construct a regular 140-sided polygon (in the sense of the *Elements*).

(k) In a circle, all the angles subtended by a chord are equal to each other.

(l) In a circle, arcs are proportional to their chords.

(m) Every circle has only one center.

(n) Every circle has only one tangent line.

(o) If two chords of a circle bisect each other, then they are both diameters.

"What do you mean, the proof is in the pudding!?"

Chapter 5

Toward Projective Geometry

Most mathematical disciplines find it necessary to incorporate the concept of infinity into their language. Euclidean geometry is no exception to this rule, and this process resulted in the beautiful structure known as projective geometry, which was first codified by the Frenchman Gérard Desargues (1591–1662). This chapter introduces and motivates some of the rudimentary ideas of this extension of plane geometry.

5.1 Division of Line Segments

The taming of geometrical infinity begins with a careful examination of geometrical ratios. Let r be a positive real number. If the point D on the line segment AB is such that

$$\frac{AD}{DB} = r \tag{1}$$

it is said that D *divides the segment* AB *internally in the ratio* r. In Figure 5.1

$$\frac{AF}{FB} = \frac{1}{3} \qquad \frac{AG}{GB} = 1 \qquad \frac{AH}{HB} = 3.$$

On the other hand, if the point D lies on the straight line $\overleftrightarrow{AB}$ but falls *outside* the line segment AB, and if Equation (1) holds again, then D *divides*

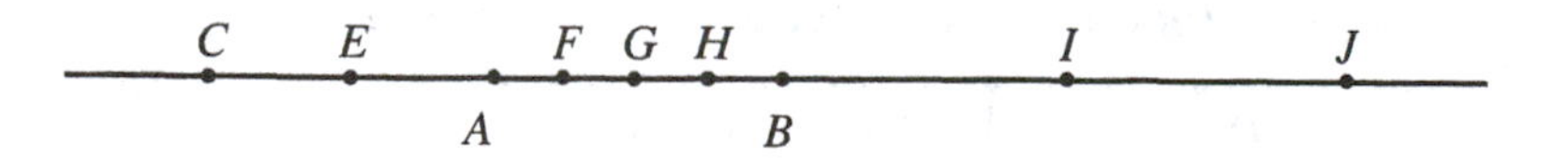

Figure 5.1. Division points

the line segment AB externally in the ratio r, or

$$\frac{AD}{DB} = -r. \tag{2}$$

In Figure 5.1

$$\frac{AC}{CB} = -\frac{1}{2} \qquad \frac{AE}{EB} = -\frac{1}{3} \qquad \frac{AI}{IB} = -2 \qquad \frac{AJ}{JB} = -\frac{3}{2}.$$

Note that when $0 < r < 1$ in Equation (2), AD must be shorter than DB so that A lies between D and B (see Fig. 5.2). On the other hand, if $r < -1$, AD is longer than DB and so B separates A and D. Exercise 16 contains more detailed information regarding the dependence of the value of AD/DB on the position of D on the line $\overleftrightarrow{AB}$.

An alternative description of the relationship between the sign of a ratio and the relative position of its points is obtained by thinking of the line segments in question as directed segments. In that case the ratio AD/DB is positive or negative according as AD and DB have the same or opposite directions.

It is important to keep in mind that the assignment of signs to ratios applies only in the case where the points A, B, D are collinear. If they are not collinear, then the ratio AD/DB is always taken to be positive. Moreover, while this definition does implicitly assume a choice of a unit of length, the actual value of the ratio AD/DB is independent of the particular choice of unit, since changing one's choice has the effect of multiplying the lengths of AD and DB by the same factor, which then disappears in the evaluation of the ratio AD/DB.

It should be mentioned that Euclid did not exhibit any interest in external division points. For that reason the correspondence between this chapter's propositions and his is somewhat tenuous.

The next proposition demonstrates that division points are unique.

Proposition 5.1.1 *Given two distinct points A and D and a real number r there exists at most one point D on* $\overleftrightarrow{AB}$ *such that* $AD/DB = r$.

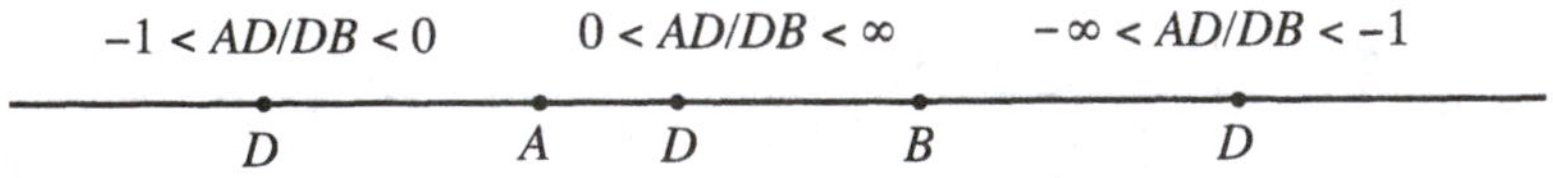

$-1 < AD/DB < 0$ $\qquad$ $0 < AD/DB < \infty$ $\qquad$ $-\infty < AD/DB < -1$

D $\quad$ A $\quad$ D $\quad$ B $\quad$ D

Figure 5.2. Division points

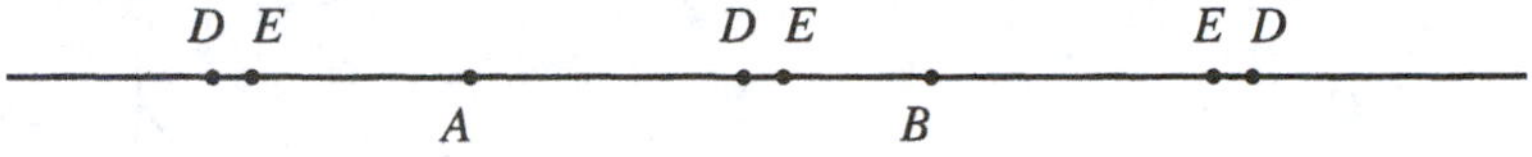

Figure 5.3. Uniqueness of division

GIVEN: Two distinct points A and B; two points D, E on AB such that (Fig. 5.3)

$$\frac{AD}{DB} = \frac{AE}{EB}.$$

TO PROVE: D and E are identical.

PROOF: Let r be the common value of the ratio AD/DB and AE/EB. Suppose first that $r > 0$. In this case both D and E are in between A and B. It follows from Proposition 3.5.5 and the proportion

$$\frac{AD}{DB} = \frac{AE}{EB}$$

that

$$\frac{AD + DB}{DB} = \frac{AE + EB}{EB} \quad \text{or} \quad \frac{AB}{DB} = \frac{AB}{EB}.$$

Hence $DB = EB$ and so, since D and E are both between A and B, they are identical.

The resolution of the other cases corresponding to $r < -1$, $r = -1$, $-1 < r < 0$, and $r = 0$ is relegated to Exercises 1–3. Q.E.D.

Before addressing the question of the existence of division points that yield arbitrary ratios, rational divisions are examined.

Proposition 5.1.2 (VI.9) *To divide a given segment, both internally and externally, in the ratio m/n, where m and n are two distinct positive integers.*

GIVEN: Line segment AB, positive integers m, n (Fig. 5.4).

TO CONSTRUCT: Points D and E on AB such that

$$\frac{AD}{DB} = \frac{m}{n} \qquad \frac{AE}{EB} = -\frac{m}{n}.$$

CONSTRUCTION: Let $AC \perp AB$ and let $A_1, A_2, \ldots, A_{m+n}$ be a sequence of distinct points on AC such that $AA_1 = A_1A_2 = \cdots = A_{m+n-1}A_{m+n}$.

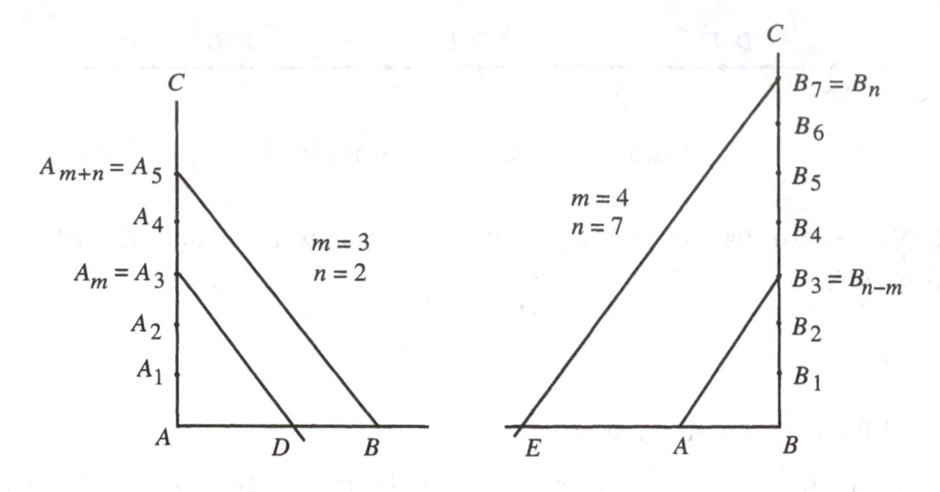

Figure 5.4. Constructing division points

Let D be the intersection of AB with the straight line through A_m that is parallel to BA_{m+n}. Turning to E, if $m < n$, then let $BC \perp AB$ and let $B_1, B_2, \ldots, B_n$ be a sequence of distinct points on BC such that $BB_1 = B_1B_2 = \cdots = B_{n-1}B_n$. Let E be the intersection of $\overleftrightarrow{AB}$ with the straight line through B_n that is parallel to AB_{n-m}. The construction of E in the case $m > n$ is relegated to Exercise 4.

PROOF: It follows from Proposition 3.5.6 that

$$\frac{AD}{DB} = \frac{AA_m}{A_mA_{m+n}} = \frac{mAA_1}{nAA_1} = \frac{m}{n}$$

and

$$\frac{AE}{EB} = \frac{B_{n-m}B_n}{B_nB} = -\frac{mBB_1}{nBB_1} = -\frac{m}{n}.$$

Q.E.D.

It is intuitively clear that given any positive real number r, there is a point D that divides AB internally in the ratio r. After all, we need simply choose a point D on AB such that

$$AD = \frac{r}{r+1}AB$$

so that

$$BD = AB - AD = AB - \frac{r}{r+1}AB = \left(1 - \frac{r}{r+1}\right)AB = \frac{1}{r+1}AB$$

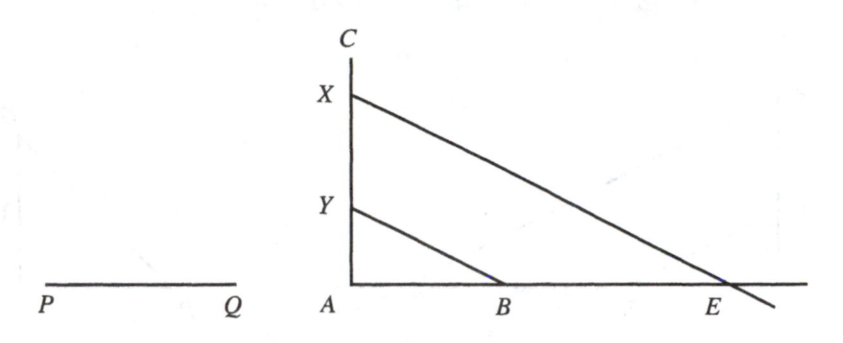

Figure 5.5. An external division point

and hence

$$\frac{AD}{DB} = \frac{r}{1} = r.$$

This, however, is merely an existential statement. The following proposition shows how this division point can be constructed within the framework of Euclid's *Elements*.

Proposition 5.1.3 (VI.10) *To divide a given line segment, both internally and externally, in any given ratio* $\neq 1$.

GIVEN: Line segment AB, line segment PQ of length $a \neq 1$ (Fig. 5.5).

TO CONSTRUCT: Points D and E such that $\frac{AD}{DB} = a$ and $\frac{AE}{EB} = -a$.

CONSTRUCTION: The construction of a point E is described for $a > 1$ only, leaving the other cases to Exercises 5 and 6. On $AC \perp AB$ let X, Y be points such that $AX = PQ$ and XY has unit length. Let E be the intersection of $\overleftrightarrow{AB}$ with the straight line through X that is parallel to BY.

PROOF: By Proposition 3.5.6

$$\frac{AE}{EB} = \frac{AX}{XY} = -\frac{a}{1} = -a.$$

Q.E.D.

In the exceptional case where the ratio is 1, the internal division point is the midpoint, which was already dealt with in Proposition 2.3.10, whereas the external division point does not exist (see Exercise 15). An alternative method for dividing a line segment in a prespecified signed ratio is described in Exercises 3.5B.6–7. The foregoing discussion is summarized by the following proposition.

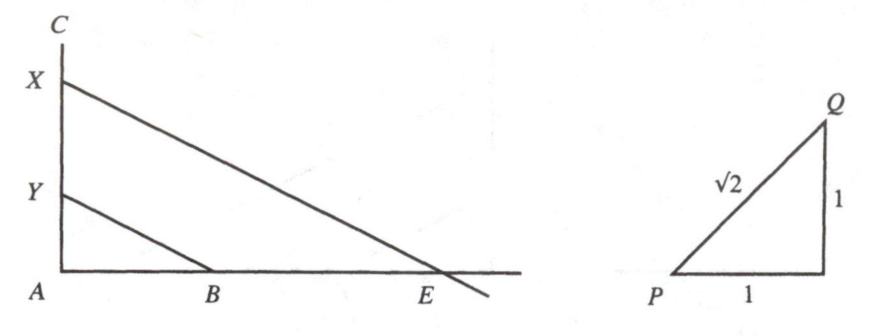

Figure 5.6. An external division

Proposition 5.1.4 *Let AB be a straight line segment and r a real number different from -1. Then there is a unique point D on $\overleftrightarrow{AB}$ such that $AD/DB = r$.* □

Example 5.1.5 To divide a line segment AB of Figure 5.6 in the ratio $-\sqrt{2}$.

CONSTRUCTION: The number $\sqrt{2}$ is represented geometrically by the hypotenuse of an isosceles right triangle whose legs are 1 unit long. The remainder of the construction follows the steps outlined in the proof of Proposition 5.1.3, again with $AX = PQ$.

Exercises 5.1

1. Prove Proposition 5.1.1 in the case $r < -1$.
2. Prove Proposition 5.1.1 in the case $-1 < r < 0$.
3. Discuss cases $r = 0, -1$ of Proposition 5.1.1.
4. Construct the point E in the case $m > n$ of Proposition 5.1.2.
5. Prove Proposition 5.1.3 in the case $1 > a > 0$.
6. Discuss Proposition 5.1.3 in the case $a = 1$.
7. Let AB be a line segment.
   (a) Divide AB internally in the ratio 4.
   (b) Divide AB internally in the ratio 1/4.
   (c) Divide AB externally in the ratio 4.
   (d) Divide AB externally in the ratio 1/4.

8. Let AB be a line segment.

   (a) Divide AB internally in the ratio $5/3$.

   (b) Divide AB internally in the ratio $3/5$.

   (c) Divide AB externally in the ratio $5/3$.

   (d) Divide AB externally in the ratio $3/5$.

9. Let AB be a line segment and n a positive integer. Divide AB into n equal segments (Proposition VI.9).

10. Let AB and PQ be line segments and X a point on the straight line PQ. Divide AB in the ratio PX/XQ.

11. Let AB be a line segment.

    (a) Divide AB internally in the ratio $\sqrt{5}$.

    (b) Divide AB internally in the ratio $1/\sqrt{5}$.

    (c) Divide AB externally in the ratio $\sqrt{5}$.

    (d) Divide AB externally in the ratio $1/\sqrt{5}$.

12. Let AB be a line segment.

    (a) Divide AB internally in the ratio $\sqrt{3}$.

    (b) Divide AB internally in the ratio $1/\sqrt{3}$.

    (c) Divide AB externally in the ratio $\sqrt{3}$.

    (d) Divide AB externally in the ratio $1/\sqrt{3}$.

13. Let AB be a line segment.

    (a) Divide AB internally in the ratio $(1+\sqrt{5})/(2+\sqrt{3})$.

    (b) Divide AB externally in the ratio $(1+\sqrt{5})/(2+\sqrt{3})$.

    (c) Divide AB externally in the ratio $(2+\sqrt{5})/(1+\sqrt{3})$.

14. Supply the details needed to complete the proof of Proposition 5.1.4.

15. Let AB be a straight line segment. Prove that there is no point E such that $AE/EB = -1$.

16. Show that if $A = (0,0)$, $B = (1,0)$, and $D = (x,0)$ in some Cartesian coordinate system, then $AD/DB = x/(1-x)$.

17. Comment on Proposition 5.1.1 in the context of the following geometries:
    (a) spherical (b) hyperbolic (c) taxicab (d) maxi

18. Comment on Proposition 5.1.4 in the context of the following geometries:
    (a) spherical (b) hyperbolic (c) taxicab (d) maxi

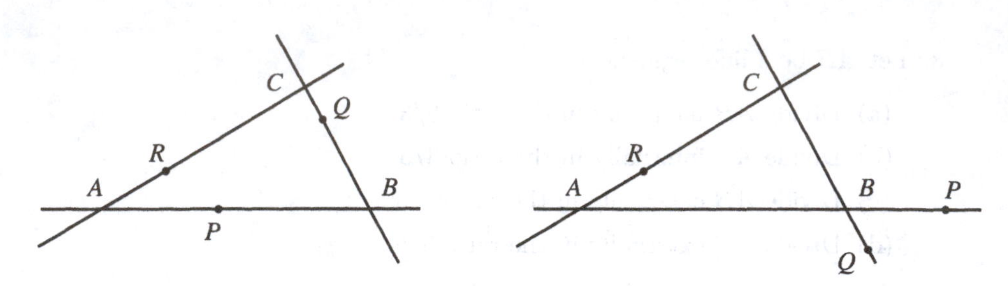

Figure 5.7. Transversals.

5.2 Collinearity and Concurrence

This section presents some key theorems that were added to the geometric lore over the centuries that followed the writing of *The Elements*. They were selected because they provide a natural transition to projective geometry.

The set of points $\{P, Q, R\}$ is said to be a *transversal* of $\triangle ABC$ if these points are distinct from A, B, C and they fall on the straight lines $\overleftrightarrow{AB}$, $\overleftrightarrow{BC}$, and $\overleftrightarrow{AC}$ respectively (see Fig. 5.7).

Proposition 5.2.1 (The Theorem of Menelaus) *Let* $\{P, Q, R\}$ *be a transversal of* $\triangle ABC$. *Then* P, Q, R *are collinear if and only if*

$$\frac{AP}{PB}\frac{BQ}{QC}\frac{CR}{RA} = -1. \tag{1}$$

GIVEN: $\triangle ABC$, P on $\overleftrightarrow{AB}$, Q on $\overleftrightarrow{BC}$, R on $\overleftrightarrow{AC}$. (Fig. 5.8).

TO PROVE: P, Q, R are collinear if and only if $\dfrac{AP}{PB}\dfrac{BQ}{QC}\dfrac{CR}{RA} = -1$.

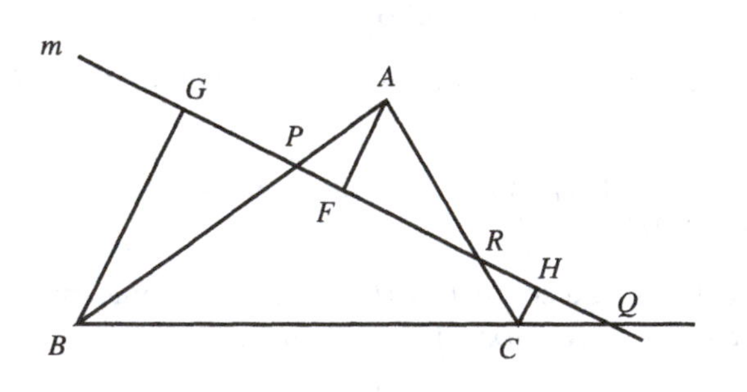

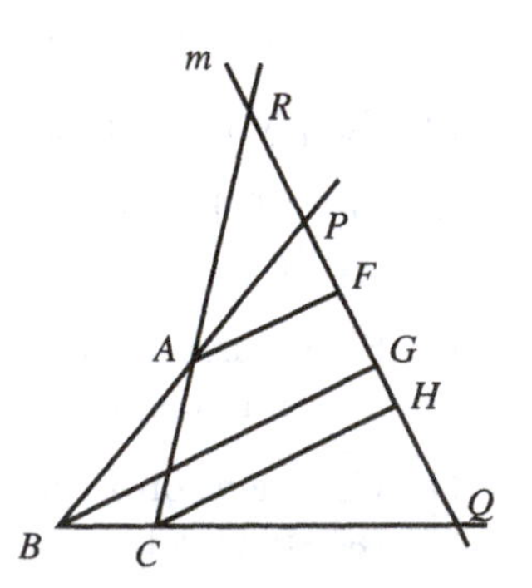

Figure 5.8.

PROOF: Assume first that P, Q, R are collinear and let m be the straight line containing them. Let F, G, H be points of the line m such that AF, BG, and CH are all perpendicular to m. Then each of the following similarities is justified by the observation that the triangles in question are all right angled and each pair either shares an acute angle or else has vertically opposite acute angles:

$$\begin{array}{llll} & \triangle APF \sim \triangle BPG, & \triangle BQG \sim \triangle CQH, & \triangle CRH \sim \triangle ARF \\ \therefore & \dfrac{AP}{PB} = \pm\dfrac{AF}{BG}, & \dfrac{BQ}{QC} = \pm\dfrac{BG}{CH}, & \dfrac{CR}{RA} = \pm\dfrac{CH}{AF}. \end{array}$$

Since m does not pass through any of the vertices of $\triangle ABC$, it cuts either 1 or 3 of its sides externally. Consequently, the product

$$\frac{AP}{PB}\frac{BQ}{QC}\frac{CR}{RA} = \left(\pm\frac{AF}{BG}\right)\left(\pm\frac{BG}{CH}\right)\left(\pm\frac{CH}{AF}\right) \tag{2}$$

contains an odd number of negative factors. After the obvious cancellations are carried out, only -1 remains in the right-hand side of Equation (2).

Conversely, suppose P, Q, R are such that Equation (1) holds (Fig. 5.9). Set $R' = \overleftrightarrow{PQ} \cap \overleftrightarrow{AC}$ (see Exercise 17). It follows from the first part of the proof that

$$\frac{AP}{PB}\frac{BQ}{QC}\frac{CR'}{R'A} = -1.$$

In combination with Equation (1) this yields

$$\frac{CR}{RA} = \frac{CR'}{R'A}.$$

By Proposition 5.1.4, $R = R'$ and so the points P, Q, R are collinear. Q.E.D.

Exercises 5.2A

1. Suppose a straight line m bisects side AB of $\triangle ABC$, and cuts BC internally into two segments one of which is double the other. Describe the two possible points where it intersects the (extended) third side.

2. Use the theorem of Menelaus to prove that the straight line joining the mid-points of two sides of a triangle is parallel to the third side.

3. Prove that if each of the bisectors of a triangle's exterior angles intersects the opposite side, then the three intersection points are collinear.

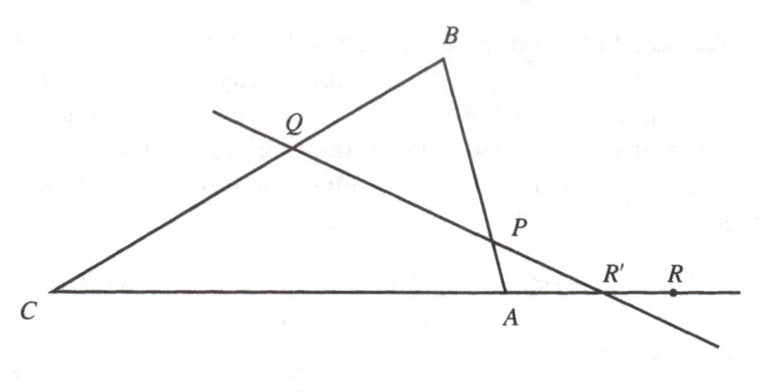

Figure 5.9.

4. Prove that if the bisector of one of the triangle's exterior angles intersects the opposite side then this intersection is collinear with the intersections of the bisectors of the interior angles at the other two vertices with the opposite sides.

5. Let $ABCD$ be a trapezoid in which the nonparallel sides AB and CD intersect in the point M and the diagonals intersect in the point N. Prove that the straight line MN bisects both of the sides BC and AD.

6. Let $ABCD$ be a trapezoid in which the nonparallel sides AB and CD intersect in the point M and let N be the midpoint of AD. Prove that if $P = BD \cap CN$ and $Q = AD \cap \overleftrightarrow{MP}$, then $AQ = 2QD$.

7. Let $ABCD$ be a trapezoid in which the nonparallel sides AB and CD intersect in the point M and Q divide AD internally in the ratio of 2. Prove that if $R = BD \cap CQ$ and $S = AD \cap \overleftrightarrow{MR}$, then $AS = 3SD$.

8. Let $ABCD$ be a trapezoid in which the nonparallel sides AB and CD intersect in the point M. (Figure 5.10). Define $A_1 = A$, $B_1 = B$, and, for each positive integer n, let $B_{n+1} = CA_n \cap BD$, and let $A_{n+1} = \overleftrightarrow{MB}_{n+1} \cap AD$. Prove that, $DA = nDA_n$ for $n \geq 1$.

9. Let A', B', C' be the respective midpoints of the sides BC, CA, AB of $\triangle ABC$. If $P = AA' \cap B'C'$ and $Q = CP \cap AB$, prove that $AB = 3AQ$.

10. Two distinct straight lines intersect the sides of $\triangle ABC$ in the transversals P, Q, R and P', Q', R', respectively. Show that the points $X = \overleftrightarrow{BC} \cap \overleftrightarrow{RP'}$, $Y = \overleftrightarrow{CA} \cap \overleftrightarrow{PQ'}$, $Z = \overleftrightarrow{AB} \cap \overleftrightarrow{QR'}$, are collinear, provided they exist.

11. Two equal segments AE and AF are taken on the sides AB and AC of $\triangle ABC$, and M is the midpoint of BC. Show that if $G = AM \cap EF$, then $GF/GE = AB/AC$.

12. The points A, B, C, D on the straight line m and A', B', C', D' on the straight line n are such that $\overleftrightarrow{AA'}$, $\overleftrightarrow{BB'}$, $\overleftrightarrow{CC'}$, $\overleftrightarrow{DD'}$ are concurrent. Prove that $(AB/BC)/(AD/DC) = (A'B'/B'C')/(A'D'/D'C')$.

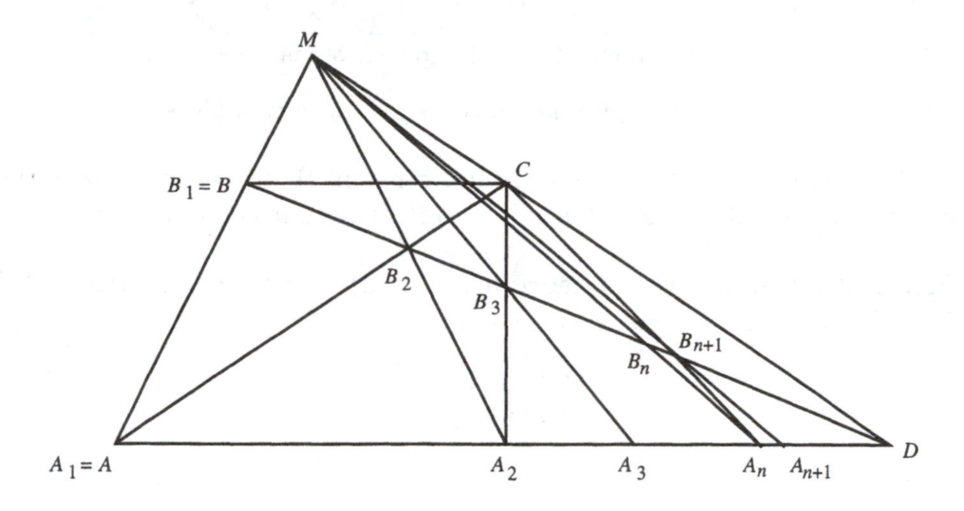

Figure 5.10.

13. Show that if each of the tangents to the circumcircle of a triangle at the vertices of the triangle intersects the extended opposite side of the triangle, then the points of intersection are collinear.

14. What happens to the theorem of Menelaus if P, Q, R are not distinct from A, B, C?

15. Let p, q, r be three circles of unequal radii each of which lies in the others' exterior. Prove that the three intersections of the common external tangents of each pair of circles are collinear.

16. Formulate and prove an analog of Exercise 15 that involves also the intersections of some common *internal* tangents.

17. Explain why the point R' in the proof of part 2 of the theorem of Menelaus exists.

18. Let P, Q, R be a spherical transversal of the spherical $\triangle ABC$. Prove that P, Q, R are spherically collinear if and only if

$$\frac{\sin AP}{\sin PB}\frac{\sin BQ}{\sin QC}\frac{\sin CR}{\sin RA} = -1.$$

(*Note:* Here AP denotes the length of the geodesic joining A and P, etc.)

19. Let P, Q, R be a hyperbolic transversal of the hyperbolic $\triangle ABC$. Prove that P, Q, R are hyperbolically collinear if and only if

$$\frac{\sinh AP}{\sinh PB}\frac{\sinh BQ}{\sinh QC}\frac{\sinh CR}{\sinh RA} = -1.$$

20. Comment on Proposition 5.2.1 in the context of taxicab geometry.

21. Comment on Proposition 5.2.1 in the context of maxi geometry.

22(C). Use a computer application to verify the theorem of Menelaus.

A *Cevian* of $\triangle ABC$ is a straight line segment that joins a vertex of the triangle to a point on the extended opposite side that is not a vertex.

Proposition 5.2.2 (The Theorem of Ceva) *The three Cevians AQ, BR, CP of $\triangle ABC$ are concurrent if and only if*

$$\frac{AP}{PB}\frac{BQ}{QC}\frac{CR}{RA} = 1.$$

See Exercise 1.

Exercises 5.2B

1. Prove the theorem of Ceva.

Use the theorem of Ceva to prove the following statements.

2. Prove that the three medians of the triangle are concurrent.

3. Prove that the bisectors of the three interior angles of a triangle are concurrent.

4. Prove that the bisector of an interior angle of $\triangle ABC$ and the bisectors of the exterior angles at the other two vertices are concurrent.

5. Prove that in a triangle, the Cevians through the points of contact of the inscribed circle are concurrent.

6. Prove that the three altitudes of every triangle are concurrent. Be sure that your proof also works for obtuse triangles.

7. Suppose AD, BE, CF are concurrent Cevians of $\triangle ABC$ and the circle through D, E, F intersects the sides $\overleftrightarrow{BC}$, $\overleftrightarrow{CA}$, $\overleftrightarrow{AB}$ again in the points D', E', F'. Prove that the Cevians AD', BE', CF' are also concurrent.

8. Let AD, BE, CF be three concurrent Cevians of $\triangle ABC$. Then the points $\overleftrightarrow{BC} \cap \overleftrightarrow{EF}$, $\overleftrightarrow{CA} \cap \overleftrightarrow{DF}$, $\overleftrightarrow{AB} \cap \overleftrightarrow{ED}$ are collinear (provided they exist).

9. Formulate and prove a converse to Exercise 8.

10. Two parallelograms $ABCD$ and $AB'C'D'$ have a common angle at A. Prove that the lines $\overleftrightarrow{BD'}$, $\overleftrightarrow{B'D}$, $\overleftrightarrow{C'C}$ are concurrent.

11. If equilateral triangles BCA', CAB', ABC' are described externally on the sides of $\triangle ABC$, then the lines $\overleftrightarrow{AA'}$, $\overleftrightarrow{BB'}$, $\overleftrightarrow{CC'}$ are concurrent.

12. If A'', B'', C'' are the centers of the equilateral triangles of the previous exercise, then the lines $\overleftrightarrow{AA''}$, $\overleftrightarrow{BB''}$, $\overleftrightarrow{CC''}$ are concurrent.

13. In the quadrilateral $ABCD$, $E = \overleftrightarrow{AC} \cap \overleftrightarrow{BD}$, $F = \overleftrightarrow{AD} \cap \overleftrightarrow{BC}$, $G = \overleftrightarrow{AB} \cap \overleftrightarrow{CD}$ and $H = \overleftrightarrow{AB} \cap \overleftrightarrow{EF}$. Prove that $AH/HB = -AG/GB$. Does your proof depend on whether E is inside or outside $ABCD$? Does it remain valid even if the cyclic ordering of the vertices of the given quadrilateral is not A, B, C, D?

14. State and prove (using spherical trigonometry) a spherical version of the theorem of Ceva.

15. Use Exercise 14 to prove that the spherical medians of a spherical triangle are concurrent.

16. Use Exercise 14 to prove that the spherical angle bisectors of a spherical triangle are concurrent.

17. State and prove (using hyperbolic trigonometry) a hyperbolic version of the theorem of Ceva.

18. Use Exercise 17 to prove that the hyperbolic medians of a hyperbolic triangle are concurrent.

19. Use Exercise 17 to prove that the hyperbolic angle bisectors of a hyperbolic triangle are concurrent.

20. Comment on Proposition 5.2.2 in the context of taxicab geometry.

21. Comment on Proposition 5.2.2 in the context of maxi geometry.

22(C). Use a computer application to verify the theorem of Ceva.

23. Formulate and prove an analog of Exercise 5.2A.15 for the common internal tangents.

Proposition 5.2.3 (The Theorem of Pappus) *If* $\{A, B, C\}$ *and* $\{A', B', C'\}$ *are two sets of collinear points, then the points* $P = \overleftrightarrow{AB'} \cap \overleftrightarrow{A'B}$, $Q = \overleftrightarrow{AC'} \cap \overleftrightarrow{A'C}$, *and* $R = \overleftrightarrow{BC'} \cap \overleftrightarrow{B'C}$ *are also collinear* (provided these intersection points all exist).

See Exercise 1.

Proposition 5.2.4 (The Theorem of Desargues) *For any* $\triangle ABC$ *and* $\triangle A'B'C'$, *the lines* $\overleftrightarrow{AA'}$, $\overleftrightarrow{BB'}$, $\overleftrightarrow{CC'}$ *are concurrent if and only if the points* $\overleftrightarrow{AB} \cap \overleftrightarrow{A'B'}$, $\overleftrightarrow{BC} \cap \overleftrightarrow{B'C'}$, $\overleftrightarrow{AC} \cap \overleftrightarrow{A'C'}$ *are collinear* (provided the intersections $\overleftrightarrow{AA'} \cap \overleftrightarrow{BB'}$, $\overleftrightarrow{BB'} \cap \overleftrightarrow{CC'}$, $\overleftrightarrow{CC'} \cap \overleftrightarrow{AA'}$, $\overleftrightarrow{AB} \cap \overleftrightarrow{A'B'}$, $\overleftrightarrow{BC} \cap \overleftrightarrow{B'C'}$, $\overleftrightarrow{AC} \cap \overleftrightarrow{A'C'}$ all exist).

See Exercises 2 and 3.

Proposition 5.2.5 (The Theorem of Pascal) *The intersections of the three pairs of opposite sides of a cyclic hexagon are collinear* (provided these intersection points all exist).

See Exercise 4.

Each of the preceding three propositions ends with a parenthetical qualification that, at a higher level, turns out to be unnecessary. An elegant reinterpretation of the elements of geometry will be offered in the next section, which indicates how such nuisances can be avoided.

Exercises 5.2C

1. Prove the theorem of Pappus.

2. Prove the first half of the theorem of Desargues: For any $\triangle ABC$ and $\triangle A'B'C'$, if the lines $\overleftrightarrow{AA'}$, $\overleftrightarrow{BB'}$, $\overleftrightarrow{CC'}$ are concurrent, then the points $P = \overleftrightarrow{AB} \cap \overleftrightarrow{A'B'}$, $Q = \overleftrightarrow{BC} \cap \overleftrightarrow{B'C'}$, $R = \overleftrightarrow{AC} \cap \overleftrightarrow{A'C'}$ are collinear (provided they exist).

3. Prove the second half of the theorem of Desargues: For any $\triangle ABC$ and $\triangle A'B'C'$, the lines $\overleftrightarrow{AA'}$, $\overleftrightarrow{BB'}$, $\overleftrightarrow{CC'}$ are concurrent if the points $\overleftrightarrow{AB} \cap \overleftrightarrow{A'B'}$, $\overleftrightarrow{BC} \cap \overleftrightarrow{B'C'}$, $\overleftrightarrow{AC} \cap \overleftrightarrow{A'C'}$ are collinear.

4. Prove the theorem of Pascal.

5. Comment on Proposition 5.2.3 in the context of the following geometries:
   (a) spherical (b) hyperbolic (c) taxicab (d) maxi

6. Comment on Proposition 5.2.4 in the context of the following geometries:
   (a) spherical (b) hyperbolic (c) taxicab (d) maxi

7. Comment on Proposition 5.2.5 in the context of the following geometries:
   (a) spherical (b) hyperbolic (c) taxicab (d) maxi

8(C). Use a computer application to verify
   (a) the theorem of Menelaus
   (b) the theorem of Desargues
   (c) the theorem of Pascal.

9. Draw nine points in the plane so that ten triples of these points are collinear.

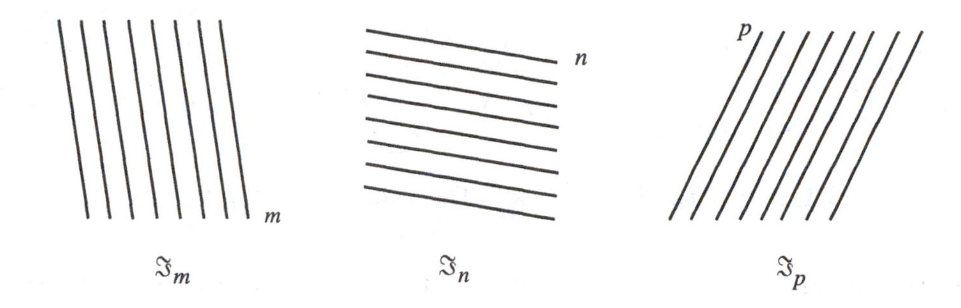

Figure 5.11. Three ideal points

5.3 The Projective Plane

It is well known that parallel lines look as though they meet in a "vanishing point" or at a "point at infinity." The edges of a long straight road look like they meet at a point on the horizon, as do adjacent railway tracks. This illusion is further supported by the fact that in the representation of such a scene on a canvas or in a photograph, the said edges do indeed meet in the plane of the representation. These suggestive and informal observations were turned by Desargues into a formal and fertile geometrical discipline, called *projective geometry*, in the mid-seventeenth century. Since then this direct descendant of Euclidean geometry acquired great depth and applicability and became an integral part of the mainstream of mathematical evolution. We will now explain how mathematicians converted the informal phrase "meet at infinity" into a formally correct statement.

An *ordinary point* is a point of the Euclidean plane. An *ordinary line* is a straight line of the Euclidean plane. The set of all the ordinary lines parallel to the ordinary line m is the *ideal point* or *point at infinity* or the *vanishing point* $\Im_m$ (Fig. 5.11). If m is any ordinary line, then *the extended line* m^* consists of all the points of m together with $\Im_m$; in other words, $m^* = m \cup \{\Im_m\}$. The *ideal line* or *line at infinity* Λ consists of the set of all the ideal points. A *projective point* is either an ordinary or an ideal point. The *projective plane* consists of all the projective points. A *projective line* is either an extended line or the ideal line.

The assignment of only a single ideal point to a straight line may seem counterintuitive, and it is commonly argued that since every line extends to infinity in two directions, each of those directions should receive its own vanishing, or ideal, point. This misconception is reinforced by the observation that when we look along the aforementioned railroad tracks first in one direction and then in the opposite, the tracks seem to meet in two

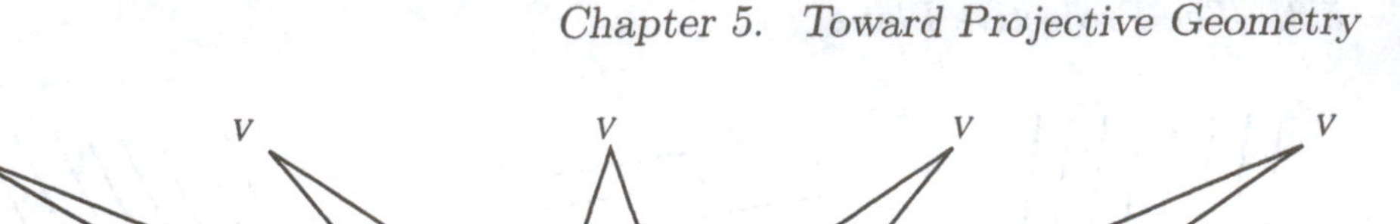

Figure 5.12. Five views of the same vanishing point

"different" ideal points. It is important to remember, however, that the vanishing point depends on the observer's point of view. In other words, the aforementioned "two" vanishing points are merely two different mansifestations of the same ideal point. Figure 5.12 indicates that there are in fact not only two but infinitely many such manifestations.

The geometry of the projective plane is very rich and elegant. Some of this elegance is manifested in the symmetry of the next two basic propositions.

Proposition 5.3.1 *Every two distinct projective points are contained in exactly one projective line.*

GIVEN: Projective points $P \neq Q$.

TO PROVE: There exists exactly one projective line that contains both P and Q.

PROOF:

Case 1: P and Q are both ordinary points. In this case P and Q are contained in exactly one ordinary line $m = \overleftrightarrow{PQ}$. Hence they are contained in exactly one extended line m^*. Since both are ordinary points they are not contained in the ideal line. Thus P and Q are contained in exactly one projective line.

Case 2: P is ordinary and Q is ideal, say $Q = \Im_n$ (Fig. 5.13). Let m be that ordinary line that belongs to Q and contains P. Then the extended line m^* contains both P and $\Im_m = \Im_n = Q$. If p^* is any extended line that contains both P and $Q = \Im_n$, then, by definition, p is an ordinary line that contains P and is parallel to n. Thus, by Playfair's postulate, $p = m$ and so $p^* = m^*$. Since the ideal line Λ consists of ideal points only, it cannot contain P. Thus, the points P and Q are contained in exactly one projective line.

Case 3: P and Q are both ideal points. Both P and Q are in the ideal line Λ. Since each extended line contains only one ideal point, no extended line contains both P and Q. Hence there is exactly one projective line that contains both P and Q. Q.E.D.

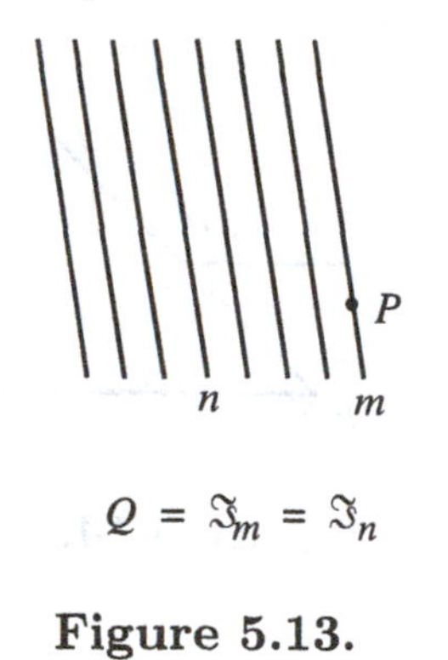

Figure 5.13.

Proposition 5.3.2 *Every two distinct projective lines intersect in exactly one projective point.*

GIVEN: Two distinct projective lines.

TO PROVE: There is exactly one projective point P on both of these lines.

PROOF: It follows from Proposition 5.3.1 that any two distinct projective lines can intersect in at most one point. Hence it suffices to show that every two projective lines intersect.

Case 1: The two lines are both extended Euclidean lines, say m^* and n^*. If $m \parallel n$, then $\Im_m = \Im_n$ and so m^* and n^* intersect in this common ideal point. If $m \nparallel n$, then m^* and n^* intersect in the ordinary point $m \cap n$.

Case 2: One of the straight lines is the ideal line Λ and the other is an extended line m^*. In this case both of the projective lines contain the point $\Im_m$. Q.E.D.

Despite the fact that the projective plane incorporates points that are seemingly infinitely far away, it is possible to extend the notion of the ratio of lengths of ordinary segments to some cases that involve ideal points in a very fruitful way. The following convention is needed to accomplish this task.

Ratio Convention 5.3.3 *Let A and B be distinct ordinary points, and let C be the ideal point on $\overleftrightarrow{AB}^*$. Then $AC/CB = CA/AB = -1$.*

Convention 5.3.3 makes it possible to convert annoying exceptions to Euclidean theorems into valid and interesting propositions. This is illustrated by a reexamination of the theorem of Menelaus in the projective plane. In the relevant figure, an ideal point is represented by three short parallel line

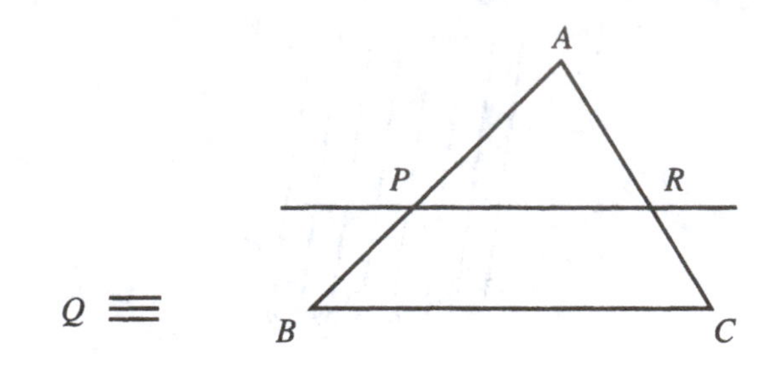

Figure 5.14.

segments that indicate the entire family of lines that constitute that ideal point.

Proposition 5.3.4 *The theorem of Menelaus is also valid when one of the transversal points is ideal.*

GIVEN: Ordinary $\triangle ABC$ with transversal $\{P, Q, R\}$, where P and R are ordinary and Q is ideal (Fig. 5.14).

TO PROVE: The points P, Q, R are collinear if and only if $\dfrac{AP}{PB}\dfrac{BQ}{QC}\dfrac{CR}{RA} = -1$.

PROOF: Suppose P, Q, R are collinear; then $PR \parallel BC$ and so it follows from Proposition 3.5.6 that

$$\frac{AP}{PB} = \frac{AR}{RC} = \frac{RA}{CR}$$

$$\therefore \quad \frac{AP}{PB}\frac{BQ}{QC}\frac{CR}{RA} = \frac{BQ}{QC} = -1.$$

Conversely, if $\dfrac{AP}{PB}\dfrac{BQ}{QC}\dfrac{CR}{RA} = -1$, then $\dfrac{AP}{PB}\dfrac{CR}{RA} = 1$ so that $\dfrac{AP}{PB} = \dfrac{AR}{RC}$. It follows from Proposition 3.5.6 that $PR \parallel BC$ and hence P, Q, R are collinear. Q.E.D.

Exercises 5.3

1. Interpret and prove the theorem of Menelaus when two or more of the transversal points are ideal.

2. Interpret and prove the theorem of Ceva in the case where the transversal point P is ideal and all the other points are ordinary.

3. Interpret and prove the theorem of Ceva in the case where the transversal points P and Q are ideal and all the other points are ordinary.

4. Interpret and prove the theorem of Pappus in the case where exactly one of the given intersection points is ideal and all the other points are ordinary.

5. Interpret and prove the theorem of Pappus in case where two or more of the given intersections are ideal and all the other points are ordinary.

6. Interpret and prove the theorem of Pascal in the case where exactly one of the given intersection points is ideal and all the other points are ordinary.

7. Interpret and prove the theorem of Pascal in the case where two or more of the given intersection points are ideal and all the other points are ordinary.

8. Interpret and prove the first half of the theorem of Desargues in the case where one of the given intersection points is ideal and all the other points are ordinary.

9. Interpret and prove the first half of the theorem of Desargues in the case where two or more of the given intersection points are ideal and all the other points are ordinary.

10. Discuss ideal points in the context of the following geometries:
(a) spherical (b) hyperbolic (c) taxicab (d) maxi

Chapter Review Exercises

In Exercises 1–5 all the points and lines are ordinary.

1. Prove that if the straight line m intersects the sides $\overleftrightarrow{AB}$, $\overleftrightarrow{BC}$, $\overleftrightarrow{CD}$, $\overleftrightarrow{DA}$ of quadrilateral $ABCD$ in the points P, Q, R, S, respectively, then $(AP/PB)\cdot(BQ/QC)(CR/RD)(DS/SA) = 1$.

2. Show that the converse of Exercise 1 is false.

3. Generalize Exercise 1 to arbitrary polygons.

*4. If all the sides of hexagon $ABCDEF$ are tangent to the same circle in its interior, then the three diagonals joining its opposite vertices are concurrent. (theorem of Brianchon).

5. A parallel to the side BC of $\triangle ABC$ meets AB in B' and AC in C'. Prove that BC' and $B'C$ intersect on the median to BC.

6. Divide a given line segment in the ratios $\pm\sqrt{2}/\sqrt{3}$ and $\pm\sqrt{3}/\sqrt{2}$.

7. Interpret and prove Exercise 1 if P is ideal and all the other points are ordinary.

8. Are the following statements true or false? Justify your answers.

(a) Given two distinct points C and D, there exists exactly one point X on $\overleftrightarrow{CD}$ such that $CX/XD = 3$.

(b) Given two distinct points C and D, there exists exactly one point X on $\overleftrightarrow{CD}$ such that $CX/XD = -3$.

(c) Given two distinct points C and D, there exists exactly one point X on $\overleftrightarrow{CD}$ such that $CX/XD = -1$.

(d) Given two distinct points C and D, there exists exactly one point X on $\overleftrightarrow{CD}$ such that $CX/XD = \sqrt{\pi}$.

(e) Given two distinct points C and D, it is possible to construct (in the sense of the *Elements*) a point X on $\overleftrightarrow{CD}$ such that $CX/XD = \sqrt{\pi}$.

(f) Given any three distinct points P, Q, R, there exists $\triangle ABC$ such that AQ, BR, CP are concurrent Cevians for that triangle.

(g) In the projective plane, every three ideal points are collinear.

(h) In the projective plane, every two ideal lines intersect.

(i) In the projective plane, every two projective lines intersect.

(j) In the projective plane, every ideal point lies on some extended line.

(k) Playfair's postulate holds in the projective plane.

(l) If C and D are two ordinary points of the projective plane and r is a real number, then there is exactly one point X on $\overleftrightarrow{CD}$ such that $CX/XD = r$.

Chapter 6

Planar Symmetries

As was noted in Chapter 2, one of the most serious deficiencies in Euclid's axiomatic development of geometry was his failure to provide an explicit discussion of motions, despite the fact that they play an important role in several of his proofs, beginning with that of Proposition 4 of Book I. These transformations were not mentioned in Hilbert's axiomatization either, where they were replaced by several congruence axioms. Other axiom systems, notably that of Mario Pieri (1860–1925), did refer to such motions explicitly. The nineteenth century also witnessed the creation of many alternative geometries, each with its own collection, or *group*, of motions. This proliferation of geometries called for their classification and in 1872 Felix Klein (1849–1925) promulgated his *Erlanger Programm*, in which he suggested that they be classified by their groups of motions.

This chapter is devoted primarily to the classification of the motions of the Euclidean plane and the allied topic of planar symmetry. Some information is also obtained about the motions of the hyperbolic plane.

6.1 Translations, Rotations, and Fixed Points

Informally speaking, a motion of the plane is a transformation that does not alter the distances between the points. More formally, a *motion* is a function f of the plane into itself such that for any two points P and Q

$$PQ = P'Q',$$

where $P' = f(P)$ and $Q' = f(Q)$. The prototypical motion is the *translation* that "slides" the plane on itself so that all straight lines remain parallel to their original positions. More precisely, given any two points A and B, the

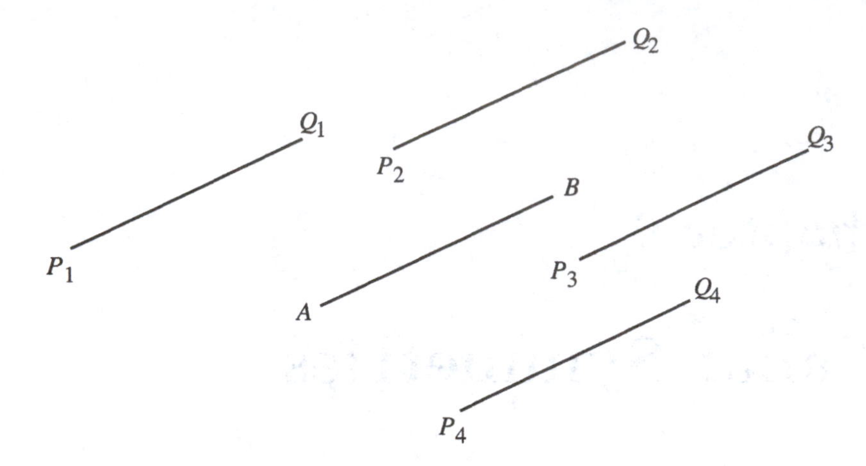

Figure 6.1. A translation

translation that carries A onto B is denoted by τ_{AB} and if P is any point then

$$\tau_{AB}(P) = Q$$

where Q is the unique point such that $AB = PQ$, $AB \parallel PQ$, and the segments AB and PQ are similarly directed. If P does not lie on $\overleftrightarrow{AB}$ then this, by virtue of Proposition 3.1.7, is tantamount to saying that the quadrilateral $ABQP$ is a parallelogram. In Figure 6.1 $\tau_{AB}(P_i) = Q_i$ for each $i = 1, 2, 3, 4$. Note that in this figure P_iP_j is both parallel and equal to Q_iQ_j whenever $i \neq j$ and hence τ_{AB} is indeed a motion.

The same translation can be represented in many different ways. Thus, the translation τ_{AB} of Figure 6.1 can also be denoted by $\tau_{P_1Q_1}$, $\tau_{P_2Q_2}$, and so on. Two motions f and g are said to be *equal* provided that

$$f(P) = g(P) \qquad \text{for all points } P \text{ in the plane.}$$

In other words, if the motion is visualized as a physical movement of the plane, then the intermediary stages of the motion are immaterial: All that matters are the final positions of the points.

This chapter's goal is the classification of all the motions of the plane, and the most important tool in this text's approach is the composition of motions. The reader is reminded that if f and g are functions of any set into itself, then their *composition* $g \circ f$ is a function of the same set into itself such that

$$g \circ f(P) = g(f(P)).$$

The *identity* transformation *Id* is defined by the equation

$$\mathrm{Id}(P) = P \qquad \text{for every point } P$$

and has the property that for any motion f,

$$f \circ \mathrm{Id} = \mathrm{Id} \circ f = f.$$

The operation of composition is associative in the sense that for any three such functions f, g, and h,

$$(f \circ g) \circ h = f \circ (g \circ h).$$

We begin with the composition of translations.

Proposition 6.1.1 *If A, B, C are any points of the plane, then*

$$\tau_{BC} \circ \tau_{AB} = \tau_{AC}.$$

PROOF: Let P be any point of the plane and set (see Fig. 6.2)

$$P' = \tau_{AB}(P) \qquad P'' = \tau_{BC}(P') = \tau_{BC} \circ \tau_{AB}(P).$$

It is necessary to show that $P'' = \tau_{AC}(P)$. However, as was noted previously, $ABP'P$ and $BCP''P'$ are both parallelograms. It follows from Proposition 3.1.8 that AP and BP' are equal to and parallel to BP' and CP'' respectively. Hence, by Proposition 3.1.7 $ACP''P$ is a parallelogram and so $P'' = \tau_{AC}(P)$. Q.E.D.

It follows from this proposition that the composition of any two translations is itself a translation. For if f and g are any translations and P is any point, then we could set $P' = f(P)$, $P'' = g(P')$ and conclude that

$$g \circ f = \tau_{P'P''} \circ \tau_{PP'} = \tau_{PP''}.$$

The *inverse* f^{-1} of the motion f is a motion such that

$$f \circ f^{-1} = f^{-1} \circ f = \mathrm{Id}.$$

It is clear that for any two points A and B, $\tau_{AB}^{-1} = \tau_{BA}$. If n is any positive integer, then f^n denotes n iterations of f, so that $f^2 = f \circ f$ and $f^3 = f \circ f \circ f$. Similarly, f^{-n} denotes n iterations of f^{-1}.

Another type of motion is the *rotation*. If C is any point of the plane and α is some directed angle, then the rotation $R_{C,\alpha}$ is the transformation that

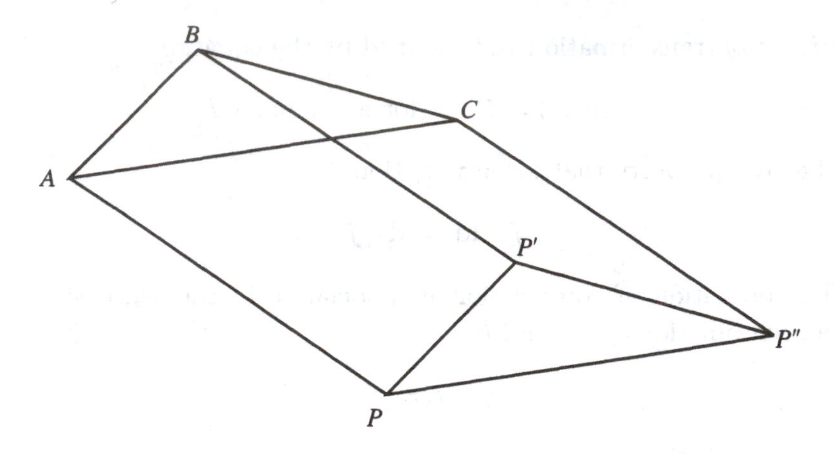

Figure 6.2. The composition of translations

moves the general point P to the point $P' = R_{C,\alpha}(P)$, where $CP' = CP$ and $\angle P'CP = \alpha$ (Fig. 6.3). Exercise 1 calls for the formal proof of the rigidity of rotations. The point C is the *pivot point* of the rotation $R_{C,\alpha}$. The angle α of the rotation is understood to be *oriented*, in the sense that it can be either positive or negative, and the rotation accordingly proceeds either counterclockwise or clockwise, respectively. Moreover, if n is any integer and $\beta = \alpha + n \cdot 360°$, then $R_{C,\beta} = R_{C,\alpha}$. Consequently, in describing any rotation $R_{C,\alpha}$ the angle will generally be chosen so that $-180° < \alpha < 360°$. Note that $R_{C,\alpha}^{-1} = R_{C,-\alpha}$.

The composition of the rotations $R_{C,\alpha}$ and $R_{C,\beta}$ is clearly $R_{C,\alpha+\beta}$, but what about the composition of $R_{C,\alpha}$ with $R_{D,\beta}$, where C and D are distinct points? In order to answer this natural question, it is first necessary to deal

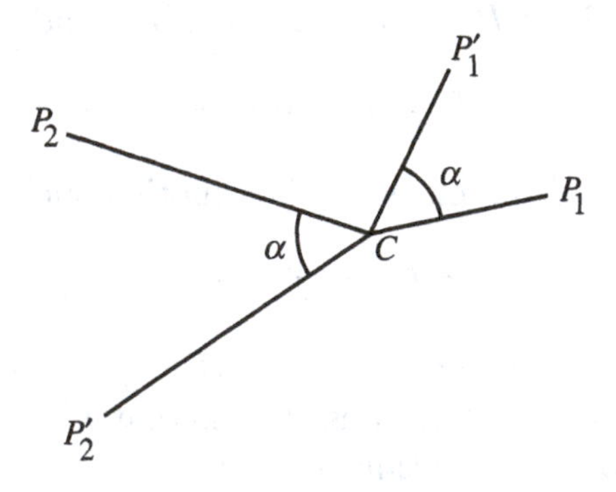

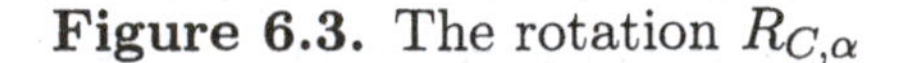

Figure 6.3. The rotation $R_{C,\alpha}$

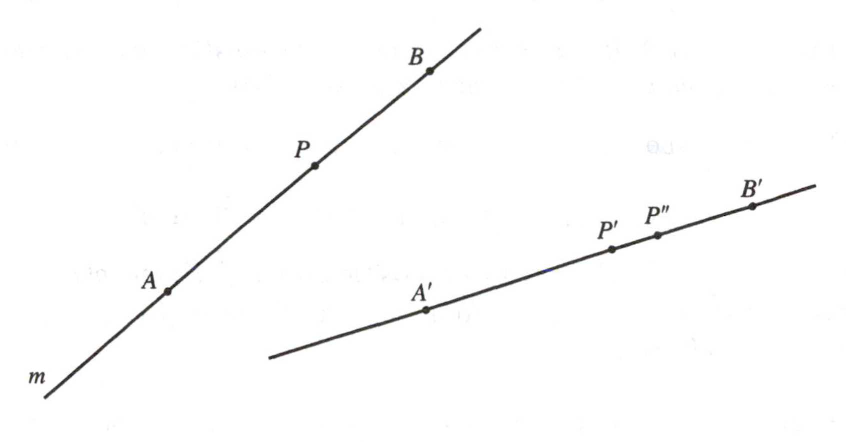

Figure 6.4.

with the issue of identifying motions in general. The following sequence of propositions aims to answer the question

> *How much information is it necessary to have about a motion before we can say that it is known?*

It will soon be seen that surprisingly little is needed.

Proposition 6.1.2 *Every motion transforms straight lines into straight lines.*

PROOF: Let f be a motion, let m be a straight line with two distinct points A and B on it, and set $A' = f(A)$ and $B' = f(B)$ (Fig. 6.4). If P is any point of m between A and B and $P' = f(P)$, then

$$A'P' + P'B' = AP + PB = AB = A'B',$$

and it follows from Proposition 2.3.25 that P' is on the line segment $A'B'$. A similar argument (Exercise 5) demonstrates that as long as P is on m, then P' is on the line $\overleftrightarrow{AB}$ even when P is not between A and B.

Conversely, let P' be any point of $\overleftrightarrow{AB}$ that lies on the line segment $A'B'$. Since $A'B' = AB$ there is a unique point P of m such that

$$AP = A'P' \quad \text{and} \quad BP = B'P'.$$

If $P'' = f(P)$, then

$$A'P'' = AP = A'P' \quad \text{and} \quad B'P'' = BP = B'P'$$

so that P' and $P'' = f(P)$ must be identical. The same holds even when P' is on $\overleftrightarrow{A'B'}$ but not between A' and B'. This means that every point of $\overleftrightarrow{A'B'}$ is covered by some point of m. In other words, $f(m) = \overleftrightarrow{A'B'}$. Q.E.D.

Proposition 6.1.3 *If two motions agree on two distinct points, then they agree at every point of the straight line joining them.*

PROOF: Let f and g be two motions and A and B two distinct points such that

$$f(A) = g(A) = A' \quad \text{and} \quad f(B) = g(B) = B'.$$

If P is any point of $\overleftrightarrow{AB}$, then, by Proposition 6.1.2, $f(P)$ and $g(P)$ are both points of $\overleftrightarrow{A'B'}$ whose distances from A' and B' are respectively equal. It follows that $f(P) = g(P)$. Q.E.D.

Theorem 6.1.4 *If two motions agree at three noncollinear points, then they agree everywhere.*

PROOF: Let f and g be two motions that agree at the three noncollinear points A, B, C. By Proposition 6.1.3, f and g agree at every point on the straight lines $\overleftrightarrow{AB}$, $\overleftrightarrow{BC}$, and $\overleftrightarrow{AC}$. If P is any point of the plane then there clearly exists a straight line through P that intersects the union of these three straight lines in some two distinct points X and Y. Since f and g agree at X and Y, it follows from Proposition 6.1.3 that they must also agree at P. Q.E.D.

Thus, in order to pin down a motion, it suffices to know how it affects some triple of noncollinear points.

A *fixed point* of the transformation f is a point P such that

$$f(P) = P.$$

It is clear that the point C is a fixed point of the rotation $R_{C,\alpha}$ and is in fact the only fixed point of that rotation. It is equally clear that, with the exception of the identity, translations have no fixed points whatsoever. On the other hand, every point is a fixed point of the identity. The following corollary is an immediate consequence of Theorem 6.1.4.

Corollary 6.1.5 *If a motion fixes three noncollinear points, then it must be the identity.*

Exercises 6.1

1. Prove that every rotation is a motion.
2. Prove that every motion transforms circles into circles.

3. Prove that if A, B, C are any three points, then $\tau_{CA} \circ \tau_{BC} \circ \tau_{AB} = \text{Id}$.
4. Let $\triangle ABC$ be a clockwise triangle with oriented interior angles α, β, γ at A, B, C, respectively. Prove that $R_{C,2\gamma} \circ R_{B,2\beta} \circ R_{A,2\alpha} = \text{Id}$.
5. Complete the proof of Proposition 6.1.2 by providing the details for the case where P is on $\overleftrightarrow{AB}$ but outside AB.
6. Let $A(a_1, a_2)$ and $B(b_1, b_2)$ be two points. Explain why the transformation $f(P) = Q$ that takes the point $P(x, y)$ to the point $Q(x', y')$ where
$$x' = x + b_1 - a_1$$
$$y' = y + b_2 - a_2$$
is in fact the translation τ_{AB}.
7. Let α be an angle. Explain why the transformation $f(P) = Q$ that maps the point $P(x, y)$ to the point $Q(x', y')$ where
$$x' = x \cos \alpha - y \sin \alpha$$
$$y' = x \sin \alpha + y \cos \alpha$$
is in fact the rotation $R_{O,\alpha}$, where O is the origin.
8. Prove that motions preserve angles. In other words, show that if f is a motion and m and n are straight lines that form an angle of measure α, then $f(m)$ and $f(n)$ are also straight lines that form an angle of measure α.

6.2 Reflections

Given a straight line m, the *reflection* ρ_m is the transformation that fixes every point of m and associates to each point P not on m the unique point P' such that m is the perpendicular bisector of PP' (see Fig. 6.5 and Exercise 28). It follows directly from the definition that $\rho_m \circ \rho_m = \text{Id}$ and hence $\rho_m^{-1} = \rho_m$. This text's classification of the motions is based on the fact that these reflections are the building blocks of all the motions in the sense that every motion can be expressed as the composition of some reflections. The next two propositions show that such is indeed the case for translations and rotations.

Proposition 6.2.1 *Let m and n be two parallel straight lines. Let AB be a directed line segment that first intersects m and then n and whose length is twice the distance between m and n. Then*

(a) $\rho_n \circ \rho_m = \tau_{AB}$
(b) $\rho_n \circ \tau_{AB} = \rho_m$
(c) $\tau_{AB} \circ \rho_m = \rho_n$

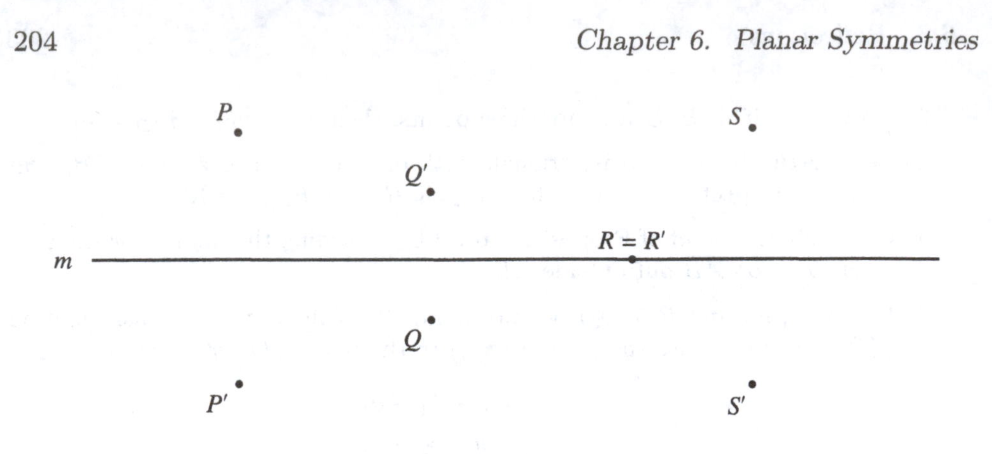

Figure 6.5.

PROOF: Let P be any point outside the infinite strip bounded by m and n such that the distance from P to m is less than the distance between m and n (Fig. 6.6). Set

$$P' = \rho_m(P) \quad \text{and} \quad P'' = \rho_n(P').$$

It is clear that P, P', and P'' are collinear and that

$$PP'' = PP' + P'P'' = 2XP' + 2P'Y = 2XY$$
$$= \text{twice the distance between } m \text{ and } n.$$

Hence

$$\rho_n \circ \rho_m(P) = \rho_n(P') = P'' = \tau_{PP''}(P). \tag{1}$$

Since it is easy to find three noncollinear positions of P that satisfy the constraints specified in the beginning of this proof, it follows that Equation

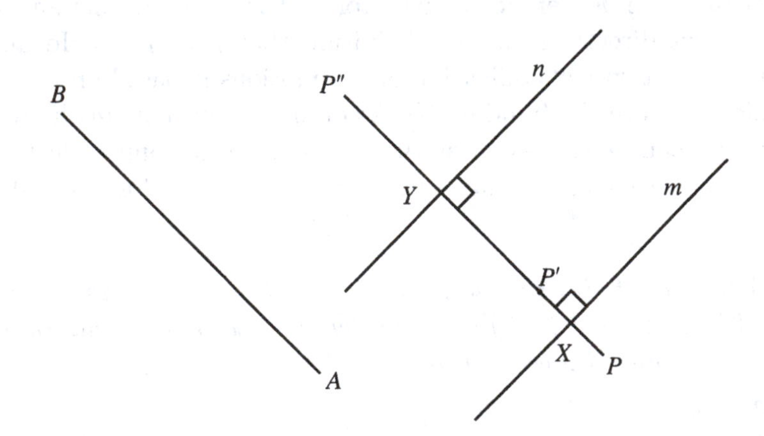

Figure 6.6.

(1) holds for three noncollinear points and hence, by Theorem 6.1.4, $\rho_m \circ \rho_n = \tau_{PP''}$. Since $AB = PP''$ and $AB \parallel PP''$, it follows that $\tau_{PP''} = \tau_{AB}$. This completes the proof of part (a). Parts (b) and (c) follow immediately, since

$$\rho_n \circ \tau_{AB} = \rho_n \circ (\rho_n \circ \rho_m) = (\rho_n \circ \rho_n) \circ \rho_m = \text{Id} \circ \rho_m = \rho_m,$$
$$\tau_{AB} \circ \rho_m = (\rho_n \circ \rho_m) \circ \rho_m = \rho_n \circ (\rho_m \circ \rho_m) = \rho_n \circ \text{Id} = \rho_n.$$

Q.E.D.

Conversely, given any translation τ_{AB}, there clearly exist two parallel straight lines that are perpendicular to $\overleftrightarrow{AB}$ and whose distance from each other equals half of AB. By the preceding proposition, either $\tau_{AB} = \rho_m \circ \rho_n$ or $\tau_{AB} = \rho_n \circ \rho_m$ and in either case the arbitrary translation τ_{AB} has been expressed as the composition of two reflections. This expression is, of course, not unique, since m can be any line that is perpendicular to AB.

Proposition 6.2.2 *Let m and n be two straight lines that intersect at a point A, and let α be the counterclockwise angle from m to n at A. Then*

$$\rho_n \circ \rho_m = R_{A,2\alpha}.$$

PROOF: Let P be a point outside $\angle BAC = \alpha$ (Fig. 6.7) but close enough to m so that $P' = \rho_m(P)$ is inside the angle. Set $P'' = \rho_n(P') = \rho_n \circ \rho_m(P)$. Then m bisects $\angle PAP'$ and n bisects $\angle P'AP''$. Consequently,

$$\angle PAP'' = 2\angle BAP' + 2\angle P'AC = 2\alpha.$$

Hence

$$R_{A,2\alpha}(P) = P'' = \rho_n \circ \rho_m(P). \tag{2}$$

Since it is easy to find three noncollinear positions of P that satisfy the constraints specified in the beginning of this proof, it follows that Equation (2) holds for three noncollinear points and hence, by Theorem 6.1.4, $\rho_n \circ \rho_m = R_{A,2\alpha}$. Q.E.D.

It was noted previously that the composition of rotations that share their pivot points is a rotation about the same point, but that the nature of the composition of rotations with distinct pivot points was unclear. We are now ready to dispose of this and other similar issues.

Proposition 6.2.3 *Let A and B be two points and let α and β be two oriented angles. Then the composition $R_{B,\beta} \circ R_{A,\alpha}$ is*

(a) *a translation if $\alpha + \beta$ is a multiple of 360°*
(b) *a rotation $R_{X,\alpha+\beta}$ if $\alpha + \beta$ is not a multiple of 360°*

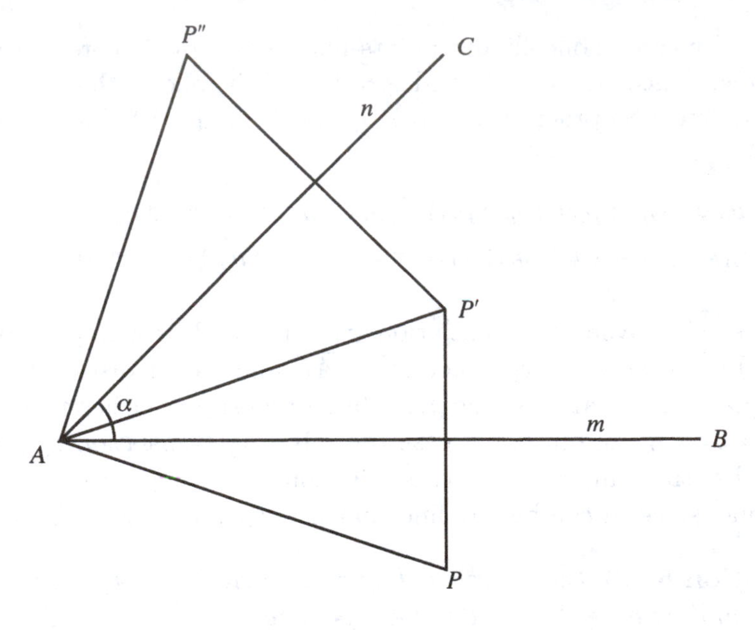

Figure 6.7.

PROOF: This is obvious if A and B are identical points as well as when either α or β is zero. It is therefore assumed that A and B are distinct, and neither α nor β is zero. Let $m = \overleftrightarrow{AB}$, let k be the line through A such that the oriented angle from k to m is $\alpha/2$, let n be the line through B such that the oriented angle from m to n is $\beta/2$ (Fig. 6.8). Then, by Proposition 6.2.2

$$R_{B,\beta} \circ R_{A,\alpha} = (\rho_n \circ \rho_m) \circ (\rho_m \circ \rho_k) = \rho_n \circ (\rho_m \circ \rho_m) \circ \rho_k = \rho_n \circ \rho_k,$$

which is either a translation or a rotation, depending on whether the lines k and n are parallel. However, these lines are parallel if and only if $\alpha/2 + \beta/2$ is a multiple of 180°, which is of course equivalent to $\alpha + \beta$ being a multiple of 360°. Hence, by Proposition 6.2.1, the composition is a translation if $\alpha + \beta$ is a multiple of 360°. When $\alpha + \beta$ is not such a multiple, then, by Proposition 6.2.2, the composition is the rotation $R_{X,\alpha+\beta}$. Q.E.D.

Example 6.2.4 Given any two points A and B, identify $R_{B,60^\circ} \circ R_{A,60^\circ}$.

It follows from Proposition 6.2.3 that this composition is a rotation $R_{C,120^\circ}$. The pivot point C is located as follows: Set (see Fig. 6.9)

$$A' = R_{C,120^\circ}(A) = R_{B,60^\circ} \circ R_{A,60^\circ}(A) = R_{B,60^\circ}(A).$$

Then C is that unique point such that $\triangle ACA'$ is isosceles with vertex angle $\angle ACA' = 120°$. In other words, C is the center of the equilateral $\triangle AA'B$.

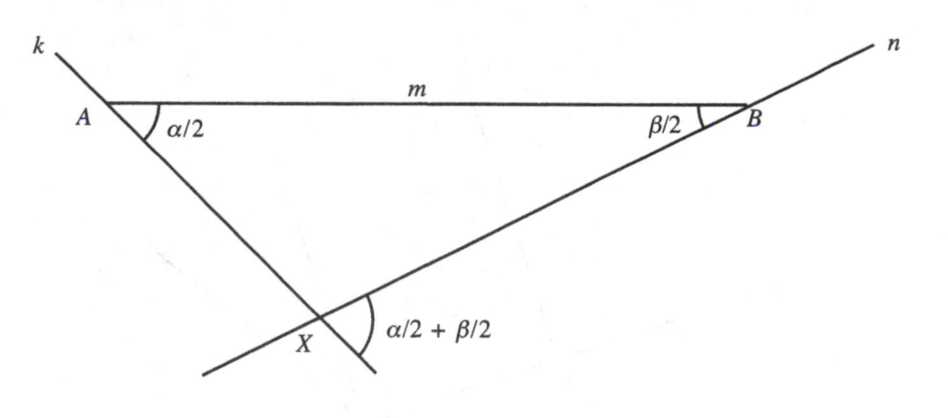

Figure 6.8.

Proposition 6.2.5 *Let R be a rotation that is not the identity and let τ be a translation. Then both $R \circ \tau$ and $\tau \circ R$ are rotations with the same angle as R.*

PROOF: Suppose $R = R_{A,\alpha}$, $A' = \tau(A)$ and let B be the midpoint of the segment AA' (Fig. 6.10). Let k and m be the lines through B and A, respectively, that are perpendicular to AA', and let n be the line through A such that the oriented angle from n to m is equal to $\alpha/2$. Then

$$\tau \circ R = (\rho_k \circ \rho_m) \circ (\rho_m \circ \rho_n) = \rho_k \circ (\rho_m \circ \rho_m) \circ \rho_n = \rho_k \circ \rho_n,$$

which is a rotation by angle α because k and n, when extended, intersect in an angle of $\alpha/2$.

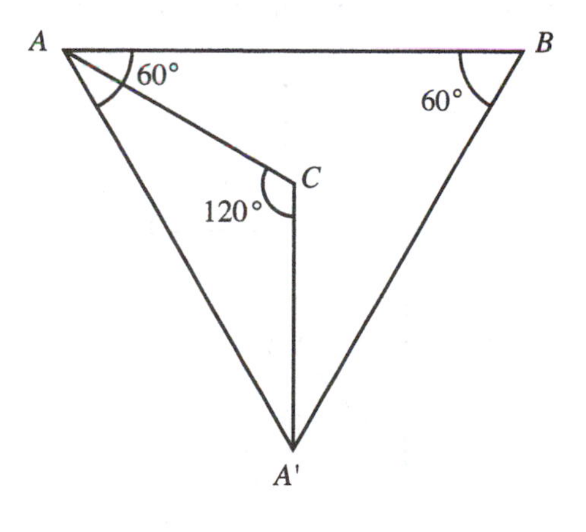

Figure 6.9.

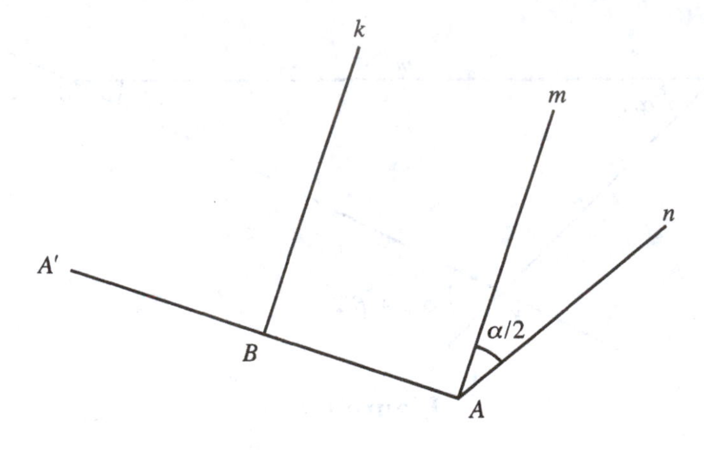

Figure 6.10.

The proof that $R \circ \tau$ is also a rotation is relegated to Exercise 25.

Q.E.D.

Example 6.2.6 For the two given points A and B of Figure 6.11, identify both $\tau_{AB} \circ R_{A,90°}$ and $R_{A,90°} \circ \tau_{AB}$.

By Proposition 6.2.5 $\tau_{AB} \circ R_{A,90°}$ is a 90° rotation $R_{X,90°}$ such that

$$R_{X,90°}(A) = \tau_{AB} \circ R_{A,90°}(A) = \tau_{AB}(A) = B.$$

It follows that the pivot point of $\tau_{AB} \circ R_{A,90°}$ is that point X such that $\triangle ABX$ is an isosceles right triangle. Similarly, $R_{A,90°} \circ \tau_{AB}$ is a 90° rotation $R_{Y,90°}$ such that

$$R_{Y,90°}(A) = R_{A,90°} \circ \tau_{AB}(A) = R_{A,90°}(B) = A'.$$

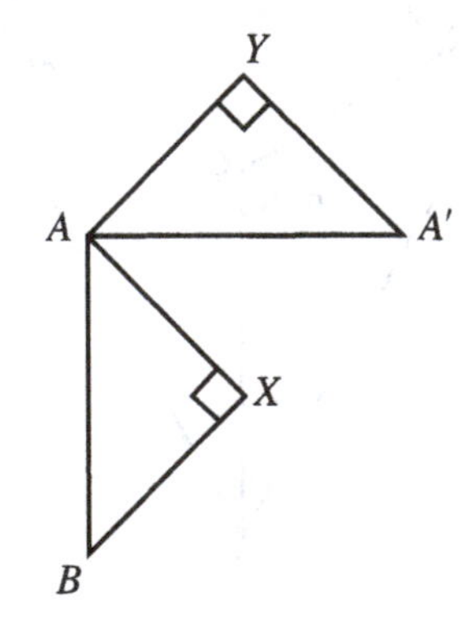

Figure 6.11.

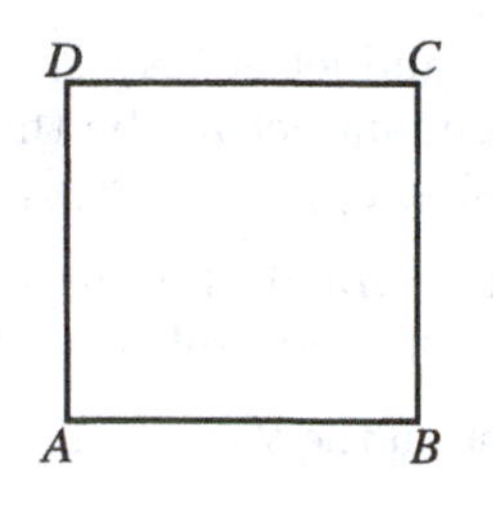

Figure 6.12.

It follows that the pivot point of $R_{A,90°} \circ \tau_{AB}$ is the point Y where $\triangle AA'Y$ is an isosceles right triangle. Note that the two compositions $\tau_{AB} \circ R_{A,90°}$ and $R_{A,90°} \circ \tau_{AB}$ are not equal. In general, motions do not commute.

Exercises 6.2

Identify the compositions of Exercises 1–18, where ABCD is the square of Figure 6.12.

1. $R_{A,90°} \circ R_{B,90°}$
2. $R_{B,90°} \circ R_{A,90°}$
3. $R_{C,180°} \circ R_{A,90°}$
4. $R_{A,90°} \circ \tau_{BC}$
5. $R_{A,90°} \circ \tau_{CA}$
6. $\tau_{CA} \circ R_{A,90°}$
7. $\tau_{BC} \circ \tau_{BA}$
8. $\tau_{BC} \circ \tau_{AD}$
9. $\tau_{DA} \circ \tau_{BC}$
10. $R_{A,270°} \circ R_{C,90°}$
11. $R_{A,180°} \circ R_{D,180°}$
12. $R_{A,45°} \circ R_{C,135°}$
13. $R_{A,45°} \circ R_{B,45°}$
14. $R_{A,60°} \circ R_{B,120°}$
15. $R_{D,120°} \circ R_{C,120°}$
16. $R_{A,90°} \circ R_{B,30°}$
17. $\tau_{AB} \circ R_{A,60°}$
18. $R_{B,60°} \circ \tau_{AB}$
19. Let $\triangle ABC$ be a clockwise triangle with oriented interior angles α, β, γ at A, B, C, respectively. Use Proposition 6.2.2 to prove that $R_{C,2\gamma} \circ R_{B,2\beta} \circ R_{A,2\alpha} = \mathrm{Id}$.
20. Let $A_1, A_2, \dots, A_n$ be the clockwise successive vertices of a polygon with n sides. If the interior angle at A_i is α_i, identify the composition $R_{A_n,2\alpha_n} \circ \cdots \circ R_{A_2,2\alpha_2} \circ R_{A_1,2\alpha_1}$.
21. Let $A_1, A_2, \dots, A_n$ be the midpoints of the successive sides of a polygon with n sides. Identify the composition $R_{A_n,\pi} \circ \cdots \circ R_{A_2,\pi} \circ R_{A_1,\pi}$ if
    (a) $n = 3$ (b) $n = 4$ (c) n is an arbitrary positive integer

22. Let n be an even integer and let $A_1, A_2, \ldots, A_n$ be the successive vertices of a regular n-sided polygon, and let m_i be the bisector of the interior angle at A_i. Identify the composition $\rho_{m_n} \circ \cdots \circ \rho_{m_2} \circ \rho_{m_1}$.
23. Let P be any point on the straight line m and let θ be any angle. Prove that both $R_{P,\theta} \circ \rho_m$ and $\rho_m \circ R_{P,\theta}$ are reflections. What are their axes?
24. Prove that if P is a point on the straight line m, then $\rho_m \circ R_{P,\theta} \circ \rho_m = R_{P,-\theta}$.
25. Complete the proof of Proposition 6.2.5 by showing that $R \circ \tau$ is also a rotation with the same angle as α.
26. Let $ABCD$ be a cyclic quadrilateral. Identify the composition $\rho_{DA} \circ \rho_{CD} \circ \rho_{BC} \circ \rho_{AB}$.
27. Let α be an angle and let m be the straight line through the origin with inclination α to the positive x axis. Explain why the transformation $f(P) = Q$ that maps the point $P(x, y)$ to the point $Q(x', y')$, where

$$x' = x \cos 2\alpha + y \sin 2\alpha$$
$$y' = x \sin 2\alpha - y \cos 2\alpha$$

is in fact the reflection ρ_m.
28. Prove that every reflection is a motion.

6.3 Glide Reflections

So far reflections have been used merely in order to explain how translations and rotations interact under compositions. We now examine how these two types interact with reflections. A special case of this issue was resolved by parts (b) and (c) of Proposition 6.2.1, wherein it was proved that the composition of a reflection with a translation whose direction is perpendicular to the direction of the translation is another reflection with an axis parallel to that of the given reflection. The composition of a reflection with a rotation whose pivot point lies on the reflection's axis is also a reflection (Exercise 6.2.23). In general, however, the composition of either a translation or a rotation with a reflection forms a new kind of motion.

Let A and B be two distinct points. The composition $\rho_{AB} \circ \tau_{AB}$ is called a *glide reflection* and is denoted by γ_{AB}. It is easily seen that the reverse composition $\tau_{AB} \circ \rho_{AB}$ also equals γ_{AB} and that the inverse of γ_{AB} is γ_{BA} (Fig. 6.13). In order to simplify the statements of some of the subsequent propositions, reflections will be considered as special cases of glide reflections. The line $\overleftrightarrow{AB}$ is called the *axis* of the glide reflection γ_{AB}, and it is easily seen that for any point P not on the axis A, the line segment joining P to $\gamma_{AB}(P)$ is bisected by $\overleftrightarrow{AB}$. (See Fig. 6.13 and Exercise 25.)

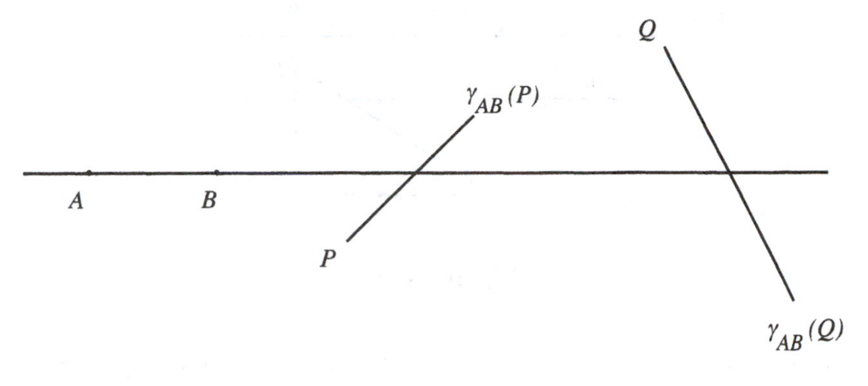

Figure 6.13.

Proposition 6.3.1 *Let τ be any translation and γ any glide reflection. Then $\gamma \circ \tau$ and $\tau \circ \gamma$ are both glide reflections.*

PROOF: Suppose $\tau = \tau_{AB}$.

If $\gamma = \rho_{AB}$, then clearly $\gamma \circ \tau = \tau \circ \gamma = \gamma_{AB}$.

If $\gamma = \rho_m$, where $m \parallel AB$, then there exist points A', B' on m such that $\tau = \tau_{AB} = \tau_{A'B'}$. Consequently, by the previous argument,

$$\gamma \circ \tau = \rho_{A'B'} \circ \tau_{A'B'} = \gamma_{A'B'} = \tau_{A'B'} \circ \rho_{A'B'} = \tau \circ \gamma.$$

If $\gamma = \rho_m$, where $m \perp AB$, then this proposition follows from Proposition 6.2.1bc.

If $\gamma = \rho_m$, where m is skew to AB, let C be a point such that $AC \parallel m$ and $BC \perp m$ (Fig. 6.14). By Proposition 6.1.1,

$$\gamma \circ \tau = \rho_m \circ \tau_{AB} = \rho_m \circ (\tau_{CB} \circ \tau_{AC}) = (\rho_m \circ \tau_{CB}) \circ \tau_{AC}.$$

By Proposition 6.2.1b, there is a line $n \perp BC$ such that $\rho_m \circ \tau_{CB} = \rho_n$, and hence

$$\gamma \circ \tau = \rho_n \circ \tau_{AC},$$

which, since $n \parallel AC$, is known to be a glide reflection. The proof that $\tau \circ \gamma$ is also a glide reflection is relegated to Exercise 23.

Finally, let γ be an arbitrary glide reflection. If $\gamma = \gamma_{CD} = \rho_{CD} \circ \tau_{CD}$ then, by Proposition 6.1.1,

$$\gamma \circ \tau = (\rho_{CD} \circ \tau_{CD}) \circ \tau_{AB} = \rho_{CD} \circ (\tau_{CD} \circ \tau_{AB}) = \rho_{CD} \circ \tau$$

for some translation τ. This, however, is known to be a glide reflection. The proof that $\tau \circ \gamma$ is also a glide reflection is relegated to Exercise 23. Q.E.D.

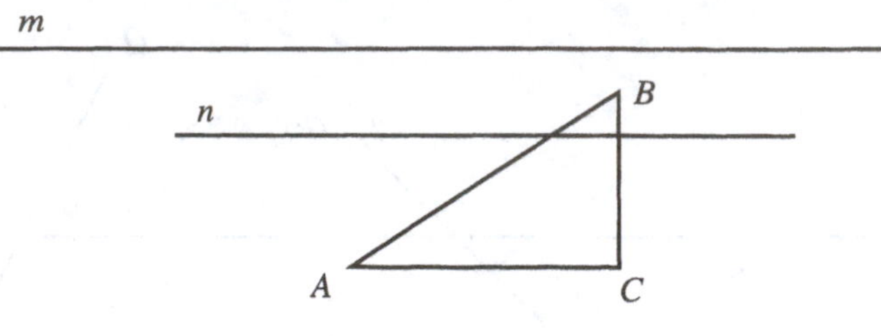

Figure 6.14.

Example 6.3.2 Identify the compositions $\gamma_{AD} \circ \tau_{AB}$ and $\tau_{AB} \circ \gamma_{AD}$, where $ABCD$ is the square of Figure 6.15.

By the previous proposition, these compositions are both glide reflections. Moreover, since

$$\gamma_{AD} \circ \tau_{AB}(A) = \gamma_{AD}(B) = B',$$

it follows that the axis of $\gamma_{AD} \circ \tau_{AB}$ must contain the midpoint M of the segment AB'. In addition,

$$\gamma_{AD} \circ \tau_{AB}(M) = \gamma_{AD}(P) = P'.$$

It follows that $\overleftrightarrow{MP}$ is the axis of $\gamma_{AD} \circ \tau_{AB}$ and in fact $\gamma_{AD} \circ \tau_{AB} = \gamma_{MP'}$.

Again,

$$\tau_{AB} \circ \gamma_{AD}(A) = \tau_{AB}(D) = C$$

and hence the axis of $\tau_{AB} \circ \gamma_{AD}$ contains the midpoint P of AC. In addition,

$$\tau_{AB} \circ \gamma_{AD}(P) = \tau_{AB}(P') = Q$$

and hence $\tau_{AB} \circ \gamma_{AD} = \gamma_{PQ}$.

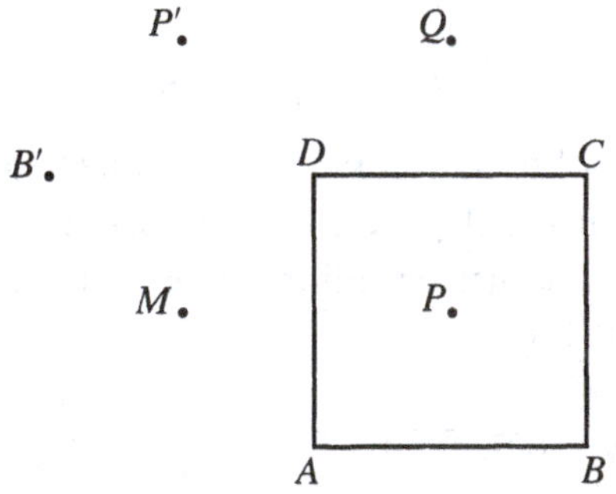

Figure 6.15.

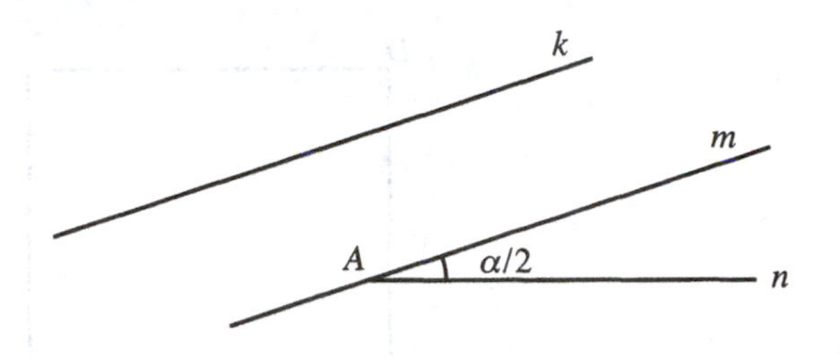

Figure 6.16.

Proposition 6.3.3 *Let R be any rotation and γ any glide reflection. Then both $\gamma \circ R$ and $R \circ \gamma$ are glide reflections.*

PROOF: Let $R = R_{A,\alpha}$ and suppose first that $\gamma = \rho_k$. Let m be the straight line through A that is parallel to k, and let n be the straight line through A such that the oriented angle from n to m is $\alpha/2$ (Fig. 6.16). Then

$$\gamma \circ R = \rho_k \circ R_{A,\alpha} = \rho_k \circ (\rho_m \circ \rho_n) = (\rho_k \circ \rho_m) \circ \rho_n.$$

Since $k \parallel m$ it follows from Proposition 6.2.1 that $\rho_k \circ \rho_m$ is a translation and hence, by Proposition 6.3.1, $\gamma \circ R = (\rho_k \circ \rho_m) \circ \rho_n$ is a glide reflection.

If γ is the arbitrary glide reflection $\tau_{CD} \circ \rho_{CD}$, then

$$\gamma \circ R = (\tau_{CD} \circ \rho_{CD}) \circ R_{A,\alpha} = \tau_{CD} \circ (\rho_{CD} \circ R_{A,\alpha}).$$

By the first part of the proof, $\rho_{CD} \circ R_{A,\alpha}$ is a glide reflection and hence it follows from Proposition 6.3.1 that $\gamma \circ R = \tau_{CD} \circ (\rho_{CD} \circ R_{A,\alpha})$ is also a glide reflection.

The proof that $R \circ \gamma$ is a glide reflection is relegated to Exercise 24.

Q.E.D.

Example 6.3.4 Identify the composition $\gamma_{AD} \circ R_{A,90^\circ}$ and $R_{A,90^\circ} \circ \gamma_{AD}$, where $ABCD$ is the square of Figure 6.17.

By the previous proposition, both of these compositions are glide reflections. Moreover,

$$\gamma_{AD} \circ R_{A,90^\circ}(A) = \gamma_{AD}(A) = D$$

so that the axis of this composition contains the midpoint X of AD. Since

$$\gamma_{AD} \circ R_{A,90^\circ}(X) = \gamma_{AD}(X') = Y,$$

it follows that $\gamma_{AD} \circ R_{A,90^\circ} = \gamma_{XY}$. Similarly,

$$R_{A,90^\circ} \circ \gamma_{AD}(A) = R_{A,90^\circ}(D) = Z$$

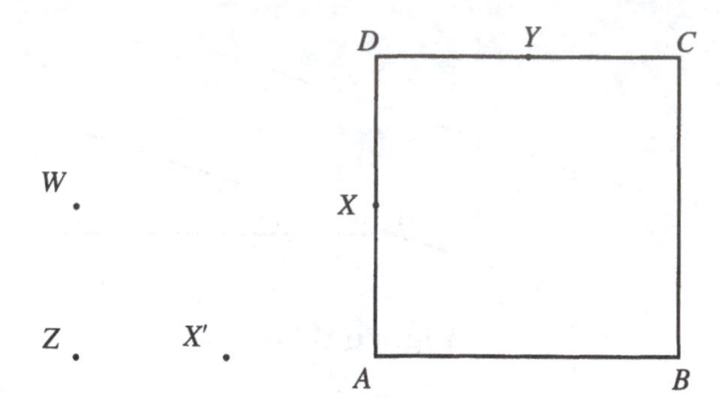

Figure 6.17.

so that the axis of this composition contains the midpoint X' of AZ. Since

$$R_{A,90^\circ} \circ \gamma_{AD}(X') = R_{A,90^\circ}(Y) = W,$$

it follows that $R_{A,90^\circ} \circ \gamma_{AD} = \gamma_{X'W}$.

Proposition 6.3.5 *Let γ_{AB} and γ_{CD} be two glide reflections. The composition $\gamma_{AB} \circ \gamma_{CD}$ is*

(a) *a translation if $AB \parallel CD$*

(b) *a rotation of angle 2α otherwise, where α is the oriented angle from $\overleftrightarrow{CD}$ to $\overleftrightarrow{AB}$*

PROOF: Note that

$$\gamma_{AB} \circ \rho_{CD} = (\tau_{AB} \circ \rho_{AB}) \circ (\rho_{CD} \circ \tau_{CD}) = \tau_{AB} \circ (\rho_{AB} \circ \rho_{CD}) \circ \tau_{CD} = \tau_{AB} \circ f \circ \tau_{CD},$$

where, by Propositions 6.2.1–2, f is a translation if $AB \parallel CD$ and a rotation by angle 2α otherwise. The desired results now follow from Proposition 6.1.1 in the first case and from Proposition 6.2.5 in the second case. Q.E.D.

Example 6.3.6 Identify the compositions $\gamma_{AD} \circ \gamma_{AB}$ and $\gamma_{AB} \circ \gamma_{CD}$, where $ABCD$ is the square of Figure 6.18.

By Proposition 6.3.5b, $\gamma_{AD} \circ \gamma_{AB}$ is a 180° rotation. Since

$$\gamma_{AD} \circ \gamma_{AB}(A) = \gamma_{AD}(B) = B',$$

it follows that the pivot point of this rotation is the midpoint M of AB'. Hence $\gamma_{AD} \circ \gamma_{AB} = R_{M,180^\circ}$. By Proposition 6.3.5a, $\gamma_{AB} \circ \gamma_{CD}$ is a translation. Since

$$\gamma_{AB} \circ \gamma_{CD}(C) = \gamma_{AB}(D) = D',$$

it follows that $\gamma_{AB} \circ \gamma_{CD} = \tau_{CD'}$.

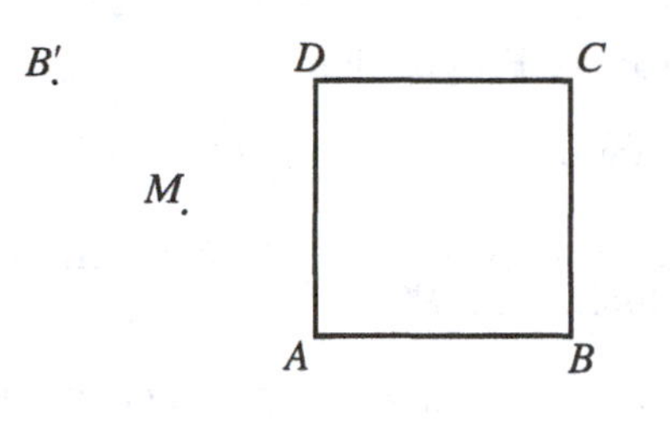

Figure 6.18.

Exercises 6.3

Identify the compositions of Exercises 1–18, where ABCD is the square of Figure 6.18.

1. $R_{D,90^\circ} \circ \gamma_{DC}$
2. $\gamma_{DC} \circ R_{D,90^\circ}$
3. $R_{P,180^\circ} \circ \gamma_{AB}$
4. $\gamma_{AB} \circ R_{P,180^\circ}$
5. $\tau_{AB} \circ \gamma_{DC}$
6. $\gamma_{CD} \circ \tau_{AB}$
7. $\tau_{AB} \circ \gamma_{BC}$
8. $\gamma_{BC} \circ \tau_{BA}$
9. $\gamma_{AD} \circ \gamma_{BC}$
10. $\gamma_{AD} \circ \gamma_{CB}$
11. $\gamma_{BA} \circ \gamma_{BC}$
12. $\gamma_{CB} \circ \gamma_{BA}$
13. $\gamma_{AC} \circ \gamma_{BD}$
14. $\rho_{AD} \circ \tau_{AB}$
15. $\rho_{AD} \circ R_{C,90^\circ}$
16. $\gamma_{CD} \circ \gamma_{BC} \circ \gamma_{AB}$
17. $\gamma_{CD} \circ \tau_{BC} \circ \gamma_{AB}$
18. $\tau_{CD} \circ \gamma_{BC} \circ \tau_{AB}$
19. If k, m, n are the perpendicular bisectors of the sides AB, BC, CA of $\triangle ABC$, respectively, show that $\rho_k \circ \rho_m \circ \rho_n$ is a reflection. What is the axis of this reflection?
20. Let A and B be any two distinct points. Prove that the composition $R_{B,180^\circ} \circ \rho_{AB} \circ R_{A,180^\circ}$ is a glide reflection and find its axis.
21. Show that the composition of the reflections in the three angle bisectors of a triangle is a reflection whose axis is perpendicular to one of the triangle's sides.
22. Let n be an odd integer and let $A_1, A_2, \ldots, A_n$ be the successive vertices of a regular n-sided polygon, and let m_i be the bisector of the interior angle at A_i. Identify the composition $\rho_{m_n} \circ \cdots \circ \rho_{m_2} \circ \rho_{m_1}$.

23. Complete the proof of Proposition 6.3.1.

24. Complete the proof of Proposition 6.3.3.

25. Prove that if γ_{AB} is a glide reflection and $\gamma_{AB}(P) = P'$, then the axis $\overleftrightarrow{AB}$ contains the midpoint of PP'.

26. Show that $\rho_k \circ \rho_m \circ \rho_n = \rho_n \circ \rho_m \circ \rho_k$ whenever the lines k, m, n are either concurrent or parallel.

27. Show that the composition of an even number of glide reflections is either a rotation or a translation.

28. Show that the composition of an odd number of glide reflections is a glide reflection.

6.4 The Main Theorems

Enough tools are now available to demonstrate that there are no Euclidean motions above and beyond those described so far.

Proposition 6.4.1 *Suppose $\triangle ABC \cong \triangle DEF$. Then there exists a sequence of no more than three reflections such that the composition of these reflections maps the points A, B, C onto the points D, E, F, respectively.*

PROOF: If the two given triangles are identical, then the composition of two identical reflections will clearly accomplish the required task.

If the two triangles share exactly two vertices, then it may be assumed that their relative position is described by Figure 6.19. In that case ρ_{AB} itself constitutes the required sequence of reflections.

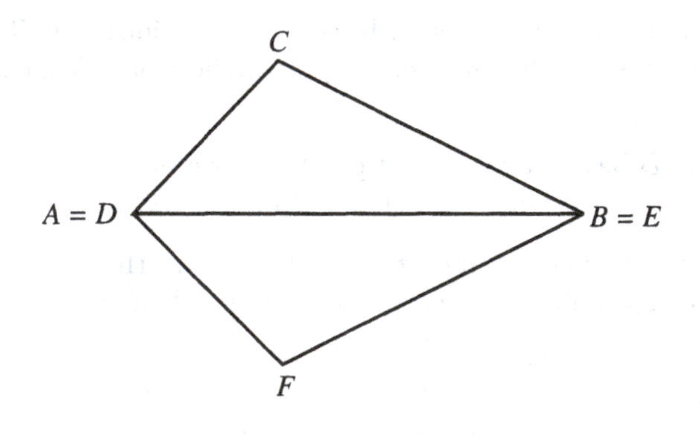

Figure 6.19.

If the two triangles share exactly one vertex, say $A = D$, let M be the midpoint of the segment BE. Then the reflection ρ_{AM} fixes A and transforms $\triangle ABC$ into $\triangle DEF'$ that shares at least two vertices with $\triangle DEF$. It follows from the previous argument that at most one more reflection will be required to transform $\triangle DEF'$ into $\triangle DEF$.

Finally, if the two triangles share no vertices, let m be the perpendicular bisector of the segment AD. The reflection ρ_m then transforms $\triangle ABC$ into $\triangle DE'F'$, which shares at least one vertex with $\triangle DEF$. By the preceding argument at most two more reflections will transform $\triangle DE'F'$ into $\triangle DEF$. It follows that at most three reflections are required to transform $\triangle ABC$ into $\triangle DEF$. Q.E.D.

The following is this chapter's main theorem.

Theorem 6.4.2 *Every motion is the composition of at most three reflections.*

PROOF: Let f be a motion, let A, B, C be three noncollinear points, and set $A' = f(A)$, $B' = f(B)$, $C' = f(C)$. Since $\triangle ABC \cong \triangle A'B'C'$, it follows from Proposition 6.4.1 that there exist at most three reflections whose composition, say g, also transforms A, B, C onto A', B', C', respectively. It follows from Theorem 6.1.4 that $f = g$. Q.E.D.

The following classification theorem is a consequence of the above. Recall that every reflection is a special kind of glide reflection.

Theorem 6.4.3 *Every motion is either a translation, a rotation, or a glide reflection.*

PROOF: The composition of no reflections is the identity that can be viewed as either a rotation $R_{A,0°}$ or a translation τ_{AA}. The composition of one reflection is a glide reflection. The composition of two reflections is, by Propositions 6.2.1–2, either a translation or a rotation. It follows that the composition of three reflections is also the composition of a reflection with either a translation or a rotation, which, by Propositions 6.3.1 and 6.3.3, is a glide reflection. Q.E.D.

Example 6.4.4 Let f and g be two motions. Prove that g is a reflection if and only if $f \circ g \circ f^{-1}$ is a reflection.

Suppose first that g is a reflection. By Theorem 6.4.3, f is either a translation, a rotation, or a glide reflection. In the first two cases it follows from Propositions 6.3.1 and 6.3.3 that $f \circ g \circ f^{-1}$ is also a glide reflection. The same conclusion can be drawn in the third case, if f is a glide reflection, but this time Proposition 6.3.5 is also needed. In order to show that $f \circ g \circ f^{-1}$ is a reflection, it suffices to show that it has a fixed point. Let P be any fixed point of the reflection g and set $P' = f(P)$. Then

$$f \circ g \circ f^{-1}(P') = f \circ g \circ f^{-1}(f(P)) = f \circ g(P) = f(P) = P',$$

so that P' is the requisite fixed point of $f \circ g \circ f^{-1}$.

Conversely, suppose $f \circ g \circ f^{-1}$ is a reflection. It then follows from the preceding that g is also a reflection because

$$f^{-1} \circ (f \circ g \circ f^{-1}) \circ (f^{-1})^{-1} = f^{-1} \circ f \circ g \circ f^{-1} \circ f = g.$$

In conclusion, we point out that the definitions of rotations and reflections as well as the proofs of Propositions 6.1.4, 6.4.1–2 are all neutral and hence they also hold for the hyperbolic plane. In particular, *every motion of the hyperbolic plane is the composition of at most three hyperbolic reflections.* These hyperbolic reflections will be described in the next chapter.

Exercises 6.4

In the following exercises, f and g denote two motions.

1. Prove that g is a glide reflection if and only if $f \circ g \circ f^{-1}$ is a glide reflection.
2. Prove that g is a rotation if and only if $f \circ g \circ f^{-1}$ is a rotation.
3. Prove that g is a translation if and only if $f \circ g \circ f^{-1}$ is a translation.
4. Prove that $f \circ g$ is a glide reflection if and only if $g \circ f$ is a glide reflection.
5. Prove that $f \circ g$ is a rotation if and only if $g \circ f$ is a rotation.
6. Prove that $f \circ g$ is a translation if and only if $g \circ f$ is a translation.
7. Is it true that $f \circ g$ is a reflection if and only if $g \circ f$ is a reflection? Justify your answer?
8. Is it true that $f \circ g$ is a translation if and only if both f and g are translations? Justify your answer.
9. Is it true that $f \circ g$ is a rotation if and only if both f and g are rotations? Justify your answer.
10. Is it true that $f \circ g$ is a reflection if and only if both f and g are reflections? Justify your answer.

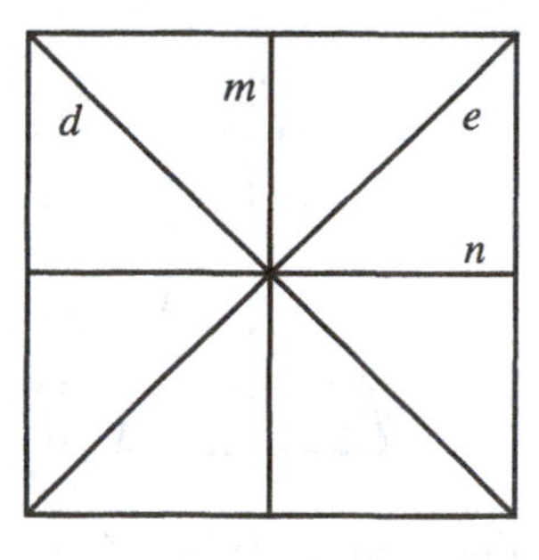

Figure 6.20. Some symmetries of the square

6.5 Symmetries of Polygons

A (*mathematical*) *symmetry* of a figure Φ is a motion f such that

$$f(\Phi) = \Phi.$$

Thus, the square of Figure 6.20 possesses the symmetries ρ_d, ρ_e, ρ_m, and ρ_n. This is the mathematical formalization of the more intuitive observation that the square is symmetrical about its diagonals and about the lines joining the midpoints of its opposite sides. However, the mathematical definition of symmetry is broader than the common usage of the term. If C denotes the geometrical center of the square, then the rotations $R_{C,90^\circ}$, $R_{C,180^\circ}$, $R_{C,270^\circ}$ (see Fig. 6.21) all rotate the square back onto itself and so they too constitute mathematical symmetries, even though they wouldn't be recognized as symmetries by the proverbial person in the street. The identity motion Id is another such symmetry. The set of all the symmetries of a figure is called its *symmetry group* or just *group*. Thus, the symmetry group of the square is

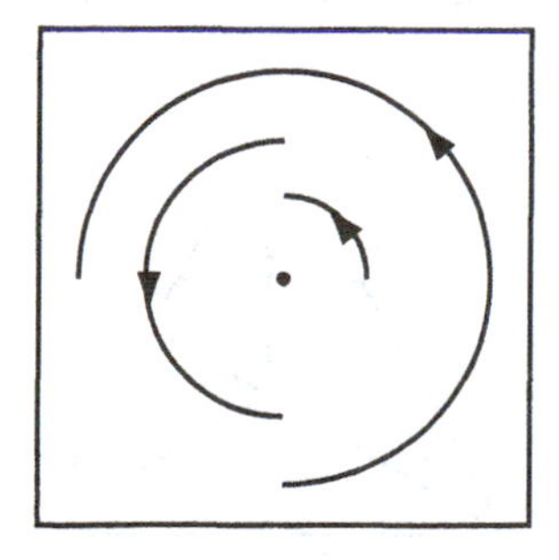

Figure 6.21. More symmetries of the square

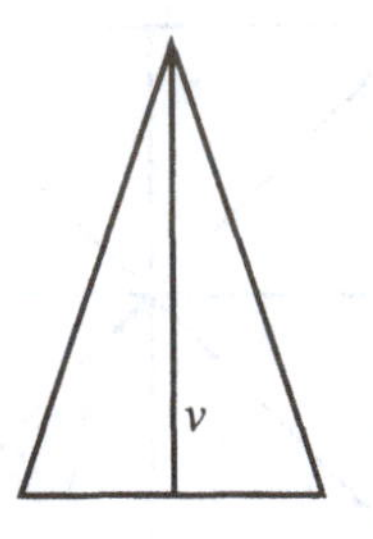

Figure 6.22. The symmetries of an isosceles triangle

$$\{\text{Id}, \rho_d, \rho_e, \rho_m, \rho_n, R_{C,90^\circ}, R_{C,180^\circ}, R_{C,270^\circ}\}.$$

By definition, every plane figure Φ has a symmetry group that contains at least the identity motion Id. The isosceles triangle of Figure 6.22 has $\{\text{Id}, \rho_v\}$ as its symmetry group whereas that of the equilateral triangle of Figure 6.23 is $\{\text{Id}, \rho_d, \rho_e, \rho_f, R_{C,120^\circ}, R_{C,240^\circ}\}$.

While figures of finite extent cannot have either translations or glide reflections as their symmetries, infinitely extended figures do admit such symmetries and a variety of interesting examples will be discussed in the next two sections. This section, however, is concerned with finite figures only, and for the symmetries of such polygons there is a useful algebraic description that is obtained by restricting attention to the action of the symmetry on the polygon's vertices. This action is described by means of the positions occupied by the vertices. Thus, if the *positions* occupied by the four vertices of the square are labeled 1, 2, 3, 4, respectively (Fig. 6.24), then any symmetry f of the square can be thought of as a function

$$f : \{1, 2, 3, 4\} \to \{1, 2, 3, 4\},$$

where, for each $i = 1, 2, 3, 4$, $f(i)$ denotes the new position of the vertex that was, prior to the execution of f, in position i. Accordingly (see Appendix

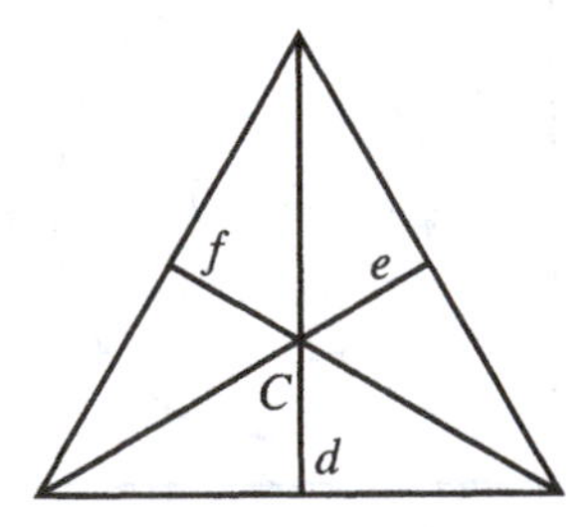

Figure 6.23. The reflections of an equilateral triangle

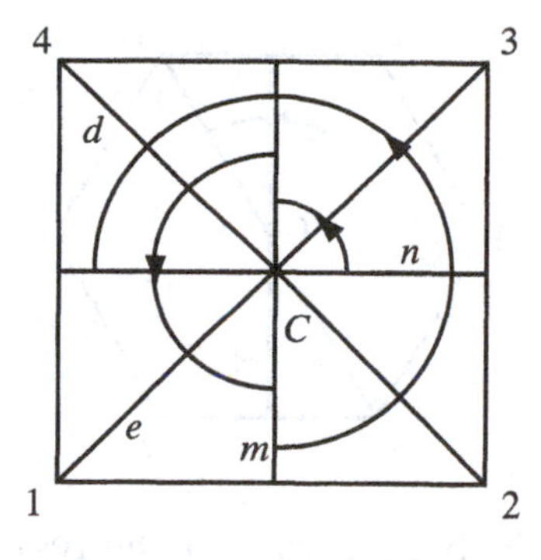

Figure 6.24. The symmetries of the square

F), the eight symmetries of the square have the following permutation representations:

$$\begin{aligned}
\text{Id} &= (1)(2)(3)(4)\\
\rho_d &= (1\ \ 3)\,(2)(4)\\
\rho_e &= (1)\,(2\ \ 4)\,(3)\\
\rho_m &= (1\ \ 2)\,(3\ \ 4)\\
\rho_n &= (1\ \ 4)\,(2\ \ 3)\\
R_{C,90^\circ} &= (1\ \ 2\ \ 3\ \ 4)\\
R_{C,180^\circ} &= (1\ \ 3)\,(2\ \ 4)\\
R_{C,270^\circ} &= (1\ \ 4\ \ 3\ \ 2)\,.
\end{aligned}$$

Note that this involves some abuse of notation as the same symbol f is being used to denote both the symmetry as it acts on the whole plane and its restriction to the vertices alone. This will lead to no difficulties and obviates the need for a new notation.

Mathematicians and physical scientists have a great interest in groups of symmetries of solids in spaces of an arbitrary number of dimensions and their classifications. The composition operation plays an important role in the classification of both the motions of the plane and the symmetry groups. The advantage of the permutation representations is that they allow for an algebraic representation of composition. Thus, since $\rho_e = (1)\,(2\ \ 4)\,(3)$ and $\rho_m = (1\ \ 3)\,(2\ \ 4)$ it follows that

$$\rho_e \circ \rho_m = (1)\,(2\ \ 4)\,(3) \circ (1\ \ 3)\,(2\ \ 4) = (1\ \ 3)\,(2)(4) = \rho_d$$

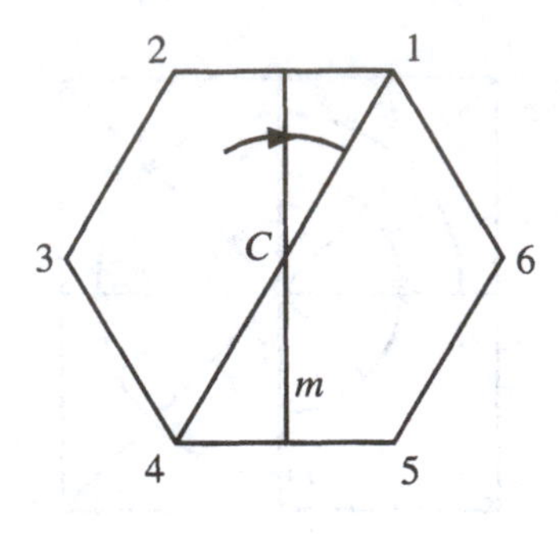

Figure 6.25. Symmetries of the regular hexagon

and since $R_{C,90^\circ} = \begin{pmatrix}1 & 2 & 3 & 4\end{pmatrix}$ it follows that

$$\rho_e \circ R_{C,90^\circ} = (1)\begin{pmatrix}2 & 4\end{pmatrix}(3) \circ \begin{pmatrix}1 & 2 & 3 & 4\end{pmatrix} = \begin{pmatrix}1 & 4\end{pmatrix}\begin{pmatrix}2 & 3\end{pmatrix} = \rho_n.$$

Similarly, the composition of the reflection $\rho_{14} = (1)\begin{pmatrix}2 & 6\end{pmatrix}\begin{pmatrix}3 & 5\end{pmatrix}(4)$ and the rotation $R_{C,-60^\circ} = \begin{pmatrix}1 & 6 & 5 & 4 & 3 & 2\end{pmatrix}$, both symmetries of the regular hexagon of Figure 6.25, is

$$\begin{aligned}\rho_{14} \circ R_{C,-60^\circ} &= (1)\begin{pmatrix}2 & 6\end{pmatrix}\begin{pmatrix}3 & 5\end{pmatrix}(4) \circ \begin{pmatrix}1 & 6 & 5 & 4 & 3 & 2\end{pmatrix} \\ &= \begin{pmatrix}1 & 2\end{pmatrix}\begin{pmatrix}3 & 6\end{pmatrix}\begin{pmatrix}4 & 5\end{pmatrix} = \rho_m.\end{aligned}$$

Exercises 6.5

1. Give the symmetry groups of the following figures:
   (a) the rectangle with unequal sides
   (b) the regular pentagon
   (c) the regular hexagon
   (d) the regular heptagon
   (e) the regular octagon

2. Identify the following compositions of the symmetries of the square of Figure 6.24. Describe them both geometrically and with permutation representations.

   (a) $\rho_m \circ \rho_e$ (b) $R_{C,90^\circ} \circ \rho_m$ (c) $\rho_n \circ \rho_e$
   (d) $\rho_e \circ \rho_d$ (e) $\rho_m \circ R_{C,180^\circ}$ (f) $\rho_n \circ R_{C,180^\circ}$
   (g) $R_{C,180^\circ} \circ \rho_e$ (h) $R_{C,90^\circ} \circ R_{C,180^\circ}$ (i) $\rho_e \circ R_{C,180^\circ}$

3. Identify the following compositions of the symmetries of the regular hexagon of Figure 6.25. Describe them both geometrically and with permutation representations.

(a)	$\rho_m \circ \rho_{14}$	(b)	$R_{C,60^\circ} \circ \rho_m$	(c)	$\rho_{14} \circ \rho_{36}$
(d)	$\rho_{25} \circ \rho_m$	(e)	$\rho_m \circ R_{C,180^\circ}$	(f)	$\rho_m \circ R_{C,120^\circ}$
(g)	$R_{C,240^\circ} \circ \rho_{25}$	(h)	$R_{C,-60^\circ} \circ R_{C,180^\circ}$	(i)	$\rho_{14} \circ R_{C,180^\circ}$

6.6 Frieze Patterns

A *frieze pattern* is a one-dimensional repeating figure. More formally, a frieze pattern is generated by a finite figure Φ, called a *block*, and a translation τ. The pattern itself consists of the union of all the figures

$$\ldots, \tau^{-2}(\Phi), \tau^{-1}(\Phi), \Phi, \tau(\Phi), \tau^2(\Phi), \ldots .$$

These frieze patterns are the mathematical idealization of such decorative designs as borders used to accent wallpapers and trim sewn or printed around a cloth (Fig. 6.26). However, unlike their physical manifestations, frieze patterns are understood to extend indefinitely in both directions, just like a straight line.

The frieze pattern created by the repetition of a block Φ is denoted by $\wp(\Phi)$ and it inherits some of the symmetries of Φ (see Exercises 1–3). This observation, however, does not account for *all* the symmetries of the frieze pattern $\wp(\Phi)$. By definition, every such pattern possesses its generating translation τ as a symmetry, since this translation shifts the infinitely extended pattern onto itself. In the case of the block Φ_1 of Figure 6.27, the frieze pattern has no other symmetries, and so its symmetry group is denoted by $\Gamma_1 = \langle \tau \rangle$.

Block Φ_2 of Figure 6.28 possesses the symmetry ρ_h (h for horizontal), which is, of course, also a symmetry of its frieze pattern. In addition, this pattern also necessarily possesses the composite glide reflection $\gamma = \rho_h \circ \tau$ as a symmetry. This frieze's symmetry group is denoted by $\Gamma_2 = \langle \tau, \rho_h, \gamma \rangle$. The symmetry ρ_v of Φ_3 of Figure 6.29 results in a multitude of symmetries of the frieze $\wp(\Phi_3)$, which are all essentially identical. It should be noted, however, that this frieze pattern possesses an additional symmetry, namely the reflection $\rho_{v'}$, which has no counterpart in the generating block Φ_3. Because of its similarity to ρ_v, the symmetry $\rho_{v'}$ is not listed in the symmetry group $\Gamma_3 = \langle \tau, \rho_v \rangle$ of this pattern. Such an additional reflection could not have appeared with the horizontal reflection of Φ_2, but similar "accidental"

Figure 6.26. Chinese ornamental friezes

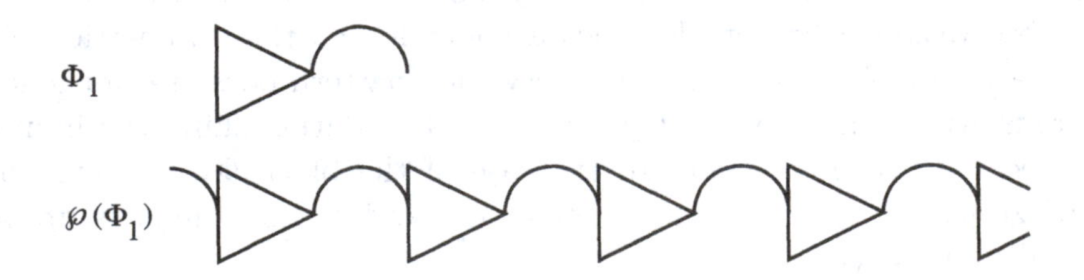

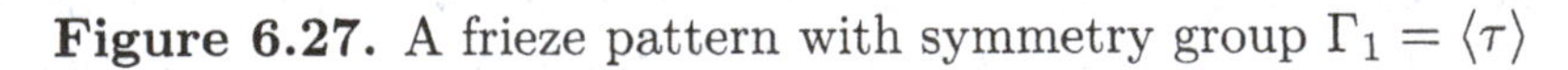

Figure 6.27. A frieze pattern with symmetry group $\Gamma_1 = \langle \tau \rangle$

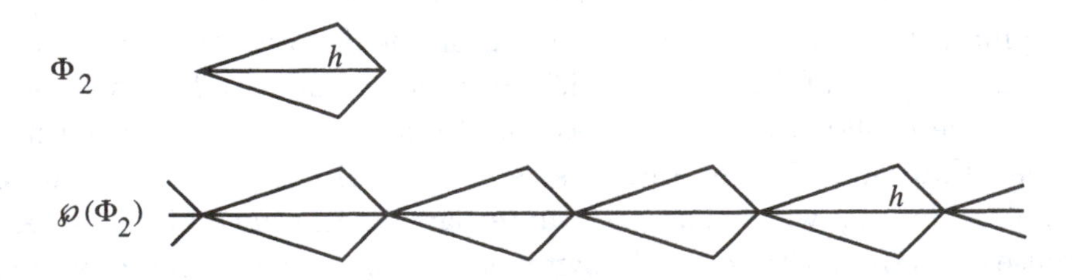

Figure 6.28. A frieze pattern with symmetry group $\Gamma_2 = \langle \tau, \rho_h, \gamma \rangle$

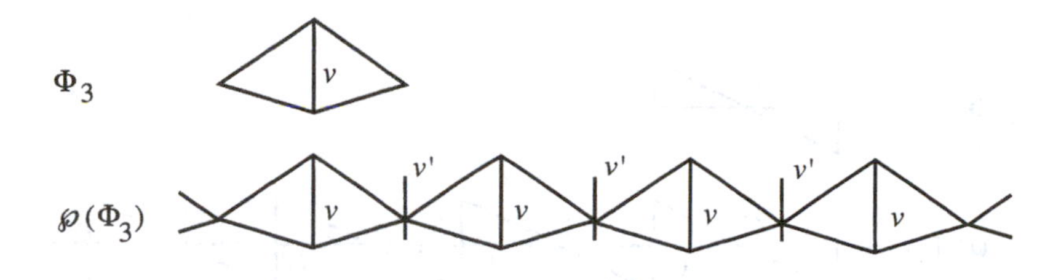

Figure 6.29. A frieze pattern with symmetry group $\Gamma_3 = \langle \tau, \rho_v \rangle$.

symmetries can arise in other cases, as will be seen. The symmetry $R_{C,180°}$ of the block Φ_4 of Figure 6.30 results in the symmetries $R = R_{C,180°}$ of the frieze pattern. Once again the frieze pattern has the additional symmetry $R_{C',180°}$. This frieze pattern's symmetry group is $\Gamma_4 = \langle \tau, R \rangle$. The block Φ_5 of Figure 6.31 possesses all the symmetries, ρ_h, ρ_v and R. Its symmetry group is $\Gamma_5 = \langle \tau, \rho_h, \rho_v, R, \gamma \rangle$. Just like Γ_2, the next two patterns of Figure 6.32 and Figure 6.33 have symmetry groups $\Gamma_6 = \langle \tau, \gamma \rangle$ and $\Gamma_7 = \langle \tau, \gamma, R \rangle$ that contain glide-reflections. However, unlike the glide-reflection of Γ_2, those of Γ_6 and Γ_7 do not have their component translation and reflection in the group.

The following theorem, attributed to Paul Niggli (1888–1953), states that the foregoing is a complete list of all the possible types of symmetry groups that frieze patterns can possess.

Proposition 6.6.1 (P. Niggli, 1926) *Every frieze pattern has a symmetry group that is identical with one of the groups*

$$\begin{aligned}
\Gamma_1 &= \langle \tau \rangle \\
\Gamma_2 &= \langle \tau, \rho_h, \gamma \rangle \\
\Gamma_3 &= \langle \tau, \rho_v \rangle \\
\Gamma_4 &= \langle \tau, R \rangle \\
\Gamma_5 &= \langle \tau, \rho_h, \rho_v, R, \gamma \rangle \\
\Gamma_6 &= \langle \tau, \gamma \rangle \\
\Gamma_7 &= \langle \tau, \gamma, R, \rho_v \rangle
\end{aligned}$$

Exercises 6.6

1. If the block Φ's frieze pattern $\wp(\Phi)$ has a reflectional symmetry with a horizontal axis, must Φ necessarily have the same symmetry?

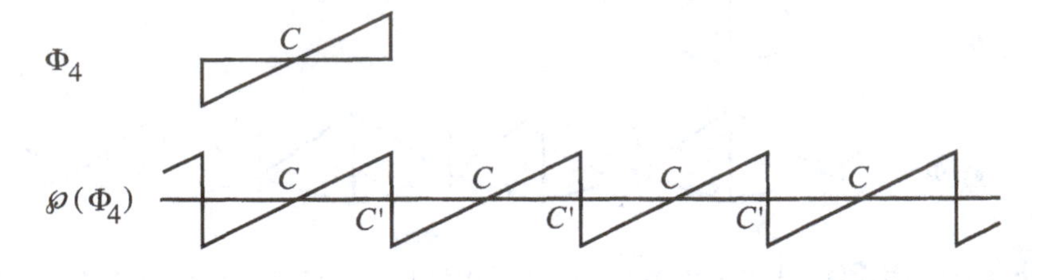

Figure 6.30. A frieze pattern with symmetry group $\Gamma_4 = \langle \tau, R \rangle$.

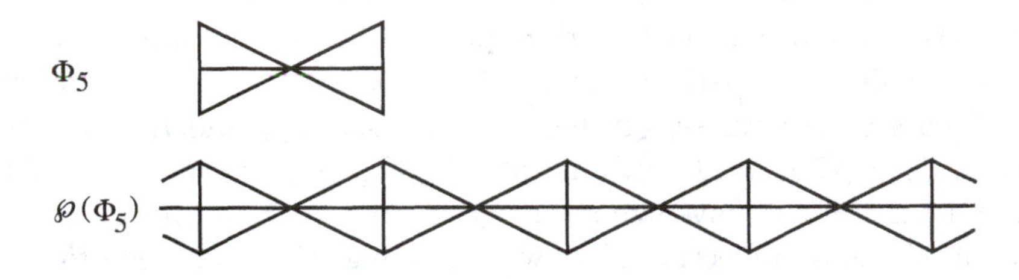

Figure 6.31. A frieze pattern with symmetry group $\Gamma_5 = \langle \tau, \rho_h, \rho_v, R, \gamma \rangle$.

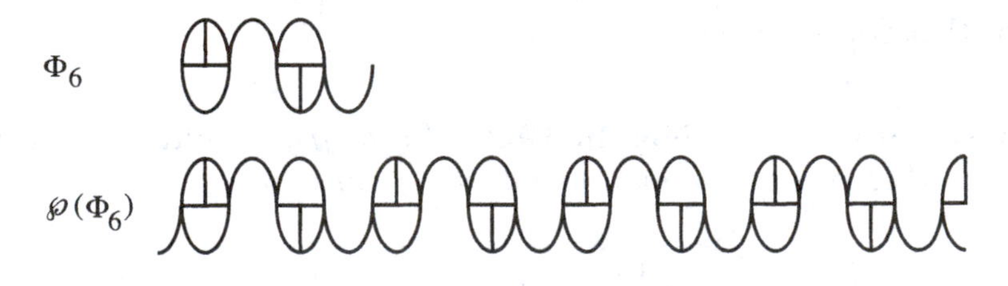

Figure 6.32. A frieze pattern with symmetry group $\Gamma_6 = \langle \tau, \gamma \rangle$.

Figure 6.33. A frieze pattern with symmetry group $\Gamma_7 = \langle \tau, \gamma, R, \rho_v \rangle$.

2. If the block Φ's frieze pattern $\wp(\Phi)$ has a reflectional symmetry with a vertical axis, must Φ necessarily have the same symmetry?

3. If the block Φ's frieze pattern $\wp(\Phi)$ has a rotational symmetry, must Φ necessarily have the same symmetry?

Identify the groups of the following frieze patterns.

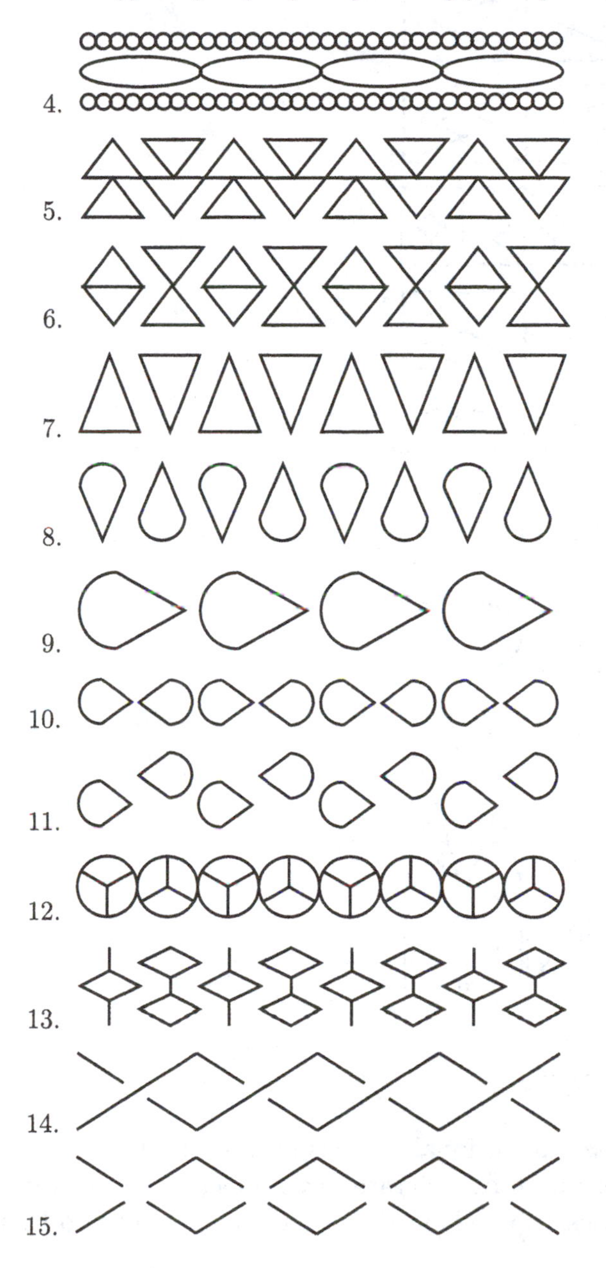

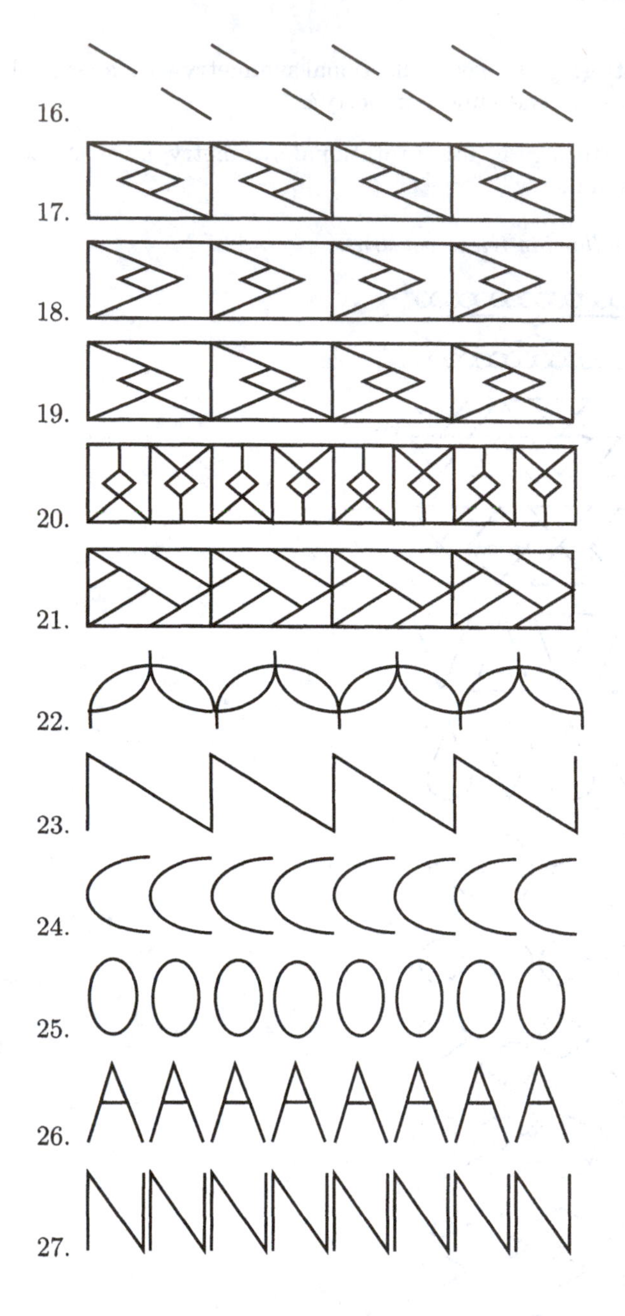

6.7 Wallpaper Designs

Wallpaper designs are the two-dimensional analogs of frieze patterns. More technically, let $\wp$ be the frieze pattern generated by a block Φ and a translation τ. If τ^* is another translation whose direction is not parallel to that of τ, then the union of the figures

$$\ldots, \tau^{*-2}(\wp), \tau^{*-1}(\wp), \wp, \tau^{*}(\wp), \tau^{*2}(\wp), \ldots$$

is the *wallpaper design* $W(\Phi)$ generated by Φ, τ, and τ^* (see Figs. 6.34 and 6.35). As their name implies, such designs are the mathematical patterns that underlie the repeating decorative artworks illustrated in Figure 6.36. Unlike the carpets and walls that carry these artworks, the mathematical wallpaper designs extend ad infinitum in all the directions of the plane.

It is clear that both τ and τ^* are symmetries of the wallpaper design they generate. As was the case for frieze patterns, the generated design $W(\Phi)$ may possess further symmetries that are not present in Φ. In contrast with the seven different groups of symmetries of frieze patterns, there are seventeen different possibilities for the symmetry groups of wallpaper designs. These, together with their labels, are exhibited in Figures 6.37–6.39. The presence of reflectional and glide-reflectional symmetries is denoted by a dashed line with a label of either ρ (for reflection) or γ (for glide reflection). The pivot points of rotational symmetries are denoted by the symbols $\Diamond$ (180°), $\triangle$ (120°), $\square$ (90°), ⬡ (60°). The chart at the end of the chapter lists the salient symmetry characteristics of each design. A rotation through an angle of $360°/n$ is said to have *order* n. A glide reflection is said to be *nontrivial* if its component translation and reflection are not symmetries of the pattern.

The symbols *p1*, *pgg*, *p31m*, ... are used to denote both a type of wallpaper design and its symmetry group. This is known as the crystallographic notation for the symmetry groups. If the second character in this symbol is an integer, it is the highest order of all the rotations in that group. The significance of the other characters is too complicated to explain here.

The following is the two-dimensional analog of Proposition 6.6.1.

Proposition 6.7.1 *There are exactly seventeen wallpaper symmetry groups.*

□

Figures 6.37–6.39 display one illustration for each of these groups.

Proposition 6.7.1 was first discovered in 1891 by Evgraf Stepanovich Fedorov (1853–1919), 35 years before Niggli stated and proved its one-dimensional analog on frieze groups (Proposition 6.6.1). Curiously, this work had been preceded by Fedorov's and Arthur Schönflies's (1853–1928) independent classifications of the 230 crystallographic groups, these being the three-dimensional analogs of the wallpaper groups. It has since then been established that there are exactly 4783 classes of such groups in four-dimensional

Φ

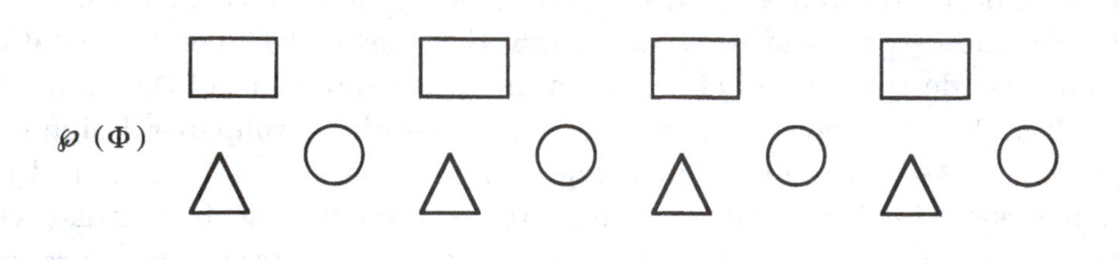

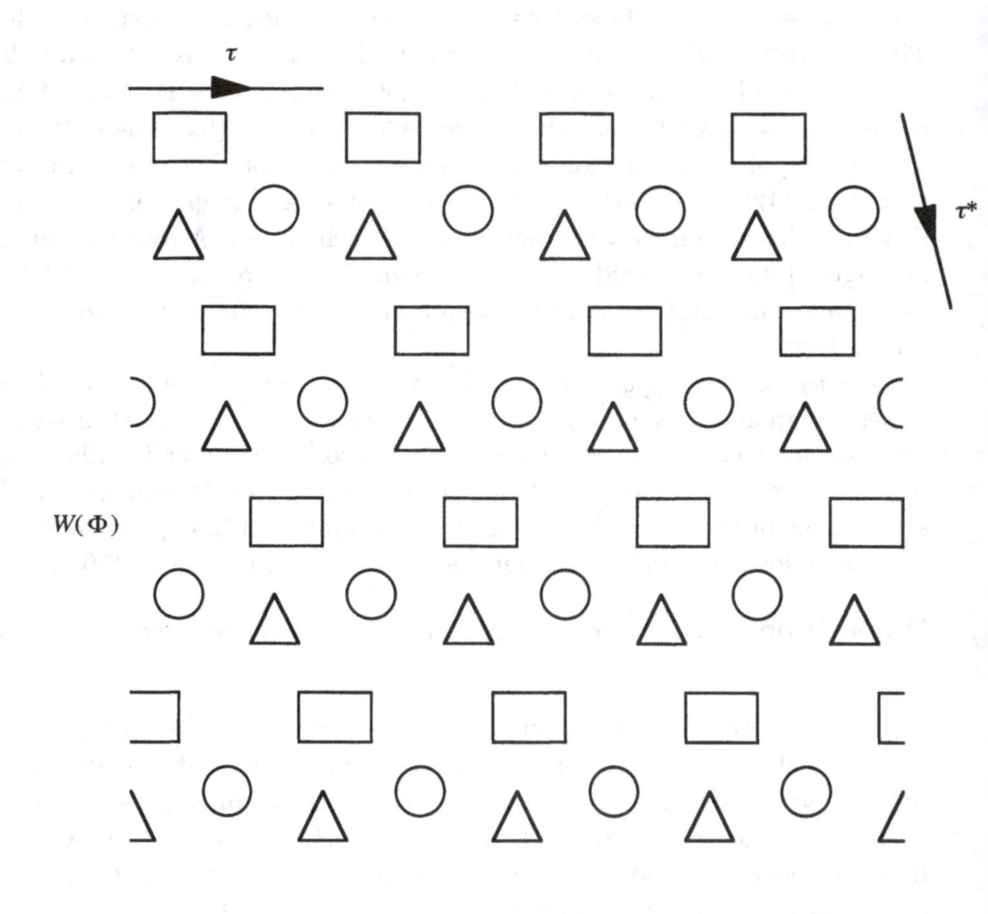

Figure 6.34. A wallpaper design

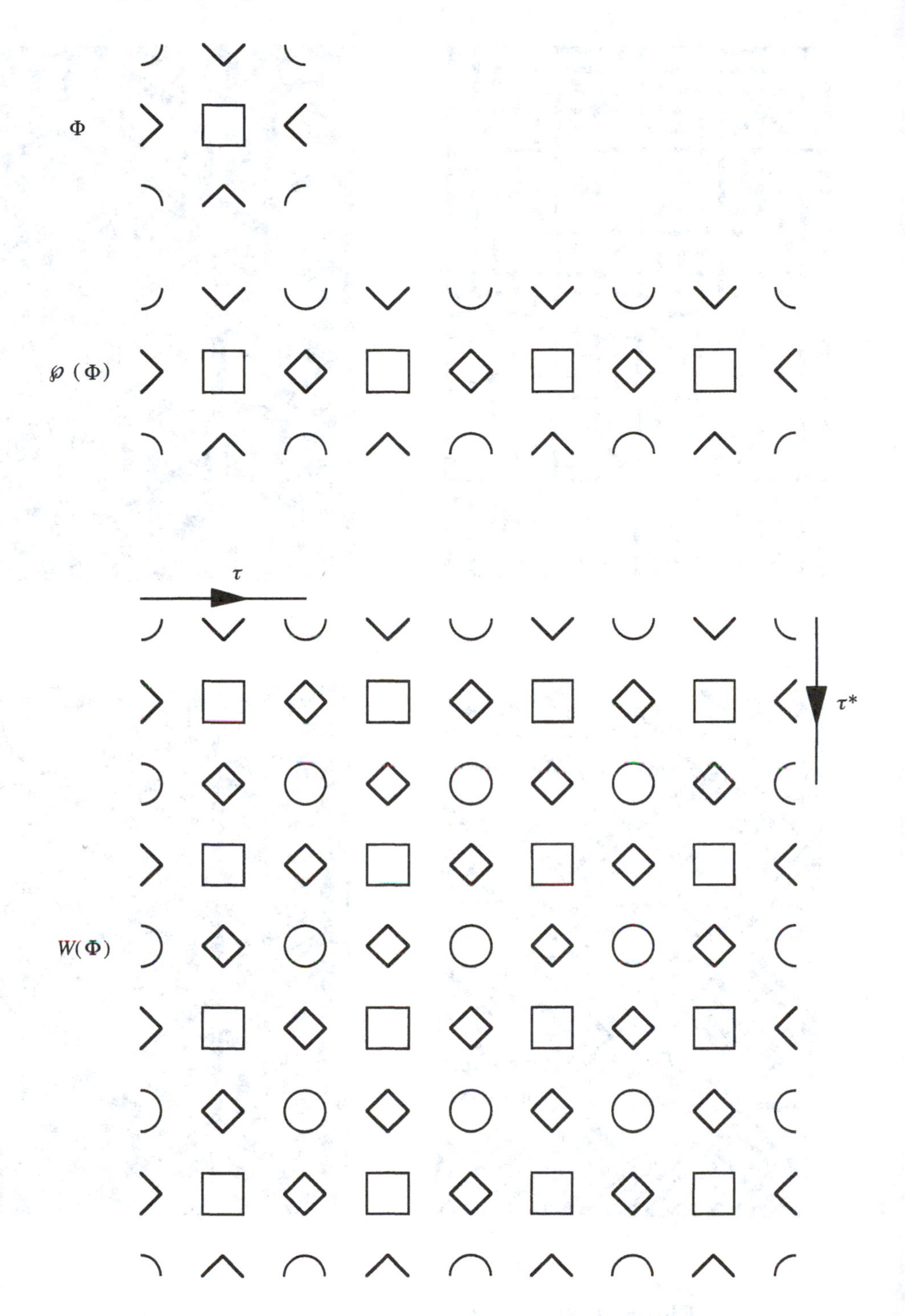

Figure 6.35. A wallpaper design

Figure 6.36. Middle Eastern ornamental designs

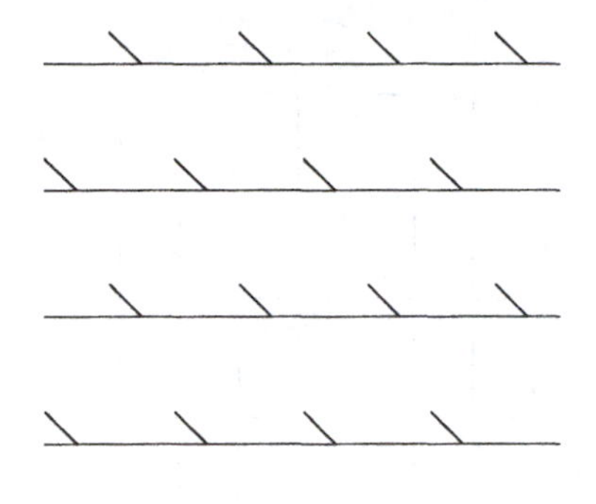

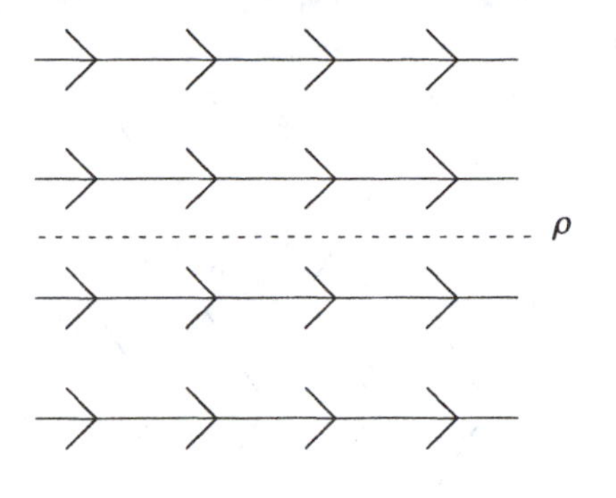

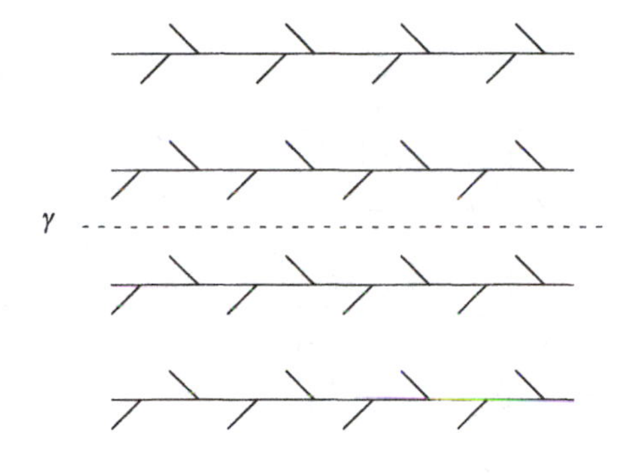

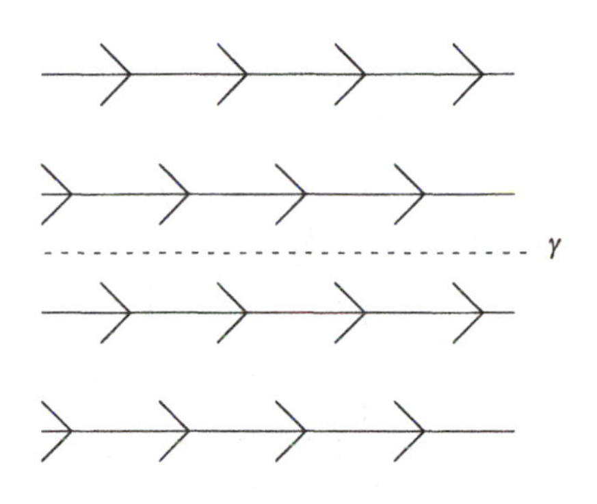

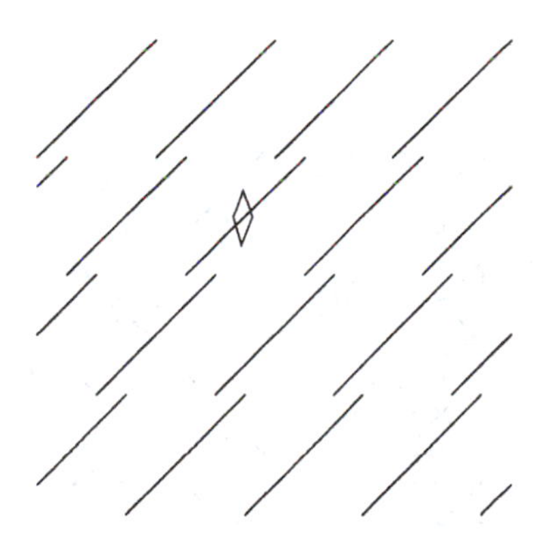

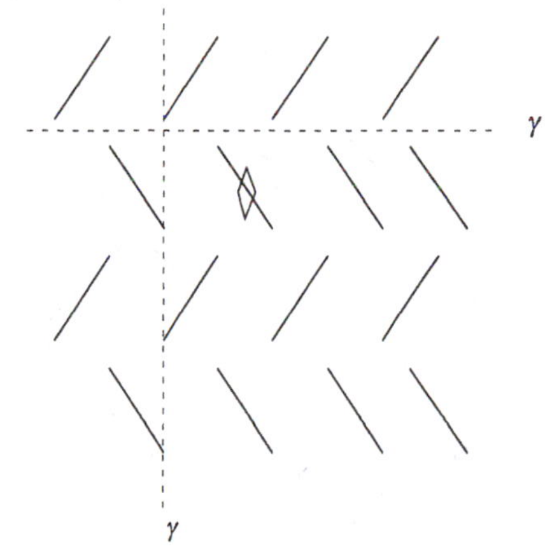

Figure 6.37.

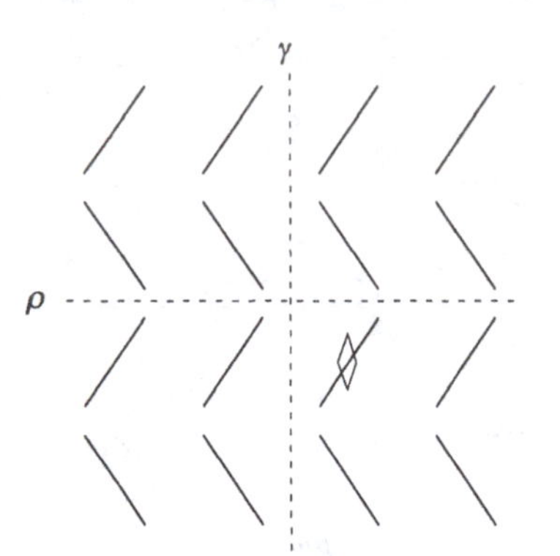

pmg

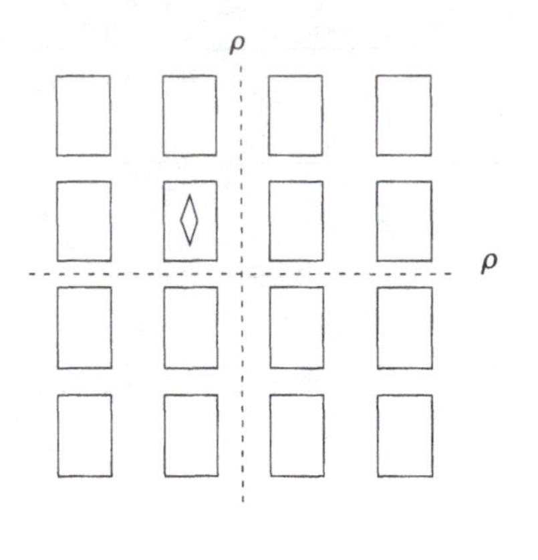

pmm

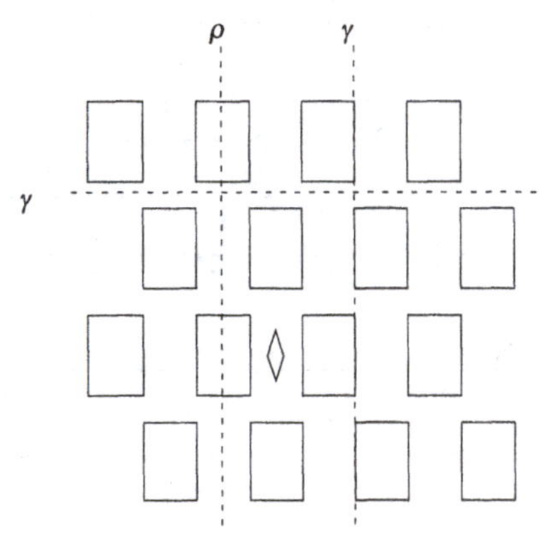

cmm

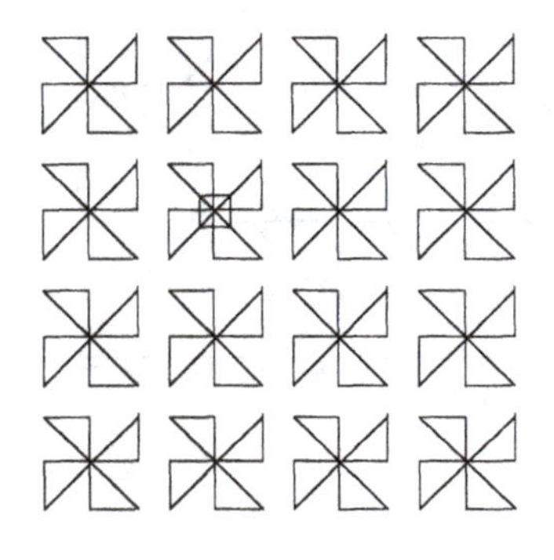

p4

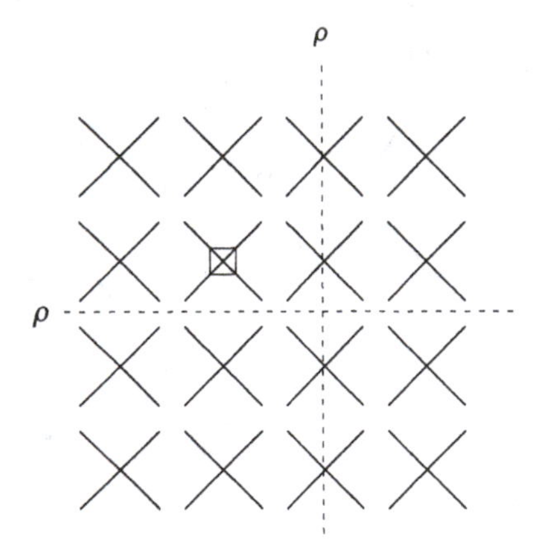

p4m

p4g

Figure 6.38.

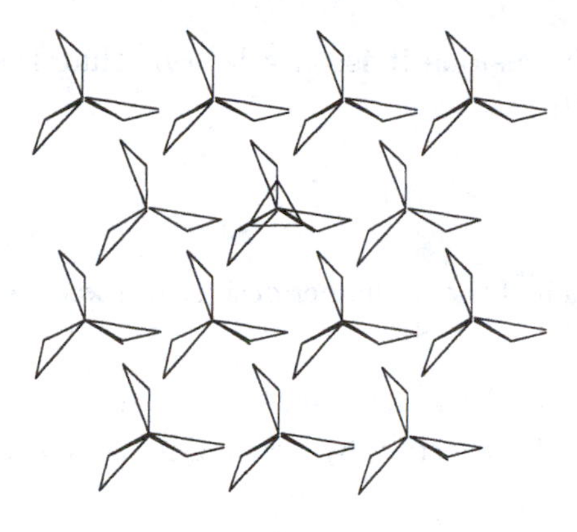

p3

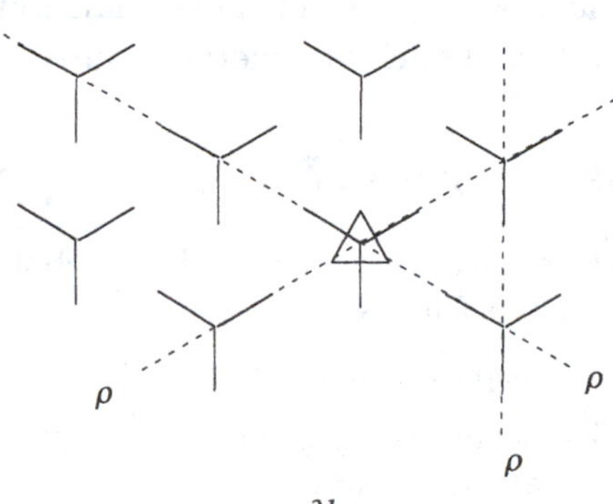

p31m

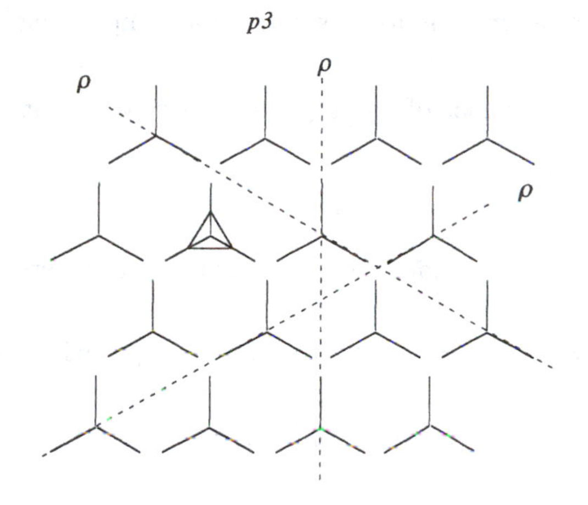

p3m1

p6

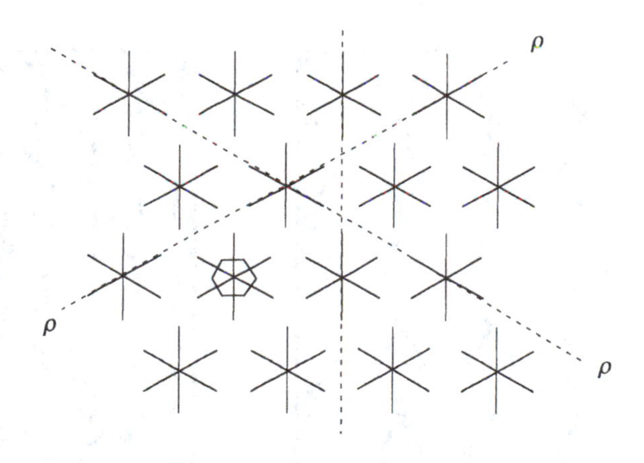

p6m

Figure 6.39.

space. For spaces of more than four dimensions it is only known that the number of such symmetry groups is finite.

Exercises 6.7

Determine the crystallographic symbol of each of the wallpaper designs of Exercises 1–34. In each case

(a) display a rotation of the highest order

(b) denote the presence of a nontrivial glide reflection by drawing its axis with an accompanying γ

(c) denote the presence of a reflection by drawing its axis with an accompanying ρ

(d) avoid redundancy by only drawing one axis of any type in any one direction

Chapter Review Exercises

1. Find the image of the line $y = x + 1$ under the translation τ_{AB}, where $A = (2,0)$ and $B = (0,2)$.
2. Find the image of the line $y = x + 1$ under the rotation $R_{A,90^\circ}$, where $A = (2,0)$.
3. Find the image of the line $y = x+1$ under the reflection ρ_{AB}, where $A = (2,0)$ and $B = (0,2)$.
4. Find the image of the line $y = x + 1$ under the glide reflection γ_{AB}, where $A = (2,0)$ and $B = (0,2)$.

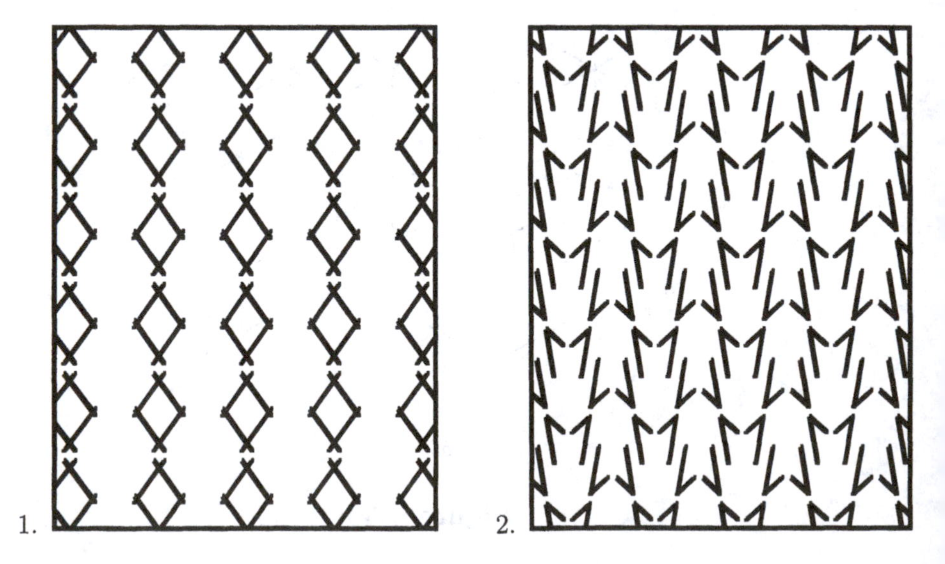

1. 2.

3.
4.
5.
6.

7.
8.
9.
10.

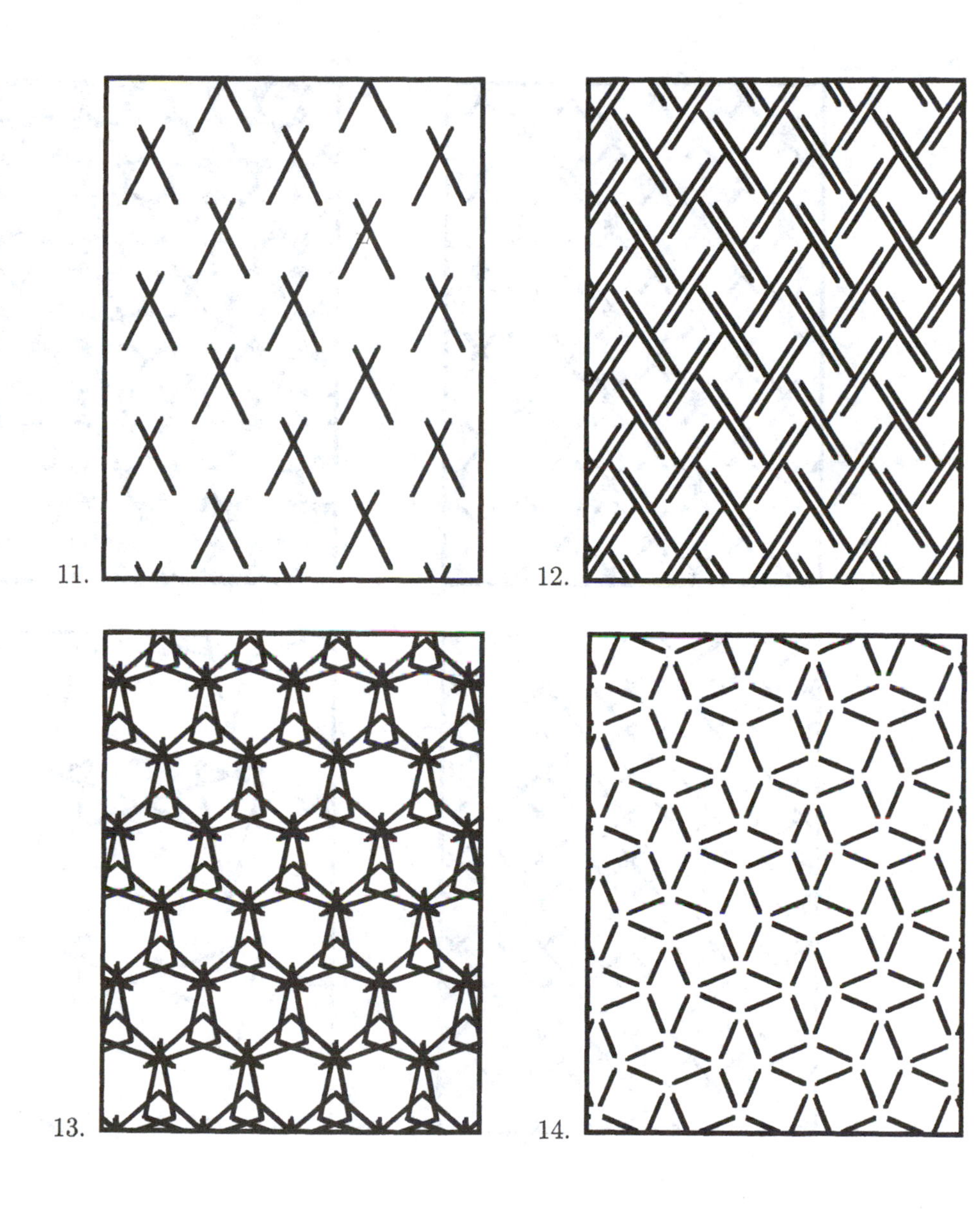
11.
12.
13.
14.

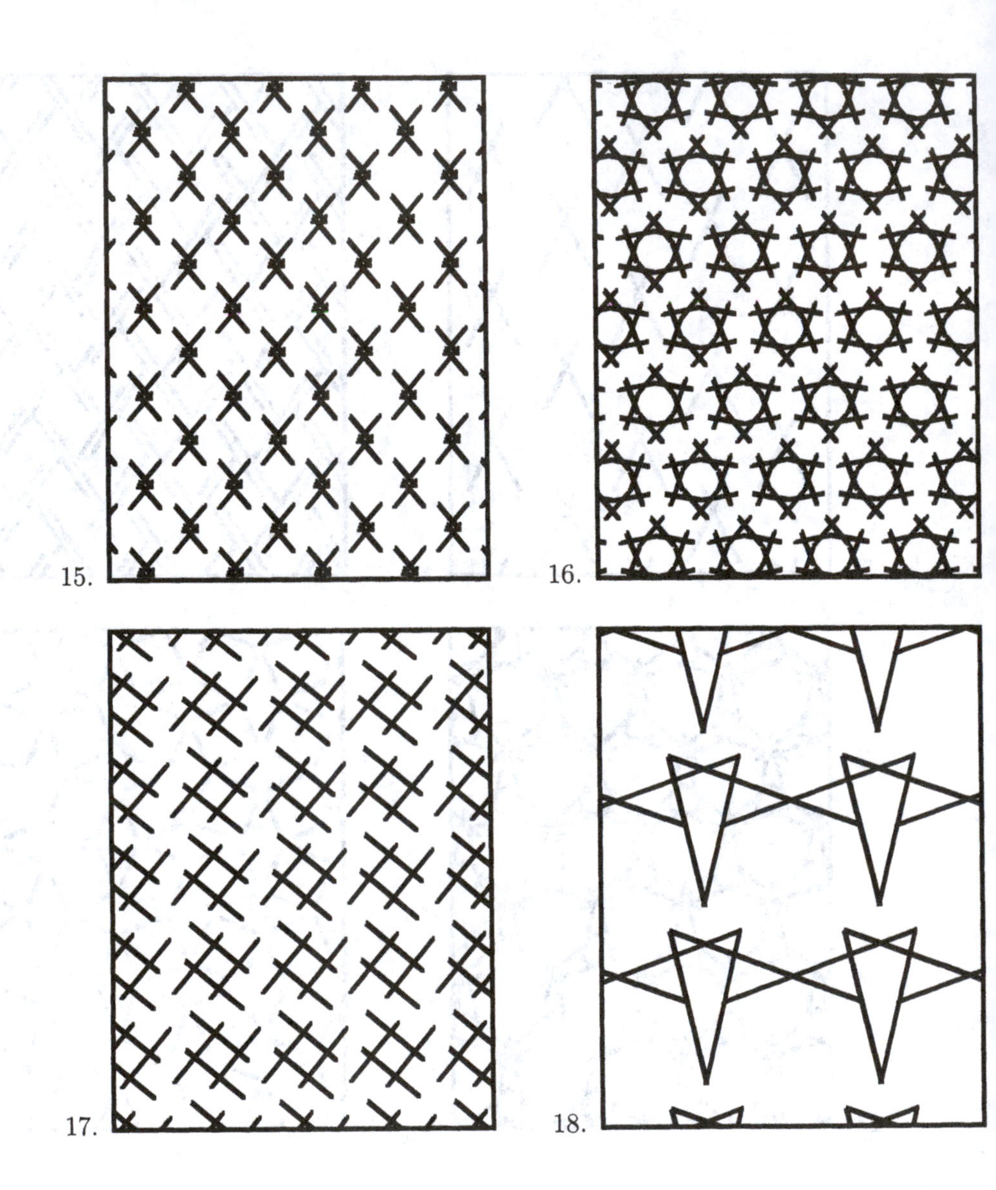
15.
16.
17.
18.

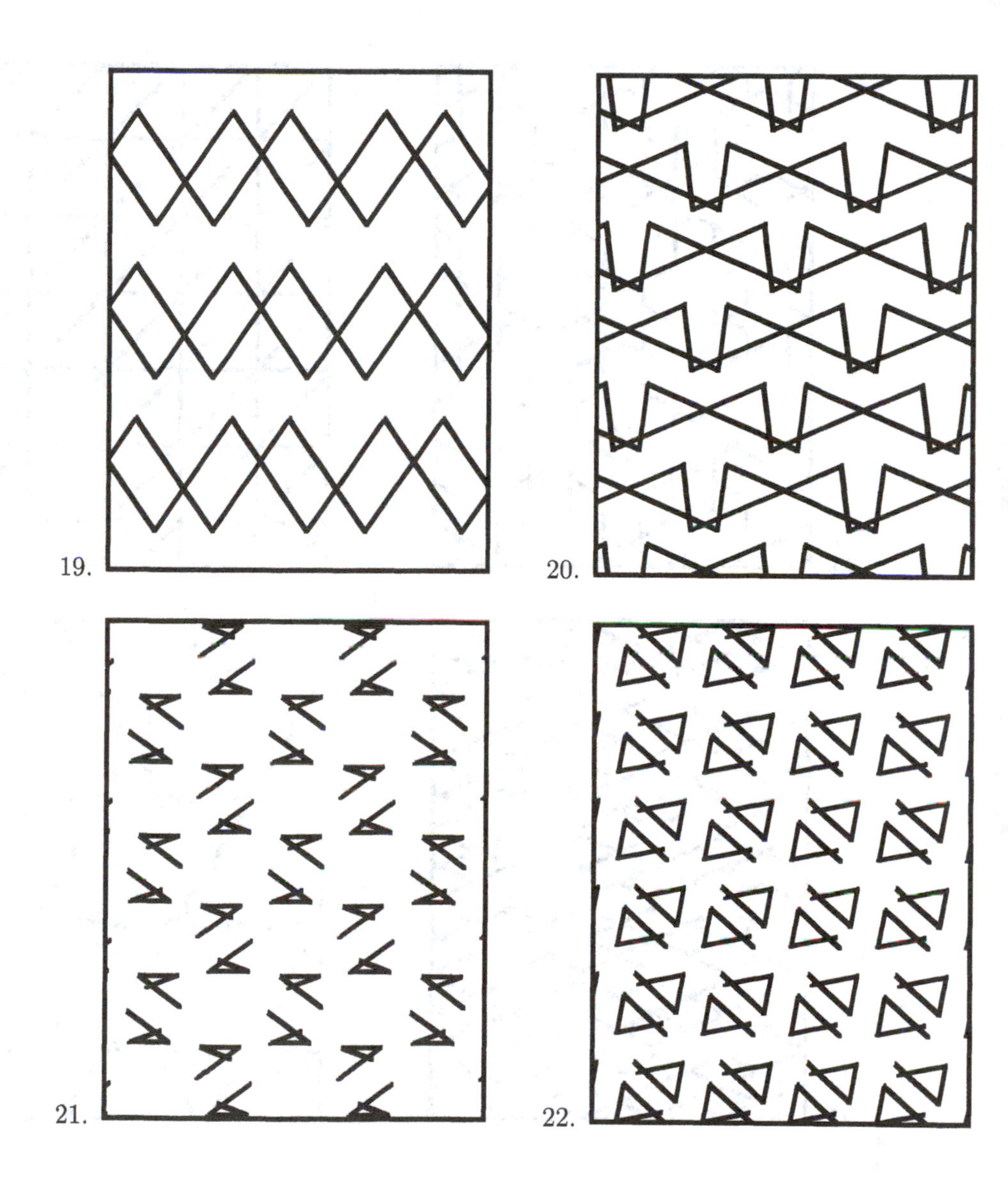

19.
20.
21.
22.

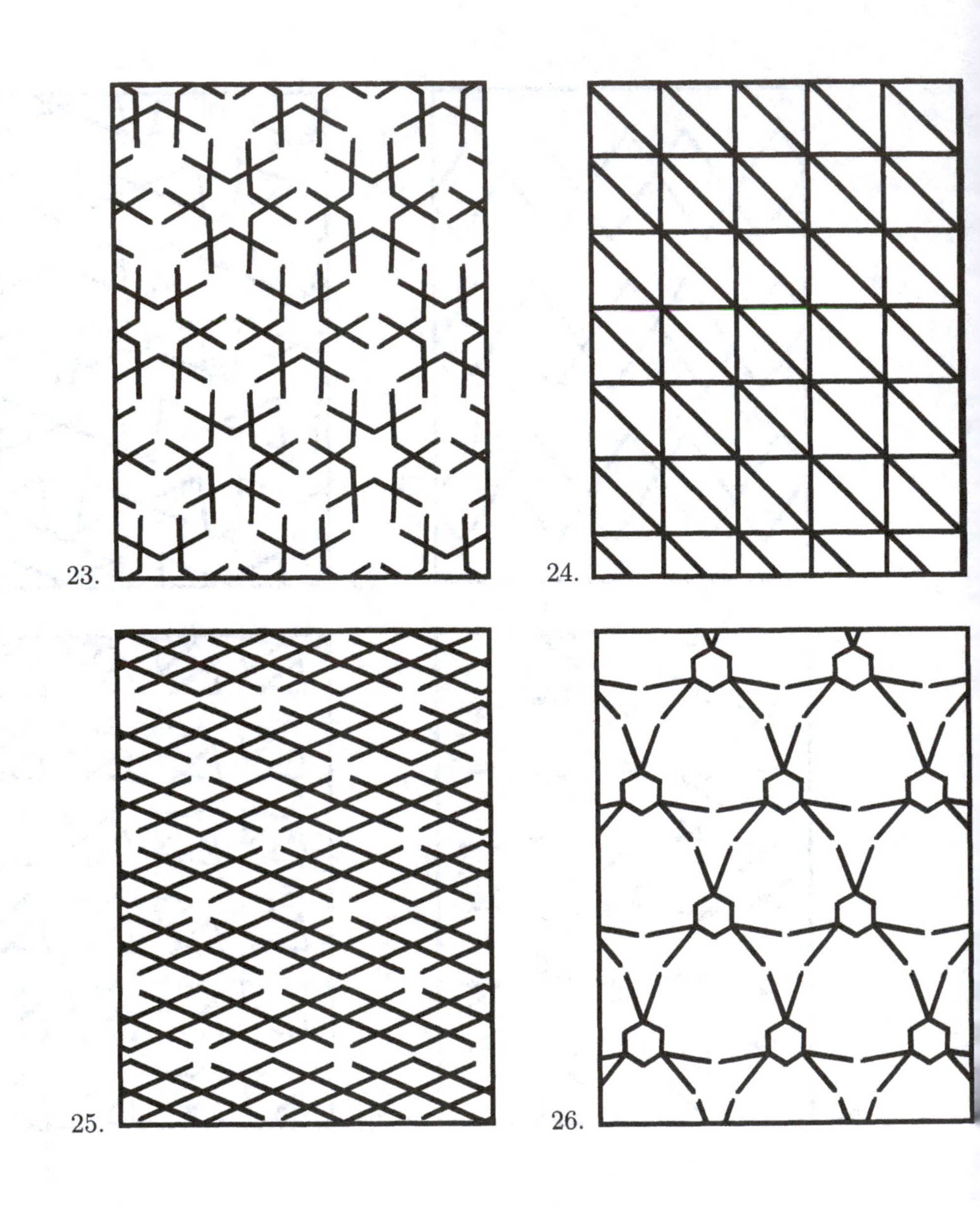
23.
24.
25.
26.

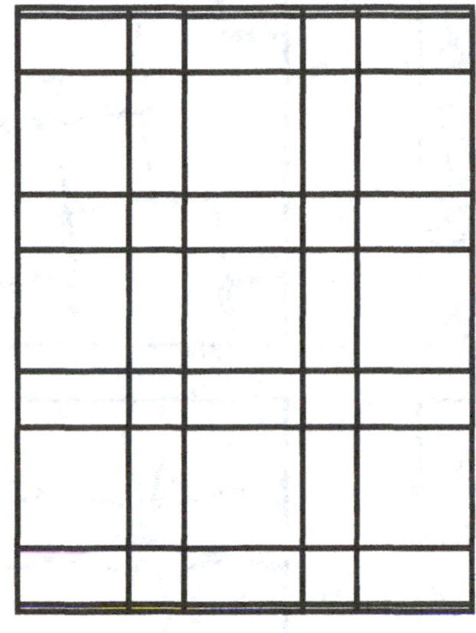

27.

28.

29.

30.

5. Assuming that $A = (2, 0)$ and $B(0, 2)$, identify the following compositions:
(a) $\tau_{AB} \circ R_{A,90^\circ}$ (b) $\tau_{AB} \circ \rho_{AB}$ (c) $\tau_{AB} \circ \gamma_{AB}$
(d) $\tau_{AB} \circ R_{A,90^\circ}$ (e) $\gamma_{AB} \circ \gamma_{AB}$ (f) $\rho_{AB} \circ R_{A,90^\circ}$
6. Suppose $A, B, \ldots, Z$ are vertices of a cyclic 26-gon. Identify the composition $\rho_{AB} \circ \rho_{BC} \circ \cdots \circ \rho_{YZ} \circ \rho_{ZA}$.
7. List the symmetries of the regular decagon.
8. Explain why the regular n-gon has $2n$ symmetries.
9. Prove that if f and g are motions, then $f \circ g \circ f^{-1} \circ g^{-1}$ is either a translation or a rotation.

10. Identify the symmetry groups of the frieze patterns in Figure 6.26.
11. Identify the symmetry groups of the wallpaper designs in Figure 6.36.
12. Are the following statements true or false in Euclidean geometry? Justify your answers.

    (a) The composition of two motions is a motion.

    (b) The composition of two translations is a translation.

    (c) The composition of two rotations is a rotation.

    (d) The composition of two reflections is a reflection.

    (e) The composition of two glide reflections is never a glide reflection.

    (f) The only motion that fixes three distinct points is the identity.

    (g) Every geometrical figure has at least one symmetry.

    (h) There exist only seven frieze patterns.

    (i) The frieze patterns have only seven pairwise distinct frieze pattern symmetry groups.

    (j) There exist only seventeen wallpaper designs.

    (k) There exist only seventeen distinct wallpaper symmetry groups.

    (l) Every frieze pattern has a translation in its symmetry group.

    (m) For each $n = 1, 2, 3, 4, 6$ there is a wallpaper design that has a rotation of order n and a nontrivial glide-reflectional symmetry.

    (n) For each $n = 1, 2, 3, 4, 6$ there is a wallpaper design that has a rotation of order n and no reflectional symmetries.

    (o) The inverse of every motion is a motion of the same type.

A Recognition Chart for Wallpaper Symmetry Groups

Type	Highest Order of Rotation	Reflections	Nontrivial Glide Reflections	Helpful Distinguishing Properties
p1	1	no	no	
p2	2	no	no	
pm	1	yes	no	
pg	1	no	yes	
cm	1	yes	yes	
pmm	2	yes	no	
pmg	2	yes	yes	parallel reflection axes
pgg	2	no	yes	
cmm	2	yes	yes	perpendicular reflection axes
p4	4	no	no	
p4m	4	yes	yes	4-fold centers on reflection axes
p4g	4	yes	yes	4-fold centers not on reflection axes
p3	3	no	no	
p3m1	3	yes	yes	all 3-fold centers on reflection axes
p31m	3	yes	yes	not all 3-fold centers on reflection axes
p6	6	no	no	
p6m	6	yes	yes	

Chapter 7

Inversions

Transformations that are not rigid can be interesting too, even though they are not as natural as the motions of the previous chapter. The inversions of this chapter are particularly appealing because they play important roles in both Euclidean and hyperbolic geometry.

7.1 Inversions as Transformations

Given a point C and a positive real number k, the *inversion* $I_{C,k}$ is a transformation of the plane that maps any point $P \neq C$ of the plane into the point $P' = I_{C,k}(P)$ such that

(a) C, P, P' are collinear with C outside the segment PP',

and

(b) $CP \cdot CP' = k^2$.

Figure 7.1 illustrates the action of a typical inversion.

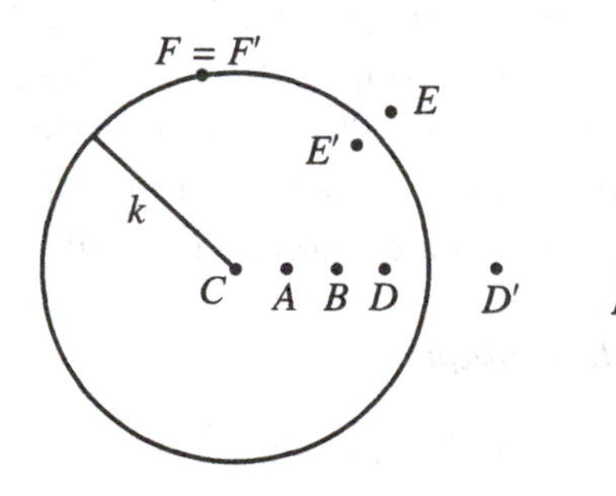

Figure 7.1. The inversion $I_{C,k}$

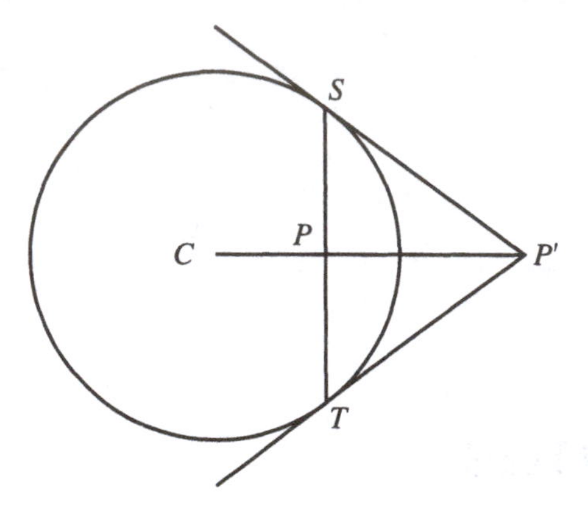

Figure 7.2.

It is clear that $I_{C,k}(P) = P'$ if and only if $I_{C,k}(P') = P$ and hence $I^2_{C,k} = \text{Id}$ and $I^{-1}_{C,k} = I_{C,k}$. Moreover, $I_{C,k}(P) = P$ if and only if P is on the circle $(C; k)$. Otherwise the point P is inside the circle $(C; k)$ if and only P' is outside it. If $g = (C; k)$, the inversion $I_{C,k}$ will also be denoted by I_g. Note that $I_{C,k}(P)$ is undefined for $P = C$ and only for C. The point C is called the *center* of the inversion $I_{C,k}$. Figure 7.2 displays the relation between P and $P' = I_{C,k}(P)$ geometrically. The circle of this figure has radius k and the lines $\overleftrightarrow{SP'}$ and $\overleftrightarrow{TP'}$ are tangent to it (see Exercise 8).

Any two points P, Q inside the circle $(C; k)$ are transformed by the inversion $I_{C,k}$ into two points P', Q' such that

$$P'Q' > PQ.$$

In fact, the closer P and Q are to the center C, the greater the discrepancy between PQ and $P'Q'$. It follows that $I_{C,k}$ is not a motion. Nevertheless, inversions do share some of the properties of motions. Motions transform straight lines into straight lines (Proposition 6.1.2) and circles into circles (Exercise 6.1.2). Inversions, too, transform straight lines and circles into straight lines and circles, although not necessarily respectively: Some straight lines are bent into circles and some circles are straightened out. The next theorem describes these phenomena in detail.

Theorem 7.1.1 *The inversion $I_{C,k}$ maps*

(a) *straight lines through C onto themselves,*

(b) *straight lines not through C onto circles through C,*

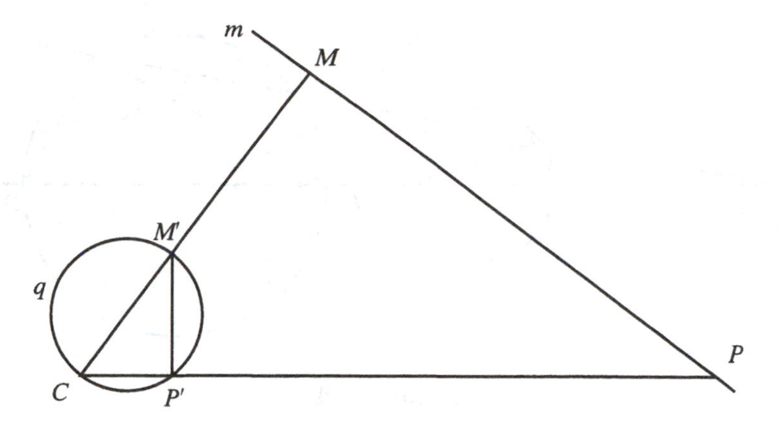

Figure 7.3.

(c) *circles through C onto straight lines not through C,*

(d) *circles not through C onto circles not through C.*

PROOF: (a) This follows directly from the definition of inversions.

(b) Let m be a straight line not through C, and let M be that point of m such that $CM \perp m$ (Fig. 7.3), whereas P is an arbitrary point of m. Set

$$M' = I_{C,k}(M) \quad \text{and} \quad P' = I_{C,k}(P).$$

Since $CM \cdot CM' = k^2 = CP \cdot CP'$, it follows that

$$\frac{CM'}{CP} = \frac{CP'}{CM}.$$

Moreover, $\angle PCM$ is common to both $\triangle CPM$ and $\triangle CM'P'$, and so it follows from Proposition 3.5.9 that these triangles are similar and consequently

$$\angle M'P'C = \angle PMC = 90^\circ.$$

Since the points C and M' are fixed (whereas P is arbitrary on m) it follows from Exercise 4.2B.5 that P' falls on the fixed circle with diameter CM'.

(c) Let q be a circle through C, let CM' be a diameter of q, set $M = I_{C,k}(M')$, and let m be the line through M perpendicular to $\overleftrightarrow{CM'}$ (Fig. 7.3). It follows from part (b) that $I_{C,k}(m) = q$ and hence

$$I_{C,k}(q) = I^2_{C,k}(m) = m.$$

(d) Let p be a circle not through C with P an arbitrary point on p. Let DE be a diameter of p whose extension contains C, and suppose the given

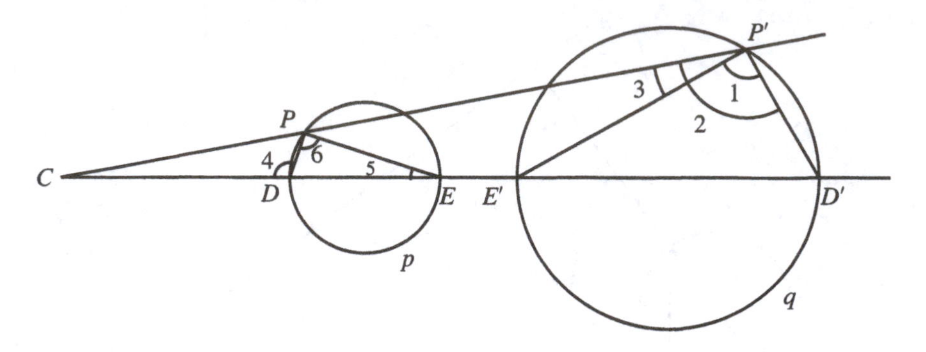

Figure 7.4.

inversion $I_{C,k}$ maps the points D, E, P onto the points D', E', P' (Fig. 7.4). Since

$$CP \cdot CP' = CD \cdot CD' = CE \cdot CE' = k^2$$

it follows that

$$\frac{CD}{CP'} = \frac{CP}{CD'} \quad \text{and} \quad \frac{CE}{CP'} = \frac{CP}{CE'}.$$

Since $\angle DCP$ is common to $\triangle DCP$, $\triangle P'CD'$, $\triangle ECP$, and $\triangle P'CE'$ it follows that the first two are similar to each other, as are the last two. Consequently,

$$\angle 4 = \angle 2 \quad \text{and} \quad \angle 5 = \angle 3$$

and

$$\angle 1 = \angle 2 - \angle 3 = \angle 4 - \angle 5 = \angle 6 = 90°.$$

Since D' and E' are fixed points, it follows that P' lies on the circle that has $D'E'$ as its diameter (see Exercise 4.2B.5). Q.E.D.

The following observations are implicit in the proof of the preceding theorem:

> *When the inversion $I_{C,k}$ transforms a circle into a circle, their centers are collinear with C. When the inversion transforms a straight line into a circle, or vice versa, the line through C perpendicular to the straight line contains the center of the circle.*

Example 7.1.2 Let $I = I_{O,6}$, where O denotes the origin, and let m denote the line $y = -6$ (Fig. 7.5). By Theorem 7.1.1b, $I(m)$ is a circle q that contains O and whose center lies on the y-axis. Since I fixes the point $(0, -6)$, this point must lie on q. It follows that q is the circle $((0, -3); 3)$.

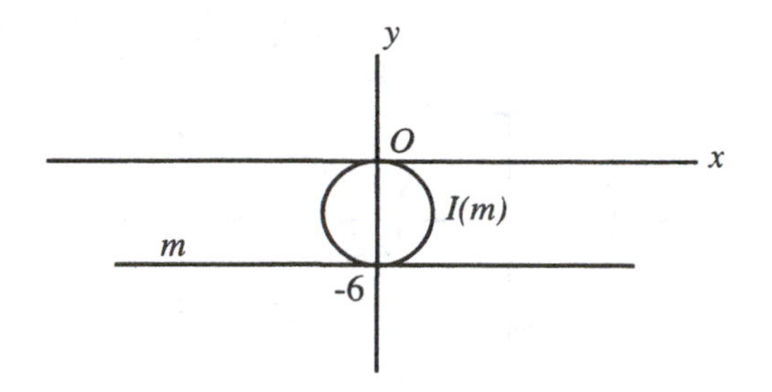

Figure 7.5.

Example 7.1.3 Let $I = I_{O,6}$ and let $q = ((4,0);1)$ (Fig. 7.6). It follows from Theorem 7.1.1d that $I(q)$ is also a circle, with center on the x-axis, which of necessity contains the points

$$I((5,0)) = \left(\left(\frac{36}{5}\right),0\right) = (7.2,0) \quad \text{and} \quad I((3,0) = \left(\frac{36}{3},0\right) = (12,0).$$

Consequently, $I(q) = ((9.6,0);2.4)$.

Example 7.1.4 Identify the inversion $I_{C,k}$ that maps the circle $(O;2)$ onto the straight line $x = 6$.

By Theorem 7.1.1c, C must be either $(2,0)$ or $(-2,0)$ (Fig. 7.7). Since C cannot lie between a point and its image, it follows that $C = (-2,0)$. Finally, since $I_{C,k}$ maps $(2,0)$ onto $(6,0)$ it follows that

$$k = \sqrt{[2-(-2)]\cdot[6-(-2)]} = 4\sqrt{2}.$$

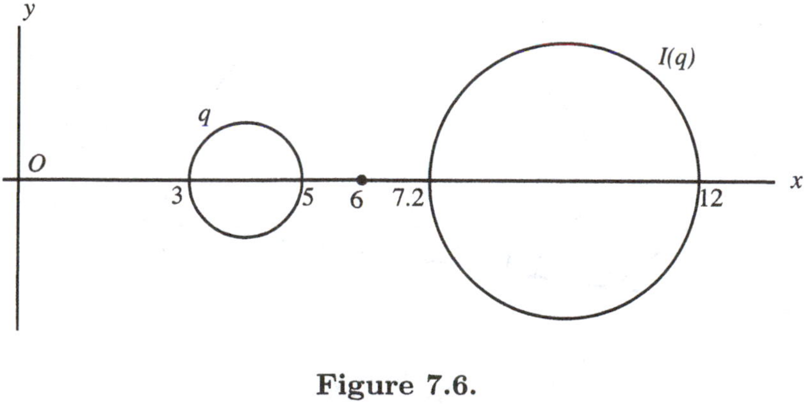

Figure 7.6.

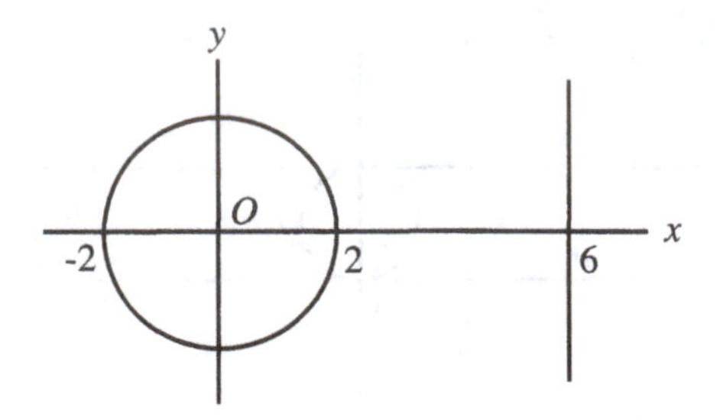

Figure 7.7.

Another trait in which inversions resemble motions is that they too preserve the measures of angles (see Exercise 6.1.8). Of course, since inversions bend straight lines, it is necessary to allow for nonrectilineal angles. The reader is reminded that the measure of the angle determined by two intersecting curves was defined to be the measure of the angle between their respective tangents at that intersection.

Proposition 7.1.5 *Inversions preserve angles (but reverse their senses).*

PROOF: Figure 7.8 describes how an inversion centered at C transforms the angle α formed by the curves h and j into the angle α' formed by the image curves $h' = I_{C,k}(h)$ and $j' = I_{C,k}(j)$. It is clear that the sense of the angle is reversed by the inversion. Since

$$\alpha = \angle 1 - \angle 2 \quad \text{and} \quad \alpha' = \angle 3 - \angle 4,$$

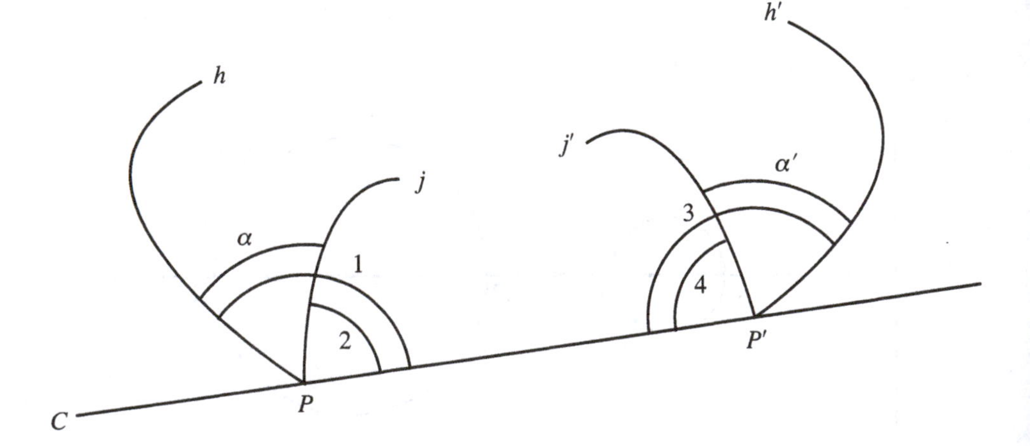

Figure 7.8.

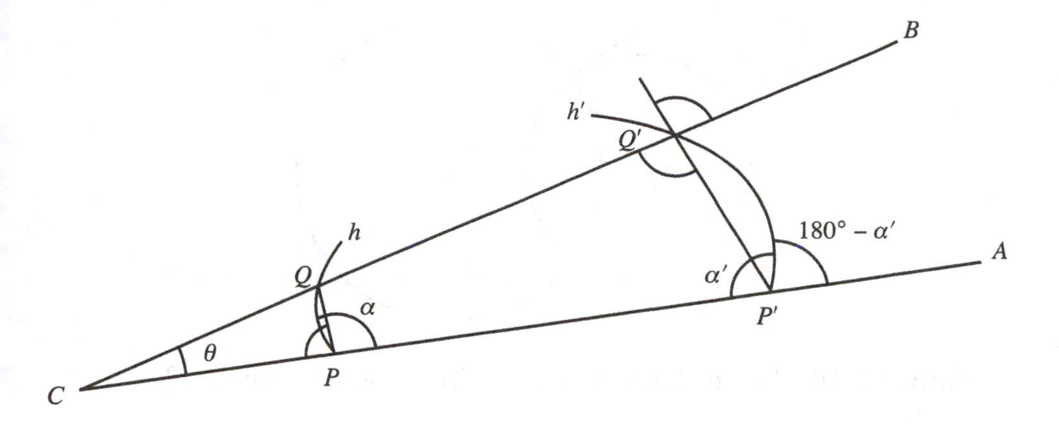

Figure 7.9.

it will suffice to prove the special case that $\angle 1 = \angle 3$, or, that $\alpha = \alpha'$ in Figure 7.9. The equation

$$CP \cdot CP' = k^2 = CQ \cdot CQ'$$

implies that

$$\triangle CQ'P' \sim \triangle CPQ$$

and hence

$$\angle CQ'P' = \angle CPQ.$$

Since the limiting values, as $\theta \to 0$, of $\angle CQ'P'$ and $\angle CPQ$ are $180° - \alpha'$ and $180° - \alpha$, respectively, it follows that $\alpha = \alpha'$. Q.E.D.

Two intersecting circles are said to be *orthogonal* if their respective tangents at the point of intersection are perpendicular to each other. However, by Proposition 4.1.4, the tangent line is perpendicular to the radius through the point of contact and hence it follows that two intersecting circles are orthogonal if and only if their tangents at the point of intersection pass through each other's centers (see Fig. 7.10). In fact, orthogonality is guaranteed by the tangent to one circle passing through the center of the other. Exercise 3 relates orthogonality to inversions.

A straight line is said to be *orthogonal* to a circle if it is perpendicular to the tangents at the intersection points. This is equivalent to saying that the straight line contains a diameter of the circle.

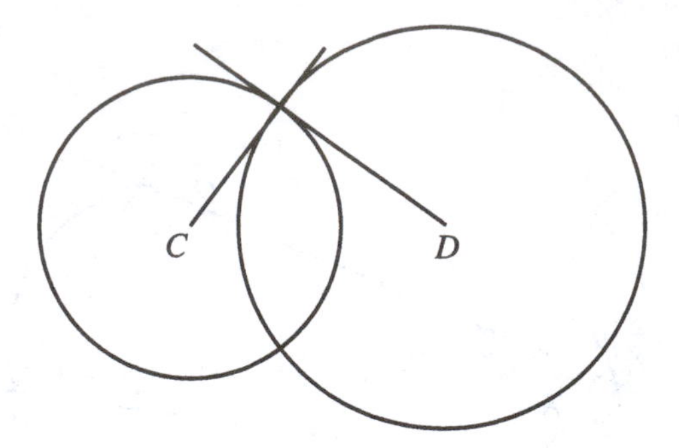

Figure 7.10. Two orthogonal circles with respective centers C and D

Exercises 7.1

1. If O denotes the origin, to what point or curve does the inversion $I_{O,4}$ transform the following sets?

   (a) The point $(3, 0)$
   (b) The point $(0, -2)$
   (c) The point $(2, 2)$
   (d) The point $(-1, 1)$
   (e) The line $y = -2x$
   (f) The line $x + y = 4$
   (g) The line $x = 4$
   (h) The line $y = -4$
   (i) The line $x = 2$
   (j) The line $y = -8$
   (k) The line $y = x + 8$
   (l) The line $y = -x + 4$
   (m) The line $y = x - 2$
   (n) The line $y = -x - 8$
   (o) The circle $(O; 3)$
   (p) The circle $(O; 8)$
   (q) The circle $((3, 0); 1)$
   (r) The circle $((3, 0);\ 6)$
   (s) The circle $((0, 8); 2)$
   (t) The circle $((0, 8);\ 4)$
   (u) The circle $((0, 8); 6)$
   (v) The circle $((0, 8); 8)$
   (w) The circle $((0, 8); 10)$
   (x) The circle $((4, 4); 4)$
   (y) The circle $((5, 5); 5)$
   (z) The circle $((5, 5); \sqrt{26})$

2. For each of the following pairs of curves, decide whether there exists an inversion that transforms one onto the other. Identify the inversion if it exists.

   (a) The x axis and the line $y = 2$
   (b) The circle $(O; 5)$ and the line $x = 3$
   (c) The circle $(O; 5)$ and the line $x = 5$

(d) The circle $(O;5)$ and the line $x = 10$

(e) The circle $(O;5)$ and the circle $(O;10)$

(f) The circle $(O;5)$ and the circle $((5,0);5)$

(g) The circle $(O;5)$ and the circle $((35,0);30))$

3. Let p be a circle, C a point, and k a positive real number. Prove that $I_{C,k}(p) = p$ if and only if the circles p and $(C;k)$ are orthogonal.

4. Let I be an inversion and let p be a circle such that $I(p)$ is also a circle. When do p and $I(p)$ have different radii?

5. Let p and q be two circles with different radii. Show that there is an inversion I such that $I(p) = q$.

6. Let m be a straight line. Characterize all the circles p such that there exists an inversion I for which $I(m) = p$.

7. Let p be a circle. Characterize all the straight lines m such that there exists an inversion I for which $I(p) = m$.

8. Prove that if the radius of the circle of Figure 7.2 is k, then $I_{C,k}(P) = P'$.

9(C). Write a script that takes a circle $(C;k)$ and a point P as input and yields $I_{C,k}(P)$ as output.

10(C). Write a script that takes a circle $(C;k)$ and a straight line m as input, and yields $I_{C,k}(m)$ as output.

11(C). Write a script that takes circles $(C;k)$ and p as input, and yields $I_{C,k}(p)$ as output.

7.2 Inversions to the Rescue

Inversions can be very useful in transforming problems about circles into simpler problems about straight lines. Two examples of this procedure are offered.

Example 7.2.1 Let two circles p and q intersect in A and B, and let the extensions of the diameters of p and q through B intersect q and p in the points C and D, respectively. Show that the line $\overleftrightarrow{AB}$ contains a diameter of the circle that circumscribes $\triangle BCD$.

Let k be any real number and suppose the inversion $I = I_{B,k}$ is applied to the given configuration of circles so that $A' = I(A)$, $C' = I(C)$, $D' = I(D)$, $p' = I(p)$, $q' = I(q)$, $r' = I(r)$ (Fig. 7.11). Since BC and BD are orthogonal to p and q, respectively, it follows from Proposition 7.1.5 that $BC' \perp p'$ and

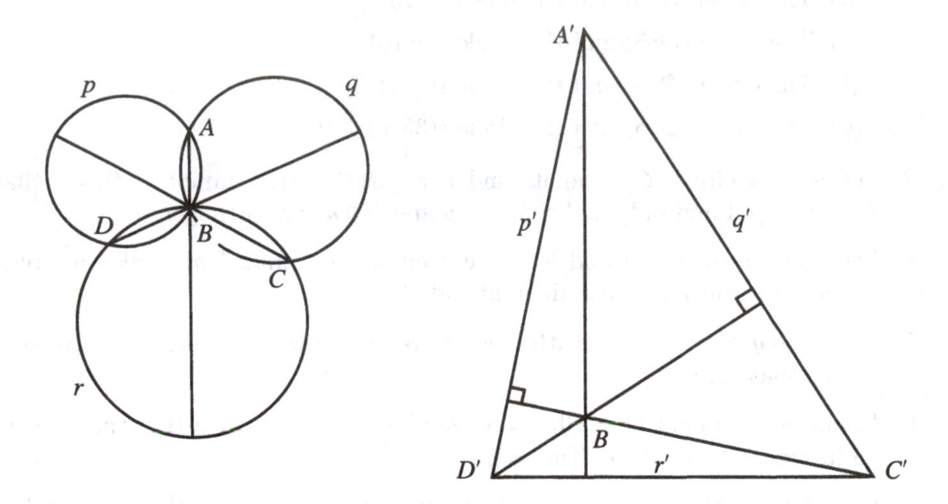

Figure 7.11.

$BD' \perp q'$. The concurrence of the three altitudes of the triangle (Exercise 4.2B.11) now implies that $BA' \perp r'$ and hence BA contains a diameter of r.

Proposition 7.2.3 was first proved by Ptolemy (ca. 85–165). It was an important tool in his construction of the table of chords, which appears in Ptolemy's definitive book on Greek astronomy, the *Almagest.* The proof of the required lemma is relegated to Exercise 1.

Lemma 7.2.2 *Suppose $P' = I_{C,k}(P)$ and $Q' = I_{C,k}(Q)$. Then*

$$P'Q' = \frac{k^2 PQ}{CP \cdot CQ}.$$

Proposition 7.2.3 *In a cyclic quadrilateral, the product of the diagonals equals the sum of the products of the two pairs of opposite sides.*

PROOF: Let $ABCD$ be a cyclic quadrilateral inscribed in a circle of diameter k (Fig. 7.12). It follows from Theorem 7.1.1b that $I_{A,k}$ inverts this circle into a tangent line that contains the points $B' = I_{A,k}(B)$, $C' = I_{A,k}(C)$, and $D' = I_{A,k}(D)$. Since $B'C' + C'D' = B'D'$, it follows from the lemma that

$$\frac{k^2 BC}{AB \cdot AC} + \frac{k^2 CD}{AC \cdot AD} = \frac{k^2 BD}{AB \cdot AD}$$

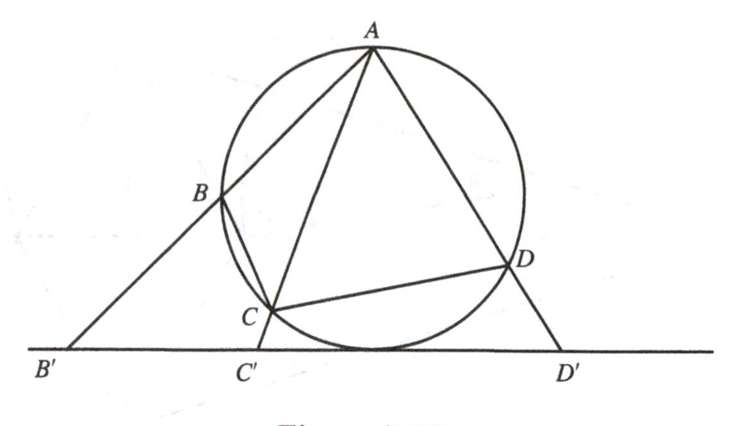

Figure 7.12.

or, upon multiplying this equation by $(AB \cdot AC \cdot AD)/k^2$,

$$BC \cdot AD + CD \cdot AB = BD \cdot AC. \qquad \text{Q.E.D.}$$

In order to illustrate one more application of inversions, we return to the issue of constructibility. In elementary geometry classes it is customary to construct figures using rulers and compasses. These two tools, however, are qualitatively different. The compass generates a circle because of its mechanical properties whereas the ruler simply allows us to copy a given straight line onto the paper. The circular analog of a ruler would be a coin or the lid of a jar. What then is the linear analog of the compass? In other words, what mechanical device, consisting of linked rods, would constrain a pencil to move so as to draw a straight line? Such devices are called *linkages*, the simplest one being a single rod AB with fixed point A. It is clear that a pencil attached at B would be constrained to draw a circle.

The utility of linkages in drawing curves has been studied for hundreds of years, but the first one capable of drawing a straight line was invented in 1864 by A. Peaucellier (1832–1913), who was an engineer in the French army. In his honor, this device, which contains seven rods, is called *Peaucellier's cell*. In 1874 Harry Hart (1848–1920) invented a five-rod linkage for drawing straight lines and it is unknown whether there are any such linkages with fewer than five rods.

Peaucellier's cell is depicted in Figure 7.13, where the points C and X are fixed, $XP = XC$, and the solid lines XP, $AP = PB = BP' = P'A < CA = CB$ denote rods that are loosely linked at their endpoints. The dashed lines denote auxiliary lines that serve only for the purpose of demonstrating the

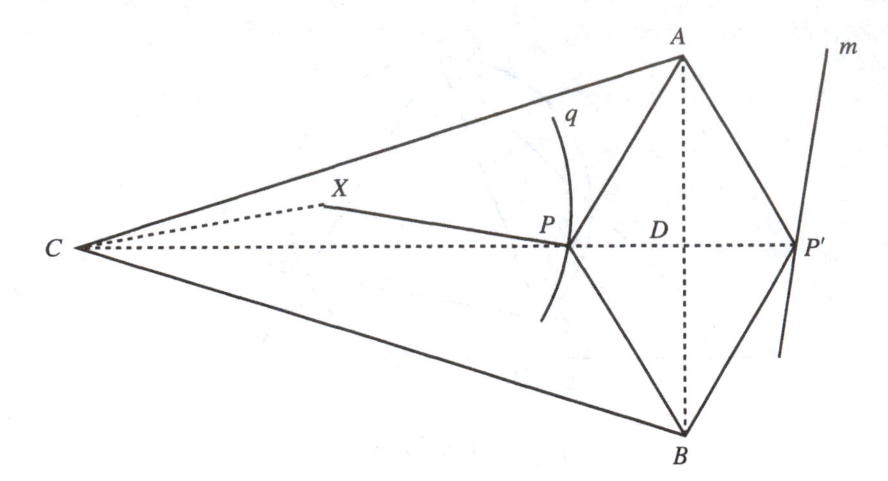

Figure 7.13. Peaucellier's cell

properties of the linkage. Since $APBP'$ is a rhombus, we have

$$\begin{aligned} CP \cdot CP' &= (CD - DP)(CD + DP) = CD^2 - DP^2 \\ &= (CD^2 + DA^2) - (DP^2 + DA^2) = CA^2 - AP^2. \end{aligned}$$

If $k = (CA^2 - AP^2)^{1/2}$, then k is constant and we have

$$P' = I_{C,k}(P).$$

Since the rod XP constrains P to move in a circle q that contains C, it follows from Proposition 7.1.1c that P' traces out a straight line m.

Exsercises 7.2

1. Prove Lemma 7.2.2.

2. Two circles p and q have a common tangent at a point T, and a variable circle through T intersects p and q orthogonally in points P and Q. Prove that $\overleftrightarrow{PQ}$ passes through a fixed point.

3. Suppose $ABCD$ is a cyclic quadrilateral. If T is the point of contact of a circle containing A and B with another circle containing C and D, show that the locus of T is a circle.

4. Prove that if $ABCD$ is a convex quadrilateral, then $BC \cdot AD + CD \cdot AB \geq BD \cdot AC$. Show that equality holds if and only if $ABCD$ is cyclic.

5. Let p be a fixed circle and let P be a fixed point not on p. Prove that there exists a point P' distinct from P such that every circle through P that is orthogonal to p also passes through P'.

6. Let p be a fixed circle and P a fixed point. Show that the locus of the centers of all the circles that pass through P and are orthogonal to p is a straight line.

7(C). Use a computer application to verify Proposition 7.2.3.

8. Given a circle p and a point Q, construct a circle q that contains Q and is orthogonal to p.

7.3 Inversions as Hyperbolic Motions

In addition to the role they play in Euclidean geometry, inversions also provide us with means for visualizing the motions of hyperbolic geometry. As was the case in Section 1.2, the exposition here is informal and no proofs are given. Instead, several examples are offered that are easily implemented on computers and substantiate the discussion. Exercises 13–17 provide an opportunity for the further exploration of the properties of the hyperbolic motions.

It was proven in Section 6.4 that every hyperbolic motion can be expressed as the composition of hyperbolic reflections. These hyperbolic reflections are now described in the context of the upper half-plane geometry.

Proposition 7.3.1 *There are two kinds of reflections of hyperbolic geometry:*

(a) *Euclidean reflections whose axes are vertical*

(b) *Inversions whose centers are on the x-axis* □

It stands to reason that Euclidean reflections with vertical axes should also double as hyperbolic reflections. After all, the distortion of lengths that was used to create this geometry depends only on the distances from the x-axis, and since these particular reflections do not change these distances, it is not surprising that they constitute hyperbolic motions. Figure 7.14 illustrates the effect of the reflections of both types on a triangle. Hyperbolic $\triangle A'B'C'$ is both the Euclidean and the hyperbolic reflection of $\triangle ABC$ in the vertical geodesic m, and $\triangle A''B''C'' = I_g(\triangle ABC)$ is the hyperbolic reflection of $\triangle ABC$ in the bowed geodesic g.

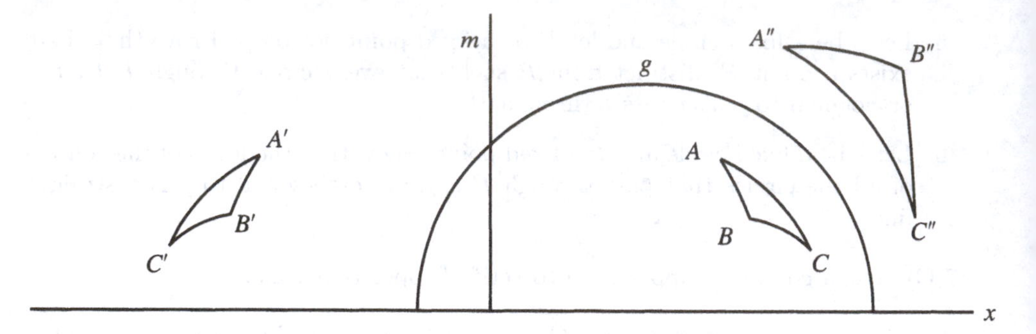

Figure 7.14. Hyperbolic reflections

Example 7.3.2 Find a hyperbolic reflection that transforms the point $P(1,1)$ to the point $Q(5,3)$ (Fig. 7.15).

The line $\overleftrightarrow{PQ}$ has equation $y-1=(1/2)(x-1)$ and intersects the x-axis at the point $C(-1,0)$. Since

$$\begin{aligned} CP \cdot CQ &= \sqrt{(1+1)^2+(1-0)^2} \cdot \sqrt{(5+1)^2+(3-0)^2} \\ &= \sqrt{5} \cdot \sqrt{45} = 15 \end{aligned}$$

it follows that the inversion $I_{C,\sqrt{15}}$ is the required hyperbolic reflection.

Example 7.3.3 Find a hyperbolic reflection that transforms the geodesic consisting of the upper half of the circle $((4,0);2)$ onto the upper half of the straight line $x=-3$ (Fig. 7.16).

Any inversion centered at $(6,0)$ will transform the given semicircle into

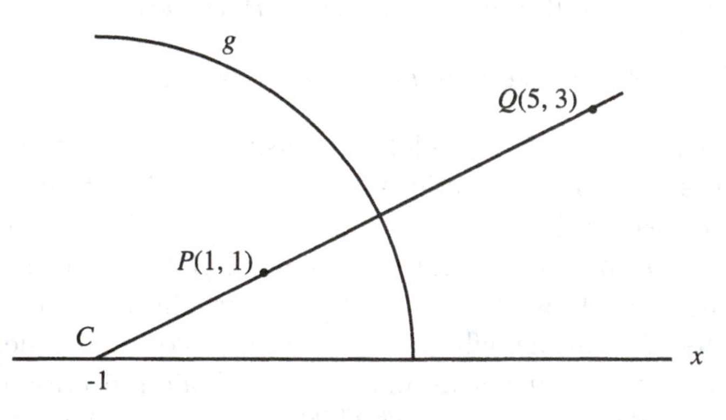

Figure 7.15.

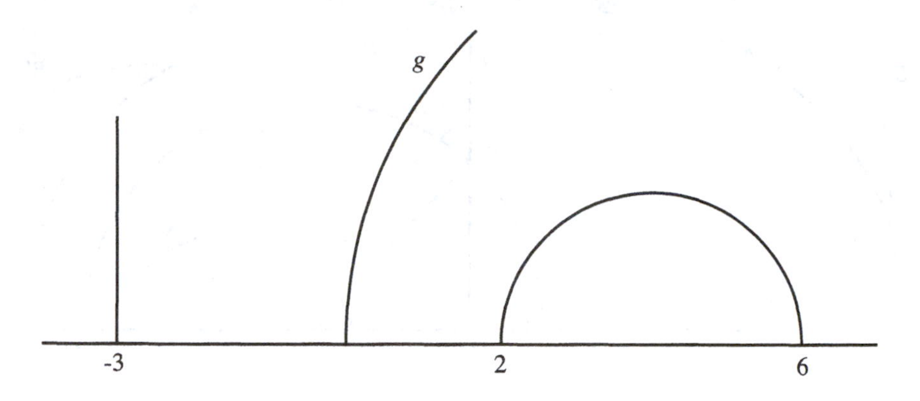

Figure 7.16. A hyperbolic reflection

a vertical ray. Since $(6-2)(6-(-3)) = 36$, it follows that the required hyperbolic reflection is the inversion $I_{(6,0),6}$ with fixed circle g.

Inasmuch as the definition of a Euclidean rotation makes no reference to parallelism or any of its consequences, it can also serve as the definition of a *hyperbolic rotation*. Unfortunately, because of the distortion of distances in the upper half-plane, this definition is not easily visualized. However, Proposition 6.2.2, which states that the composition of two reflections with intersecting axes is a rotation about the point of intersection, is neutral, and so it applies here as well. Hence, if I_g denotes the inversion whose fixed circle is g (Fig. 7.17), and we recall that ρ_m is also a hyperbolic reflection, then $\rho_m \circ I_g$ is a hyperbolic rotation R about the point X. Note that

$$R(\triangle ABC) = \rho_m(I_g(\triangle ABC)) = \rho_m(\triangle A'B'C') = \triangle A''B''C''.$$

If $g = (O; 3)$ and m is the line $x = -2$, it follows from Exercise 12 that

$$\angle(g, m) = \angle XOM = \cos^{-1}\left(\frac{2}{3}\right) = 48.2°.$$

Consequently, $R = R_{X,96.4°}$. Figure 7.18 illustrates a hyperbolic rotation $R = R_{Y,60°}$. In this figure $P_{i+1} = R(P_i)$ for $i = 1, 2, \cdots, 5$ and $P_1 = R(P_6)$. Obviously, these points all lie on a hyperbolic circle centered at Y. Exercise 15(C) examines this issue further.

The definition of a Euclidean translation does involve parallelism and is therefore of no use in the hyperbolic context. Instead, motivated by Proposition 6.2.1, a *hyperbolic translation* is defined as the composition of two hyperbolic reflections whose axes do not intersect. If the axes of both

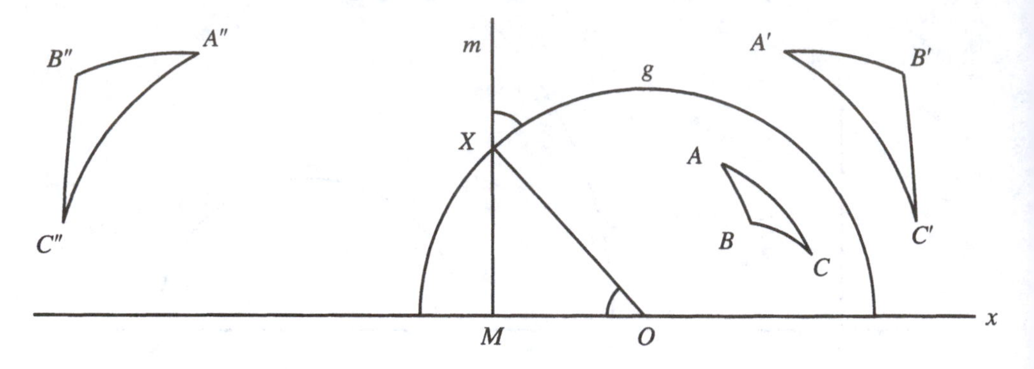

Figure 7.17. Hyperbolic reflections

the hyperbolic reflections are straight geodesics, then their composition is a horizontal Euclidean translation. This is illustrated in Figure 7.19, where $\tau = \rho_n \circ \rho_m$ and $\tau(P_i) = P_{i+1}$ for $i = 1, 2, 3, 4$. Such horizontal Euclidean translations constitute hyperbolic motions for the same reason that the Euclidean reflections ρ_m and ρ_n do. They do not alter the distances of points from the x-axis.

Figure 7.20 illustrates the composition of hyperbolic reflections of mixed

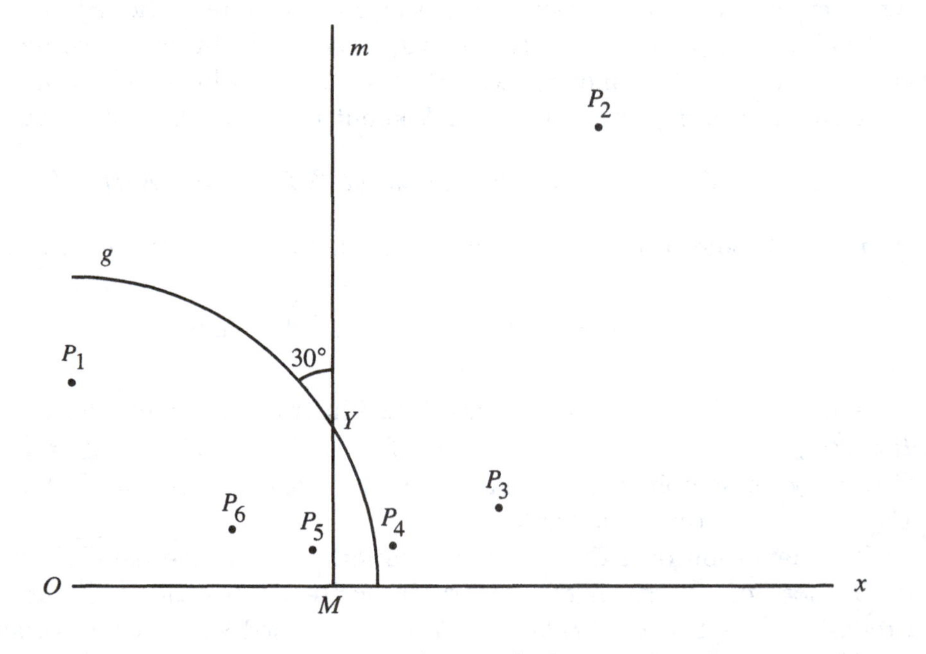

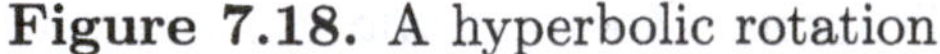
Figure 7.18. A hyperbolic rotation

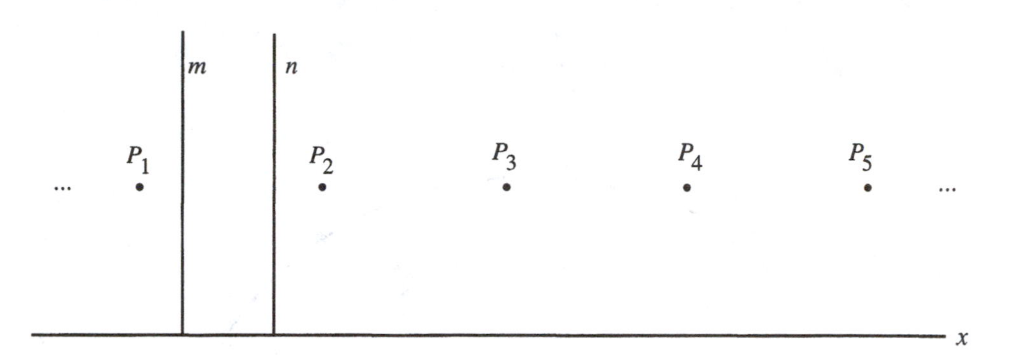

Figure 7.19. Both a Euclidean and a hyperbolic translation

types with nonintersecting axes. If $\tau = I_g \circ \rho_m$, then $P_{i+1} = \tau(P_i)$ for $i = 1, 2, 3, 4$. Note that the orbit of this translation τ does go out to hyperbolic infinity, just as Euclidean translations go out to Euclidean infinity. This stands in marked contrast with the orbit of the hyperbolic rotation R of Figure 7.18.

Figure 7.21 displays an orbit of the *hyperbolic glide reflection* $\gamma = \tau \circ I_{O,5}$, where τ is the horizontal shift $(x, y) \rightarrow (x + 2, y)$.

Since the proof of Theorem 6.4.2 is neutral, it follows that every hyperbolic motion is the composition of no more than three hyperbolic reflections. The group of all the motions of the upper half-plane is one of the best studied structures of advanced mathematics. It has also proved to be an indispensable tool in such diverse areas as geometry (of course), number

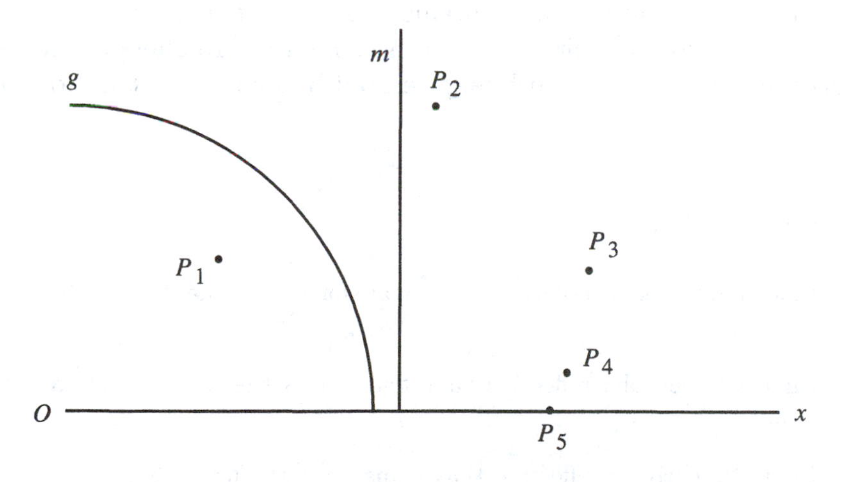

Figure 7.20. A hyperbolic translation

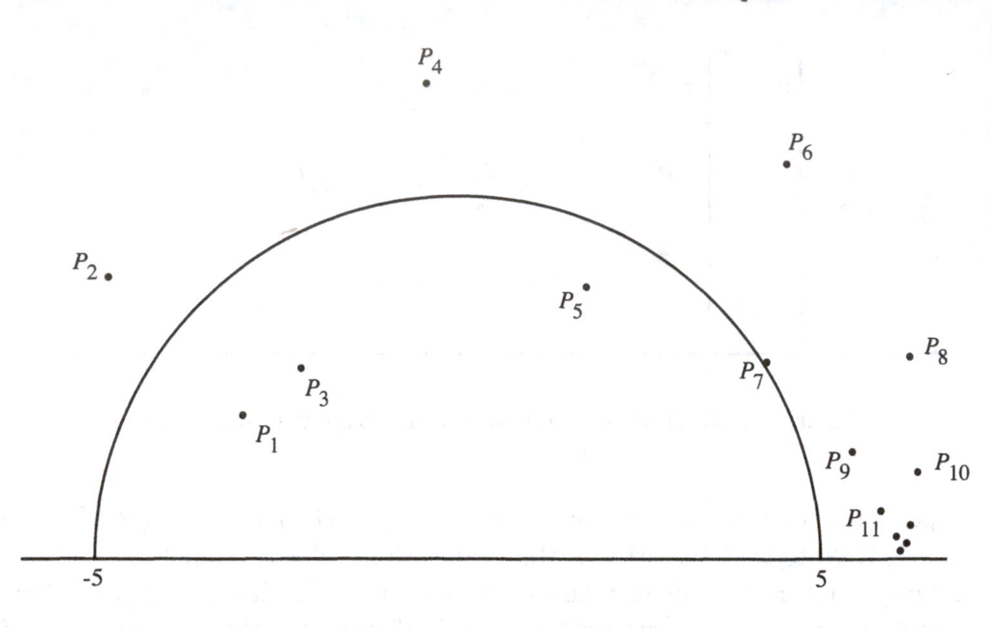

Figure 7.21. A hyperbolic glide reflection

theory, and analysis. An elementary discussion of this topic and its surprising connection with complex numbers can be found in the author's *The Poincaré Half-Plane: A Gateway to Modern Geometry.*

Given a point P and a transformation f, the *orbit* of P generated by f is the sequence of points $P, f(P), f^2(P), f^3(P), \ldots$. Thus, the sequences $P_1, P_2, P_3, \ldots$ of Figures 7.18–21 are all orbits. These orbits help visualize the action of the transformation that generated them. It is clear that orbits of Euclidean translations are contained in straight lines whereas orbits of Euclidean rotations are circular in nature. Exercise 15 offers the reader the opportunity to explore the orbits generated by hyperbolic transformations.

Exercises 7.3

1. Find a hyperbolic reflection that transforms the point $(1, 1)$ to the point $(-3, 5)$.

2. Find a hyperbolic reflection that transforms the point $(1, 1)$ to the point $(-3, 1)$.

3. Find a hyperbolic reflection that transforms the upper half of $x = 4$ to the upper half of the circle $(O; 4)$.

4. Find a hyperbolic reflection that transforms the upper half of $x = 4$ to the upper half of the circle $(O; 2)$.

5. Find a hyperbolic reflection that transforms the upper half of $x = 4$ to the upper half of the circle $(O; 5)$.

6. Find a hyperbolic reflection that transforms the upper half of $x = 4$ to the upper half of $x = 17$.

7. Find a hyperbolic reflection that transforms the upper half of $(O; 5)$ to the upper half of $(O; 3)$.

8. Find a hyperbolic reflection that transforms the upper half of $(O; 5)$ to the upper half of $((9, 0); 2)$.

9. Prove that given any two points of the upper half-plane, there is a hyperbolic reflection that transforms one onto the other.

10. Prove that given any two intersecting geodesics of the upper half-plane, there is a hyperbolic reflection that transforms one onto the other.

11. Prove that given any two nonintersecting geodesics of the upper half-plane, there is a hyperbolic reflection that transforms one onto the other.

12. Prove that in Figure 7.17 $\angle(g, m) = \angle XOM$.

13(C). Write a script that will reflect any two points of the upper half-plane in any bowed geodesic. Use the script of Exercise 1.2.17 to substantiate the claim that this transformation is indeed a hyperbolic motion.

14(C). Write a script that will reflect any triangle of the upper half-plane in any bowed geodesic.

15(C). Let P be an arbitrary point of the upper half-plane, m an arbitrary vertical straight line, and g an arbitrary circle centered on the x-axis. Write a script that takes P, m, and g as its input and yields five iterations of the action of $\rho_m \circ I_g$ on P. Use this script to explore the following questions:

    (a) When m and g intersect, what is the geometrical nature of the orbits of $\rho_m \circ I_g$?

    (b) What does a hyperbolic circle look like to a Euclidean observer?

    (c) When m and g do not intersect, what is the geometrical nature of the orbits of $\rho_m \circ I_g$? To be specific, what is the nature of each orbit of $\rho_m \circ I_g$ and how do these orbits relate to each other?

16(C). Use a computer application to model hyperbolic glide reflections.

*17(C). Use a computer application to explore the notion of a hyperbolic inversion.

Chapter Review Exercises

1. Prove that if f is any Euclidean motion and I is any inversion, then $f \circ I \circ f^{-1}$ is also an inversion.
2. Suppose f is any Euclidean motion and I is any inversion. Is $I \circ f \circ I^{-1}$ necessarily a Euclidean motion?
3. Is the composition of two inversions ever an inversion?
4. When is the composition of two inversions a Euclidean motion?
5. Are the following statements true or false? Justify your answers.

   (a) Every inversion is a motion of Euclidean geometry.

   (b) Some inversions are motions of Euclidean geometry.

   (c) Every inversion is a motion of hyperbolic geometry.

   (d) Some inversions are motions of hyperbolic geometry.

   (e) Every motion of hyperbolic geometry is an inversion.

   (f) Some motions of hyperbolic geometry are inversions.

   (g) Inversions transform circles into circles.

   (h) Given a straight line, there exists no inversion that will transform it into another, distinct, straight line.

   (i) Peaucellier's cell was a notorious torture chamber in the Bastille.

   (j) Given any two circles, there exists either a Euclidean motion or an inversion that transforms one into the other.

THAT SPECIAL GIFT FOR MATHEMATICIANS

Chapter 8

Symmetry in Space

The focus now shifts to a discussion of some symmetries in three (and more) dimensions. Attention is restricted mostly to the symmetries of the five regular solids. The chapter concludes with a discussion of some recent discoveries in group theory.

8.1 Regular and Semiregular Polyhedra

Proposition 1 of Book I of the *Elements* states that it is possible to construct equilateral triangles. The thirteenth and last of these books is concerned exclusively with the constructibility of the three dimensional analogs of the regular polygons. The symmetries and other properties of these solids are the subject matter of this chapter. Euclid's decision to both begin and end his text with a discussion of highly symmetrical figures was in all likelihood conscious and testifies to a concern with esthetic issues that goes back to the roots of geometry in Pythagorean mysticism.

A *polyhedron* is a solid body of finite extent whose surface consists of several polygons, called *faces*. The sides and vertices of these polygonal faces are, respectively, the *edges* and *vertices* of the polyhedron. The vertices, edges, and faces of a polyhedron are collectively referred to as its *cells*.

A *regular polyhedron* is a polyhedron whose cells satisfy the following constraints:

1. *All the faces are congruent to the same regular polygon.*

2. *All the vertices are equivalent in the sense that for any two vertices u and v there is a rotation of the polyhedron that replaces the vertex*

Table 8.1.

	v = vertices	e = edges	f = faces	$v-e+f$
Cube	8	12	6	$8-12+6=2$
Tetrahedron	4	6	4	$4-6+4=2$
Octahedron	6	12	8	$6-12+8=2$
Dodecahedron	20	30	12	$20-30+12=2$
Icosahedron	12	30	20	$12-30+20=2$

u with the vertex v and also replaces all the edges emanating from u with the edges emanating from v.

As proved by Euclid, there are five regular polyhedra. The easiest regular polyhedron to visualize is, of course, the *cube* (Fig. 8.1) whose faces consist of 6 congruent squares. It has 12 edges and 8 vertices. Almost as immediate as the cube is the *tetrahedron*, a triangle-based pyramid, whose faces consist of 4 equilateral triangles. It has 6 edges and 4 vertices. The *octahedron*, a double square-based pyramid, has 8 equilateral triangles as its faces. It has 12 edges and 6 vertices. The *dodecahedron* has 12 regular pentagons as its faces, 30 edges and 20 vertices. The *icosahedron* has 20 equilateral triangles as its faces, 30 edges and 12 vertices. These counts are tabulated in Table 8.1.

It is commonly accepted that the Pythagoreans were aware of all five regular polyhedra. Theaetetus (415?–369? B.C.) is credited with being the first mathematician to formally prove their existence. While the existence of the cube, tetrahedron, and octahedron hardly requires justification, the existence of the dodecahedron and icosahedron is much less obvious. One way of demonstrating the existence of all these polyhedra is by means of coordinates. Assume that space has been endowed with a Cartesian coordinate system so that each point is described by a triple (x, y, z) and let

$$\mathbb{R}^3 = \{(x, y, z) \text{ such that } x,\ y, \text{ and } z \text{ are real numbers}\}$$

denote space. Then the following coordinates describe the vertices of the regular polyhedra.

Cube

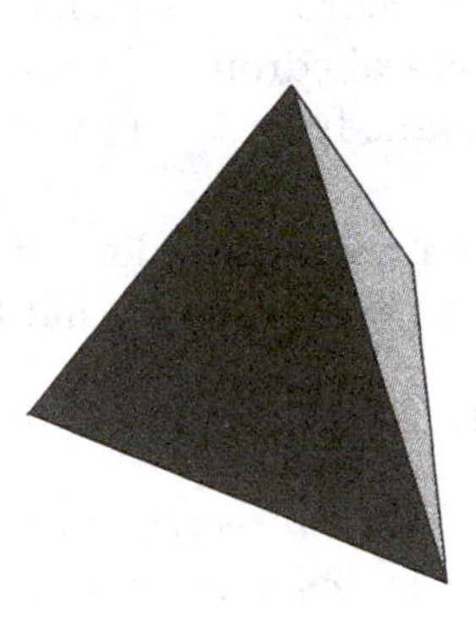

Tetrahedron

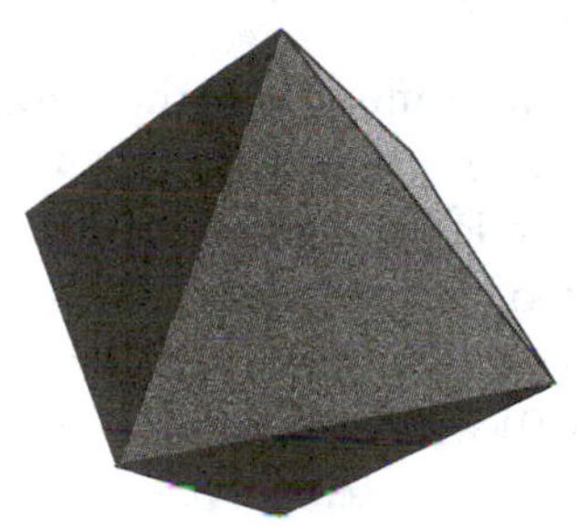

Octahedron

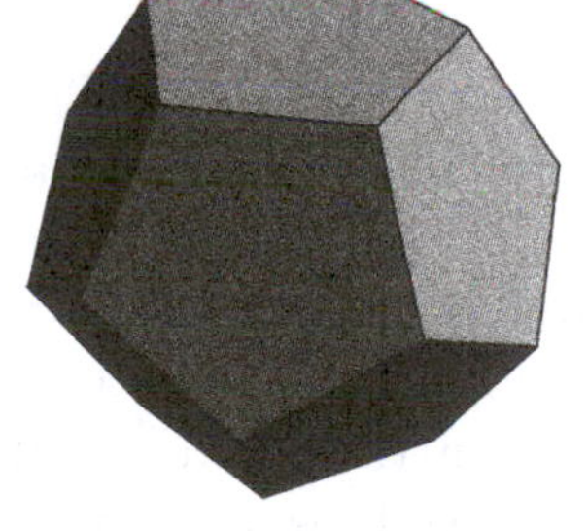

Dodecahedron

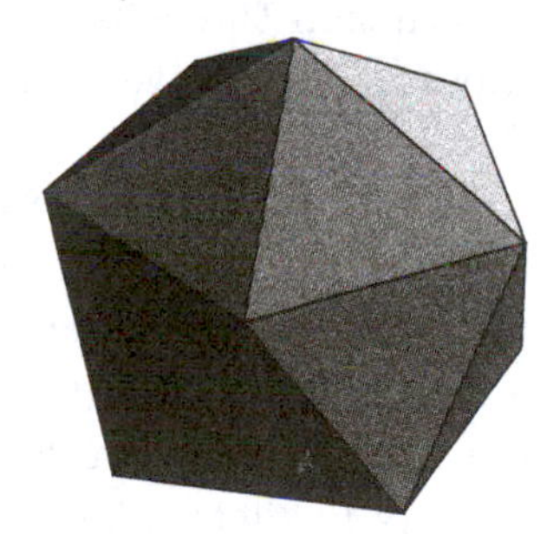

Icosahedron

Figure 8.1. The Platonic or regular polyhedra

Cube:	$(\pm 1, \pm 1, \pm 1)$
Tetrahedron:	$(1,1,1)$, $(1,-1,-1)$, $(-1,1,-1)$, $(-1,-1,1)$
Octahedron:	$(\pm 1,0,0)$, $(0,\pm 1,0)$, $(0,0,\pm 1)$
Dodecahedron:	$(0,\pm\tau,\pm 1/\tau)$, $(\pm 1/\tau,0,\pm\tau)$, $(\pm\tau,\pm 1/\tau,0)$, $(\pm 1,\pm 1,\pm 1)$
Icosahedron:	$(\pm\tau,0,\pm 1)$, $(0,\pm 1,\pm\tau)$, $(\pm 1,\pm\tau,0)$

where τ is the golden ratio $(\sqrt{5}-1)/2$ of Proposition 3.4.1.

A polyhedron is said to be *semiregular* if

1. *All the faces are regular polygons;*

2. *All the vertices are equivalent in the sense that for any two vertices u and v there is a rotation of the polyhedron that replaces the vertex u with the vertex v and also replaces all the edges emanating from u with the edges emanating from v.*

The semiregular polyhedra differ from the regular ones only in that their faces need not all be the same regular polygon. While one of the semiregular polyhedra was mentioned by Plato, their first serious study is attributed to Archimedes. They consist of the *prisms* and *antiprisms* (Fig. 8.2) as well as the *Archimedian polyhedra.* His work on this topic was lost and it was Johannes Kepler (1571–1630) who once again constructed all the semiregular polyhedra and discussed their relation to the regular polyhedra. Some of these semiregular polyhedra can be derived by truncating the corners of the regular polyhedra. This process is demonstrated here for the cube. In this description the cells of the original cube are referred to as the *old* cells and those of the derived solid as the *new* ones.

Truncated cube I: All the corners of the cube are cut off by planes that meet at the midpoints of the old edges (see Fig. 8.4). There are 12 new vertices, one for each of the old edges. Each of the 8 old vertices contributes 3 new edges, for a total of 24. Each of the 6 old square faces has been trimmed down to a smaller new square face and each of the 8 truncated corners has left a new triangular face, for a total of 14 new faces. The new polyhedron will be recognized as the cuboctahedron of Figure 8.3.

Truncated cube II: Again all the corners of the cube are cut off, but this time the cutting planes do not meet at the midpoints of the edges. Instead, a central portion of the old edge is left, whose length equals that of the edge of the new triangular face created by the truncation process (see Fig. 8.5). There are 24 new vertices, two for each of the old edges. Each of the 8 old vertices contributes 3 new edges, and there are also the 12 remnants of the

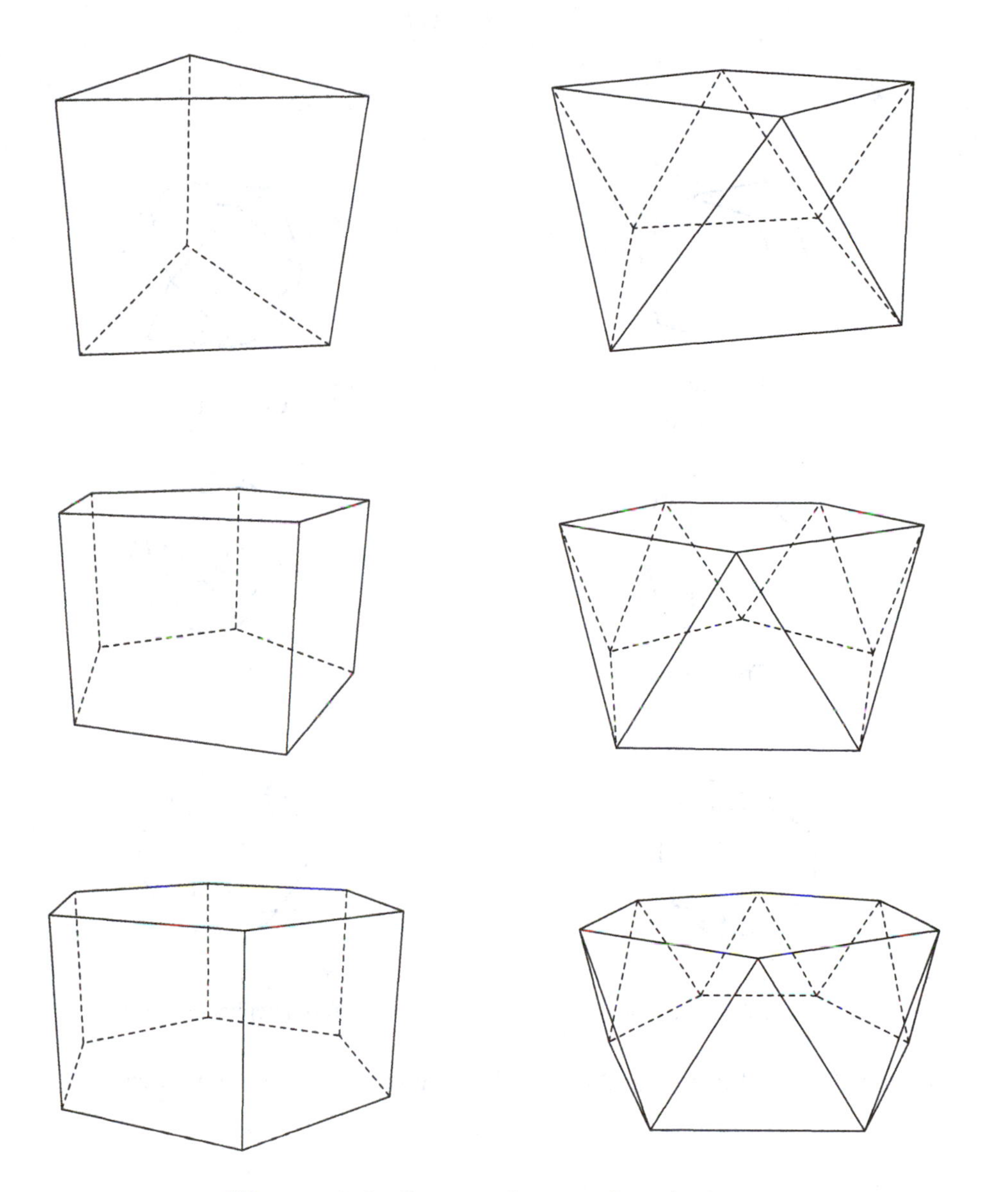

Figure 8.2. Some prisms and antiprisms

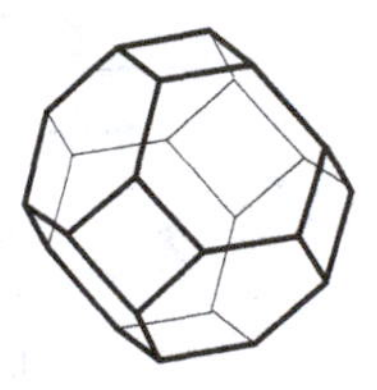

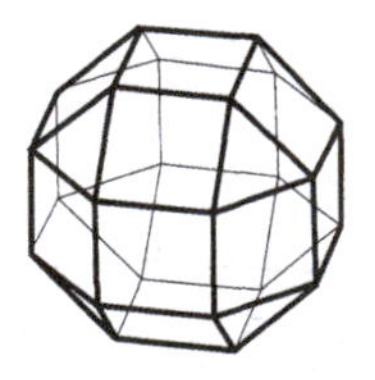

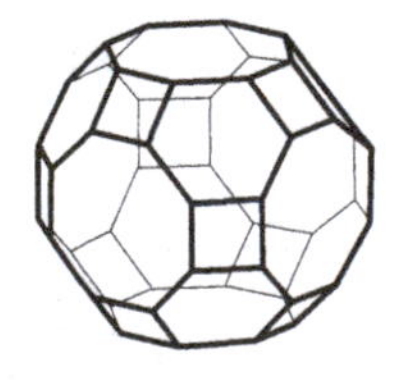

Figure 8.3. (a) Six of the Archimedian or semiregular polyhedra.

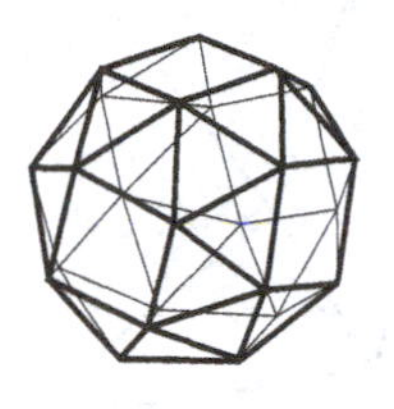

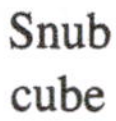

Snub cube

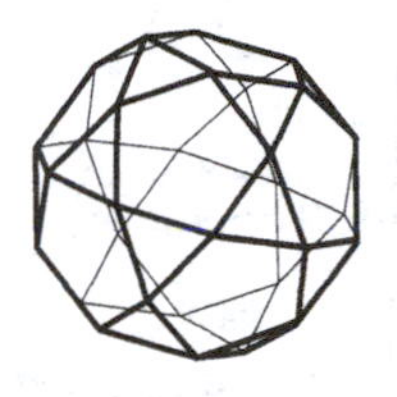

Icosidodecahedron

Truncated dodecahedron

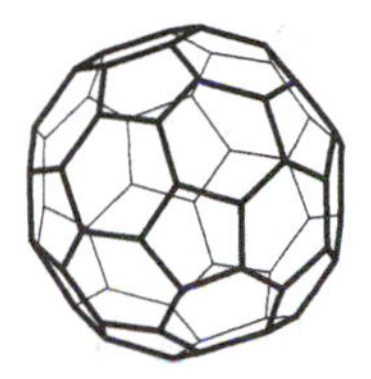

Truncated icosahedron

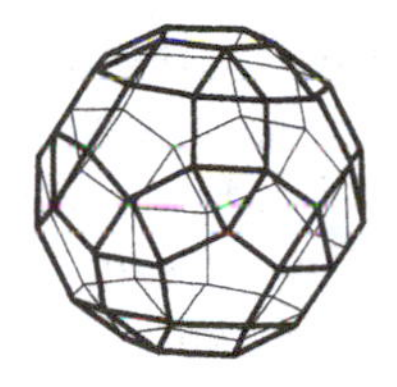

Small rhombicosidodecahedron

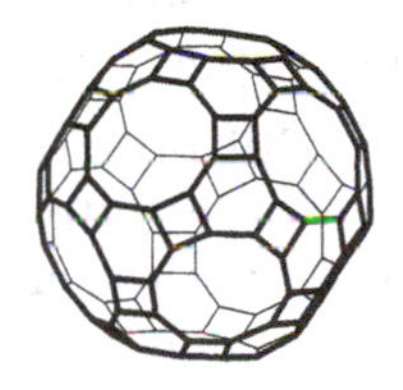

Great rhombicosidodecahedron

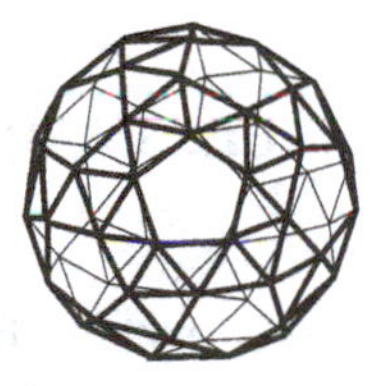

Snub dodecahedron

Figure 8.3. (b) Seven of the Archimedean or semiregular polyhedra

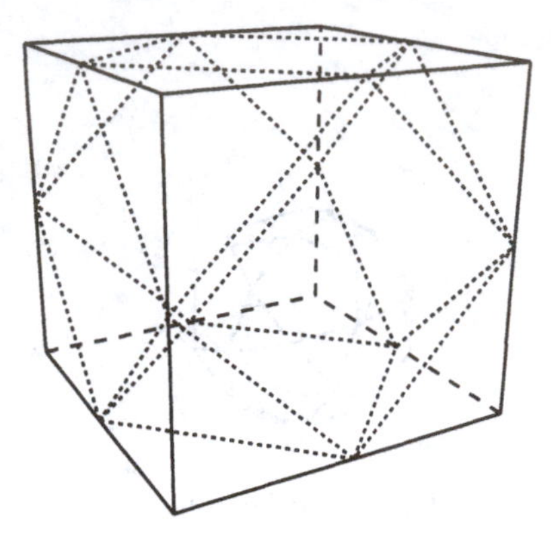

Cutting off corners

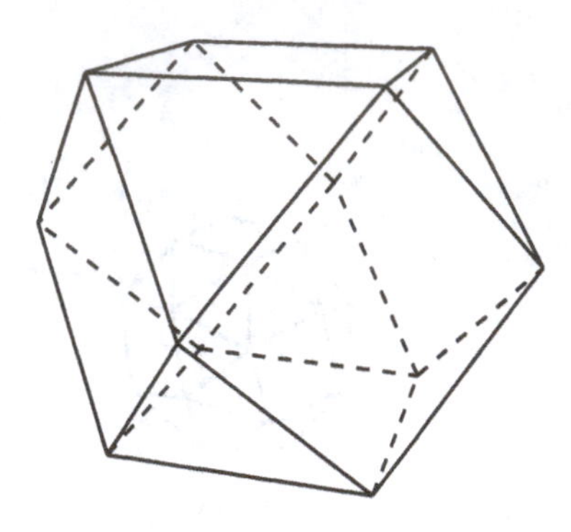

The truncated cube I

Figure 8.4.

old edges. These add up to a total of 36 new edges. Each of the 6 old square faces has been trimmed down to an octagon, and each of the 8 truncated corners has left a triangular face. Hence, this polyhedron has 14 faces and is the *truncated cube* in Figure 8.3.

The rightmost column of the tally of the cells of the regular polyhedra (Table 8.1) indicates that these counts are subject to a very simple and surprising relationship. This relationship actually holds for all polyhedra provided that their definition excludes the possibility of such troublesome features as the hole in the torus, the point juncture illustrated in Figure 8.6, and the edge juncture of Figure 8.7. For such trouble-free polyhedra, which we call *spherelike* and whose precise definition falls outside the bounds of this text, the following proposition holds.

Proposition 8.1.1 (Euler's equation, 1758) *For any spherelike polyhedron,* $v - e + f = 2$.

This equation is named after its discoverer, Euler, who has already been mentioned several times in this text. The discovery of this equation initiated a flourishing branch of mathematics now known as topology. Note that the cells of the aforementioned truncated cubes also satisfy Euler's equation. The pyramid and the prism that are based on n-sided polygons are offered as further examples of nonregular polyhedra for which Euler's equation holds. The pyramid has $n + 1$ vertices, $2n$ edges, and $n + 1$ faces so that

$$(n + 1) - 2n + (n + 1) = 2$$

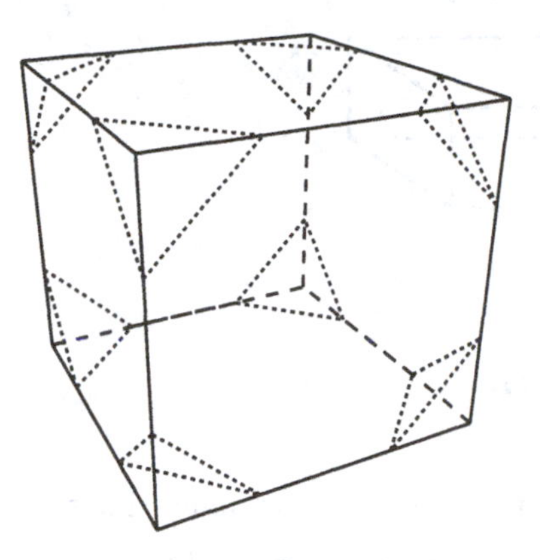

Cutting off corners

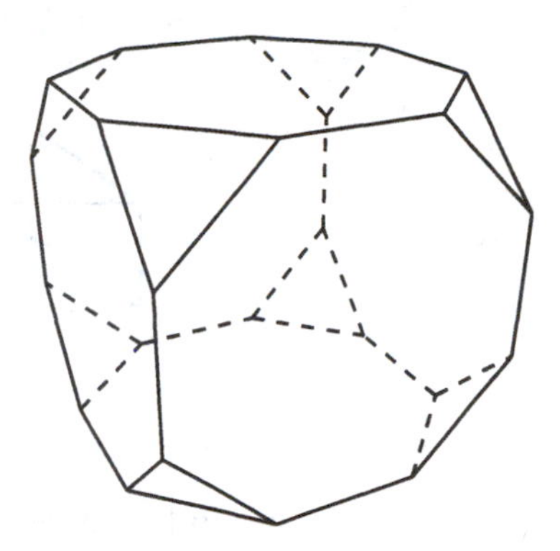

The truncated cube II

Figure 8.5.

and the prism has $2n$ vertices, $3n$ edges, and $n + 2$ faces so that again

$$2n - 3n + (n + 2) = 2.$$

Exercises 8.1

1. Answer the following questions for each of the two polyhedra obtained from the octahedron by the two truncation methods described in Figures 8.4 and 8.5 (parts a, b, c are to be answered without reference to Euler's equation).

   (a) How many vertices does it have?

   (b) How many edges does it have?

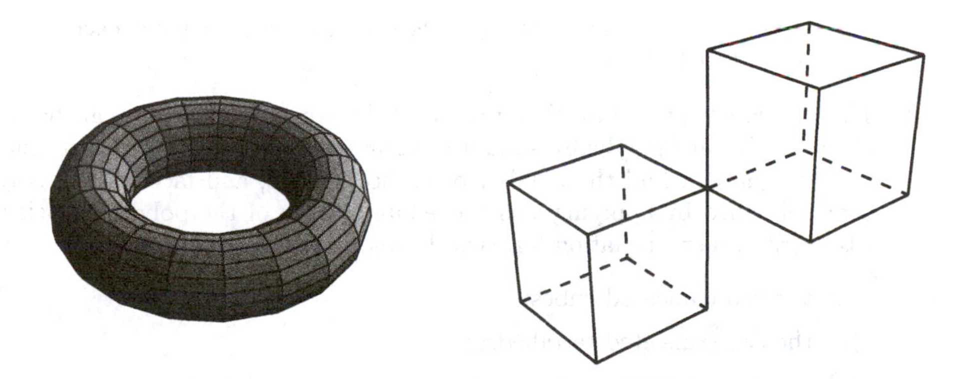

Figure 8.6. Counterexamples to Euler's equation

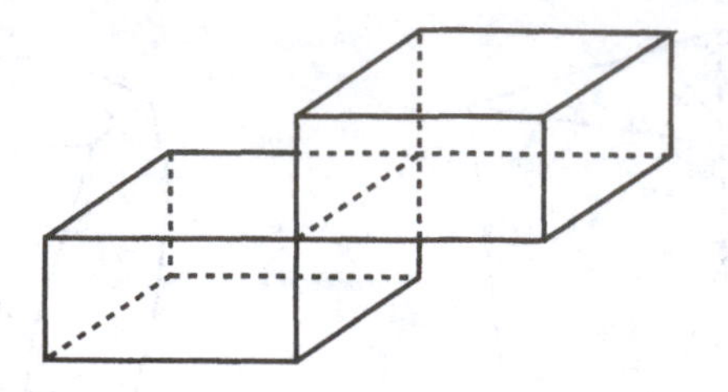

Figure 8.7.

(c) What regular polygons appear as its faces and how many times?

(d) Identify the truncated polyhedron in Figure 8.1 or in Figure 8.3.

(e) Verify that the cells of this polyhedron satisfy Euler's equation.

2. Repeat Exercise 1 for the tetrahedron.

3. Repeat Exercise 1 for the dodecahedron.

4. Repeat Exercise 1 for the icosahedron.

5. The truncation procedure that produced the truncated cube I can be applied to arbitrary polyhedra so as to obtain new polyhedra. Without using Euler's equation, find the number of vertices, edges, and faces of the polyhedra obtained by applying this procedure to each of the polyhedra below. Also verify Euler's equation for each derived polyhedron.

   (a) the two truncated cubes;

   (b) the two truncated tetrahedra;

   (c) the two truncated octahedra;

   (d) the two truncated dodecahedra;

   (e) the two truncated icosahedra;

   (f) a polyhedron with v vertices, e edges, and f faces, in which each vertex is incident to 3 edges.

6. The truncation procedure that produced the truncated cube II can be applied to arbitrary polyhedra so as to obtain new polyhedra. Without using Euler's equation, find the number of vertices, edges, and faces of the polyhedra obtained by applying this procedure to each of the polyhedra below. Also verify Euler's equation for each derived polyhedron.

   (a) the two truncated cubes;

   (b) the two truncated tetrahedra;

   (c) the two truncated octahedra;

   (d) the two truncated dodecahedra;

(e) the two truncated icosahedra;

(f) a polyhedron with v vertices, e edges, and f faces, in which each vertex is incident to d edges.

7. Use your favorite mathematical computer application to draw the following regular polyhedron from the coordinates given in this section:

   (a) tetrahedron

   (b) octahedron

   (c) cube

   (d) dodecahedron

   (e) icosahedron.

8. Show that there are infinitely many polyhedra all of whose faces are congruent squares.

9. Show that the cell counts of the polyhedron in Figure 8.7 do not satisfy Euler's equation. Explain why this is not a counterexample to this equation.

10. Show that the cell counts of the polyhedron in Figure 8.8 do not satisfy Euler's equation. Explain why this is not a counterexample to this equation.

11. Show that the cell counts of the polyhedron in Figure 8.9 do satisfy Euler's equation. Explain why this is not a counterexample to this equation.

12. Show that the cell counts of the polyhedron in Figure 8.10 do not satisfy Euler's equation. Explain why this is not a counterexample to this equation.

13. Each of the corners of the pyramid based on an n-sided polygon is truncated (in the usual two ways). Derive the number of vertices, edges, and faces of the resultant polyhedra without using Euler's equation and verify that this equation does indeed hold.

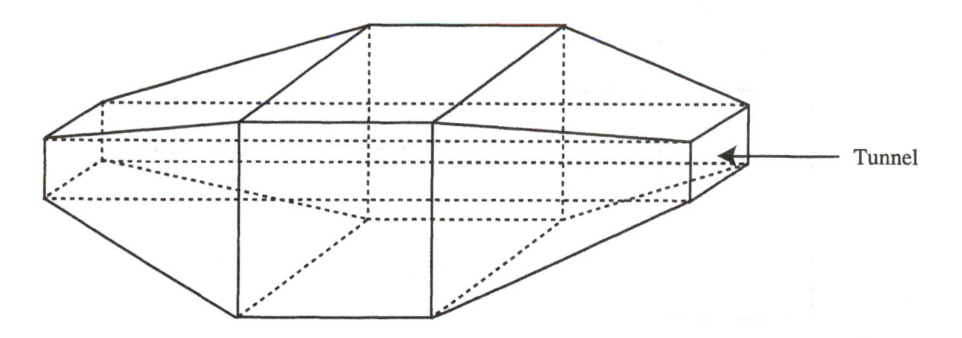

Figure 8.8.

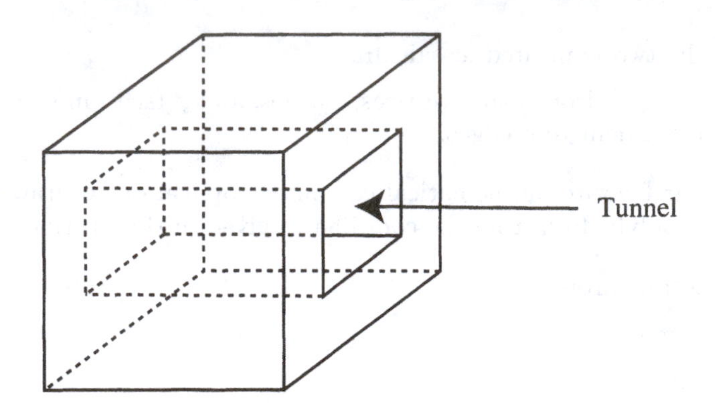

Figure 8.9.

14. Each of the corners of the prism based on an n-sided polygon is truncated (in the usual two ways). Derive the number of vertices, edges, and faces of the resultant polyhedra without using Euler's equation, and verify that this equation does indeed hold.

15. A diagonal of a polyhedron is a line segment joining two of its vertices. A polyhedron is said to be convex if it contains all of its diagonals either in its interior or on its faces. Show that there is only one convex polyhedron all of whose faces are congruent squares.

16. Construct a cube using the medium of your choice.

17. Construct a tetrahedron using the medium of your choice.

18. Construct an octahedron using the medium of your choice.

19. Construct a dodecahedron using the medium of your choice.

20. Construct an icosahedron using the medium of your choice.

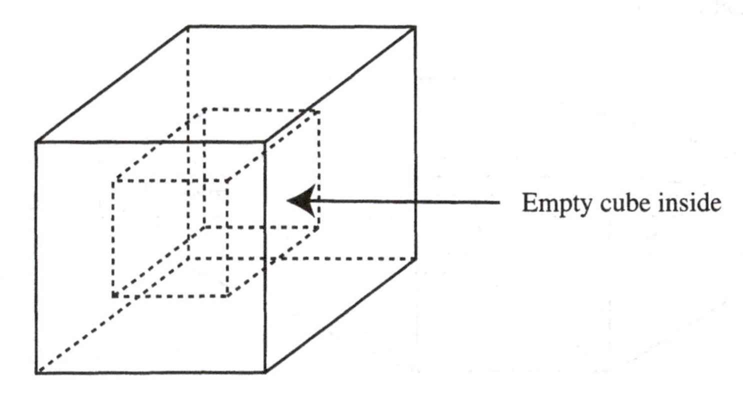

Figure 8.10.

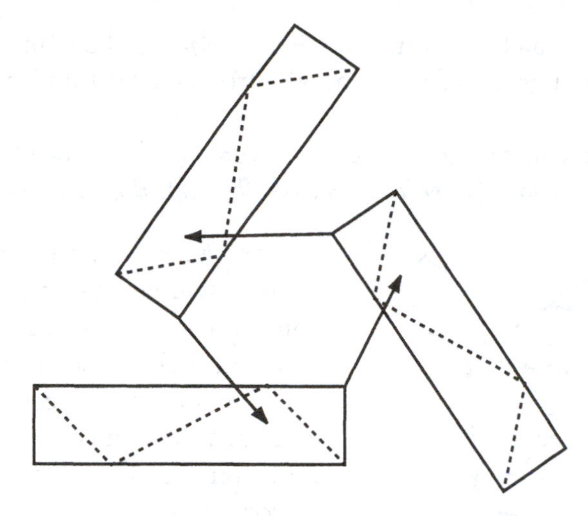

Figure 8.11.

21. A paper model of the dodecahedron can be constructed from thirty square sheets of paper ($8.5'' \times 8.5''$ is easy to work with). Each piece should be folded in half, and then each half is to be folded in half again, accordion fashion. Next, crease each piece along the dashed lines indicated in Figure 8.11, where the two corners are isosceles right triangles. These last three folds should all bend toward you. The pieces are to be tucked into each other as indicated in Figure 8.11.

22. Construct all thirteen semiregular polyhedra using the medium of your choice.

23. Note that the torus of Figure 8.6 has 19 "meridians." Assuming that it has n "equators," derive the number of vertices, edges, and faces (in terms of n) and show that Euler's equation is not satisfied.

8.2 Rotational Symmetries of Regular Polyhedra

A *symmetry* of a polyhedron is a motion of its ambient space that leaves the polyhedron in its original position. If the transformation is a rotation, then the symmetry is said to be a *rotational symmetry.* During the rotation the polyhedron may very well pass through nearby parts of space that it did not occupy initially, but when the rotation is accomplished the solid's position must coincide exactly with its initial position. Spatial rotations are denoted by the symbol $R_{A,\alpha}$, where A indicates the axis and α the magnitude and sense of the rotation.

The set of symmetries of a polyhedron is its *symmetry group* and the set of rotational symmetries is its *rotation group*. A theorem of Euler implies that the rotation group of every polyhedron is closed under composition.

Theorem 8.2.1 *If the axes of two rotations of* $\mathbb{R}^3$ *intersect, then their composition is a rotation whose axis passes through the intersection point.* □

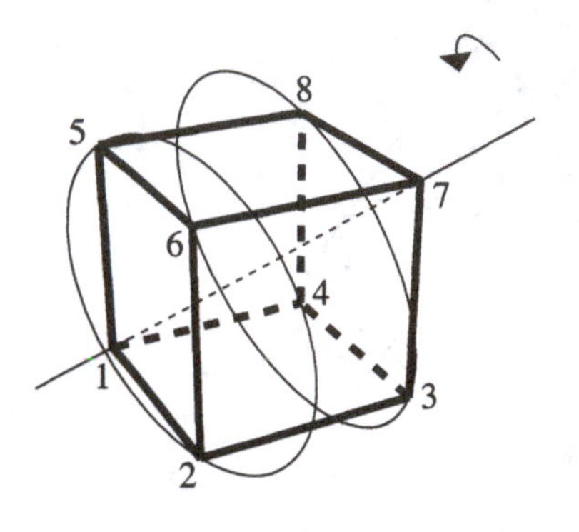

Figure 8.12. $R_{7,120^\circ} = R_{1,-120^\circ}$

The axes of those rotations that constitute symmetries of a polyhedron are constrained by the fact that they must pass through a vertex, the midpoint of an edge, or else the center of some face. This observation can be used to devise a notation for the rotational symmetries. Accordingly, $R_{7,\alpha}$ denotes a rotation of the cube whose axis passes through the vertex at 7 (Fig. 8.12), $R_{26,\alpha}$ denotes a rotation of the cube whose axis passes through the midpoint of the edge 26 (Fig. 8.13), and $R_{1265,\alpha}$ denotes a rotation of the cube whose axis passes through the center of the square 1265 (Figs. 8.14–8.15). This notation is subject to some redundancy. Thus, for the cube of Figures 8.12–8.15, the rotation $R_{7,\alpha}$ can also be written as $R_{1,-\alpha}$, the rotation $R_{26,\alpha}$ can also be written as $R_{48,-\alpha}$, and the rotation $R_{1265,\alpha}$ can also be written as $R_{3487,-\alpha}$.

The symbol $R_{A,\alpha}$ denotes a rotation by the oriented angle α, where the orientation is understood to be determined by an observer positioned outside the polyhedron near A. Thus, $R_{1265,90^\circ}$ (Fig. 8.14) denotes the 90° rotation of the cube about the axis that passes through the centers of the faces 1265 and 3487, counterclockwise from the point of view of an observer situated outside the cube near the face 1265. Note that since −90° denotes

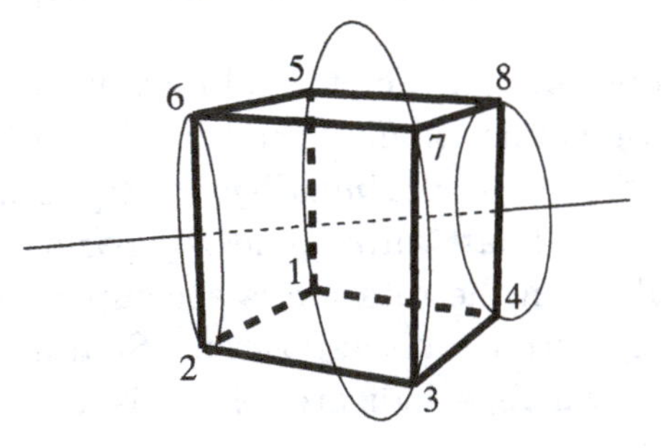

Figure 8.13. $R_{26,180^\circ} = R_{48,180^\circ}$

a clockwise rotation, it follows that $R_{1265,90^\circ} = R_{3487,-90^\circ}$. The circles in the illustrations are meant to help visualize the rotation; they are the "tracks" in which the vertices move to their new positions.

Just like their two-dimensional counterparts, spatial symmetries also have permutation representations. Thus, since the rotation $R_{1265,90^\circ}$ cycles the vertices in positions 1, 2, 6, 5 and also the vertices in positions 3, 7, 8, 4, it has the permutation representation $(1\ \ 2\ \ 6\ \ 5)$ $(3\ \ 7\ \ 8\ \ 4)$. Similarly, the 180° rotation $R_{1265,180^\circ}$ has the permutation representations $(1\ \ 6)(2\ \ 5)(7\ \ 4)(8\ \ 3)$.

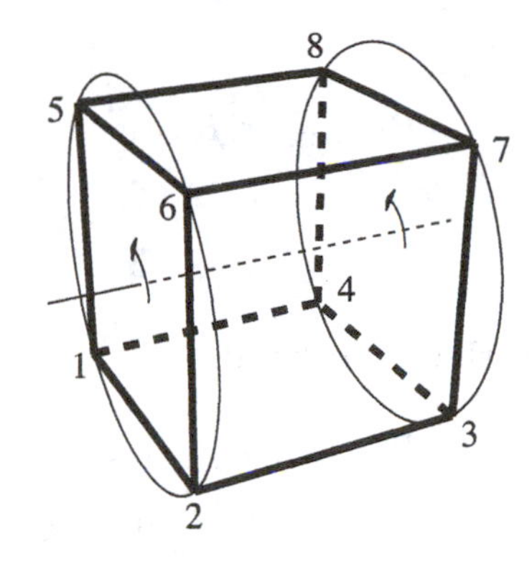

Figure 8.14. $R_{1265,90^\circ} = R_{3487,-90^\circ}$

A qualitatively different symmetry of the cube is obtained by a 180° rotation about an axis that passes through the midpoints of two diametrically opposite edges of the cube. Such, for example, is the rotation $R_{26,180^\circ} = R_{48,180^\circ}$ (Fig. 8.13). It has the permutation representation $(1\ \ 7)(2\ \ 6)(3\ \ 5)(4\ \ 8)$. While this permutation looks very much like that of $R_{1265,180^\circ}$, there is a significant geometrical difference between them. The permutation representation $(1\ \ 7)(2\ \ 6)(3\ \ 5)(4\ \ 8)$ of $R_{26,180^\circ}$ has cycles that are in fact cells of the cube, namely $(2\ \ 6)$ and $(4\ \ 8)$. On the other hand, none of the cycles of the permutation representation $(1\ \ 6)\ (2\ \ 5)\ (3\ \ 8)\ (4\ \ 7)$ of $R_{1265,180^\circ}$ are cells of the cube.

The permutation representation of the rotation $R_{7,120^\circ}$ of Figure 8.13 is $(1)(2\ \ 4\ \ 5)(3\ \ 8\ \ 6)(7)$.

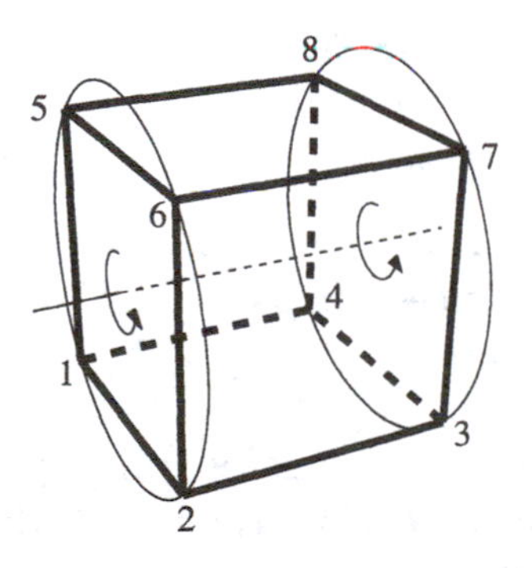

Figure 8.15. $R_{1265,270^\circ} = R_{3487,-270^\circ}$

The compositions of rotational symmetries are easily computed by means of their permutation representations. Accordingly, since $R_{7,120^\circ} = (1)(2\ 4\ 5)(3\ 8\ 6)(7)$ and $R_{1265,90^\circ} = (1\ 2\ 6\ 5)(3\ 7\ 8\ 4)$, it follows that

$$\begin{aligned}
&R_{1265,90^\circ} \circ R_{7,120^\circ} \\
&\quad = (1\ 2\ 6\ 5)(3\ 7\ 8\ 4) \circ (1)(2\ 4\ 5)(3\ 8\ 6)(7) \\
&\quad = (1\ 2\ 3\ 4)(5\ 6\ 7\ 8) = R_{1234,-90^\circ}.
\end{aligned}$$

On the other hand,

$$\begin{aligned}
&R_{7,120^\circ} \circ R_{1265,90^\circ} \\
&\quad = (1)(2\ 4\ 5)(3\ 8\ 6)(7) \circ (1\ 2\ 6\ 5)(3\ 7\ 8\ 4) \\
&\quad = (1\ 4\ 8\ 5)(2\ 3\ 7\ 6) = R_{1485,-90^\circ}.
\end{aligned}$$

Similarly, since $R_{26,180^\circ} = (1\ 7)(2\ 6)(3\ 5)(4\ 8)$ and $R_{34,180^\circ} = (1\ 7)(2\ 8)(3\ 4)(5\ 6)$, it follows that

$$\begin{aligned}
&R_{26,180^\circ} \circ R_{34,1800^\circ} \\
&\quad = (1\ 7)(2\ 6)(3\ 5)(4\ 8) \circ (1\ 7)(2\ 8)(3\ 4)(5\ 6) \\
&\quad = (1)(2\ 4\ 5)(3\ 8\ 6)(7) = R_{7,120^\circ}.
\end{aligned}$$

The number of symmetries in a group is that group's *order.*

Proposition 8.2.2 *The rotation group of the cube has order* 24 *and its rotations are classified as*

Id
8 *rotations of the type* $R_{\text{vertex},120^\circ}$
6 *rotations of the type* $R_{\text{edge},180^\circ}$
6 *rotations of the type* $R_{\text{face},90^\circ}$
3 *rotations of the type* $R_{\text{face},180^\circ}$

SKETCH OF PROOF: This follows from the fact that the cube has four axes that join opposite vertices, six axes that join opposite edges, and three axes that join opposite faces. Q.E.D.

Whereas the axis of any rotational symmetry of the cube joins the midpoints of cells of the same dimension, the tetrahedron presents us with a new alternative. The axis of the symmetry $R_{2,-120^\circ}$ joins the vertex 2 to the center of the triangular face with vertices 1, 3, and 4 (Fig. 8.16). This rotation has the permutation representation $(1\ \ 3\ \ 4)(2)$. The only qualitatively different rotational symmetry of the tetrahedron is the 180° rotation about the line joining the midpoints of two opposite edges. Such, for example, is $R_{24,180^\circ} = (1\ \ 3)(2\ \ 4)$ (Fig. 8.17). Note that

$$R_{2,-120^\circ} \circ R_{24,180^\circ} = (1\ \ 3\ \ 4)(2) \circ (1\ \ 3)(2\ \ 4) = (1\ \ 4\ \ 2)(3)$$
$$= R_{3,-120^\circ}.$$

Proposition 8.2.3 *The rotation group of the tetrahedron has order* 12 *and its rotations are classified as*

Id

4 *rotations of each of the types* $R_{\text{vertex},120^\circ}$ *and* $R_{\text{vertex},240^\circ}$

3 *rotations of the type* $R_{\text{edge},180^\circ}$

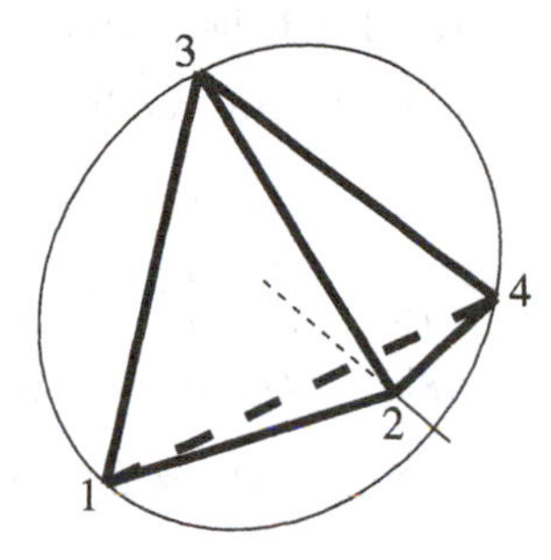

Figure 8.16. $R_{3,240^\circ}$

The verification of the analogous proposition regarding the octahedron is relegated to Exercise 1.

Proposition 8.2.4 *The rotation group of the octahedron has order* 24 *and its rotations are classified as*

Id

6 *rotations of the type* $R_{\text{vertex},90^\circ}$

3 *rotations of the type* $R_{\text{vertex},180^\circ}$

6 *rotations of the type* $R_{\text{edge},180^\circ}$

8 *rotations of each of the types* $R_{\text{face},120^\circ}$

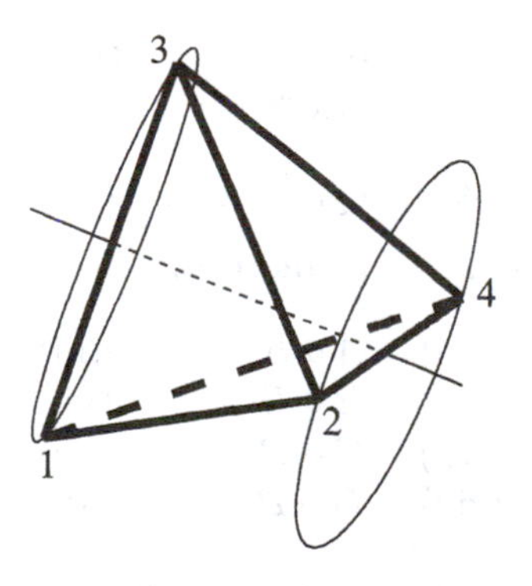

Figure 8.17. $R_{13,180^\circ}$

Exercises 8.2

1. Classify the rotational symmetries of the octahedron.
2. Classify the rotational symmetries of the dodecahedron.
3. Classify the rotational symmetries of the icosahedron.
4. Suppose $A = R_{1,120^\circ}$, $B = R_{26,180^\circ}$, $C = R_{2376,90^\circ}$, $D = R_{1234,180^\circ}$ are symmetries of the cube of Figures 8.12–8.15.

   (a) Find the permutation representations of A, B, C, D.

   (b) Identify the following symmetries:

(i)	$A \circ A$	(ii)	$A \circ B$	(iii)	$A \circ C$	(iv)	$A \circ D$
(v)	$B \circ A$	(vi)	$B \circ B$	(vii)	$B \circ C$	(viii)	$B \circ D$
(ix)	$C \circ A$	(x)	$C \circ B$	(xi)	$C \circ C$	(xii)	$C \circ D$
(xiii)	$D \circ A$	(xiv)	$D \circ B$	(xv)	$D \circ C$	(xvi)	$D \circ D$

5. Repeat Exercise 4 with $A = R_{2,-120^\circ}$, $B = R_{15,180^\circ}$, $C = R_{3487,270^\circ}$, $D = R_{5678,180^\circ}$.
6. Suppose $A = R_{1,120^\circ}$, $B = R_{24,180^\circ}$, $C = R_{2,240^\circ}$, $D = R_{14,180^\circ}$ are symmetries of the tetrahedron of Figures 8.16–8.17.

   (a) Find the permutation representations of A, B, C, D.

   (b) Identify the following symmetries:

(i)	$A \circ A$	(ii)	$A \circ B$	(iii)	$A \circ C$	(iv)	$A \circ D$
(v)	$B \circ A$	(vi)	$B \circ B$	(vii)	$B \circ C$	(viii)	$B \circ D$
(ix)	$C \circ A$	(x)	$C \circ B$	(xi)	$C \circ C$	(xii)	$C \circ D$
(xiii)	$D \circ A$	(xiv)	$D \circ B$	(xv)	$D \circ C$	(xvi)	$D \circ D$

7. Repeat Exercise 6 with $A = R_{3,120^\circ}$, $B = R_{23,180^\circ}$, $C = R_{3,240^\circ}$, $D = R_{12,180^\circ}$.
8. Suppose $A = R_{1,90^\circ}$, $B = R_{25,180^\circ}$, $C = R_{235,240^\circ}$, $D = R_{4,180^\circ}$ are symmetries of the octahedron of Figure 8.18.

   (a) Find the permutation representations of A, B, C, D.

   (b) Identify the following symmetries:

(i)	$A \circ A$	(ii)	$A \circ B$	(iii)	$A \circ C$	(iv)	$A \circ D$
(v)	$B \circ A$	(vi)	$B \circ B$	(vii)	$B \circ C$	(viii)	$B \circ D$
(ix)	$C \circ A$	(x)	$C \circ B$	(xi)	$C \circ C$	(xii)	$C \circ D$
(xiii)	$D \circ A$	(xiv)	$D \circ B$	(xv)	$D \circ C$	(xvi)	$D \circ D$

9. Repeat Exercise 8 with $A = R_{1,180^\circ}$, $B = R_{46,180^\circ}$, $C = R_{345,120^\circ}$, $D = R_{6,90^\circ}$.

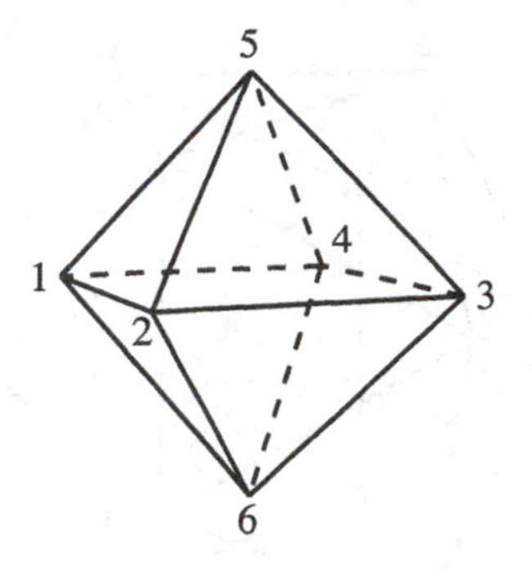

Figure 8.18. Octahedron

10. Suppose $A = R_{5,120°}$, $B = R_{57,180°}$, $C = R_{57jbf,72°}$, $D = R_{1d5f2,144°}$ are symmetries of the dodecahedron of Figure 8.19.

    (a) Find the permutation representations of A, B, C, D.

    (b) Identify the following symmetries:

(i)	$A \circ A$	(ii)	$A \circ B$	(iii)	$A \circ C$	(iv)	$A \circ D$
(v)	$B \circ A$	(vi)	$B \circ B$	(vii)	$B \circ C$	(viii)	$B \circ D$
(ix)	$C \circ A$	(x)	$C \circ B$	(xi)	$C \circ C$	(xii)	$C \circ D$
(xiii)	$D \circ A$	(xiv)	$D \circ B$	(xv)	$D \circ C$	(xvi)	$D \circ D$

11. Repeat Exercise 10 with $A = R_{3,240°}$, $B = R_{bf,180°}$, $C = R_{68iae,144°}$, $D = R_{2fbch,72°}$.

12. Suppose $A = R_{1,72°}$, $B = R_{4a,180°}$, $C = R_{349,240°}$, $D = R_{4,144°}$ are symmetries of the icosahedron of Figure 8.20.

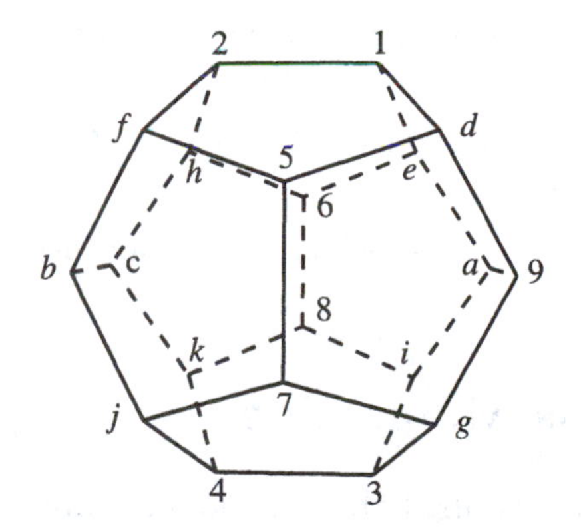

Figure 8.19. Dodecahedron

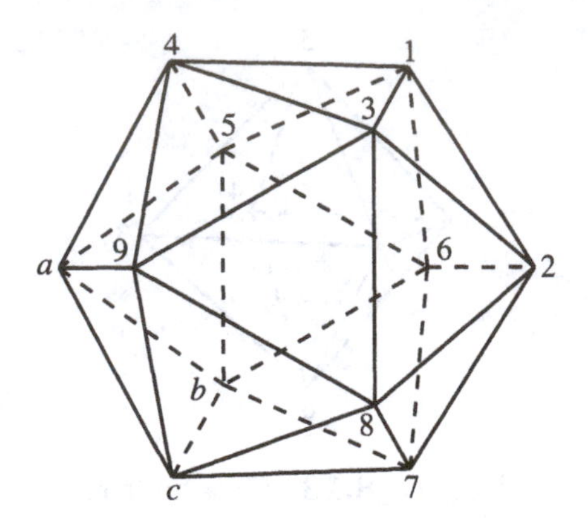

Figure 8.20. Icosahedron

(a) Find the permutation representations of A, B, C, D.

(b) Identify the following symmetries:

(i)	$A \circ A$	(ii)	$A \circ B$	(iii)	$A \circ C$	(iv)	$A \circ D$
(v)	$B \circ A$	(vi)	$B \circ B$	(vii)	$B \circ C$	(viii)	$B \circ D$
(ix)	$C \circ A$	(x)	$C \circ B$	(xi)	$C \circ C$	(xii)	$C \circ D$
(xiii)	$D \circ A$	(xiv)	$D \circ B$	(xv)	$D \circ C$	(xvi)	$D \circ D$

13. Repeat Exercise 12 with $A = R_{6,144^\circ}$, $B = R_{89,180^\circ}$, $C = R_{126,120^\circ}$, $D = R_{c,144^\circ}$.

14. How many rotational symmetries does a rectangular box have if all of its dimensions are different?

15. How many rotational symmetries does a rectangular box have if exactly two of its dimensions are the same?

16. A triangular prism has an equilateral base. How many rotational symmetries does it have?

17. A triangular prism has a base with sides 6, 6, 4. How many rotational symmetries does it have?

18. A prism has a base that is a regular n-gon. How many rotational symmetries does it have?

8.3 Monstrous Moonshine

The previous section described the rotational symmetries of the cube and the octahedron separately. Since the cube's symmetries are permutations of 8 vertices whereas those of the octahedron are permutations of 6 vertices, these polyhedra's rotation groups look quite different. Nevertheless, there is a natural sense in which these two groups are identical. Observe that in Figures 8.21–8.23 a cube has been placed inside an octahedron so that each of the vertices of the first is the center of a triangular face of the latter. The feasibility of this placement implies that every R_{face}-symmetry of the cube is also an R_{vertex}-symmetry of the octahedron (Fig. 8.21). Similarly, every R_{edge}-symmetry of the cube is also an R_{edge}-symmetry of the octahedron (Fig. 8.22) and every R_{vertex}-symmetry of the cube is an R_{face}-symmetry of the octahedron (Fig. 8.23). Thus, the rotation groups of the cube and the octahedron are now revealed as being identical. Technically, they are said to be *isomorphic*. It is clear that isomorphic groups must have the same order, and so the rotation groups of the cube and the tetrahedron are not isomorphic. Groups of the same order need not be isomorphic either. This is demonstrated by the symmetry group of the regular 12-gon. This polygon has 24 symmetries and the cube has 24 rotational symmetries. However, the 30° rotation of this polygon has order 12 and no rotational symmetry of the cube has such an order. Hence the symmetry group of the regular 12-gon and the rotation group of the cube are not isomorphic, even though they have the same orders (see Exercise 6).

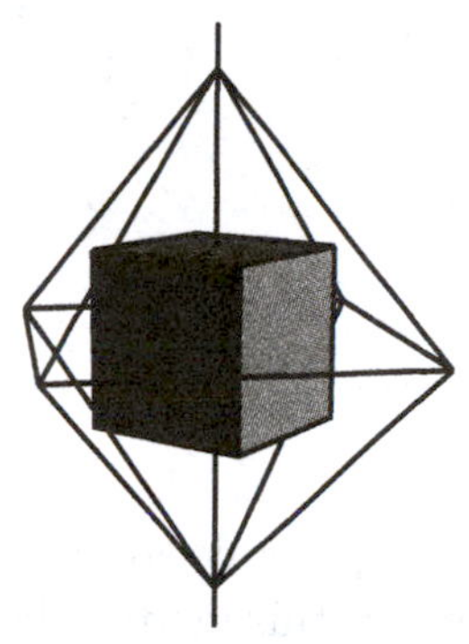

Figure 8.21.

Group theory, the mathematical theory of symmetry, has its origins in the work of Joseph Louis Lagrange (1736–1813) on the theory of equations. It was later used by Niels Henrik Abel (1802–1829) and Évariste Galois (1811–1832) to settle the question of which equations can be solved by explicit algebraic formulas and which can only be solved by means of successive approximations. The subsequent investigations of Felix Klein (1849–1925) and Henri Poincaré (1854–1912) pointed out the central role that group theory also plays in geometry.

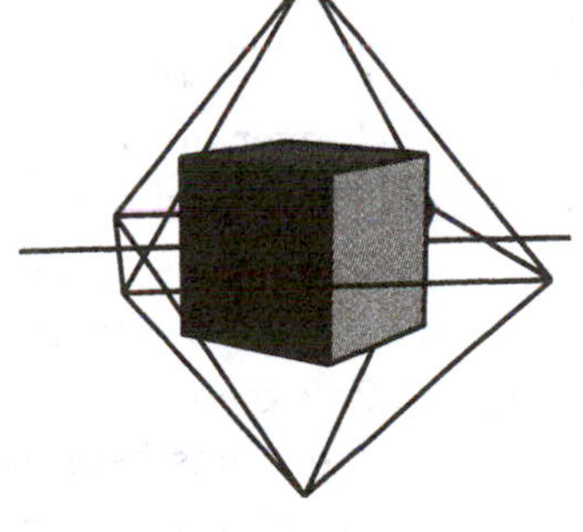

Figure 8.22.

One of the main goals of group theory is the classification of all groups up to isomorphism. While there is no expectation that this goal will be achieved in the foreseeable future, a significant milestone was passed less than twenty years ago when the *finite simple groups* were completely classified. There is

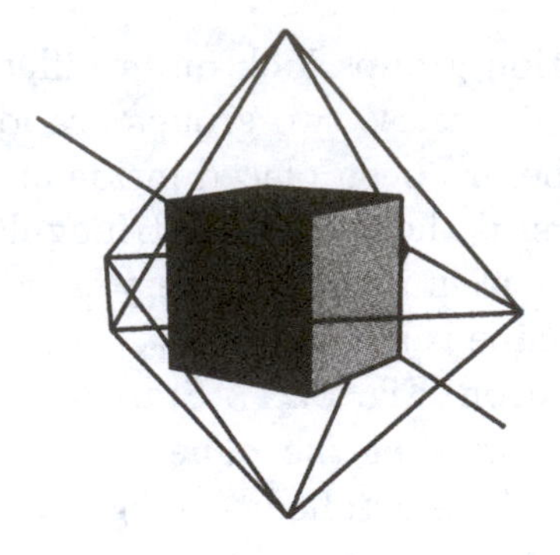

Figure 8.23.

nothing simple about the simple groups, nor is it possible to characterize them in this text. The symmetries of the icosahedron (and the dodecahedron) constitute a simple group whereas those of the cube and tetrahedron are not simple, but this difference does not have a geometrical interpretation. Algebraically, though, the difference is extremely important. The simplicity of the dodecahedral group turns out to be responsible for the nonexistence of a formulaic solution for the general fifth-degree equation

$$ax^5 + bx^4 + cx^3 + dx^2 + ex + f = 0.$$

Conversely, the nonsimplicity of the symmetry group of the tetrahedron is ultimately tantamount to the existence of such a formula for the fourth-degree equation

$$ax^4 + bx^3 + cx^2 + dx + e = 0.$$

The classification of the finite simple groups constitutes the most monumental task ever accomplished by mathematicians. Its proof is spread over 500 articles comprising more than 14,000 journal pages written by hundreds of researchers. This classification asserts that the finite simple groups fall into two categories: several infinite families of groups that possess clear patterns and 26 exceptional groups, known as the *sporadic groups*, for which no general pattern has been found.

The first of the sporadic groups was discovered in 1861 and the last two almost simultaneously in 1980. The largest of these was nicknamed MONSTER because of its order, which is

808,017,424,794,512,875,886,459,904,961,710,757,005,754,368,000,000,000.

MONSTER, discovered by Bernd Fischer and Robert L. Griess, is the group of symmetries of a (nonregular) polyhedron in 196,883 dimensions.

When word of this discovery reached John McKay he pointed out the remarkable coincidence that 196,884 is one of the coefficients in the series

$$\begin{aligned} j = q^{-1} &+ 744 + 196{,}884q + 21{,}493{,}760q^2 + 864{,}299{,}970q^3 \\ &+ 20{,}245{,}856{,}256q^4 + 333{,}202{,}640{,}600q^5 + 4{,}252{,}023{,}300{,}096q^6 \\ &+ 44{,}656{,}994{,}071{,}935q^7 + 401{,}490{,}886{,}656{,}000q^8 + \cdots , \end{aligned}$$

which arises in the seemingly unrelated context of doubly periodic functions on the hyperbolic plane. For the purposes of this exposition, these functions can be thought of as patterns that underlie hyperbolic wallpaper designs. As mathematicians were at a loss to explain this conjunction, they dubbed the following equation as *McKay's formula:*

$$196{,}884 = 1 + 196{,}883.$$

Shortly thereafter John Thompson noted that actually much more was true. The number 196,883 is the second one of an important sequence of 194 integers, the *degrees of the characters* of MONSTER, of which the first twelve are

$$\begin{gathered} 1 \\ 196{,}883 \\ 21{,}296{,}876 \\ 842{,}609{,}326 \\ 18{,}538{,}750{,}076 \\ 19{,}360{,}062{,}527 \\ 293{,}553{,}734{,}298 \\ 3{,}879{,}214{,}937{,}598 \\ 36{,}173{,}193{,}327{,}999 \\ 125{,}510{,}727{,}015{,}275 \\ 190{,}292{,}345{,}709{,}543 \\ 222{,}879{,}856{,}734{,}249 \end{gathered}$$

and the last one is

$$258{,}823{,}477{,}531{,}055{,}064{,}045{,}234{,}375.$$

The significance of these integers is that, except for 1, each denotes the number of dimensions required by a new polyhedron whose symmetry group is isomorphic to MONSTER. Thompson discovered that, with the exception of 744, all the other early coefficients in the j-series also have simple expressions in terms of MONSTER'S degrees. For example,

$$21{,}493{,}760 = 1 + 196{,}883 + 21{,}296{,}876$$
$$864{,}299{,}970 = 2 \cdot 1 + 2 \cdot 196{,}883 + 21{,}296{,}876 + 842{,}609{,}326.$$

John H. Conway assigned the name *Moonshine* to these and other related unexplained phenomena in 1979, "intending the word to convey our feelings that they are seen in a dim light, and that the whole subject is rather vaguely illicit." It should be remembered that at that time the existence of MONSTER had only been conjectured so that even the aforementioned list of degrees was questionable, not to mention their purported relation with the coefficients of the j-series.

As was noted previously, the existence of MONSTER was conclusively demonstrated in 1980 by R. Griess, who tried, unsuccessfully, to have its name changed to *The Friendly Giant.* Monstrous Moonshine mathematics was finally explained by Richard E. Borcherds, who found the connection in the theory of vertex algebras, a discipline developed recently for the purpose of providing a mathematical foundation to the new superstring theory of physics. For this work Borcherds received the 1998 Fields Medal, the most prestigious award bestowed by the mathematical community.

Exercises 8.3

In the following exercises, $f_1 = 1$, $f_2 = 196{,}883$, $f_3 = 21{,}296{,}876$, ... denote the degrees of the characters of MONSTER and a_k denotes the coefficient of q^k in the j-series, so that $a_{-1} = 1$, $a_0 = 744$, $a_1 = 196{,}884$, $a_2 = 21{,}493{,}760$, ...

1. Find integers $x_1, x_2, \ldots, x_6$ such that $a_4 = x_1 f_1 + x_2 f_2 + \cdots + x_6 f_6$.
2. Find integers $x_1, x_2, \ldots, x_7$ such that $a_5 = x_1 f_1 + x_2 f_2 + \cdots + x_7 f_7$.
3. Find integers $x_1, x_2, x_3, \ldots$ such that $a_6 = x_1 f_1 + x_2 f_2 + x_3 f_3 + \cdots$.
4. Find integers $x_1, x_2, x_3, \ldots$ such that $a_7 = x_1 f_1 + x_2 f_2 + x_3 f_3 + \cdots$.
5. Find integers $x_1, x_2, x_3, \ldots$ such that $a_8 = x_1 f_1 + x_2 f_2 + x_3 f_3 + \cdots$.
6. Show that the rotation groups of the cube and the prism whose base is a regular 12-gon are not isomorphic even though they have the same orders.

Chapter Review Exercises

In the following exercises, for each integer $n \geq 3$, P_n denotes the pyramid obtained by joining the vertices of a regular n-gon to a point outside the plane of the n-

gon and lying directly above its center. The polyhedron PP_n is the double pyramid obtained by fitting together two copies of the pyramid P_n at their base.

1. Verify that for each $n \geq 3$, Euler's equation holds for the pyramid P_n and for both of the polyhedra obtained from it by truncation.
2. Verify that for each $n \geq 3$, Euler's equation holds for the double pyramid PP_n and for both of the polyhedra obtained from it by a type II truncation.
3. Describe the rotational symmetries of the pyramid P_3. Note that there are two cases to be considered.
4. Describe the rotational symmetries of the pyramid P_n, $n \geq 4$.
5. Describe the rotational symmetries of the double pyramid PP_4. Note that there are two cases to be considered.
6. Describe the rotational symmetries of the double pyramid PP_n , $n \neq 4$.
7. Explain why the regular n-gon and the double pyramid PP_n have isomorphic rotation groups for $n \geq 5$ (when embedded in $\mathbb{R}^3$).
8. Prove that the icosahedron and dodecahedron have isomorphic rotation groups.
9. Are the following statements true or false? Justify your answers.
   (a) Every polyhedron has at least one symmetry.
   (b) Every regular polyhedron has at least two symmetries.
   (c) Every regular polyhedron has at least as many symmetries as vertices.
   (d) Every regular polyhedron has at least as many symmetries as edges.
   (e) The composition of every two rotations of $\mathbb{R}^3$ is also a rotation.
   (f) If two rotations of $\mathbb{R}^3$ have intersecting axes, then their composition is also a rotation.
   (g) All the semiregular polyhedra can be obtained from the regular polyhedra by truncation.
   (h) Euler discovered the regular polyhedra.
   (i) 196,883 is an interesting integer.
   (j) MONSTER is a symmetry group.
   (k) MONSTER was discovered by John H. Conway.
   (l) Every two symmetry groups are isomorphic.
   (m) The rotation groups of the cube and the octahedron are isomorphic.
   (n) The rotation groups of the cube and the tetrahedron are isomorphic.
   (o) There are at least 193 distinct polyhedra whose symmetry groups are isomorphic to MONSTER.

Chapter 9

Informal Topology

As was indicated in Chapter 1, there are many variants of Euclidean geometry. Each of these involves some modification of the notion of distance. Having relativized the notion of distance, it was only a question of time before mathematicians began to consider the possibility of a geometry that is independent of the notion of distance and is concerned instead with such issues as connectedness and contiguity. This distanceless geometry came into its own in the twentieth century when it came to be known as *topology*.

Consider, by way of an example, Euler's theorem (Proposition 8.1.1), which states that if v, e, and f are the respective counts of the vertices, edges, and faces of a solid, then

$$v - e + f = 2. \tag{1}$$

Large as the class of solids that this equation governs is, it can be widened even further. Euler's equation remains valid even after the solids are subjected to distortions which result in the curving of their edges and faces (see Fig. 9.1). We need simply relax the definition of edges and faces so as to allow for any non-self-intersecting curves and surfaces. Soccer and volley balls, together with the patterns formed by their seams, are examples of such curved solids to which Euler's equation applies. Moreover, it is clear that the equation still holds after the balls are deflated.

Topology is the study of those properties of geometric figures that remain valid even after the figures are subjected to distortions. This is commonly expressed by saying that topology is *rubber-sheet geometry*. Accordingly, our necessarily informal definition of a *topological space* identifies it as any subset of space from which the notions of straightness and length have been abstracted; only the aspect of contiguity remains. Points, arcs, loops, triangles, solids (both straight and curved), and the surfaces of the latter are all

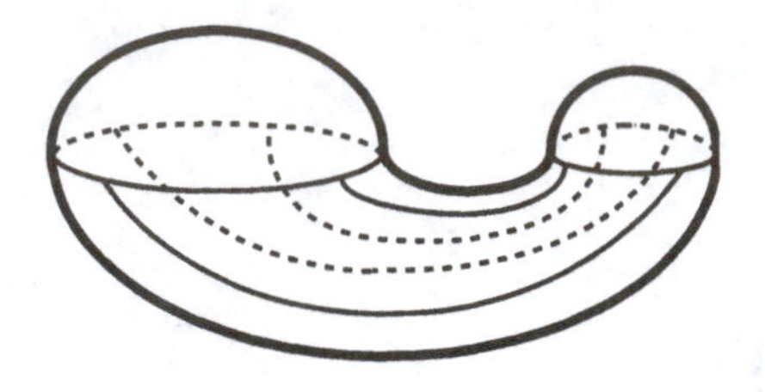

Figure 9.1. A curved cube

examples of topologic spaces. They are, of course, also geometric objects, but topology is only concerned with those aspects of their geometry that remain valid despite any translations, elongations, inflations, distortions, or twists.

Another topological problem investigated by Euler, somewhat earlier, in 1736, is known as *The Bridges of Koenigsberg.* At that time this Prussian city straddled the two banks of a river and also included two islands, all of which were connected by seven bridges in the pattern indicated in Figure 9.2. On Sunday afternoons the citizens of Koenigsberg entertained themselves by strolling around all of the city's parts, and eventually the question arose as to whether an excursion could be planned that would cross each of the seven bridges exactly once. This is clearly a geometric problem since its terms are defined visually, and yet the exact distances traversed in such excursions are immaterial (so long as they are not too excessive, of course). Nor are the precise contours of the banks and the islands of any consequence. Hence, this is a topological problem. Theorem 10.2.2 will provide us with a tool for easily resolving this and similar questions.

The notorious four-color problem, which asks whether it is possible to color the countries of every geographical map with four colors so that adjacent countries sharing a border of nonzero length receive distinct colors, is also of a topological nature. Maps are clearly visual objects, and yet

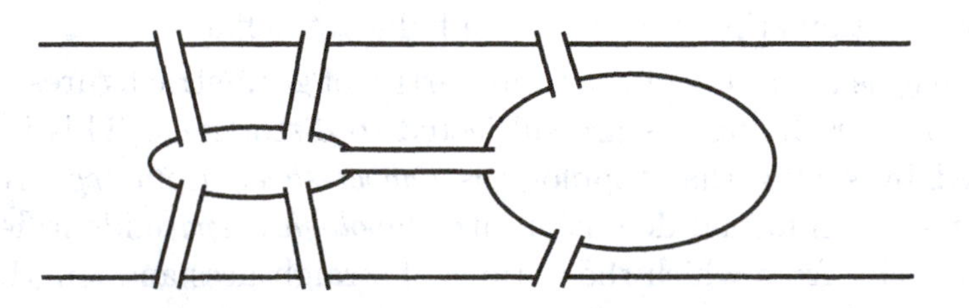

Figure 9.2. The city of Koenigsberg

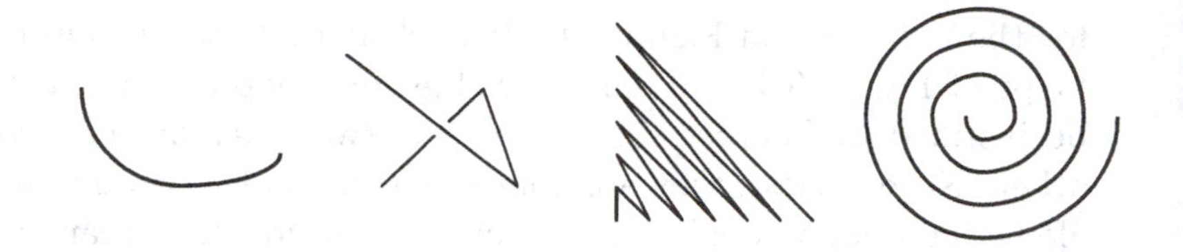

Figure 9.3. Homeomorphic open arcs

the specific shapes and sizes of the countries in such a map are completely irrelevant. Only the adjacency patterns matter.

Every mathematical discipline deals with objects or structures and most will provide a criterion for determining when two of these are identical, or equivalent. The equality of real numbers can be recognized from their decimal expansions, and two vectors are equal when they have the same direction and magnitude. Topological equivalence is called *homeomorphism*. The surface of a sphere is homeomorphic to those of a cube, a hockey puck, a plate, a bowl, and a drinking glass. The reason for this is that each of these objects can be deformed into any of the others. Similarly, the surface of a donut is homeomorphic to those of an inner tube, a tire, and a coffee mug. On the other hand, the surfaces of the sphere and the donut are not homeomorphic. Our intuition rejects the possibility of deforming the sphere into a donut shape without either tearing a hole in the middle or else stretching the sphere out and juxtaposing and pasting its two ends together. Tearing, however, destroys some contiguities, whereas juxtaposition introduces new contiguities where there were none before, and so neither of these transformations is topologically admissible.

The easiest way to establish the homeomorphism of two spaces is to describe a deformation of one onto the other that involves no tearing or juxtapositions. Such a deformation is called an *isotopy*. Whenever isotopies are used in the sequel, their existence will be clear and will require no formal justification. Such is the case, for instance, for the isotopies that establish the homeomorphisms of all the open arcs in Figure 9.3, all the loops in Figure 9.4, and all the ankh-like configurations of Figure 9.5. Note that whereas the page on which all these curves are drawn is two dimensional, the context is definitely three dimensional. In other words, all our curves (and surfaces) reside in Euclidean 3-space $\mathbb{R}^3$ and the isotopies may make use of all three dimensions.

The concept of isotopy is insufficient to describe all homeomorphisms. There are spaces that are homeomorphic but not isotopic. Such is the case for the two loops in Figure 9.6. It is clear that loop b is isotopic to all the loops of Figure 9.4, and it is plausible that loop a is not, a claim that will be justified in Chapter 12. Hence, the two loops are not isotopic to each other. Nevertheless, they are homeomorphic in the sense that ants crawling along these loops would experience them in identical manners. To express this homeomorphism somewhat more formally, it is necessary to resort to the language of functions. First, however, it should be pointed out that the word *function* is used here in the sense of an association, or an assignment, rather than the end result of an algebraic calculation. In other words, a function $f : S \to T$ is simply a rule that associates to every point of S a unique point of T. In this text most of the functions will be described visually rather than algebraically.

Given two topological spaces S and T, a *homeomorphism* is a function $f : S \to T$ such that

1. f matches all the points of S to all the points of T (distinct points of S are matched with distinct points of T and vice versa);
2. f preserves contiguity.

The vagueness of the notion of *contiguity* prevents this from being a formal definition. Since any two points on a line are separated by an infinitude of other points, this concept is not well defined. The homeomorphism of S and T is denoted by $S \approx T$. The homeomorphism of the loops of Figure 9.6 can now be established by orienting them, labeling their lowest points A and B, and matching points that are at equal distances from A and B where the distance is measured along the oriented loop (Fig. 9.7). Of course, the positions of A and B can be varied without affecting the existence of the homeomorphism.

A similar function can be defined so as to establish the homeomorphism of any two loops as long as both are devoid of self-intersections. Suppose two such loops c and d, of lengths γ and δ, respectively, are given (Fig. 9.8). Again begin by specifying orientations and initial points C and D on

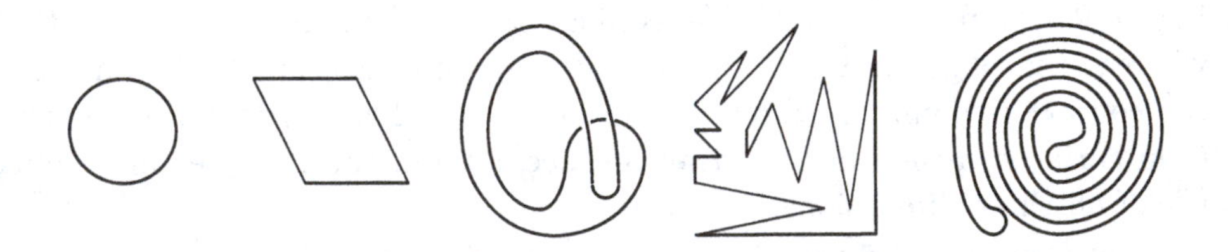

Figure 9.4. Homeomorphic loops

Figure 9.5. Homeomorphic ankhs

the two loops. Then, for every real number $0 \leq r < 1$, match the point at distance $r\gamma$ from C along c with the point at distance $r\delta$ from D along d.

Figure 9.9 contains another instructive example. Each of its three topological spaces consists of a band of, say, width 1 and length 20. They differ in that e is untwisted, f has one twist, and g has two twists.

Band f differs from the other two in that its border is in fact one single loop whereas bands e and g have two distinct borders each. It therefore comes as no surprise that band f is not homeomorphic to either e or g. These last two, however, are homeomorphic to each other. To describe this homeomorphism a coordinate system is established on each of the bands as follows. For each number $0 \leq r \leq 1$, let $L_{e,r}$ and $L_{g,r}$ denote the oriented loops of length 20 that run along the band at a constant distance of r away from the bottom borders of e and g respectively. Choosing start lines as described in Figure 9.10, the coordinate pair (r, s), $0 \leq r \leq 1$, $0 \leq s < 20$, describes those points on the loops $L_{e,r}$ and $L_{g,r}$ at a distance s from the respective starting line. The required homeomorphism simply matches up points of e and g that have the same coordinate pairs. The reason this wouldn't work for band f is that for this band the coordinatization process fails (see Fig. 9.11).

As mentioned previously, f is not homeomorphic to e and g because it has a different number of borders. In general, borders and other extremities are a good place to look for differences between topological spaces. For example,

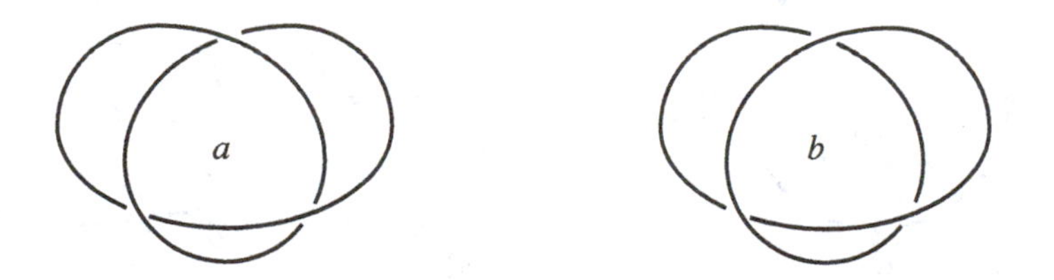

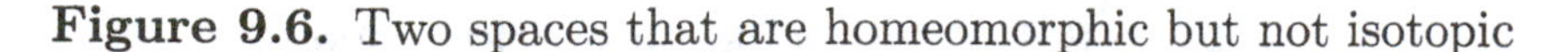

Figure 9.6. Two spaces that are homeomorphic but not isotopic

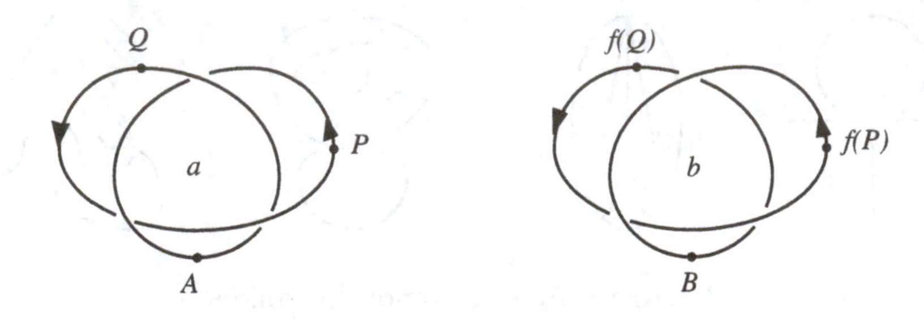

Figure 9.7. A homeomorphism of two loops

every two of the spaces in Figure 9.12 are nonhomeomorphic because they each have a different number of extremities. The number of components of a space can also serve as a tool for distinguishing between homeomorphism types. All the spaces in Figure 9.13 have the same number of extremities, but they are nevertheless nonhomeomorphic because each has a different number of components: 1, 2, 3, and 4 respectively.

Another method for distinguishing between spaces is to examine what remains when an equal number of properly selected points is deleted from each. For instance, both spaces of Figure 9.14 have one component and neither has extremities. Nevertheless, they are not homeomorphic because the removal of the two endpoints of the diameter of the θ-like space results in a space with three components whereas the removal of any two points of the circle leaves only two components. In general, a *topological property* of a space is a property that is shared by all the spaces that are homeomorphic to it. The number of endpoints and the number of components are both such topological properties. On the other hand, neither the length of an interval nor the area of a region are topological properties.

The foregoing discussion of topological spaces, homeomorphisms, and isotopies is informal and must remain so at this level. The precise definitions

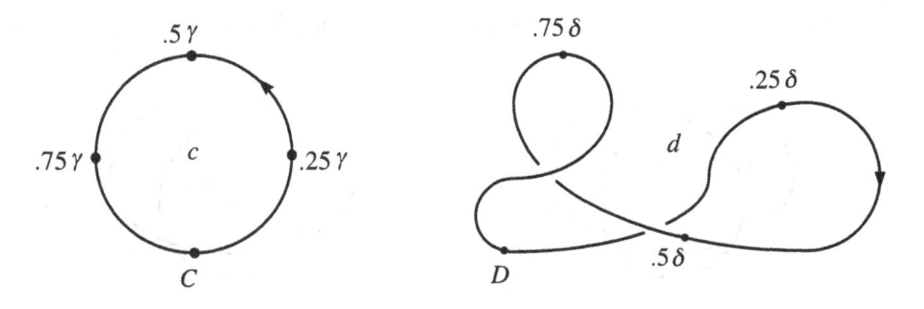

Figure 9.8. A homeomorphism of two loops

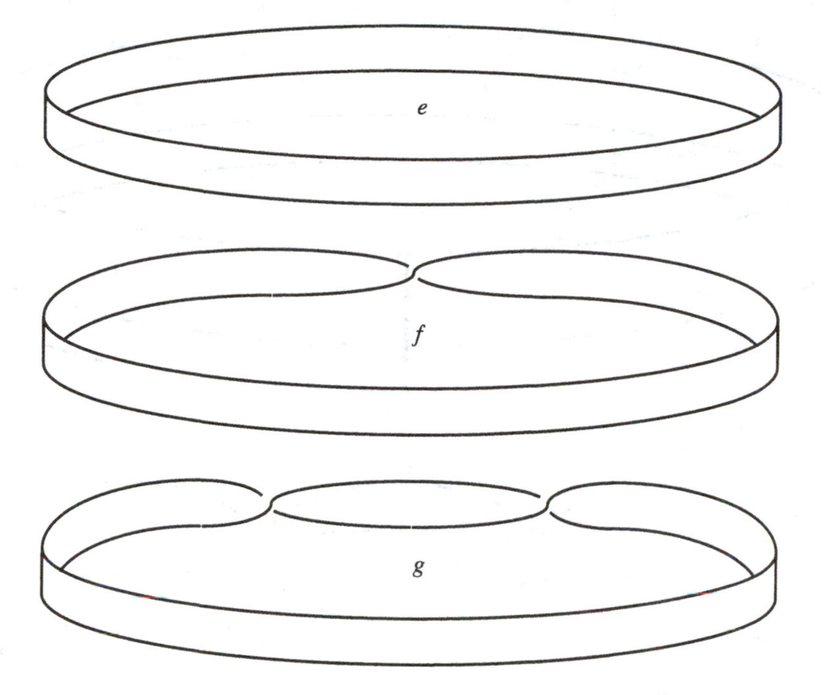

Figure 9.9. Three bands

of these terms, which can be found in the Bibligraphy, require so prohibitive an amount of preparatory work that it would be impractical to include them in this survey text. Moreover, experience indicates that this lack of precision will not hamper readers in their comprehension of the subsequent chapters.

In working out the exercises, the readers may find it useful to note that both homeomorphism and isotopy are *equivalence relations* in the sense that they satisfy the following three conditions:

Reflexivity: Every topological space is homeomorphic (isotopic) to itself.

Symmetry: If S is homeomorphic (isotopic) to T then T is homeomorphic (isotopic) to S.

Transitivity: If R is homeomorphic (isotopic) to S and S is homeomorphic (isotopic) to T then R is homeomorphic (isotopic) to T.

Exercises 9

1. Which of the letters in Figure 9.15 are homeomorphic?

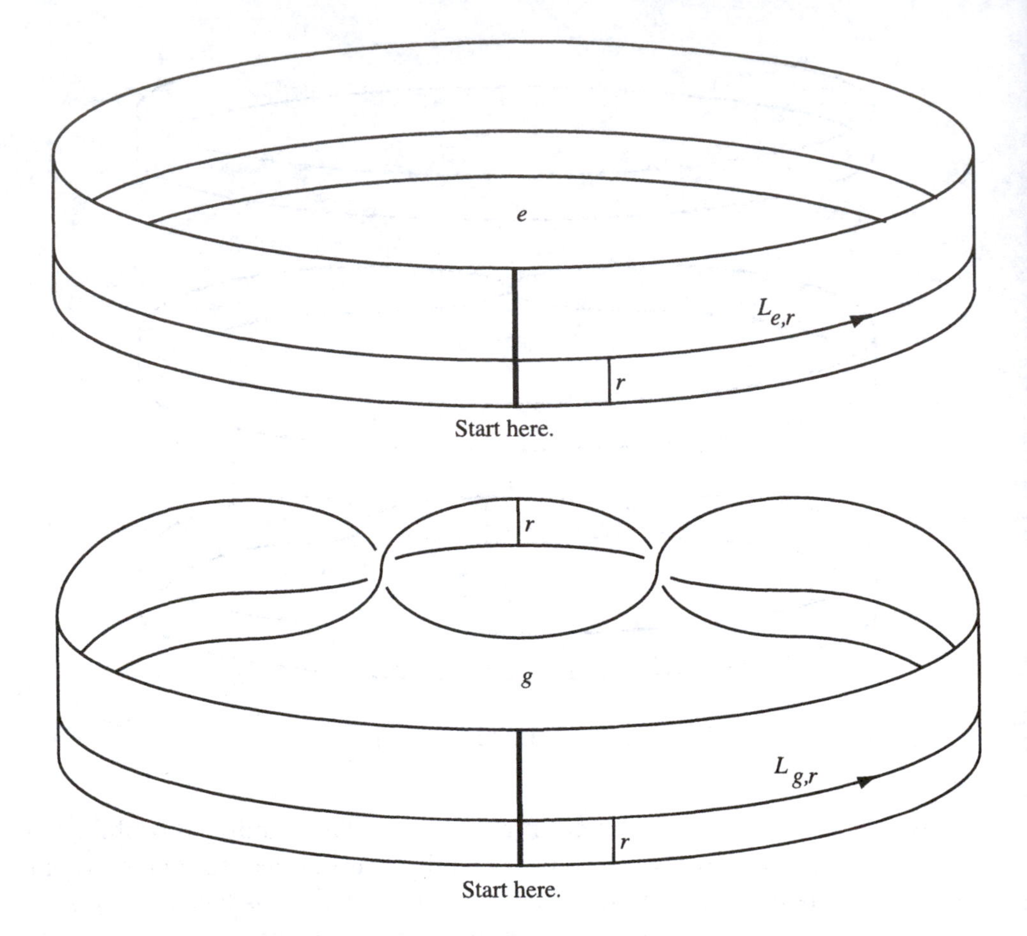

Figure 9.10. The homeomorphism of two bands

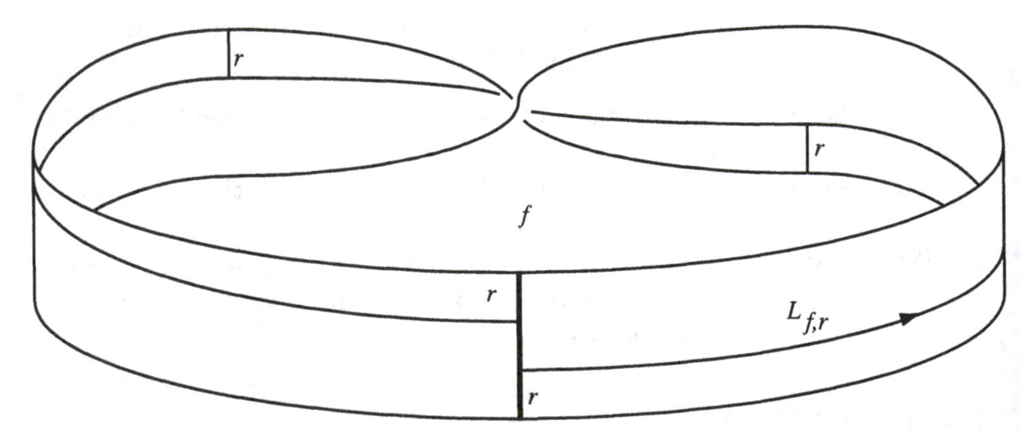

Figure 9.11. A failed homeomorphism

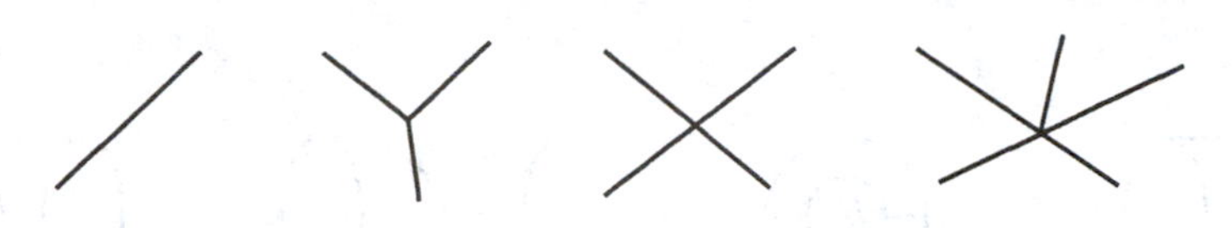

Figure 9.12. Four nonhomeomorphic spaces

Figure 9.13. Four nonhomeomorphic spaces

2. Which of the topological spaces in Figure 9.16 are homeomorphic?
3. Which of the topological spaces in Figure 9.17 are homeomorphic?
4. Which of the topological spaces of Figure 9.18 are isotopic?
5. Are the following statements true or false? Justify your answers.
   (a) If two topological spaces are homeomorphic, then they are also isotopic.
   (b) If two topological spaces are isotopic, then they are also homeomorphic.
   (c) Topological equivalence is synonymous with homeomorphism.
   (d) Topological equivalence is synonymous with isotopy.
   (e) Every two loops are isotopic.
   (f) Every two loops are homeomorphic.

Figure 9.14. Two nonhomeomorphic spaces

A B C D E F G H I J K L M
N O P Q R S T U V W X Y Z

Figure 9.15. Twenty-six topological spaces

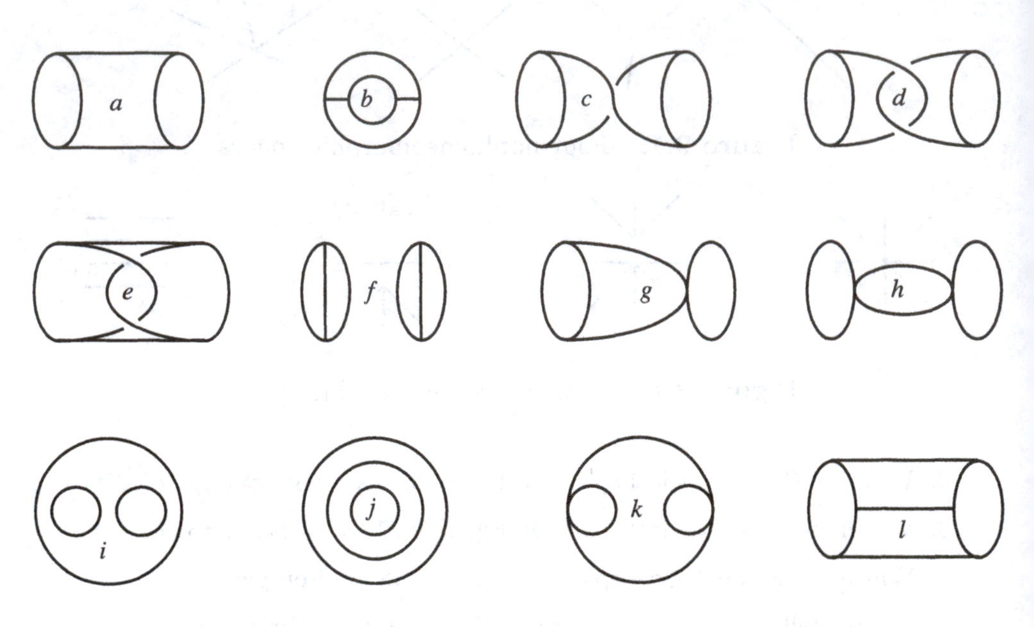

Figure 9.16. Some one-dimensional topological spaces

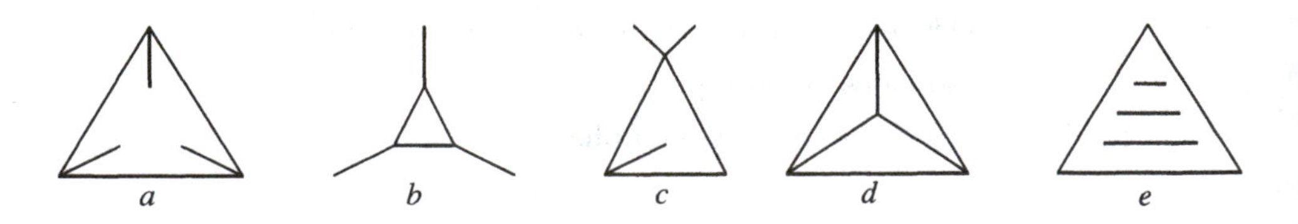

Figure 9.17. Some one-dimensional topological spaces

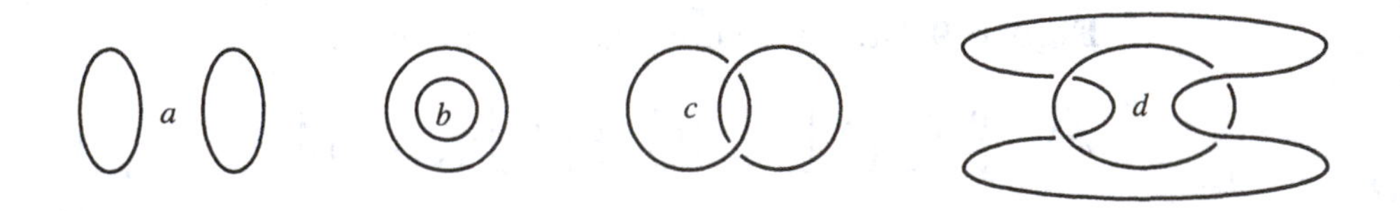

Figure 9.18. Some one-dimensional topological spaces

Chapter 10

Graphs

The one-dimensional topological objects are arcs. Graphs are created by the juxtaposition of a finite number of arcs, and they underlie many applications of mathematics as well as popular riddles. Traversability, planarity, colorability, and homeomorphisms of graphs are discussed.

10.1 Nodes and Arcs

An *open arc* is any topological space that is homeomorphic to a line segment (Fig. 9.3) and a *loop* is any topological space that is homeomorphic to a circle (Fig. 9.4); both are collectively referred to as *arcs*. A graph G consists of a set of points $N(G) = \{v_1, v_2, \ldots, v_p\}$, called the *nodes* of G, and a set of arcs $A(G) = \{a_1, a_2, \ldots, a_q\}$. The endpoints of each open arc are nodes, and every loop is assumed to begin and end at a node; other than that, arcs contain no nodes. Two nodes are said to be *adjacent* in G if they are the endpoints of a common arc of G. The adjacency of u and v is denoted by $u \sim v$. The number of arcs of which a node v is an endpoint, with each loop counted twice, is called the *degree* of v in G and is denoted by $\deg_G v$, or simply $\deg v$. For example, in the graph G of Figure 10.1, $\deg u = 4$, $\deg v = 4$, $\deg w = 7$, $\deg x = 4$, $\deg y = 1$, $\deg t = 2$, $\deg z = 0$. The location of nodes of degree at least 3 is, of course, indicated by the conjunction of the arcs at that node. Nodes of degree 1 are equally conspicuous. Nodes of degrees 2 and 0 have their location marked by a solid dot. If $v_1, v_2, \ldots, v_p$ is a listing of the nodes of G such that

$$\deg v_1 \geq \deg v_2 \geq \cdots \geq \deg v_p$$

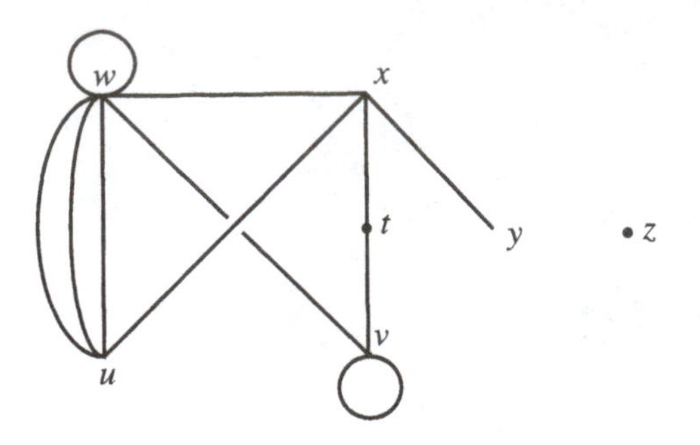

Figure 10.1. A graph

then $(\deg v_1, \deg v_2, \ldots, \deg v_p)$ is the *degree sequence* of G. Thus, the degree sequence of the graph of Figure 10.1 is (7, 4, 4, 4, 2, 1, 0). The number of nodes and arcs of the graph G are denoted by $p(G)$ and $q(G)$, or simply p and q, respectively.

The first proposition describes a simple and useful relation between the degrees of the nodes of a graph and the number of its arcs.

Proposition 10.1.1 *In any graph G, $\sum_{v \in N(G)} \deg v = 2q(G)$.*

PROOF: Each open arc of G contributes 1 to the degrees of each of its two endpoints and each loop contributes 2 to the degree of its only node. Hence each arc contributes 2 to $\sum_{v \in N(G)} \deg v$. The desired equation now follows immediately. Q.E.D.

This proposition immediately eliminates (3, 3, 2, 2, 1) as a possible degree sequence since the sum of the degrees must be even.

Distinct arcs that join the same endpoints are said to be *parallel*, and a graph that contains neither loops nor parallel arcs is said to be *simple*. Of the three graphs in Figure 10.2, only the middle one is simple.

If n is any positive integer, then the *complete graph* K_n is the simple graph with n nodes in which every pair of distinct nodes is joined by an arc. It is clear that each of the n nodes of K_n has degree $n-1$ and that K_n has $\binom{n}{2} = n(n-1)/2$ arcs. The graph K_1, which consists of a single node and no arcs, is also known as the *trivial* graph. If $m \geq n$ are two positive integers, then the *complete bipartite graph* $K_{m,n}$ is formed as follows. The node set $N(K_{m,n})$ is the union of two disjoint sets $A = \{u_1, u_2, \ldots, u_m\}$ and $B = \{v_1, v_2, \ldots, v_n\}$, and the arc set consists of mn open arcs that join each u_i to each v_j. In general, given positive integers $n_1 \geq n_2 \geq \cdots \geq n_k$ the *complete n-partite* (simple) *graph* $K_{n_1,n_2,\ldots,n_k}$ has the union $A_1 \cup A_2 \cup \cdots \cup A_k$ as

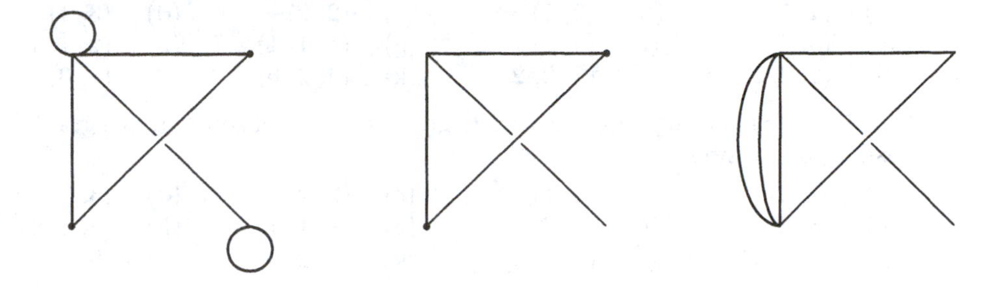

Figure 10.2. Simple and nonsimple graphs

its node set, where the sets $A_1, A_2, \ldots, A_k$ have cardinalities $n_1, n_2, \ldots, n_k$, respectively, and are pairwise disjoint. The arcs of $K_{n_1,n_2,\ldots,n_k}$ join all node pairs u, v where u and v belong to distinct A_i's. Three examples of such complete graphs appear in Figure 10.3.

If G and G' are graphs such that $N(G) \supset N(G')$ and $A(G) \supset A(G')$, then G' is said to be a subgraph of G. If G' is a subgraph of G, then $G - G'$ is the subgraph of G with node set $N(G)$ and arc set $A(G) - A(G')$ (i.e., all the arcs of G that are not arcs of G').

Exercises 10.1

1. Compute the degree sequences of the following graphs.

   (a) K_5 (b) K_n (c) $K_{3,4}$ (d) $K_{m,n}$
   (e) $K_{3,2,1}$ (f) $K_{l,m,n}$ (g) $K_{7,6,5,4}$ (h) $K_{n_1,n_2,\ldots,n_k}$

2. Which of the following sequences are degree sequences of some graph? Justify your answer.

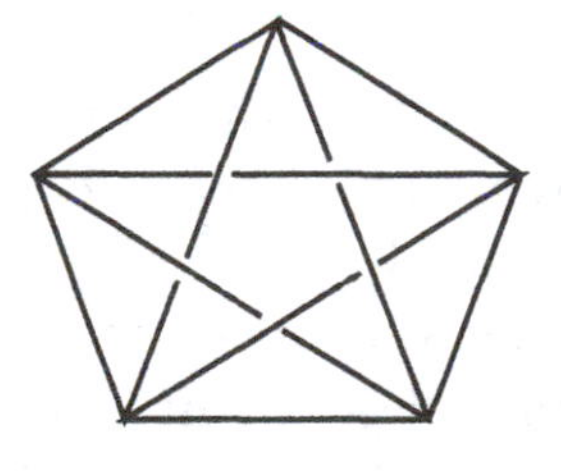

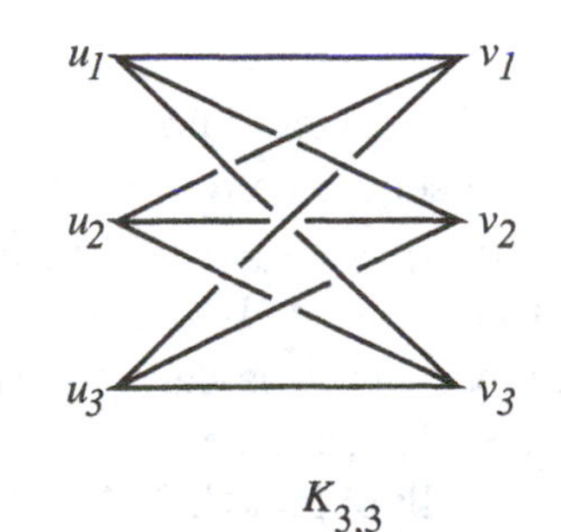

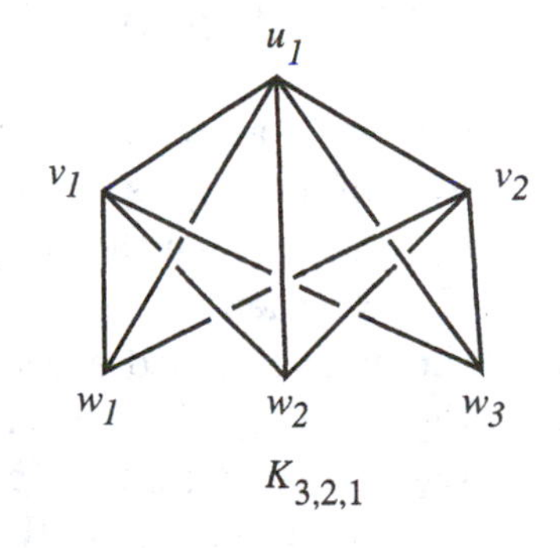

Figure 10.3. Three simple graphs

(a)	(1, 1)	(b)	(2, 1)	(c)	(2, 2)	(d)	(3, 1)
(e)	(3, 2)	(f)	(3, 3)	(g)	(1, 1, 1)	(h)	(2, 1, 1)
(i)	(2, 2, 1)	(j)	(2, 2, 2)	(k)	(4, 2, 0)	(l)	(8, 0)

3. Which of the following sequences are degree sequences of some simple graph? Justify your answer.

(a)	(1, 1)	(b)	(2, 1)	(c)	(2, 2)	(d)	(3, 1)
(e)	(3, 2)	(f)	(3, 3)	(g)	(1, 1, 1)	(h)	(2, 1, 1)
(i)	(2, 2, 1)	(j)	(2, 2, 2)	(k)	(4, 2, 0)	(l)	(8, 0)

4. For which values of n are the following sequences degree sequences of some graph? Justify your answer.

    (a) $(n, n-1, \dots, 1)$ (b) $(n, n, n-1, n-2, \dots, 1)$
    (c) $(n, n-1, n-2, \dots, 0)$ (d) $(n, n, n-1, \dots, 0)$
    (e) $(n, n, n-1, n-1, \dots, 0, 0)$ (f) $(k, k, \dots, k)$ (n k's)

5. For which values of n are the following sequences degree sequences of some simple graph? Justify your answer.

    (a) $(n, n-1, \dots, 1)$ (b) $(n, n, n-1, n-2, \dots, 1)$
    (c) $(n, n-1, n-2, \dots, 0)$ (d) $(n, n, n-1, \dots, 0)$
    (e) $(n, n, n-1, n-1, \dots, 0, 0)$ (f) $(k, k, \dots, k)$ (n k's)

6. Prove that in any graph G, the number of nodes with odd degree is even.

7. Compute the number of arcs of $K_{n_1, n_2, \dots, n_k}$ as well as its degree sequence.

*8. Prove that if $a_1 \geq a_2 \geq \cdots \geq a_p \geq 0$ are all integers, then $(a_1, a_2, \dots, a_p)$ is the degree sequence of some graph if and only if $\sum_{i=1}^{p} a_i$ is even.

*9. Prove that if $a_1 \geq a_2 \geq \cdots \geq a_p \geq 0$ are all integers, then $(a_1, a_2, \dots, a_p)$ is the degree sequence of some loopless graph if and only if $\sum_{i=1}^{p} a_i$ is even and $a_1 \leq a_2 + a_3 + \cdots + a_p$.

*10. Prove that every simple graph can be placed in $\mathbb{R}^3$ so that its arcs are straight line segments.

10.2 Traversability

If u and v are two (not necessarily distinct) nodes of the graph G, then a u-v *walk* of *length* n is an alternating sequence $u = v_0, a_1, v_1, a_2, v_2, \dots, v_{n-1}, a_n, v_n = v$ in which the endpoints of a_k are v_{k-1} and v_k for each $k = 1, 2, \dots, n$. If $u = v$ then the walk is said to be *closed*. If all the arcs of a walk are distinct, the walk is called a *trail*. A closed trail is called a *circuit*. For example, in Figure 10.4, W_1: $u, c, x, g, w, i, w, b, u, a, w, g, x$ is a u-x walk of length 6, and W_2: v, h, v, d, w, d, v is a closed walk of length 3. Neither W_1 nor W_2 is a trail. On the other hand, W_3: $u, c, x, g, w, i, w, b, u, a, w$ is a u-w trail, and W_4: $v, d, w, i, w, g, x, e, v$ is a circuit. A trail $v_0, a_1, v_1, a_2, v_2, \dots, v_{n-1}, a_n, v_n$ all

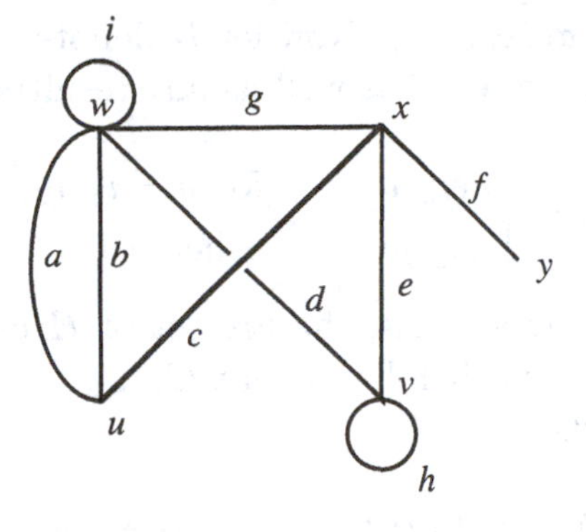

Figure 10.4. A graph with labeled arcs and nodes

of whose nodes are distinct is said to be a *path*. A circuit $v_0, a_1, v_1, a_2, v_2, \ldots,$ v_{n-1}, a_n, v_0 in which $v_0, v_1, \ldots, v_{n-1}$ are all distinct is called a *cycle*. The aforementioned trail W_3 is not a path, nor is W_4 a cycle. However, $u, c, x, g, w, d,$ is a path and u, c, x, g, w, b, u is a cycle. When no ambiguities can arise, walks will be abbreviated by listing only their vertices. Thus, the short form of the walk W_5: $y, f, x, e, v, d, w, i, w, g, x, c, u$ of Figure 10.4 is y, x, v, w, w, x, u.

A graph in which every two nodes can be joined by a trail is said to be *connected*. The maximal connected subgraphs of a graph are its *components*. A circuit that contains all the arcs of the graph G is an *Eulerian* circuit of G. A graph that possesses an Eulerian circuit is said to be *Eulerian*. Thus, K_3 and K_5 are clearly Eulerian whereas trial and error shows that K_4 is not. It turns out that Eulerian graphs are easily recognized by the fact that all their nodes have even degrees. This characterization of Eulerian graphs is one of the earliest theorems of graph theory. Its statement is preceded by a lemma.

Lemma 10.2.1 *Suppose G is a graph all of whose nodes have positive even degree. Then there is a sequence $C_1, C_2, \ldots, C_m$ of cycles of G such that every arc of G is in exactly one of these cycles.*

PROOF: We proceed by induction on the number of arcs of G. The lemma clearly holds if G has two arcs. Let n be a fixed integer, let G have n arcs, and suppose the lemma holds for all graphs that satisfy the evenness hypothesis and have less than n arcs. Let P: $v_0, a_1, v_1, a_2, v_2, \ldots, a_k, v_k$ be a path of G that has maximum length. Since $\deg_G v_k$ is at least 2, there is another arc a at v_k that is distinct from a_k. It follows from the maximality of P that this arc joins v_k to v_i for some $i = 1, 2, \ldots, k$, thus creating a cycle

C: $v_i, a_{i+1}, v_{i+1}, \ldots, a_k, v_k, a, v_i$. Now let H denote the graph obtained from G by deleting all the arcs of C as well as any resulting isolated nodes. Note that

$$\deg_H v = \begin{cases} \deg_G v - 2 & \text{for } v = v_i, v_{i+1}, \ldots, v_k \\ \deg_G v & \text{otherwise.} \end{cases}$$

By the induction hypothesis, the arc set of H can be decomposed into cycles $C_1, C_2, \ldots, C_m$ and it follows that $C_1, C_2, \ldots, C_m, C$ constitutes the desired partition of the arcs of G. Q.E.D.

Theorem 10.2.2 (Euler 1736) *A nontrivial connected graph is Eulerian if and only if each of its nodes has even degree.*

PROOF: Suppose the nontrivial connected graph G is Eulerian so that it possesses an Eulerian circuit C. Let v be any node of G. If $\ldots a_{i_1}, v, a_{i_2}, \ldots, a_{i_3}, v, a_{i_4}, \ldots, a_{i_{2n-1}}, v, a_{i_{2n}}, \ldots$ constitute all the occurrences of v in C, then, because C contains each of the arcs of G exactly once, $\deg_G v = 2n$, which is even. Hence each node of G has even degree.

The converse is proved by induction on the number of arcs in G. Let G be a connected graph in which every node has even degree. If G has only one arc, then it consists of a single loop and so is Eulerian. Let n be a fixed positive integer and suppose that all connected graphs whose nodes all have even degrees are Eulerian whenever they have fewer than n arcs. Suppose next that G is such a graph with n arcs. By the lemma the arcs of G can be partitioned into a set of disjoint cycles $C_1, C_2, \ldots, C_m$. If H is the graph obtained from G by deleting the arcs in C_m, then H is a possibly disconnected graph with components, say, $H_1, H_2, \ldots, H_c$ (some of these components may be trivial). For all the nodes v of G

$$\deg_H v = \begin{cases} \deg_G v - 2 & \text{if } v \text{ is on } C_m \\ \deg_G v & \text{otherwise.} \end{cases}$$

It follows that the nodes of H all have even degrees. By the induction hypothesis each of the nontrivial components H_i has an Eulerian circuit. These Eulerian circuits and the trivial components can be combined by means of C_m into an Eulerian circuit of G. Hence, G is Eulerian. Q.E.D.

The resolution of the Koenigsberg bridges problem of Chapter 9 is now an easy matter (see Exercise 8). The simplicity and efficacy of Euler's theorem may mislead the beginner into the expectation that all graph theoretical problems lend themselves to such easy resolutions. To counter this

Figure 10.5. A Hamiltonian and a non-Hamiltonian graph

misconception, a superficially similar question is posed whose satisfactory resolution is still lacking, and not for lack of trying.

A *Hamiltonian* cycle of the graph G is a cycle that contains all the nodes of G. A graph that possesses a Hamiltonian cycle is also said to be *Hamiltonian.* It is clear that K_n is Hamiltonian for all $n \geq 3$ whereas K_2 is not. Other interesting Hamiltonian and non-Hamiltonian graphs appear in the exercises. Being Hamiltonian is not a topological property of graphs. Figure 10.5 displays two homeomorphic graphs, only one of which is Hamiltonian. Given a specific graph, it is of course possible in principle to check on all of the permutations of its nodes to see whether one of them yields a Hamiltonian cycle. In practice, however, this method is impractical as it is much too time consuming. While better and more sophisticated algorithms have been found, they too are essentially impractical. Efficient methods for deciding whether a graph is Hamiltonian would also help resolve many other problems in both pure and applied mathematics and have been the subject of much research.

A characterization of Hamiltonian graphs that is analogous to that of Eulerian graphs is as yet unavailable and probably nonexistent. Instead, a sufficient condition for a graph to be Hamiltonian is offered. If u and v are nodes of a graph G, then $G + uv$ denotes the graph obtained by adding to G an arc that joins u and v (but is otherwise disjoint from G).

Proposition 10.2.3 (O. Ore 1960) *If G is a simple graph with at least three nodes such that for all nonadjacent nodes u and v,*

$$\deg u + \deg v \geq p,$$

then G is Hamiltonian.

PROOF: We proceed by contradiction. Suppose the theorem is false. Of all the simple non-Hamiltonian graphs with p nodes that satisfy the hypothesis of the theorem, let G be one with the maximum number of arcs. Since $p \geq 3$ and every complete graph K_p with $p \geq 3$ is Hamiltonian, it follows that G is not complete. Let u and v be some two nonadjacent nodes of G. The aforementioned maximality of G implies that $G+uv$ contains a Hamiltonian

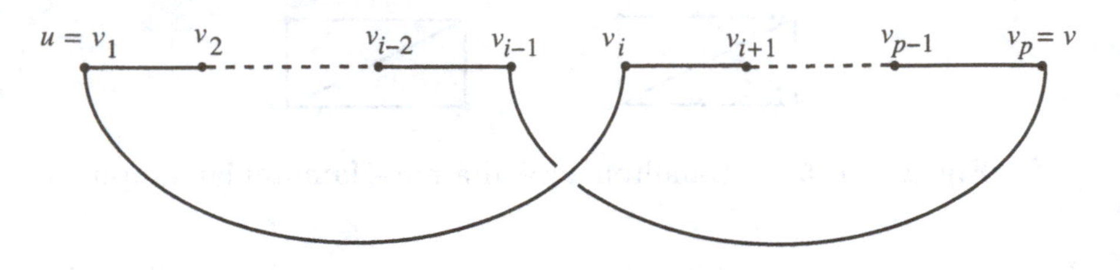

Figure 10.6. A would-be Hamiltonian cycle of G

cycle C, which, since G is not Hamiltonian, contains the new arc joining u and v. Thus, G has a u-v path P: $u = v_1, v_2, \ldots, v_p = v$ that contains every node of G.

For every $i = 2, \ldots, p$, either $v_1 \nsim v_i$ or $v_{i-1} \nsim v_p$ since otherwise

$$v_1, v_i, v_{i+1}, \ldots, v_p, v_{i-1}, v_{i-2}, \ldots, v_1$$

would constitute a Hamiltonian cycle of G (see Fig. 10.6). Hence, if v_{i_1}, v_{i_2}, ..., v_{i_d} are the nodes of G that are adjacent to $u = v_1$, then v_{i_1-1}, v_{i_2-1}, ..., v_{i_d-1} are all not adjacent to $v = v_p$. Hence

$$\deg v \leq p - 1 - d = p - 1 - \deg u,$$

which contradicts the assumption that $\deg u + \deg v \geq p$. It follows that G is Hamiltonian. Q.E.D.

Exercises 10.2

1. Which of the diagrams in Figure 10.7 can be drawn without lifting the pencil, retracing any lines, and so that the start and finish point are the same?

2. Which of the floor plans of Figure 10.8 can be traversed so that each door is used exactly once and the tour both starts and ends in the yard?

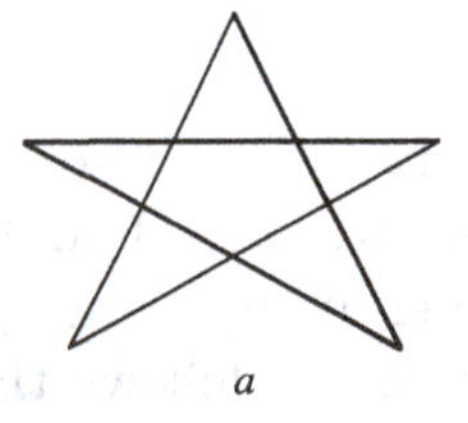

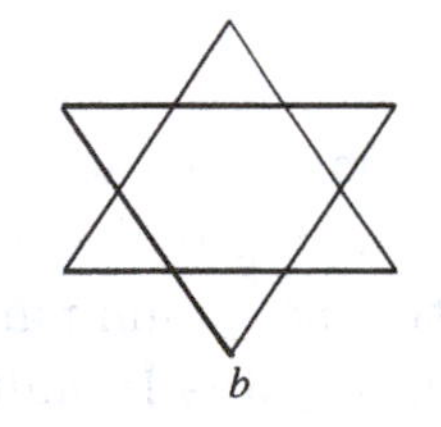

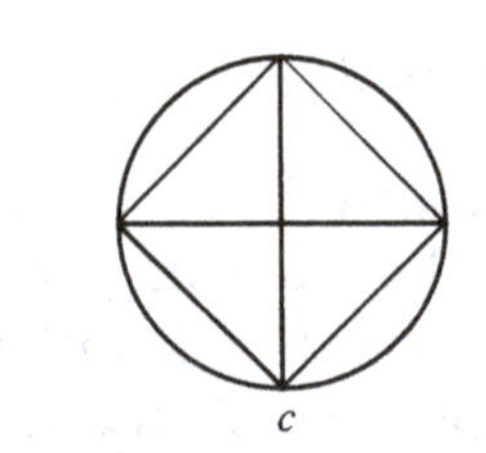

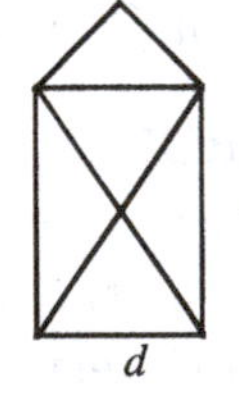

Figure 10.7. Drawing figures

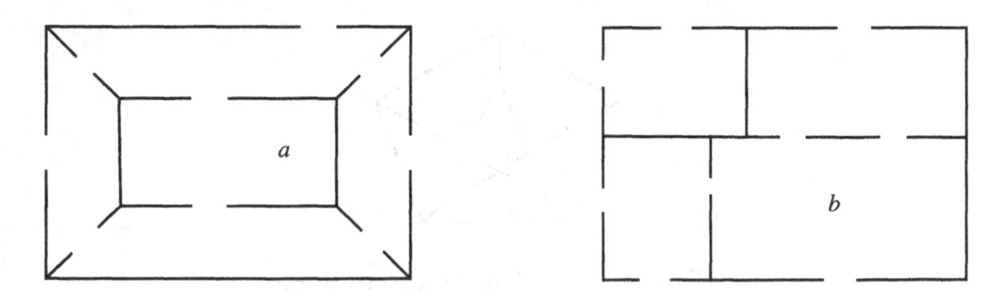

Figure 10.8. Floor plans

3. Which of the following graphs is Eulerian?

(a)	K_7	(b)	K_8	(c)	$K_{3,4}$	(d)	$K_{4,4}$
(e)	$K_{3,5}$	(f)	$K_{4,5}$	(g)	$K_{1,2,2}$	(h)	$K_{1,2,3}$
(i)	$K_{2,2,2}$	(j)	$K_{1,2,3,4}$	(k)	$K_{2,2,2,2}$	(l)	$K_{3,3,3,3}$

4. For which values of n is K_n Eulerian?

5. For which values of n_1, n_2 is K_{n_1,n_2} Eulerian?

6. For which values of n_1, n_2, n_3 is K_{n_1,n_2,n_3} Eulerian?

7*. For which values of $n_1, n_2, \dots, n_m$ is $K_{n_1,n_2,\dots,n_m}$ Eulerian?

8. Solve the Koenigsberg bridges problem of Chapter 9.

9. A graph is said to be *traversable* if it has a walk that contains each edge exactly once. Prove that a connected graph is traversable if and only it has either two or no nodes of odd degree.

10. Which of the diagrams in Figure 10.7 can be drawn without lifting the pencil and without retracing any lines?

11. Prove that if G is a graph with $p \geq 3$ nodes such that $\deg v \geq p/2$ for each node v, then G is Hamiltonian.

12. Which of the following graphs is Hamiltonian?

(a)	K_7	(b)	K_8	(c)	$K_{3,4}$	(d)	$K_{4,4}$
(e)	$K_{3,5}$	(f)	$K_{4,5}$	(g)	$K_{1,2,2}$	(h)	$K_{1,2,3}$
(i)	$K_{2,2,2}$	(j)	$K_{1,2,3,4}$	(k)	$K_{2,2,2,2}$	(l)	$K_{3,3,3,3}$

13. For which values of n_1, n_2 is K_{n_1,n_2} Hamiltonian?

14. For which values of n_1, n_2, n_3 is K_{n_1,n_2,n_3} Hamiltonian?

15. For which values of $n_1, n_2, \dots, n_m$ is $K_{n_1,n_2,\dots,n_m}$ Hamiltonian?

16. Prove that the Petersen graph of Figure 10.9 is not Hamiltonian.

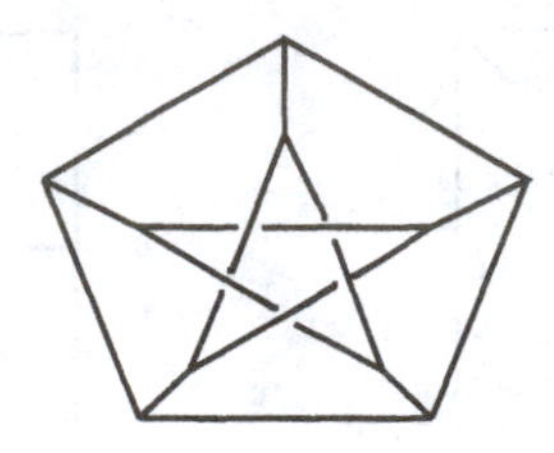

Figure 10.9. The Petersen graph

17. The graph P_n has node set $\{u_1, u_2, \ldots, u_n, v_1, v_2, \ldots, v_n\}$. For each $i = 1, 2, \ldots, n$, u_i is adjacent to u_{i-1}, u_{i+1}, v_i, and v_i is adjacent to v_{i-2}, v_{i+2}, u_i (addition modulo n). For which values of n is the graph P_n Hamiltonian? Prove your answer.

18. Prove that the Tutte graph of Figure 10.10 is not Hamiltonian.

19. Prove that if a is an arc of the connected graph G, then $G - a$ has either one or two components.

10.3 Colorings

A graph G is said to be *k-colorable* if each of its nodes can be labeled with one of the numbers (also called *colors*) $1, 2, \ldots, k$ in such a manner that adjacent nodes receive distinct colors (Fig. 10.11).

It is easy to see that cycles of even length are 2-colorable whereas cycles of odd length are not. Since even and odd cycles are homeomorphic, this means that k-colorability is *not* a topological property of graphs. The complete graph K_n is k-colorable only if $k \geq n$. Graphs with loops are never colorable. A graph is 1-colorable if and only if it has no arcs. The following theorem

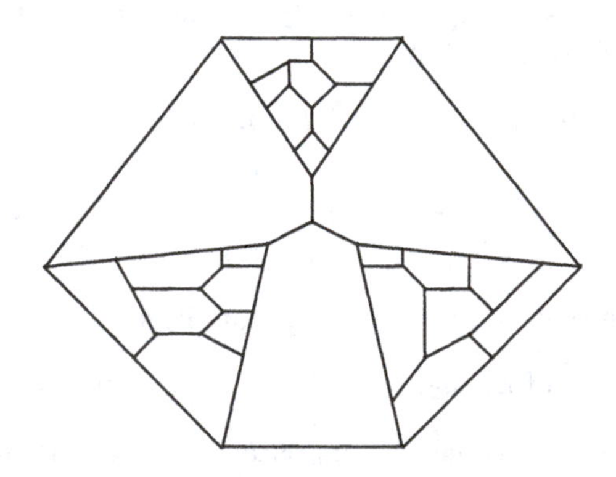

Figure 10.10. The Tutte graph

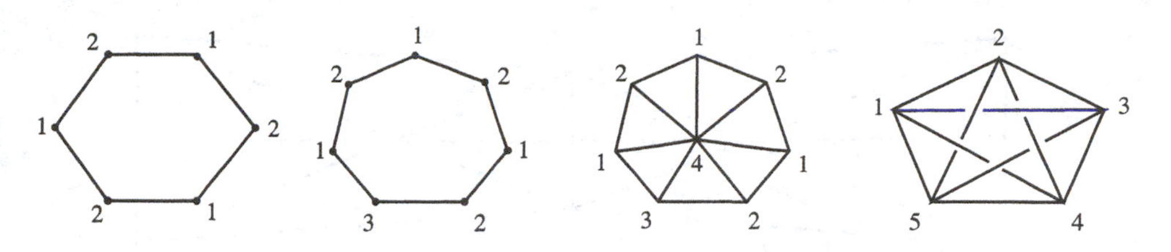

Figure 10.11. Four graph colorings

characterizes 2-colorable graphs. No characterization of k-colorable graphs exists for any $k > 2$.

Theorem 10.3.1 *The graph G is 2-colorable if and only if it has no odd cycles.*

PROOF: Suppose G has been 2-colored and C: $v_1, v_2, \ldots, v_n$ is a cycle of G. It may be assumed without loss of generality that the nodes v_1 and v_2 are colored 1 and 2, respectively. It follows that for each $1 \leq i \leq n$ the node v_i is colored 1 or 2 according as i is odd or even. Since the node v_n is adjacent to v_1, it is necessarily colored 2 and hence n is even. Thus, every cycle of G is even.

Conversely, suppose G contains no odd cycles. It clearly suffices to show that each component of G is 2-colorable and so it may be assumed that G is connected. Let v be any node of G and set $V_1 = \{v\}$. Assuming that the node set V_n, $n \geq 1$, has been defined, V_{n+1} is defined to be the set of all those nodes of G that are not in $V_1 \cup V_2 \cup \cdots \cup V_n$ and are adjacent to some nodes of V_n. It follows from this definition that a node of V_n can be adjacent only to nodes of $V_{n-1} \cup V_n \cup V_{n+1}$. It is next demonstrated by contradiction that for each n, nodes of V_n cannot be adjacent to each other. For suppose u and w are adjacent nodes of some V_n. Let P: $v = u_1, u_2, \ldots, u_n = u$ and Q: $v = w_1, w_2, \ldots, w_n = w$ be v-u and v-w paths of length n whose existence is guaranteed by the definition of V_n. Let $x = u_m = w_m$ be the last node shared by these paths. Then $x = u_m, u_{m+1}, \ldots, u_n = u$, $w = w_n, w_{n-1}, \ldots, x = w_m$ is an odd cycle (of length $2(n-m)+1$) of G. This contradicts the hypothesis, and hence for each n the nodes of V_n are adjacent only to nodes of $V_{n-1} \cup V_{n+1}$. A 2-coloring of the nodes of G is now obtained by assigning to the nodes of each V_n either 1 or 2 according as n is either odd or even. Q.E.D.

Given any graph G and any positive integer k, it is, in principle, easy to determine whether or not G is k-colorable. Every *k-coloring* can be viewed as

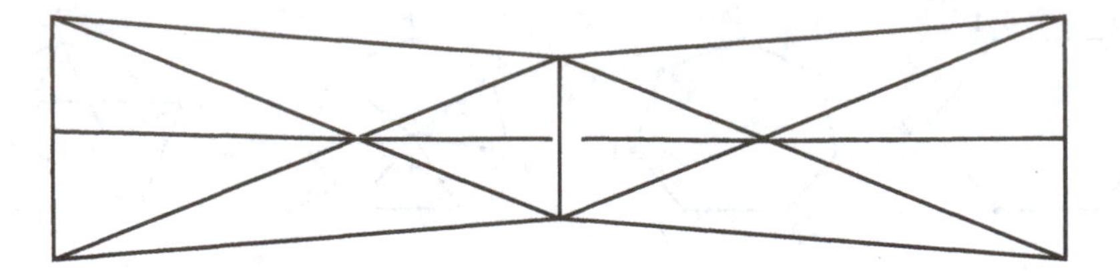

Figure 10.12. A 3-degenerate graph

a function that assigns to each node of G a number in $\{1, 2, \ldots, k\}$. Hence we need merely methodically examine all the k^p functions from $V(G)$ to $\{1, 2, \ldots, k\}$ and see whether there is at least one such function that assigns distinct colors to adjacent nodes. In practice, however, this method is so time consuming, even for relatively small graphs, that it is ineffective. Two theoretical partial answers to this question are now provided. The first is a sufficient condition guaranteeing colorability under certain circumstances, and the second is a necessary condition.

Let k be a positive integer. A graph G is said to be *k-degenerate* if there is a listing $v_1, v_2, \ldots, v_p$ of its nodes such that for each $i = 1, 2, \ldots, p$, the node v_i is adjacent to at most k of the nodes $v_1, v_2, \ldots, v_{i-1}$. For example, the graph of Figure 10.12 is 3-degenerate, even though it has several nodes of degree greater than three. Any listing of the nodes that begins with the four central nodes constitutes a proof of this fact.

Proposition 10.3.2 *Every loopless k-degenerate graph is* $(k+1)$*-colorable.*

PROOF: Let G be a k-degenerate graph and $v_1, v_2, \ldots, v_p$ a listing of its nodes such that each node is adjacent to at most k of its predecessors. The nodes can now be colored by an inductive process. Assign the color 1 to v_1. Assume that $v_1, v_2, \ldots, v_i$ have been assigned colors in $\{1, 2, \ldots, k+1\}$ so that adjacent nodes have different colors. Since v_{i+1} is adjacent to at most k of the previous nodes, it can be assigned a color from $\{1, 2, \ldots, k+1\}$ that is distinct from its neighbors' colors. This will eventually result in a $(k+1)$-coloring of G. Q.E.D.

It follows from this proposition that the graph of Figure 10.12 is 4-colorable. It also follows that if $\Delta(G)$ denotes the *maximum degree* of the graph G, then every loopless graph G is $(\Delta(G)+1)$-colorable. Necessary conditions for colorability are even harder to come by than sufficient conditions. Only the following obvious criterion can be offered here.

Proposition 10.3.3 *If the graph G contains the complete graph K_{k+1} as a subgraph, then G is not k-colorable.*

Since the graph of Figure 10.12 contains a K_4 in its center, it is not 3-colorable.

Exercises 10.3

1. For which values of k are the following graphs k-colorable? Justify your answers.

   (a) K_7 (b) $K_{3,4}$ (c) $K_{m,n}$ (d) $K_{m,n,r}$ (e) $K_{n_1,n_2,\ldots,n_m}$

   (f) The Petersen graph of Figure 10.9

   (g) The Tutte graph of Figure 10.10

   (h) The 3-degenerate graph of Figure 10.12

*2. A graph is said to be *acyclic* if it contains no cycles. Prove that every acyclic graph is 1-degenerate and conclude that it is 2-colorable.

3. Let G be a graph formed by a plane polygon together with a collection of nonintersecting diagonals. Prove that G is 2-degenerate and conclude that it is 3-colorable.

*4. A simple graph is said to be a *series-parallel network* if it contains no subgraph that is homeomorphic to K_4. Prove that every series-parallel network is 2-degenerate and hence is 3-colorable.

5. Prove that if the loopless graph G has less than $k(k+1)/2$ arcs, then G is k-colorable.

6. The wheel-like graph W_n is formed by joining every node of the cycle C_n to a new node w. For each positive integer n, determine those k's for which W_n is k-colorable.

*7. Prove that if G is neither a complete graph nor an odd cycle, then G is $\Delta(G)$-colorable.

8. Prove that every graph is homeomorphic to a 2-colorable graph.

9. Prove that the graph G is 2-colorable if and only if it is a subgraph of some complete bipartite graph.

10.4 Planarity

A *plane graph* is one that is drawn in the plane in its entirety, without any spurious intersections. In other words, any two of the arcs in the drawing intersect only in their common endpoints. A *planar graph* is a graph that

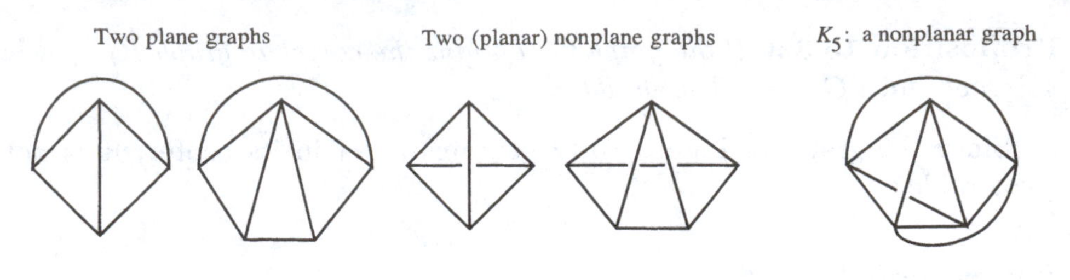

Figure 10.13. Plane, planar, and nonplanar graphs

can be drawn as a plane graph. A *nonplanar graph* is one that cannot be drawn in the plane (Fig. 10.13). The graph K_5 is in a sense the smallest nonplanar graph (Exercise 12).

This section deals with the characterization of planar graphs. The discussion begins with a closer examination of plane graphs. The arcs of a plane graph divide the ambient plane into two-dimensional pieces called *regions*. Note that each plane graph has one region of infinite extent, which is called the *exterior region*. The two plane graphs of Figure 10.13 have four and five regions, respectively. The set of arcs of G that delimit a region is called its *perimeter*. In the two plane graphs of Figure 10.13 every arc is on the perimeter of two regions. That need not always be the case. Both the arcs a and b in Figure 10.14 are on the perimeter of a single region each. When the perimeters of two distinct regions share an arc, they are said to *abut* along that arc. The regions R and R' in Figure 10.14 are said to *self-abut* along a and b, respectively.

Most topological investigations of the plane as well as much of Euclidean geometry rely on the fact that every plane loop has a well-defined interior and exterior. Despite the apparent simplicity of this proposition, both its precise formulation and proof are surprisingly difficult and evolved over the span of more than a century. A limited form of this observation appeared in Section 2.2 as Postulate S. It is now stated somewhat more formally and

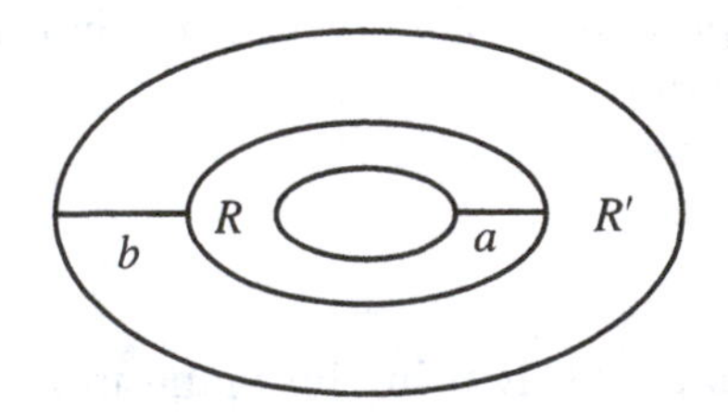

Figure 10.14. Two self-abutting regions

will be used repeatedly and implicitly in the sequel. In this general form it is attributed to Camille Jordan (1838–1922).

Theorem 10.4.1 (Jordan curve theorem) *Every plane loop divides the plane into two regions such that any arc joining a point of one region to a point of the other must intersect the dividing loop.* □

The following is one of the earliest of all topological theorems. It is as central to topology as the theorem of Pythagoras is to Euclidean geometry. It was first discovered by Euler as Proposition 8.1.1, and the exact relationship between the two versions is the subject of Exercises 16 and 17.

Theorem 10.4.2 (Euler's equation, 1758) *If G is a connected plane graph with p nodes, q arcs, and r regions, then $p - q + r = 2$.*

PROOF: We proceed by induction on p. If $p = 1$, then every arc of G is a loop and it follows from the Jordan curve theorem that $r = q + 1$. Hence,

$$p - q + r = 1 - q + (q + 1) = 2.$$

If $p > 1$, assume that the equation holds for all connected plane graphs with less than p nodes. Since G is connected, it has an open arc a. Let G' be the plane graph obtained from G by contracting the arc a to a point (see Figure 10.15). Then G' is a connected plane graph such that $p(G') = p - 1$, $q(G') = q - 1$, and $r(G') = r$ because all the regions of G are passed on to G', albeit one or two with a slightly shorter perimeter. It follows that

$$\begin{aligned} p - q + r &= (p(G') + 1) - (q(G') + 1) + r(G') \\ &= p(G') - q(G') + r(G') = 2. \end{aligned}$$

This completes the induction process. Q.E.D.

The following lemma and corollaries provide some necessary conditions for planarity.

Lemma 10.4.3 *Let G be a plane graph with q arcs and for each $k =$ $1, 2, 3, \ldots$ let r_k denote the number of regions whose perimeter consists of a closed walk of length k. Then*

$$2q = r_1 + 2r_2 + 3r_3 + \cdots.$$

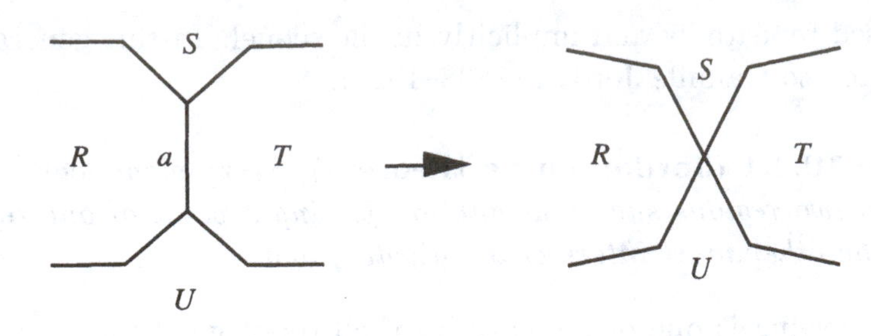

Figure 10.15. The contraction of the arc a

PROOF: Since G is a plane graph, every arc of G has two sides in the drawing of G. The total number of such sides is clearly $2q$. For every fixed positive integer k, the r_k regions whose perimeters have length k contribute kr_k sides to $2q$. Hence the desired equation. Q.E.D.

Corollary 10.4.4 *Let G be a simple planar graph with p nodes and q arcs. Then*

1. $q \leq 3p - 6$.
2. $q \leq 2p - 4$ *if G is 2-colorable.*

PROOF: Suppose G is drawn in the plane and the number of regions whose perimeter has length k is r_k. Since G is simple, $r_1 = r_2 = 0$. Hence, by Lemma 10.4.3,

$$2q = 3r_3 + 4r_4 + 5r_5 + \cdots \geq 3(r_3 + r_4 + r_5 + \cdots) = 3r.$$

By Euler's equation,

$$2 = p - q + r \leq p - q + 2q/3 = p - q/3$$

and the first inequality follows.

If G is 2-colorable, then it contains no 3-cycles so that $r_3 = 0$. Hence, by Lemma 10.4.3,

$$2q = 4r_4 + 5r_5 + 6r_6 + \cdots \geq 4(r_4 + r_5 + r_6 + \cdots) = 4r.$$

By Euler's equation,

$$2 = p - q + r \leq p - q + q/2 = p - q/2$$

and the second inequality follows immediately. Q.E.D.

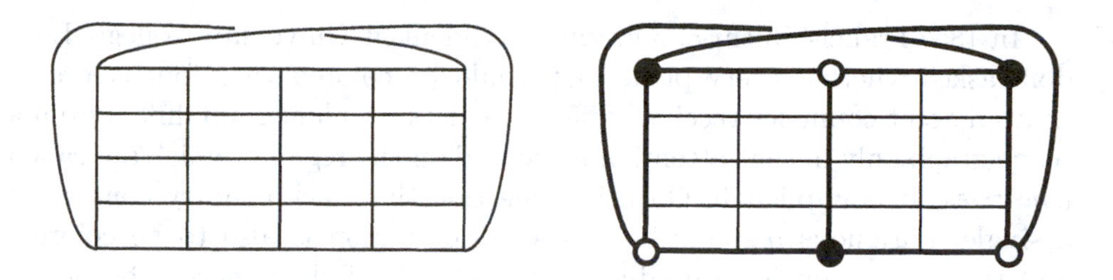

Figure 10.16. A nonplanar graph

Proposition 10.4.5 *The graphs K_5 and $K_{3,3}$ are not planar.*

PROOF: For K_5, $p = 5$ and $q = 10$. Since

$$10 \not\leq 3 \cdot 5 - 6,$$

it follows from Corollary 10.4.4.1 that K_5 is not planar.

For the 2-colorable $K_{3,3}$, $p = 6$ and $q = 9$. Since

$$9 \not\leq 2 \cdot 6 - 4,$$

it follows from Corollary 10.4.4.2 that $K_{3,3}$ is not planar. Q.E.D.

It turns out that K_5 and $K_{3,3}$ can be used to state a succinct description of the difference between planar and nonplanar graphs. The proof of this theorem, while minuscule in comparison with some of the graph theoretical proofs produced in the last three decades, is still too long to be included in this text. The theorem itself is attributed to Kasimierz Kuratowski (1896–1980), who wrote one of the first books on topology.

Theorem 10.4.6 (Kuratowski 1930) *A graph is nonplanar if and only if it contains a subgraph that is homeomorphic to either K_5 or $K_{3,3}$.* □

Example 10.4.7 The left graph in Figure 10.16 is nonplanar because, as is demonstrated on the right, it contains a subgraph homeomorphic to $K_{3,3}$ (drawn with a bold line).

As a rule of thumb, a nonplanar graph is more likely to contain a copy of $K_{3,3}$ than K_5. Efficient methods for finding such subgraphs are known. One good informal strategy for finding a copy of $K_{3,3}$ in a graph is to locate a long cycle six of whose nodes can be connected by either arcs or paths in the pattern indicated by Figure 10.16.

In 1852 Francis Guthrie, a graduate student at University College, London, asked whether every plane map could be colored with four colors so that adjacent countries received different colors. A plane map differs from a plane graph only in that attention is focused on the regions, which are called *countries*. It is implicit in Guthrie's question that each country consists of a single contiguous region, that the exterior region is also to be colored, and that two countries are adjacent if and only if they share a border of positive length. This question, which became known as the *four-color problem*, turned out to be very difficult, but, because of the simplicity of its statement, it attracted the attention of many mathematicians, of both the professional and amateur varieties. More erroneous proofs were produced in pursuit of its resolution than in that of any other mathematical problem. In 1976, Kenneth Appel and Wolfgang Haken used a localized version of Euler's equation, developed by Henri L. Lebesgue (1875–1941) and H. Heesch, to prove that four colors do indeed suffice. Their intricate reasoning was supplemented by over 1200 hours of computerized calculations. Another, somewhat shorter proof, was produced recently by Neil Robertson, Paul D. Seymour, and Robin Thomas. No proof that is suitable for inclusion in a textbook has appeared so far. A proof of the 6-colorability of all planar maps is offered next.

Lemma 10.4.8 *Every simple plane graph is 6-colorable.*

PROOF: It follows from Proposition 10.1.1 and Corollary 10.4.4.1 that if G is a simple plane graph, then

$$\sum_{v\in V(G)} \deg v = 2q \leq 6p - 12$$

and hence every simple plane graph contains a node of degree at most 5. Since every subgraph of a plane graph is also plane, it may be concluded by a straightforward induction that every simple plane graph is 5-degenerate. Hence, by Proposition 10.3.2, every simple plane graph is 6-colorable. Q.E.D.

Proposition 10.4.9 *Every plane map is 6-colorable.*

PROOF: Let M be a plane map. Associate to it a simple graph G_M by selecting a node inside each country, two distinct new nodes being joined by a new arc if and only if their countries are adjacent (in the example of Fig. 10.17 the arcs of G_M are dashed). The graph G_M is, by its construction,

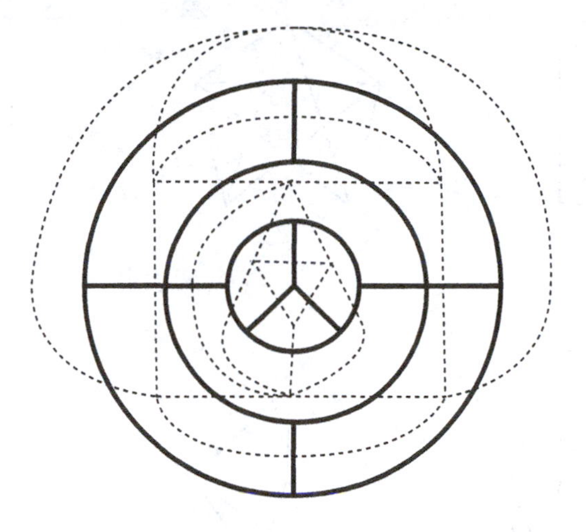

Figure 10.17. Associating a map to a plane graph

simple and plane. It follows from Lemma 10.4.8 that G_M is 6-colorable, and any such 6-coloring can clearly be construed as a 6-coloring of the map M.
Q.E.D.

Exercises 10.4

1. Which of the graphs of Figure 10.18 are planar and which are nonplanar? Justify your answers.

2. Which of the graphs of Figure 10.19 are planar and which are nonplanar? Justify your answers.

3. Prove that the regions of a Hamiltonian plane graph form a 4-colorable plane map.

4. A plane graph G is formed by drawing a polygon in the plane and adding a set of nonintersecting diagonals in its interior. Prove that both G and the map formed by its regions are 3-colorable.

5. Prove that the Tutte graph is 3-colorable and the map formed by its regions is 4-colorable. Is the map 3-colorable?

6. Find a 2-coloring of the maps in Figure 10.20.

7. For which values of n_1, n_2 is K_{n_1,n_2} planar? Justify your answer.

8. For which values of n_1, n_2, n_3 is K_{n_1,n_2,n_3} planar? Justify your answer.

9. For which values of n_1, n_2, n_3, n_4 is K_{n_1,n_2,n_3,n_4} planar? Justify your answer.

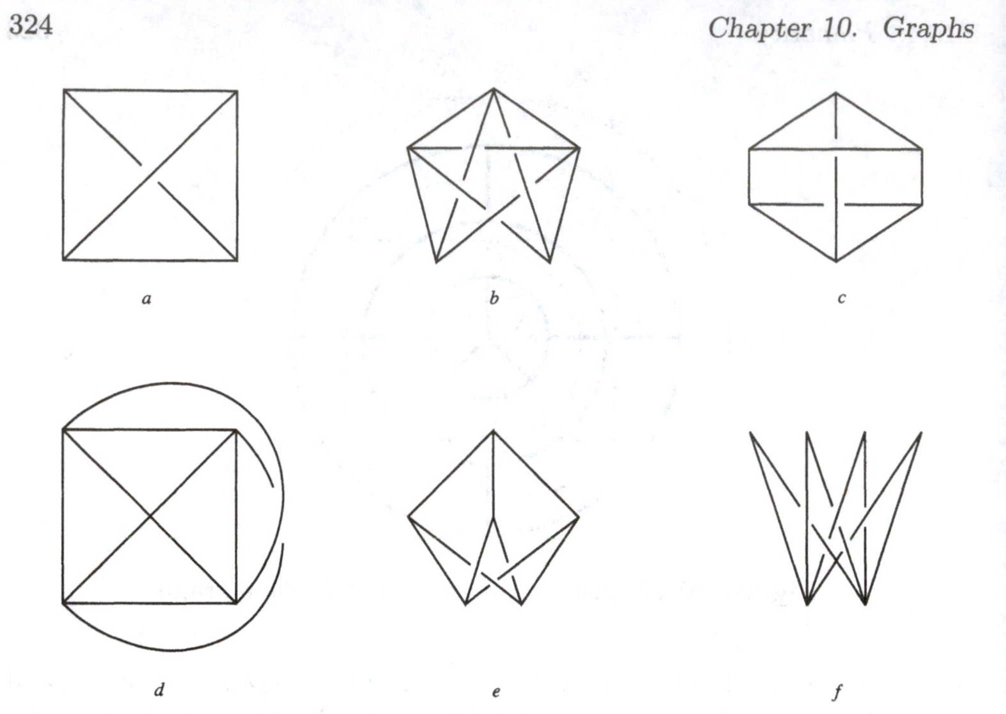

Figure 10.18.

10. A map is said to be *Eulerian* if the graph that consists of the map's borderlines is Eulerian. Prove that every Eulerian map is 2-colorable.

11. Prove that if a plane map is 2-colorable, then it is Eulerian.

12. Prove that K_5 is the only simple nonplanar graph with five or fewer nodes.

13. Prove that the plane map formed by a finite number of infinitely extended straight lines is 2-colorable.

14. Is it true that every loopless plane graph is 6-colorable?

15. Prove that the plane map formed by a finite number of circles is 2-colorable.

16. Use Theorem 10.4.2 to prove Proposition 8.1.1.

17. Are Theorem 10.4.2 and Proposition 8.1.1's version of Euler's equation logically equivalent?

18. Formulate and prove a generalization of Theorem 10.4.2 that holds for disconnected graphs.

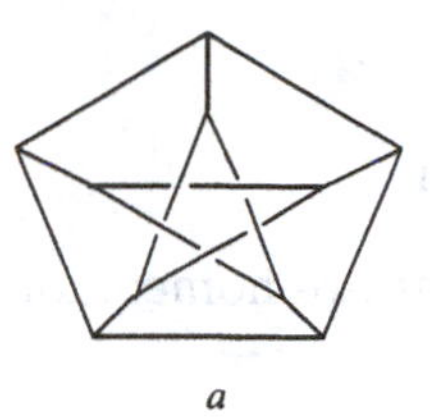

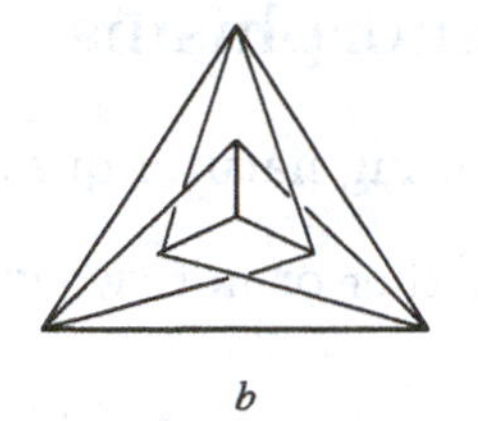

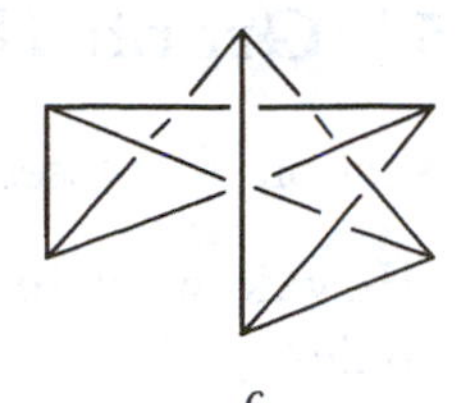

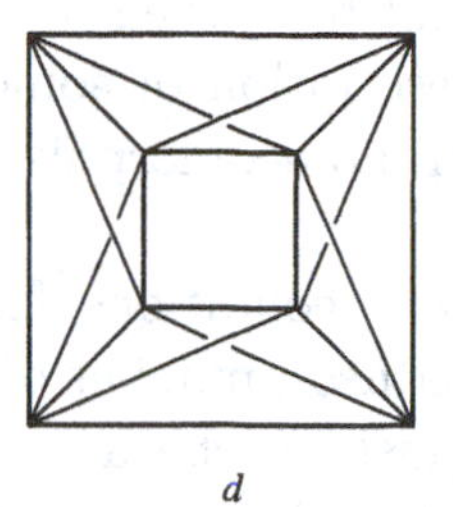

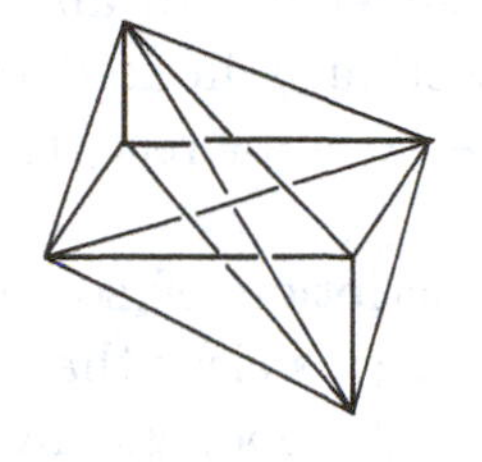

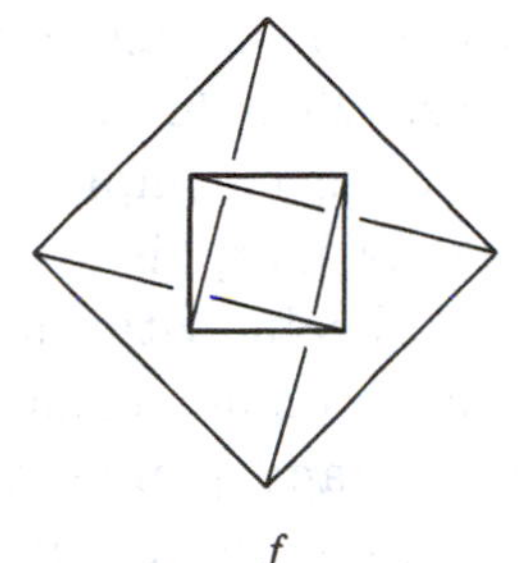

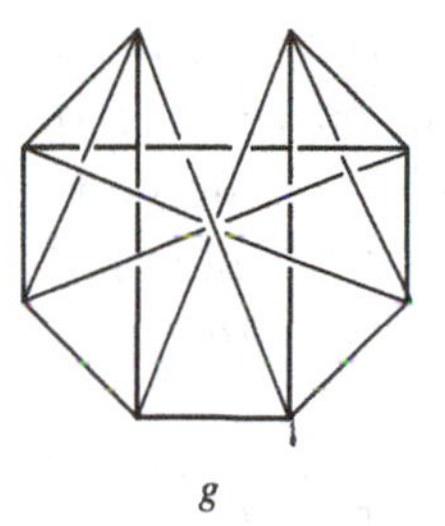

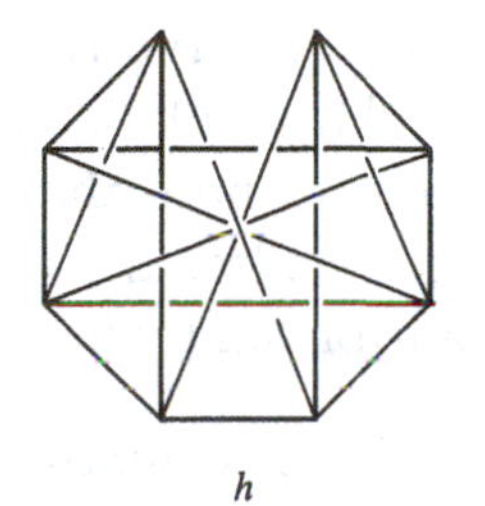

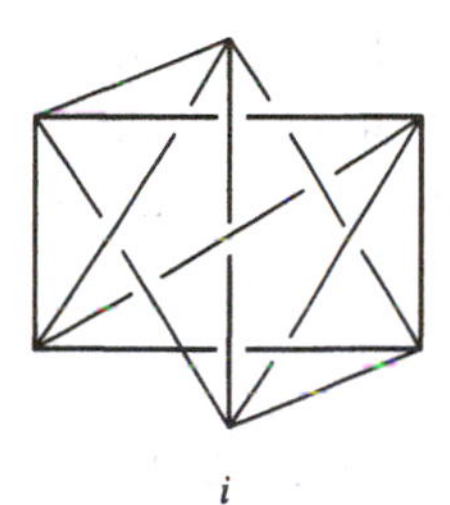

Figure 10.19. Planar and nonplanar graphs

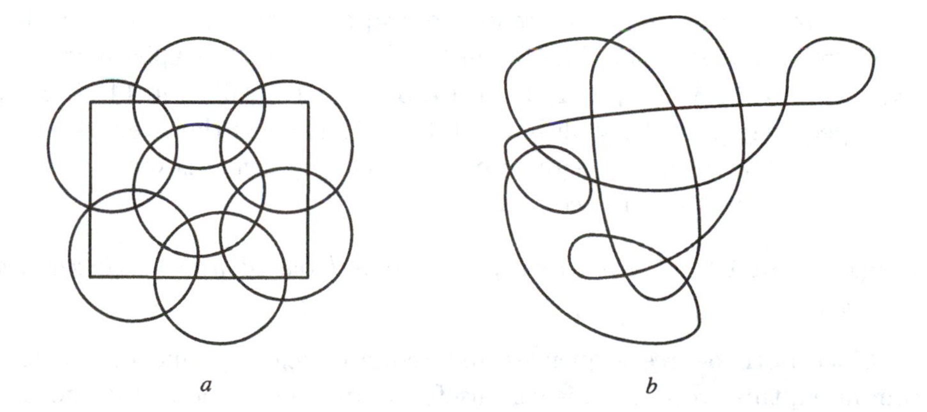

Figure 10.20. 2-Colorable maps

10.5 Graph Homeomorphisms

This section addresses the following natural question:

> How do we determine whether or not two graphs are homeomorphic?

There is an algorithm that will answer this question. Unfortunately, its full statement is somewhat awkward, and what's worse, its execution is so time consuming as to render it into an impractical tool, even for small graphs. Therefore, this section is limited to the description of some easier, though incomplete, methods for recognizing both homeomorphism and nonhomeomorphism.

We begin with the description of some necessary conditions for two graphs to be homeomorphic. It is clear that connectedness, number of components, and planarity are all topologically distinguishing characteristics. A connected graph and a disconnected graph cannot be homeomorphic to each other, and neither can a planar and a nonplanar graph be homeomorphic. The degree sequence can also serve the same purpose, but some care must be exercised in the formulation of this criterion, as vertices of degree 2 complicate the issue. To see this, note that all cycles are homeomorphic, regardless of the number of their nodes. In view of the discussion in Chapter 9, the following lemma is plausible and will be taken for granted.

Lemma 10.5.1 *If $f : G \to H$ is a homeomorphism of graphs, then for every node v of G with $\deg_G v \neq 2$, $f(v)$ is a node of H and*

$$\deg_G v = \deg_H f(v). \qquad \square$$

The *reduced degree sequence* of the graph G consists of those terms of its degree sequence that differ from 2. Thus, if three graphs have degree sequences (4, 3, 3, 2, 2, 2, 1, 1, 0), (4, 3, 3, 2, 1, 1, 0), and (4, 3, 3, 1, 1, 0) respectively, then they all have (4, 3, 3, 1, 1, 0) as their reduced degree sequence. The reduced degree sequence of every cycle is empty. Lemma 10.5.1 yields the following corollary.

Proposition 10.5.2 *Homeomorphic graphs have identical reduced degree sequences.*

Since both degree sequences and reduced degree sequences are easily computed, this proposition is very useful in proving the nonhomeomorphism of graphs. Two other techniques for demonstrating topological differences are illustrated in the next example.

Figure 10.21. Are these graphs homeomorphic?

Figure 10.22. Two nonhomeomorphic graphs

Example 10.5.3 The two graphs of Figure 10.21 are both connected, planar, and have the same reduced degree sequences. Nevertheless, they are not homeomorphic. One difference is that a has two distinct arcs joining the same two vertices, which is not the case for b. Alternately, if one of the nodes of each graph is deleted, the remainders pictured in Figure 10.22 are clearly not homeomorphic.

One way to prove the homeomorphism of two graphs is to find a labeling of their nodes that makes the matching of points obvious, as has been done for the two graphs in Figure 10.23. The homeomorphism can then be proved by (somewhat laboriously) verifying that similarly labeled pairs of vertices are either both adjacent or both nonadjacent in the two graphs. Unfortunately, even when they exist, such corresponding labelings are not easily found.

This homeomorphism problem (better known as the graph isomorphism problem, which is its combinatorial formulation) has many implications for

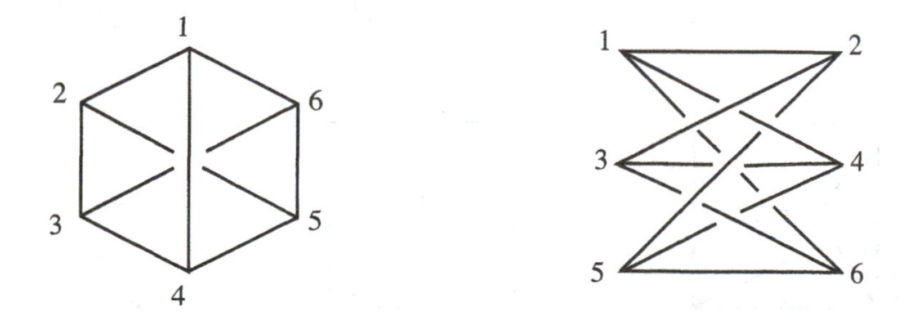

Figure 10.23. Two homeomorphic graphs

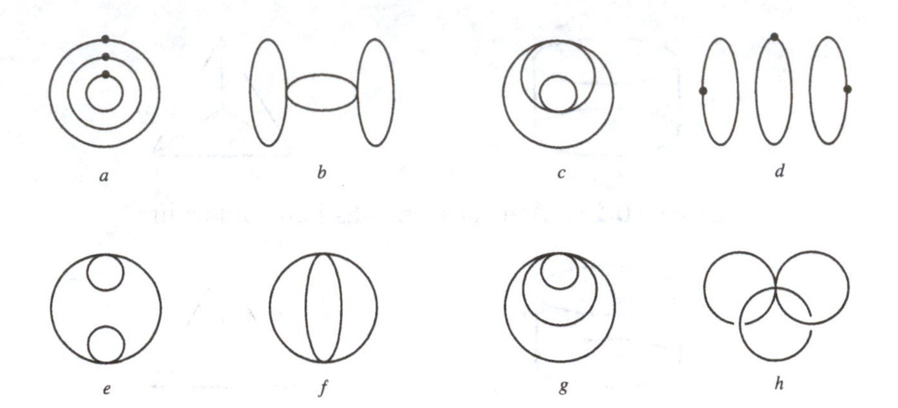

Figure 10.24. Some graphs

applied mathematics, and a considerable amount of effort has been expended on the derivation of efficient algorithms for its resolution. The consensus is that so far no such algorithm has been found.

Exercises 10.5

1. Which of the graphs of Figure 10.24 are homeomorphic?
2. Which of the graphs of Figure 10.25 are homeomorphic?
3. Which of the four graphs of Figure 10.26 are homeomorphic to each other?
4. Which of the graphs of Figure 10.27 are homeomorphic to each other?

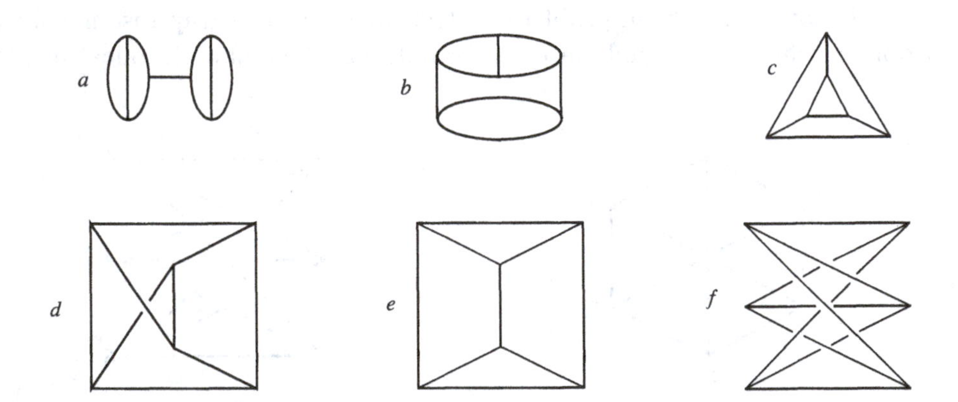

Figure 10.25. Some homeomorphic and nonhomeomorphic graphs

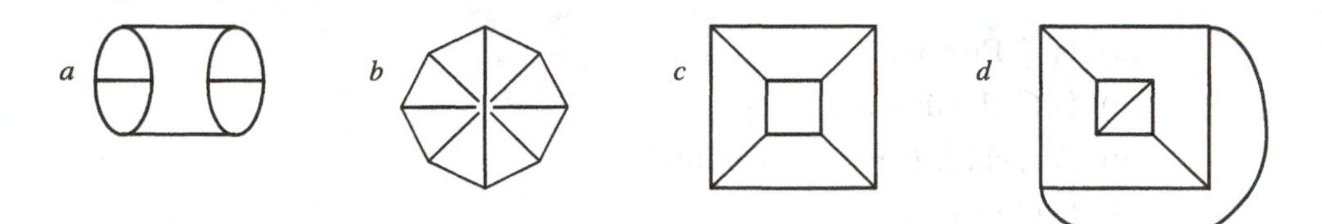

Figure 10.26. Some graphs

Chapter Review Exercises

1. Let G be the graph formed by the edges of the truncated cube I of Figure 8.4. Answer the following questions and justify your answers.

   (a) Is G Eulerian?

   (b) Is G Hamiltonian?

   (c) For which k is G k-colorable?

   (d) Is G planar?

2. Let G be the graph formed by the edges of the truncated cube II of Figure 8.5. Answer the following questions and justify your answers.

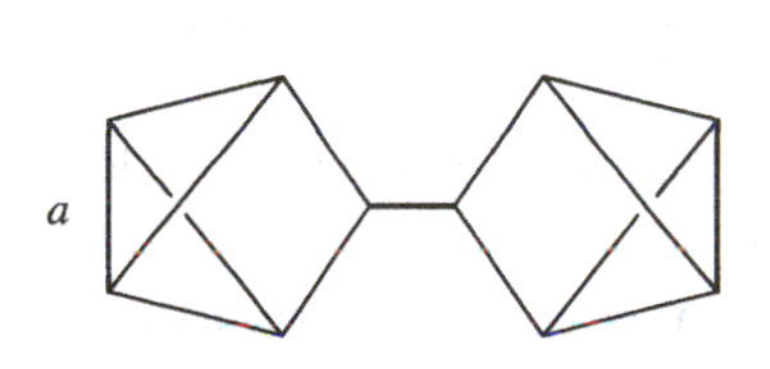

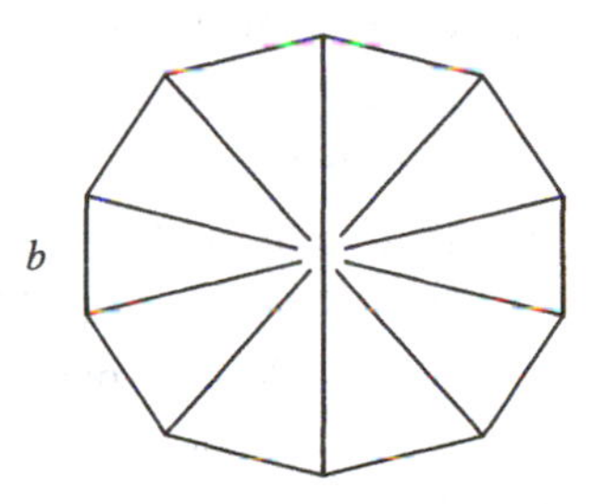

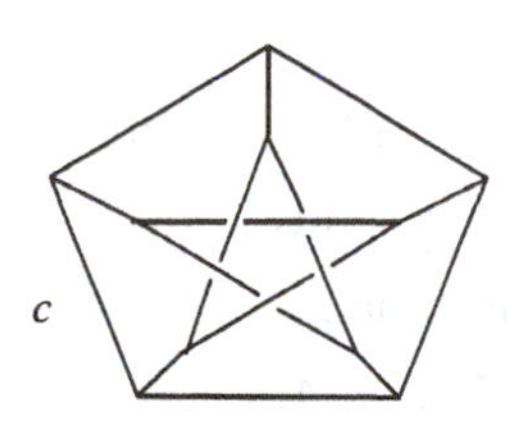

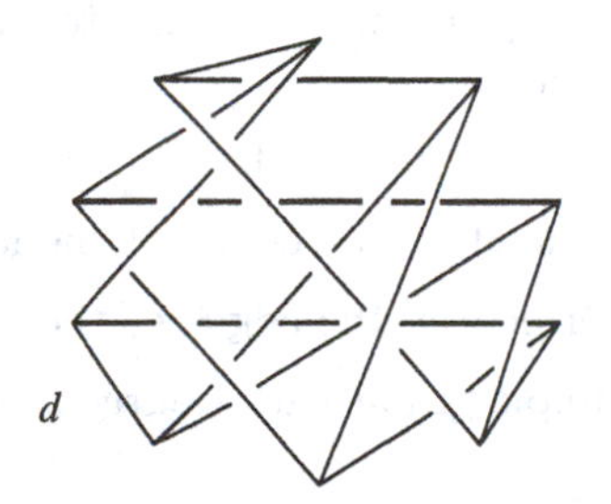

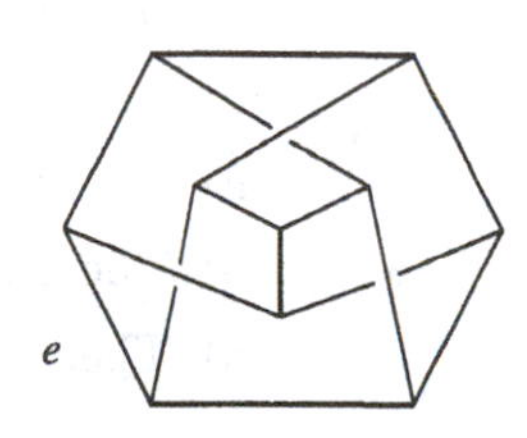

Figure 10.27. Some graphs

(a) Is G Eulerian?

(b) Is G Hamiltonian?

(c) For which k is G k-colorable?

(d) Is G planar?

3. Let G be the graph formed by the edges of the tunneled solid of Figure 8.8. Answer the following questions and justify your answers.

   (a) Is G Eulerian?

   (b) Is G Hamiltonian?

   (c) For which k is G k-colorable?

   (d) Is G planar?

4. Are the following statements true or false?

   (a) Every graph is k-colorable for some integer k.

   (b) There is an integer k such that every graph is k-colorable.

   (c) Every complete graph with at least three vertices is Eulerian.

   (d) Every complete graph with at least three vertices is Hamiltonian.

   (e) Every Eulerian graph is Hamiltonian.

   (f) Every Hamiltonian graph is Eulerian.

   (g) Every planar Eulerian graph is also Hamiltonian.

   (h) Every simple planar graph that is both Hamiltonian and Eulerian is also 3-colorable.

   (i) If a connected plane graph has the same number of nodes and arcs, then it has exactly two regions.

   (j) If a 2-colorable planar graph has twice as many arcs as nodes, then it is Hamiltonian.

   (k) The graph formed by the edges of the tetrahedron is homeomorphic to K_4.

   (l) The graph $K_{2,2,2}$ is homeomorphic to the graph formed by the edges of the octahedron.

   (m) The graph formed by the edges of the dodecahedron is Hamiltonian.

   (n) The graph formed by the edges of the icosahedron is 3-colorable.

   (o) The graph formed by the edges of the cube is Eulerian.

   (p) There is a simple graph with degree sequence (5, 5, 4, 4, 3).

TOPOLOGIST'S DILEMMA:
MOBIUS LETTERHEAD
S.HARRIS

Chapter 11

Surfaces

The two-dimensional topological objects are surfaces. While the study of the metric properties of surfaces goes back to the ancient Greeks, their topological aspects only came under scrutiny in the 1800s. In the early part of that century Abel, Carl G. J. Jacobi (1804–1851), and later Georg F. B. Riemann (1826–1866), pointed out that *elliptic integrals*, then at the forefront of mathematical research, as well as more complicated integrals, became more tractable if their independent variables were assumed to be complex rather than real. Moreover, the proper domain of each integrand should consist of a stack of several copies of the complex plane that are so interconnected as to form new mathematical objects, which later became known as *Riemann surfaces*. The topological classification of these surfaces turned out to be crucial for the proper understanding of the aforementioned integrals. While some earlier work of Euler, Gauss, and others could also be viewed as topological in nature, it was these concerns with analytic issues that brought topology into the mainstream of mathematics.

11.1 Polygonal Presentations

The prototypical surface is the Euclidean plane. Today it is customary to understand by this a flat surface that extends infinitely far in all directions. As was noted in Chapter 2, however, Euclid thought of it as a flat surface of finite, though arbitrarily large, extent. For our purposes here it is convenient to adopt this earlier point of view. A *disk* refers to any plane circle with its interior, or any topological space homeomorphic to it, such as a triangle (with its interior) or indeed any polygon whatsoever. It should be stressed that the polygon's perimeter is devoid of self-intersections (i.e., it is a loop).

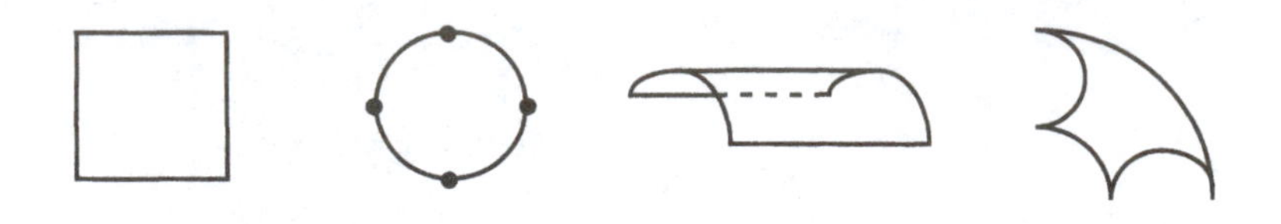

Figure 11.1. Disks

The context being topology, curved isotopes of polygons are also disks, and the requirement that the edges be straight line segments is relaxed into the stipulation that they can be any open arcs (Fig. 11.1).

The Cartesian graph of a continuous single-valued function $z = f(x, y)$, $a \leq x \leq b$, $c \leq y \leq d$ is also a disk because it is homeomorphic to a rectangle. This homeomorphism is established by matching every point (x, y) in the given rectangular domain with the point $(x, y, f(x, y))$ on the surface.

The sphere (i.e., surface of a ball) is an example of a surface that is not homeomorphic to the disk. The latter has a rim, or border, which is lacking in the sphere. Surfaces, such as the sphere, that are completely lacking in such borders are said to be *closed.* Now, while the sphere is homeomorphic to neither the plane nor the disk, it can be cut into two hemispheres each of which is homeomorphic to a disk. This decomposability into polygons has been chosen to be the defining characteristic of all surfaces. The open-ended cylinder of Figure 11.2 can be "unrolled" into the plane if it is first cut lengthwise along the arc a. The direction of the cut is recorded by means of an arrowhead just in case the original surface needs to be reconstructed from its flattened version (this will come up later). The sphere can also be flattened into a single polygon by means of a cut (see Fig. 11.3). The *torus* (i.e., the surface of a donut) is transformed in Figure 11.4 into a rectangle by means of two cuts a and b or else into two rectangles (Fig. 11.5) by means of the four cuts a, b, c, d. The *double torus* of Figure 11.6 can be transformed into two octagons by means of the seven cuts a, b, c, d, e, f, g.

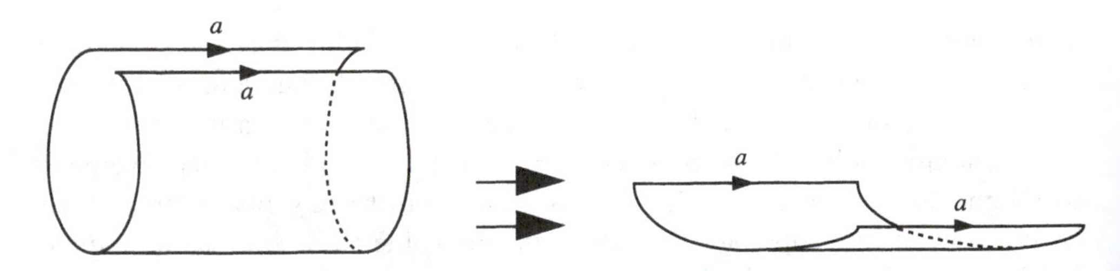

Figure 11.2. Flattening the cylinder

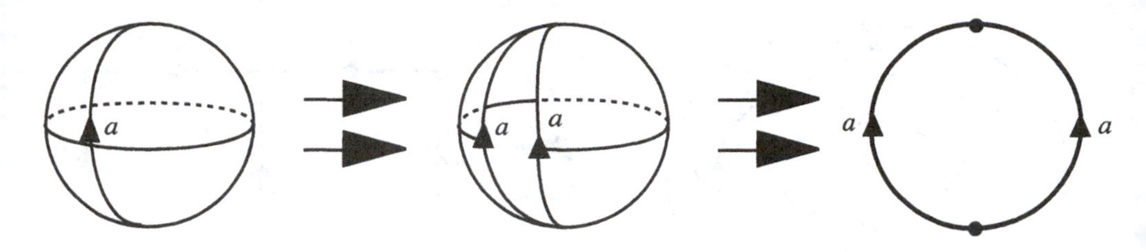

Figure 11.3. Flattening the sphere

Note that in all of the foregoing "flattenings" each cut leaves a trace consisting of two directed edges on the perimeters of the resultant polygons. Thus, if the unit square $\{(x,y) \mid 0 \leq x \leq 1,\ 0 \leq y \leq 1\}$ is cut along the vertical straight line segment joining (0.5, 0) to (0.5, 1), then the result consists of the two rectangles $\{(x,y) \mid 0 \leq x \leq 0.5,\ 0 \leq y \leq 1\}$ and $\{(x,y) \mid 0.5 \leq x \leq 1, 0 \leq y \leq 1\}$.

The definition of closed surfaces reverses the aforementioned flattening process. A *pasting* is the reversal of a cut. The pasted arcs will always be directed, and the pasting must be consistent with these directions. In other words, if the directed edge AB is to be pasted with the directed edge CD, then A and B are pasted to C and D, respectively, with the rest of the points

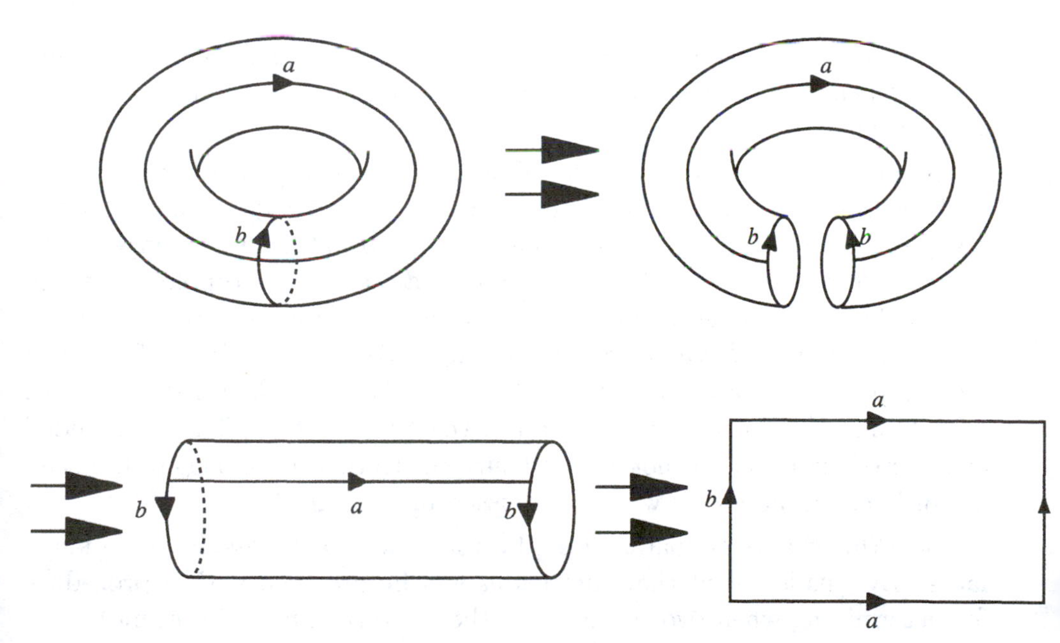

Figure 11.4. Flattening the torus

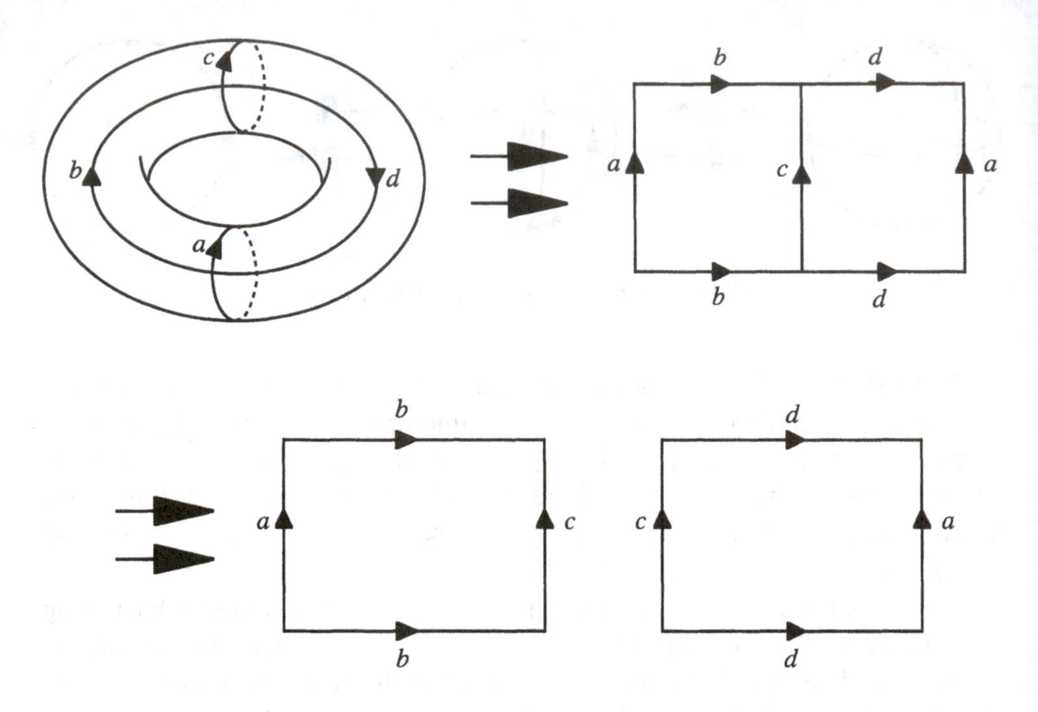

Figure 11.5. Flattening the torus into two polygons

of the edges following suit in a continuous manner. This (directed) pasting operation can also be visualized by means of an imaginary zipper that joins the two edges.

A *polygonal presentation* is a set of plane polygons together with a pairing of their directed edges, such that the pasting of paired edges connects all the polygons. The edges of these polygons are labeled so that paired edges carry the same letter (see Figs. 11.7 and 11.8 for examples). The topological space obtained by performing the indicated pastings is a *closed surface*. (Wherever possible, the term *closed surface* will be abbreviated to *surface*.) The set of labeled polygons obtained by the flattening of surfaces depicted in Figures 11.3–11.6 is a polygonal presentation of the given surface. The open-ended cylinder of Figure 11.1 is not a closed surface. Rather, it belongs to the class of bordered surfaces that will be discussed in Section 11.4.

It is this chapter's goal to describe the topological classification of surfaces. By this is meant that the reader will be provided with a procedure for determining when two surfaces are the same (i.e., homeomorphic).

The vertices and (pasted) edges of a polygonal presentation Π constitute a graph called the *skeleton* of Π and denoted by $G(\Pi)$. For example, the

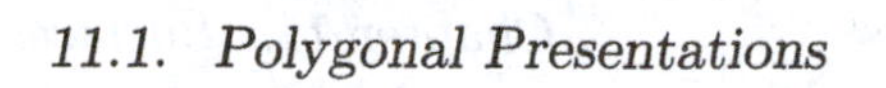

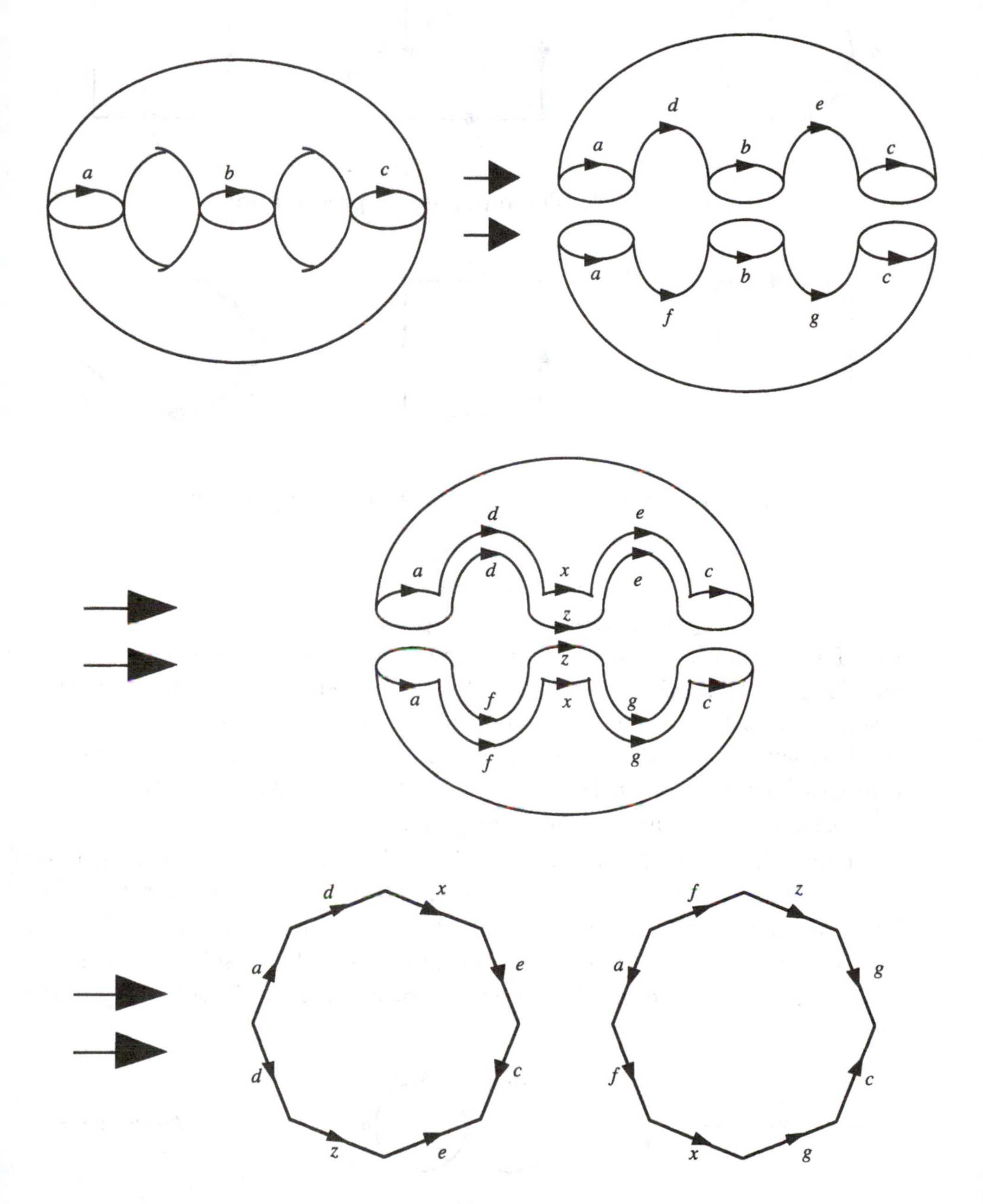

Figure 11.6. Flattening the double torus

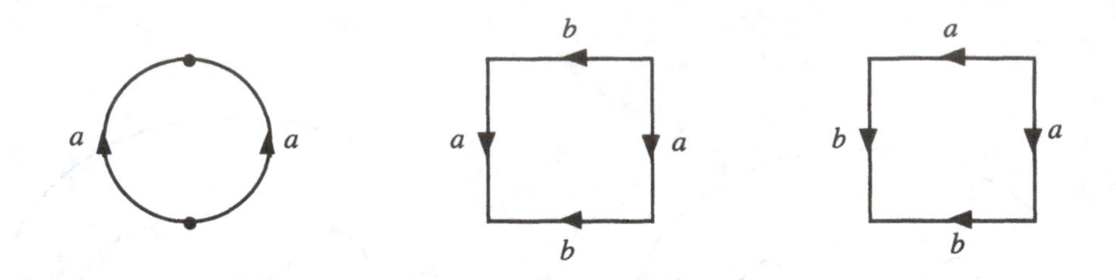

Figure 11.7. Three polygonal presentations

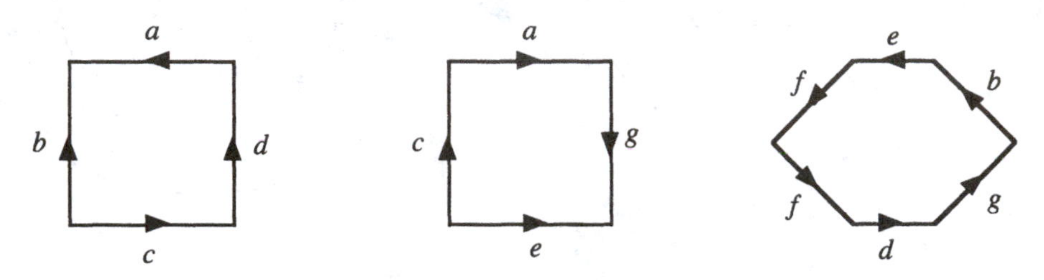

Figure 11.8. A presentation with three polygons

graph G_1 of Figure 11.9 is the skeleton of the presentation of the sphere (Fig. 11.3), and G_2 is the skeleton of the presentation of the torus in Figure 11.4. The graph G_3 of the same figure is the skeleton of the rightmost polygon of Figure 11.7, which, as will be demonstrated momentarily, is another presentation of the sphere. A method for deriving the skeleton of any presentation will be described in Section 11.2. The stipulation that the pairing of labels of a presentation Π connects all of the polygons is tantamount to the skeleton $G(\Pi)$ being a connected graph.

There arises now the natural question of identifying (or realizing) the surface that corresponds to a polygonal presentation as a subset of Euclidean 3-space. The leftmost of the three presentations of Figure 11.7 will be recognized as the end result of the flattening of the sphere illustrated above (Fig. 11.3). The middle presentation of Figure 11.7 looks very much

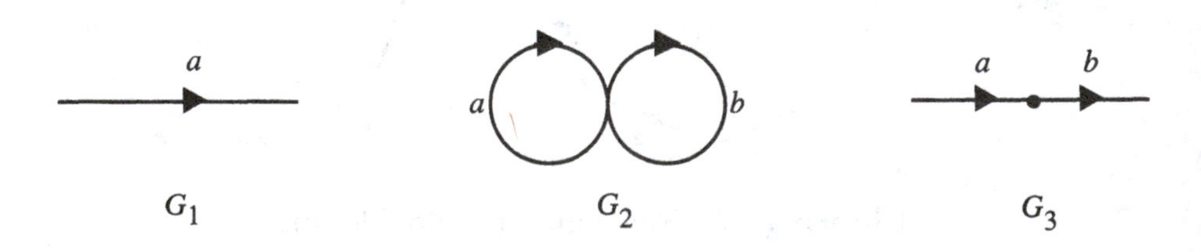

Figure 11.9. Some skeletons

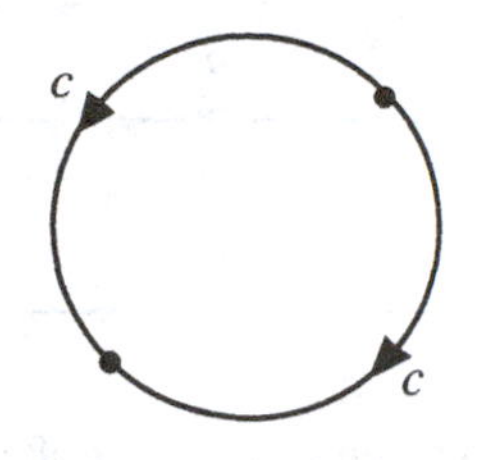

Figure 11.10. A polygonal presentation

like the flattening of the torus depicted in Figure 11.4 except that the labels and directions are different. The different labeling is, of course, immaterial, nor do the differing directions matter. As a subset of Euclidean space, every arc is identical to its reverse and so the reversal of two arrows that carry the same label does not affect the presented surface. Consequently, the surface that corresponds to the middle presentation is indeed the torus. The rightmost presentation of Figure 11.7 also bears some resemblance to the flattening of the torus, but this similarity is misleading. If the union of the edges a and b is replaced by a single arc c, we obtain the presentation if Figure 11.10, which is clearly identical to the leftmost presentation of Figure 11.7. The rightmost presentation therefore also yields a sphere.

But what about the presentation of Figure 11.11, which greatly resembles that of the torus? We could begin reconstructing the surface by first pasting the two edges labeled a so as to obtain a cylinder. This would then be followed by a pasting of the b's, perhaps by keeping the cylinder's left end fixed in place while bending the rest of the cylinder so as to bring the right end flush against the fixed left end (Figure 11.12). Unfortunately, as is indicated in the diagram, the directions of the two versions of b are such as to make the required pasting impossible. This difficulty is inescapable and the two b's must wind up facing each other with inconsistent orientations no matter how the cylinder is twisted through space. To see this, note

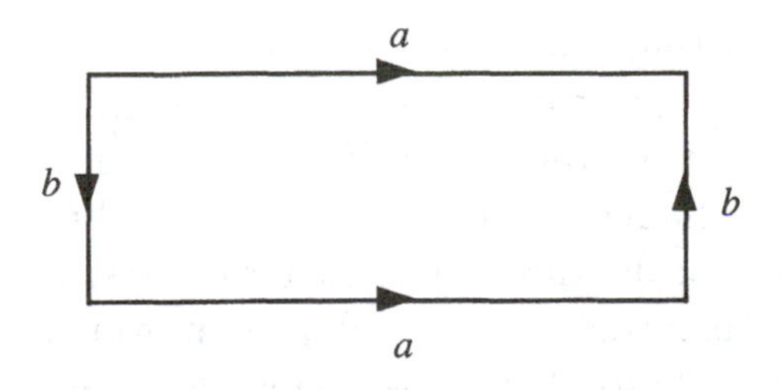

Figure 11.11. A mystery polygonal presentation

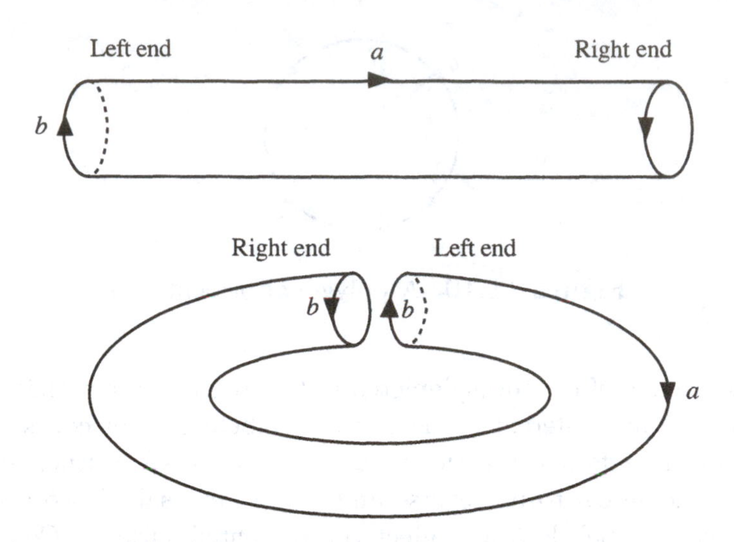

Figure 11.12. A seemingly impossible pasting

that when the cylinder was straight, each of the b's was endowed with the counterclockwise orientation from the point of view of an observer who is positioned at the geometrical center of the border loop and is looking into the cylinder from the outside. No amount of bending or twisting will alter these relative orientations and, when the two b's are eventually brought together, the observers must stand back-to-back so that the orientations of the b's are inconsistent.

The impossibility of carrying out the pasting of Figure 11.12 in 3-space notwithstanding, there is a surface, called the *Klein bottle*, that corresponds to it and that can be realized in four-dimensional Euclidean space. To demonstrate this, it is helpful to think of time as the fourth dimension. Begin as before by converting the presentation into a cylinder, fix its left end and, bending the cylinder, pass the right end *through* the cylinder's surface and bring this end up to the left end through the cylinder's *interior* (Fig. 11.13). Note that the orientations of the two b's are now consistent so that there is no obstacle to pasting them together. Of course, this pasting involves a "cheat"—the self-crossing c of the cylinder, which is not allowed for in the polygonal form. However, this self-crossing can be eliminated if each point on the bent cylinder is endowed with an additional time coordinate above and beyond its usual three spatial coordinates, which time coordinate is assumed to be smoothly distributed in conformance with the pattern indicated by the Now-Yesterday-Now-Tomorrow labels in the diagram. The two parts of the cylinder do not really cross each other in the loop c because

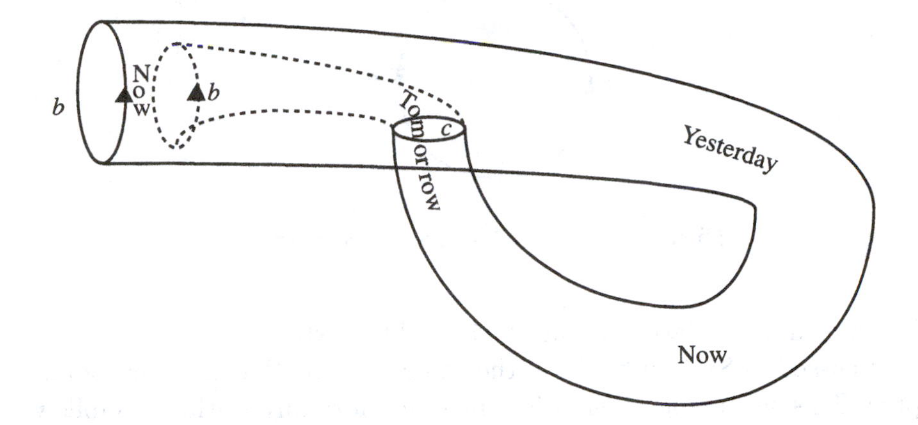

Figure 11.13. Constructing a four-dimensional realization of the Klein bottle

they pass through it at different times.

Four-dimensional Euclidean space can be formally represented by the abstract mathematical object

$$\mathbb{R}^4 = \{(x, y, z, w) \text{ such that } x, y, z, w \text{ are real numbers}\}.$$

Nevertheless, the informal space-time continuum point of view is quite useful and will arise again.

Another surface that is unrealizable in $\mathbb{R}^3$ is the *projective plane* whose presentation appears in Figure 11.14 and that underlies the alternate geometry discussed in Section 5.3. Just like the Klein bottle, this surface is realizable in $\mathbb{R}^4$. The realization process is depicted in Figure 11.15 and also makes use of an illusory self-crossing. The top left-hand circle describes some points on the "perimeter" of the projective plane that need to be pasted. This disk is first deepened into a bowl, which is further distorted as in the central figure. Next, a time coordinate is attached to every point on the surface of the distorted bowl. This time coordinate is assumed to vary smoothly with values between -4 and 4. Note that each of the two occurrences of each of the points A, B, C, D, E, F, G, H has been assigned the same time coordinate: 0 to A, -2 to B, -4 to C, -2 to D, 0 to E, 2 to F, 4 to G, and 2 to H. Consequently, the two halves of the distorted lip can be identified as in the bottom figure, where what looks like a self-crossing is no such thing at all—the points B and H occupy the same location at different times -2 and 2, and so on.

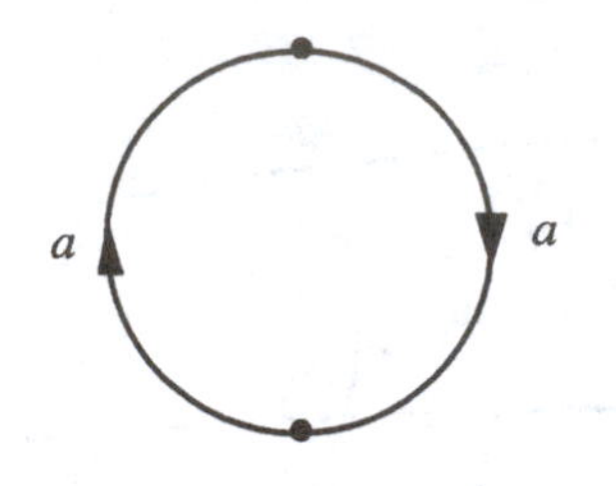

Figure 11.14. The projective plane

The reason the surface of Figure 11.14 and the extension of the Euclidean plane discussed in Section 5.3 have the same name is that they are homeomorphic. This surprising observation may be informally justified as follows. Visualize the ideal points of projective plane of Section 5.3 as lying on an infinitely far circle that surrounds the Euclidean plane. This allows us to view the extended plane as a disk (albeit one with an infinite radius). The stipulation that only one ideal point is associated with every pair of parallel Euclidean straight lines forces us to identify diametrically opposite points on the perimeter of this disk. This identification is realized by the pasting of the arcs labeled a in Figure 11.14.

There are many other such surfaces that cannot be realized in $\mathbb{R}^3$, and they will be discussed later. These observations call for some new terminology. A topological space S is said to be *embeddable* or *realizable* in the topological space T if S is homeomorphic to a subset of T. Thus, a graph is planar if and only if it is embeddable in a disk. Every graph is embeddable in $\mathbb{R}^3$ and the projective plane and the Klein bottle are not embeddable in $\mathbb{R}^3$. They are, however, embeddable in $\mathbb{R}^4$. In fact, every closed surface is embeddable in $\mathbb{R}^4$.

Exercises 11.1

1. The sphere is decomposed in two ways by means of the cuts indicated in Figure 11.16. In each case describe the resulting polygons.

2. The sphere is decomposed in two ways by means of the cuts indicated in Figure 11.17. In each case describe the resulting polygons.

3. The torus is decomposed in two ways by means of the cuts indicated in Figure 11.18. In each case describe the resulting polygons.

4. The torus is decomposed in two ways by means of the cuts indicated in Figure 11.19. In each case describe the resulting polygons.

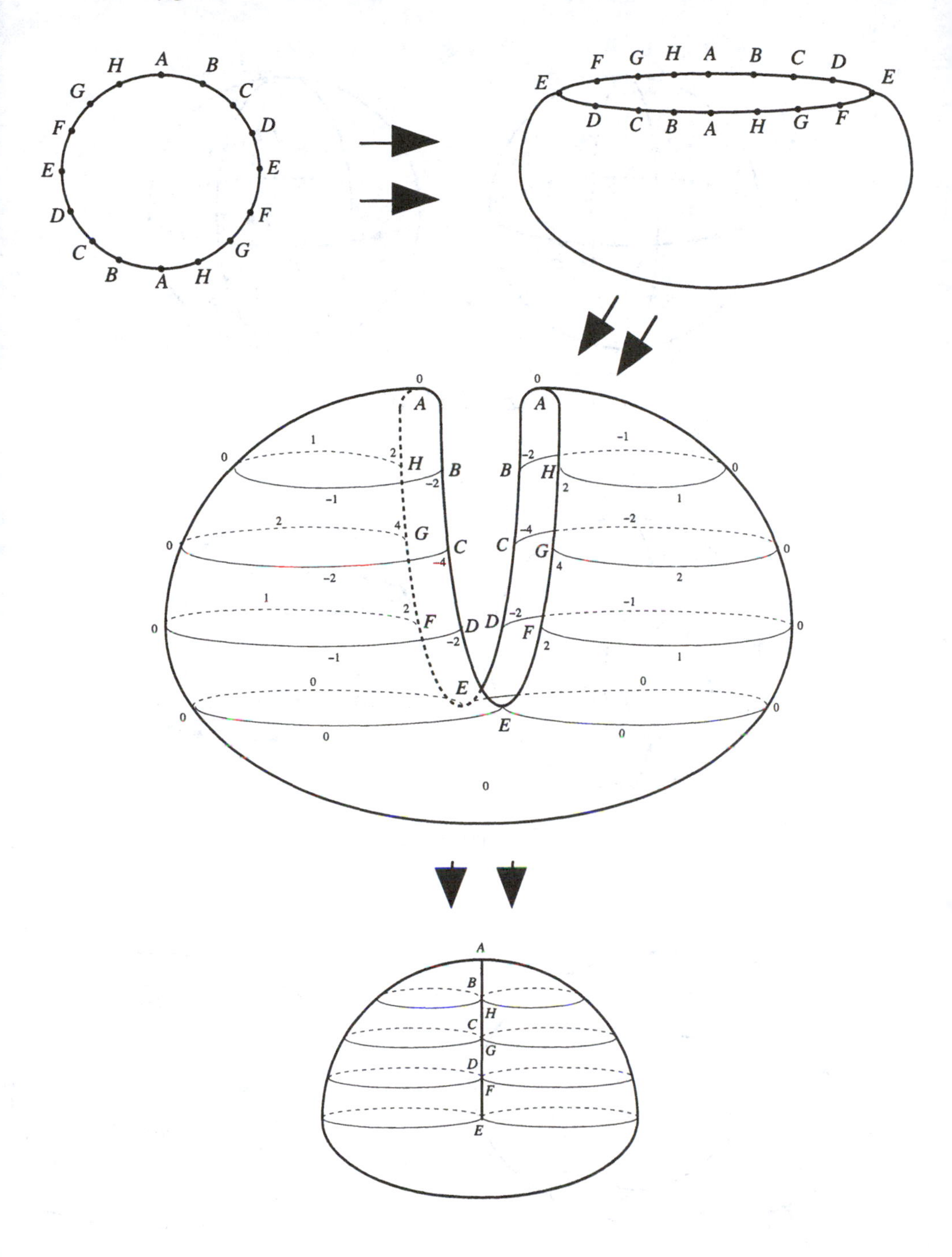

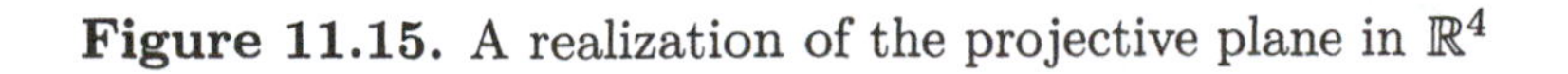
Figure 11.15. A realization of the projective plane in $\mathbb{R}^4$

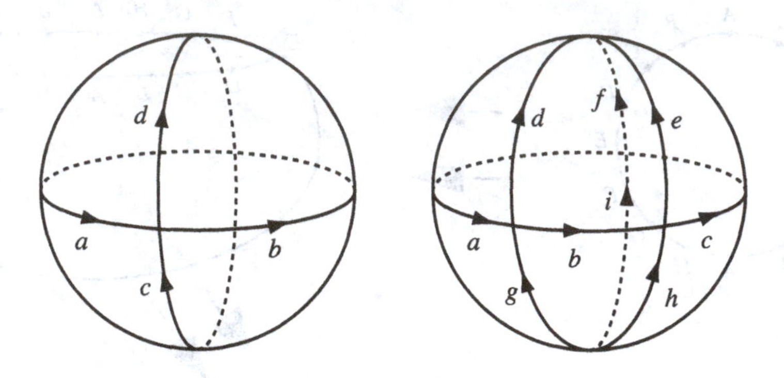

Figure 11.16.

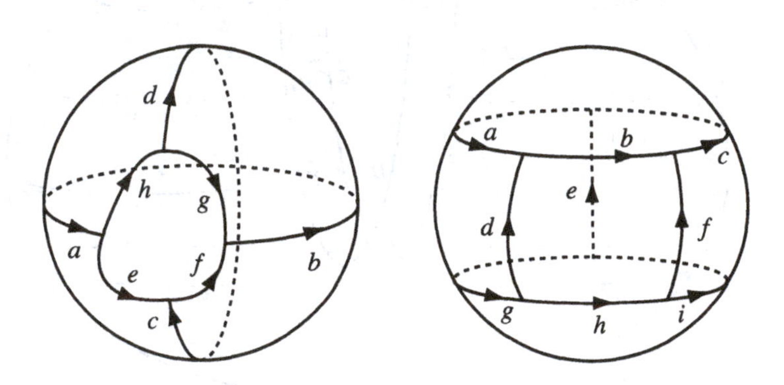

Figure 11.17.

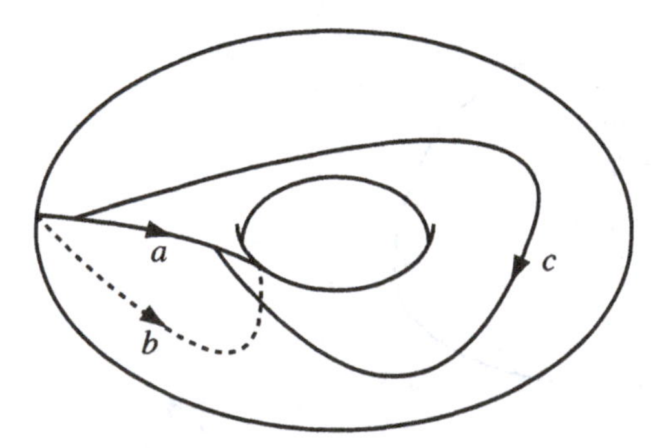

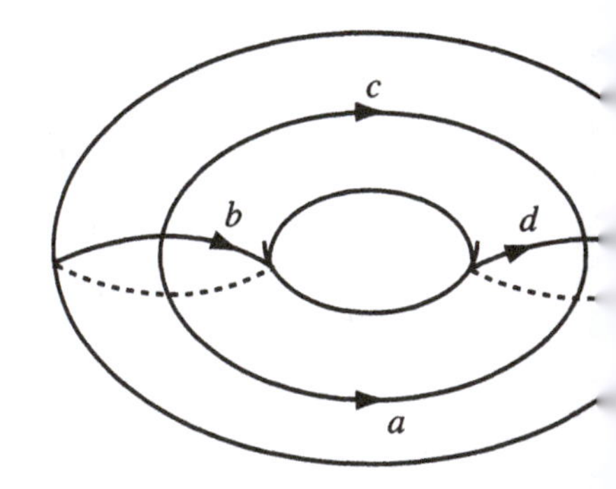

Figure 11.18.

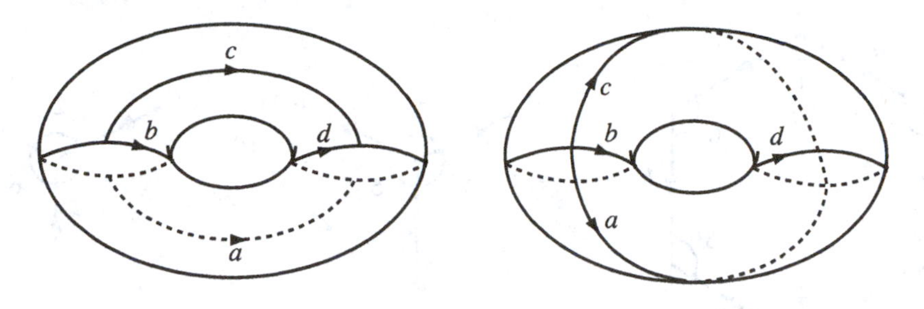

Figure 11.19.

5. Explain how the surface A of Figure 11.20 can be flattened into one polygon. Describe the resulting polygonal presentation.

6. Explain how the surface B of Figure 11.20 can be flattened into one polygon. Describe the resulting polygonal presentation.

7. Explain how the surface C of Figure 11.20 can be flattened into one polygon. Describe the resulting polygonal presentation.

8. Explain how the surface D of Figure 11.20 can be flattened into one polygon. Describe the resulting polygonal presentation.

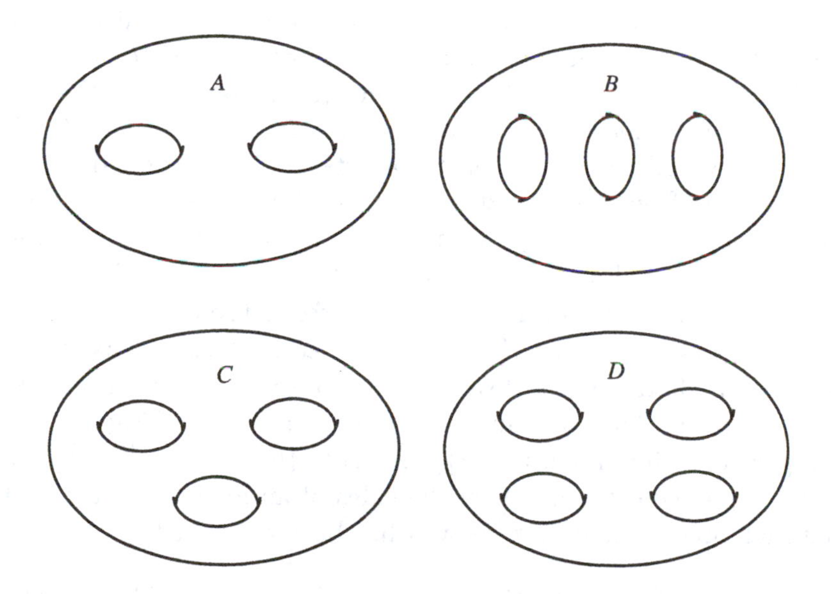

Figure 11.20.

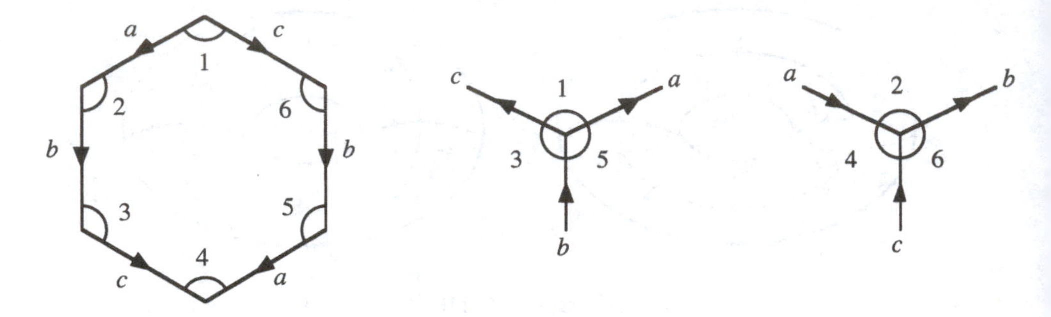

Figure 11.21. A presentation and the vicinities of its two nodes

11.2 Closed Surfaces

This section is concerned with the topological classification of all the surfaces defined by polygonal presentations. The task begins with an examination of the vertices of the constituent polygons. As is indicated by the surfaces of Figures 11.3–11.6 and their polygonal presentations, different vertices of the constituent polygons may in fact designate *occurrences* of the same node of the skeleton $G(\Pi)$—in other words, they may correspond to the same point on the surface that is being presented. If the vertex of a polygon is an occurrence of a node, then the vertex is also said to *belong* to the node. Two vertices that belong to the same node are said to be *equivalent*.

It will be necessary to ascertain the actual number of distinct nodes of $G(\Pi)$ that appear on the perimeters of the polygons of Π, and a method for deriving this number can be based on the examination of the surface in the vicinity of its nodes. If Π is a presentation of the surface S, then the vicinity of any node u of $G(\Pi)$ on S consists of a disk that is divided by the arcs emanating from u into sectors (Fig. 11.21). These sectors correspond to the interior angles of the polygons of Π, and each node can be identified by listing all the angles of the polygons of Π that form its vicinity. For this purpose it is convenient to denote the *reverse* of the oriented arc a by a^{-1} with the understanding that $(a^{-1})^{-1} = a$ (see Fig. 11.22). When an *angle* at a vertex is described, its two sides will be written down as emanating from that vertex. Thus, the angles 1–6 of the polygon of Figure 11.21 have the respective pairs of sides $\{c, a\}$, $\{a^{-1}, b\}$, $\{b^{-1}, c\}$, $\{c^{-1}, a^{-1}\}$, $\{a, b^{-1}\}$, and $\{b, c^{-1}\}$. The order in which the sides of an angle are written down is immaterial; angle 2 could just as well has been described as $\{b, a^{-1}\}$.

Example 11.2.1 Consider the 1-polygon presentation of Figure 11.21. We begin with $\angle 1 = \{c, a\}$ and observe that because of the a-pasting this angle

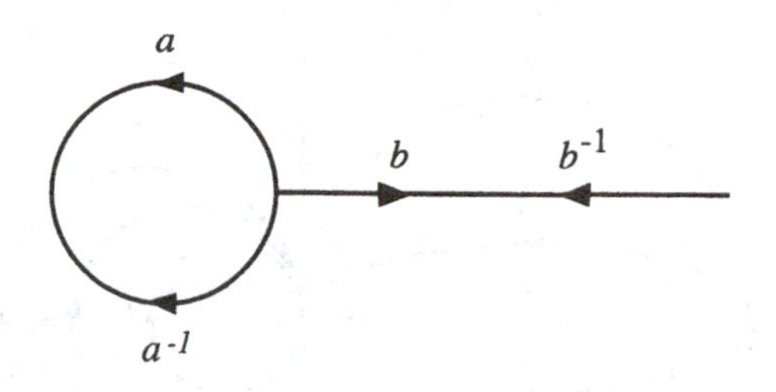

Figure 11.22. Arcs and their inverses

must abut on the actual surface with $\angle 5 = \{a, b^{-1}\}$. Next, because of the b-pasting, comes $\angle 3 = \{b^{-1}, c\}$. The c-pasting brings us back to $\angle 1$. This means that there is a node of the presentation, call it A, whose vicinity consists of the angles 1, 3, and 5. The process that leads to the identification of this node is symbolized by the list

$$A: \qquad c, a, b^{-1}, c.$$

The other three angles constitute the vicinity of another node B, where

$$B: \qquad a^{-1}, b, c^{-1}, a^{-1}.$$

Hence the presentation of Figure 11.21 has two nodes.

Example 11.2.2 The polygonal presentation of Figure 11.23 has three nodes u, v, w that are given by the following lists:

$$u: \qquad d, a, f^{-1}, g^{-1}, h, d.$$
$$v: \qquad c, c^{-1}, d^{-1}, e^{-1}, h^{-1}, f, e, g, i^{-1}, i, a^{-1}, b, c.$$
$$w: \qquad b^{-1}, b^{-1}.$$

The skeleton of this presentation is therefore the graph of Figure 11.24.

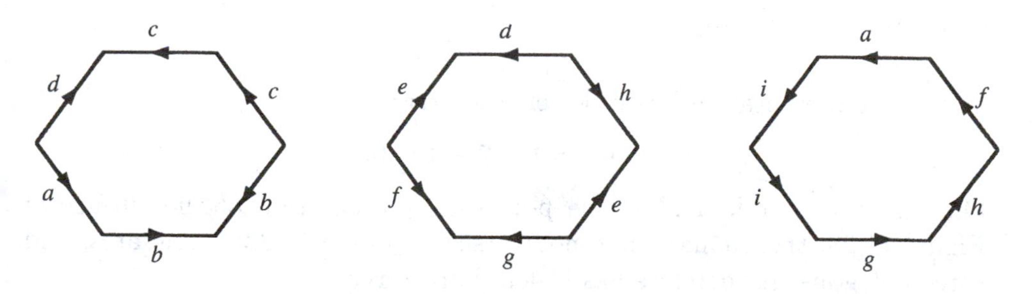

Figure 11.23. A polygonal presentation

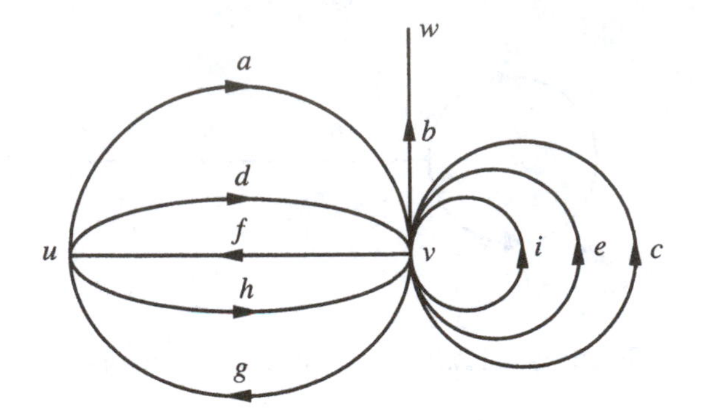

Figure 11.24. A skeleton

Presentations are classified by two attributes: the numerical Euler characteristic and the geometric orientability character.

Euler characteristic: Let the presentation Π have r polygons and suppose $G(\Pi)$ has p nodes and q arcs. Then the quantity $p-q+r$ is called the *Euler characteristic* of Π and is denoted by $\chi(\Pi)$.

Example 11.2.3 If Π is the presentation of the torus in Figure 11.4, then it has the two arcs a and b, a single node (where a and b intersect), and a single rectangular polygon. Hence its Euler characteristic is

$$\chi(\Pi) = 1 - 2 + 1 = 0.$$

By a similar reasoning, the toroidal presentation that is the end product of Figure 11.5 has Euler characteristic

$$\chi(\Pi) = 2 - 4 + 2 = 0.$$

If Π is the presentation of the Klein bottle given in Figure 11.11, then, since the underlying surface is not given, it is necessary to determine the number of nodes by the method of Example 11.2.2. Accordingly, there is a unique node given by the list

$$b, a, b^{-1}, a^{-1}, b$$

and so the presentation has Euler characteristic

$$\chi(\Pi) = 1 - 2 + 1 = 0.$$

On the other hand, if Π is the polygonal presentation of the surface of Figure 11.23, then it has three nodes (see Example 11.2.2), nine arcs, and three polygons. It therefore has Euler characteristic

$$\chi(\Pi) = 3 - 9 + 3 = -3.$$

Example 11.2.4 The faces of every one of the regular, semiregular, as well as other spherelike polyhedra discussed in Section 8.1 can be considered as a polygonal presentation of the surface of that solid, a presentation whose polygons have not been separated yet. Using the notational conventions of that chapter, it follows from Euler's theorem (Proposition 8.1.1) that every such presentation Π has Euler characteristic

$$\chi(\Pi) = p(\Pi) - q(\Pi) + r(\Pi) = p - q + r = 2.$$

Example 11.2.5 Figure 8.6 can be construed as a polygonal presentation Π of the torus, one whose rectangular polygons are still to be separated. Note that this drawing has 19 "meridians," and assume that it has n "equators." Then

$$p(\Pi) = r(\Pi) = 19n$$

and it follows from Proposition 10.1.1 that

$$q(\Pi) = \frac{1}{2} \cdot 4 \cdot 19n = 38n.$$

Thus,

$$\chi(\Pi) = 19n - 38n + 19n = 0$$

regardless of the value of n.

The following theorem summarizes the most important information regarding the Euler characteristic. While it cannot be proved here, it is supported by the preceding examples.

Theorem 11.2.6 *Polygonal presentations of homeomorphic surfaces have the same Euler characteristics.* □

In particular, presentations of the same surface S also have the same characteristic. This common value of the characteristics of all the presentations of a surface is called the *Euler characteristic of the surface* and is denoted by $\chi(S)$. Example 11.2.3 demonstrates that nonhomeomorphic surfaces can have the same characteristic.

Orientability character: An *orientation* of a single polygon consists of a sense of traversal of its perimeter. Every polygon therefore has two orientations, which, when the polygon is visualized as drawn on a page, are referred to as clockwise and counterclockwise (from the point of view of the reader). In the case of the rectangle of Figure 11.25, these orientations are

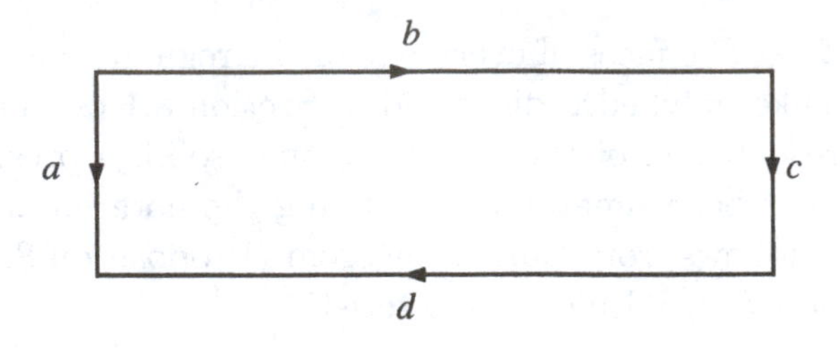

Figure 11.25. A polygon to be oriented

$a^{-1}bcd$ and $ad^{-1}c^{-1}b^{-1}$, respectively. A directed edge a on the perimeter of a polygon P is said to be *consistent* with an orientation of P provided that its direction agrees with the orientation. Otherwise a is *inconsistent* with the orientation. Equivalently, the directed edge a is *consistent* with an orientation when it appears with exponent 1 (i.e., no exponent) when that orientation is written out as a succession of labels, and the edge is *inconsistent* if it appears with exponent -1 when the orientation is written out. In Figure 11.25, edge a is consistent with the counterclockwise orientation $ad^{-1}c^{-1}b^{-1}$, whereas the other edges are consistent with the clockwise orientation $a^{-1}bcd$. Orientations are usually described by a circular arc in the center of the oriented polygon (see Fig. 11.26).

An *orientation* of a polygonal presentation Π consists of an assignment of an orientation to each of its polygons. Such an orientation of Π is said to be *coherent* if, of the two occurrences of every arc on the oriented polygon perimeters, one is consistent and the other is inconsistent. The orientation of the polygonal presentation of Figure 11.26 is coherent since the indicated traversals are $a^{-1}bcb^{-1}$ and $ac^{-1}dd^{-1}$. It is clear that the coherence of an orientation of a presentation is tantamount to every arc label appearing with both a^1 and a^{-1} as exponents. A presentation is said to be *orientable* if it possesses a coherent orientation. The presentations of Figures 11.21 and 11.26 are both orientable. On the other hand, a minimal amount of trial

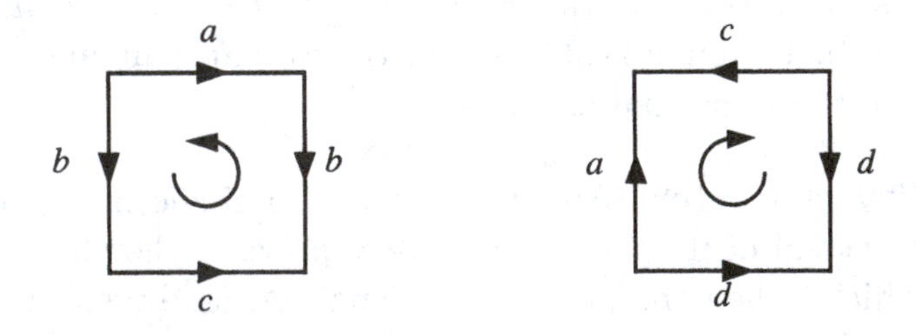

Figure 11.26. A coherently oriented presentation

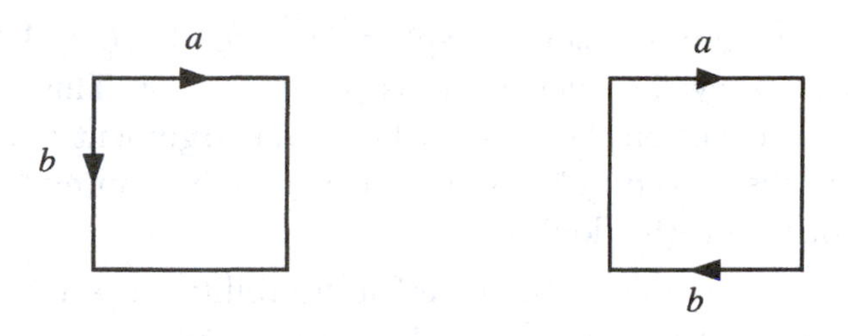

Figure 11.27. A nonorientable presentation

and error demonstrates that the rectangular presentation of the Klein bottle (Fig. 11.11) is not orientable, nor can the partially labeled presentation of Figure 11.27 be completed to a coherent orientation. Such polygonal presentations are said to be *nonorientable.*

To explain the significance of orientability we begin by demonstrating some of the undesirable consequences of nonorientability. Imagine that in Figure 11.28 the circle 1 drawn on a presentation of the nonorientable Klein bottle glides to the left so that it crosses the edge a (position 2) and then reappears at the right end. The apparent change in the positions of B and C is dictated by the arrows on the two edges marked a. Then, still moving to the left, bring the circle to position 3, close to its starting point. Note that notwithstanding the fact that the circle has been subjected to a *smooth* or *continuous* sliding motion, its sense has been reversed. The cyclic order of A, B, C is clockwise in position 1 but counterclockwise in position 3. Such a reversal is impossible on the sphere. For suppose the three points A, B, C on a circle on the sphere have a clockwise cyclic order from the point of view of an observer who is stationed at a distance of, say, one unit outside the sphere. No matter where this circle glides, as long as the observer moves

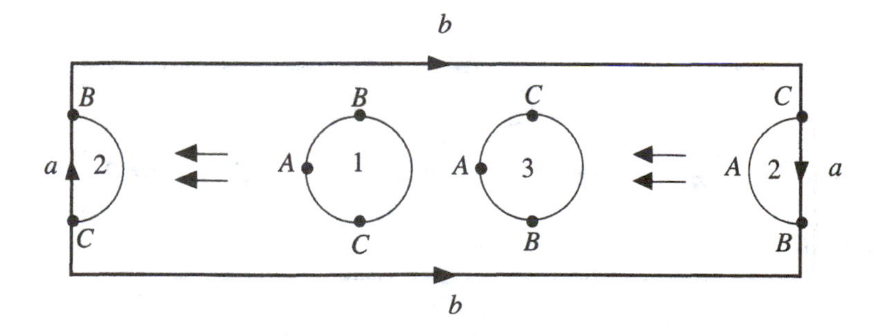

Figure 11.28. Movement on a nonorientable surface

with it, he will remain outside the sphere and so A, B, and C will always stay in a clockwise cyclic order from his point of view. Thus, no reversal of cyclic orders can occur on the sphere. The same argument clearly applies to any surface that is placed in $\mathbb{R}^3$ with a clearly defined interior and exterior, such as the torus and the double torus.

This makes it possible to interpret orientability in terms of the surface rather than any of its presentations. A surface is orientable when the ambiguity displayed in Figure 11.28 cannot arise. In other words, it is possible to specify a rotational sense on the surface at each point so that if an oriented circle glides arbitrarily on the surface, the orientation of the circle always agrees with the rotational sense of the surface in its location (see Fig. 11.29). Consequently, it is possible to visualize an oriented surface as one with several arched arrows such that whenever one arrow is slid toward another, their rotational senses agree (see Fig. 11.30). Note that when two such arcs are placed next to each other, they must pass through the point of contact in *opposite* directions. That is the analogous concept to the coherence of an orientation of a presentation.

We digress here to discuss some alternative terminology. It is also customary to refer to orientable surfaces as *two sided* and nonorientable ones as *one sided*. This terminology can be misleading. Unlike a sheet of paper, a polygon consists of a single layer of points. If we were to write on a polygon with a blue pen, the color and the writing would be equally visible no matter which of the two sides of the polygon was examined. Nevertheless, it is not altogether unreasonable to speak of the sphere and the torus as two-sided since each clearly has an *interior* and an *exterior*. This "separation" property is a feature of all closed surfaces embedded in $\mathbb{R}^3$. Moreover, it can be demonstrated that all orientable surfaces are embeddable in $\mathbb{R}^3$ with well-defined interiors and exteriors, and hence they can all be said to be two

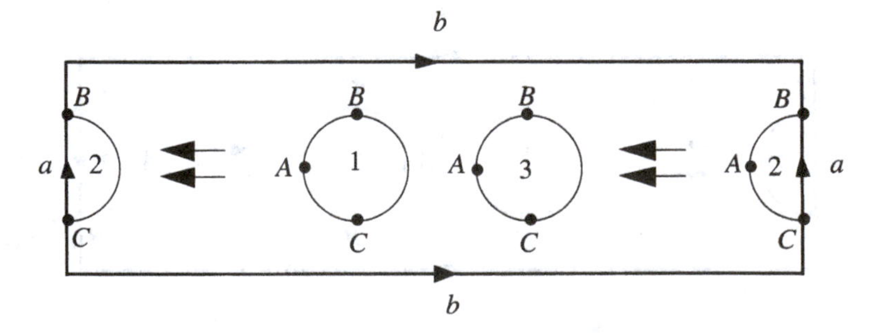

Figure 11.29. Movement on an orientable surface

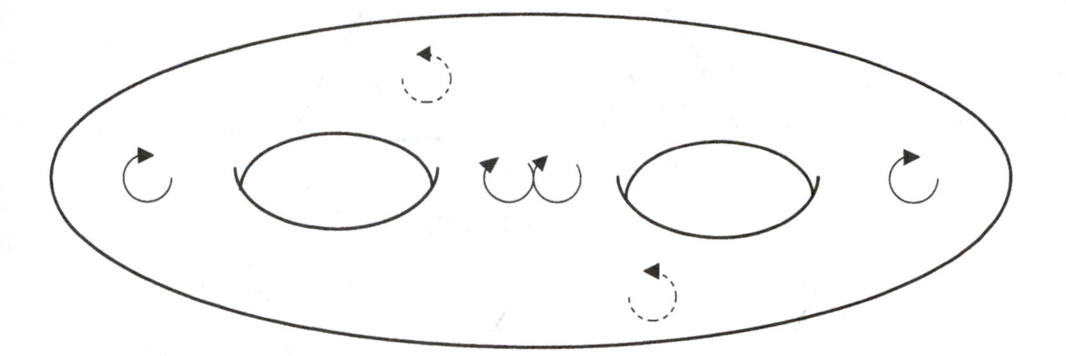

Figure 11.30. An oriented double torus (dotted lines lie on the back side)

sided. Nonorientable surfaces, on the other hand, can be at best embedded in $\mathbb{R}^4$. There they fail to separate the ambient space, much as any loop fails to separate $\mathbb{R}^3$.

It should be mentioned here that the equivalence of orientability to two-sidedness holds only in the context of embeddings in $\mathbb{R}^3$. In higher dimensions the two notions are independent of each other.

The orientability character of a presentation is easily determined by trying to choose orientations of the constituent polygons one at a time in such a manner that each choice is coherent with the previous choices.

Example 11.2.7 Neither the counterclockwise $abca^{-1}b^{-1}c$ nor the clockwise $c^{-1}bac^{-1}b^{-1}a^{-1}$ orientations of the polygon of Fig 11.31 is coherent (c is the troublesome edge). The presentation is therefore nonorientable.

Example 11.2.8 The presentation of Figure 11.32 is orientable because we can choose the indicated counterclockwise abc and clockwise $c^{-1}a^{-1}b^{-1}$ traversals.

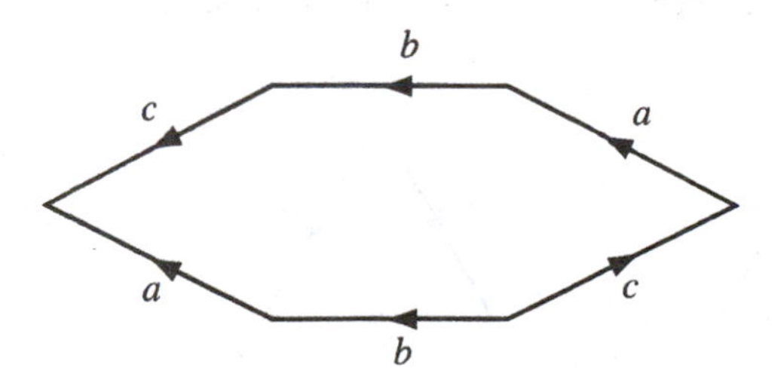

Figure 11.31.

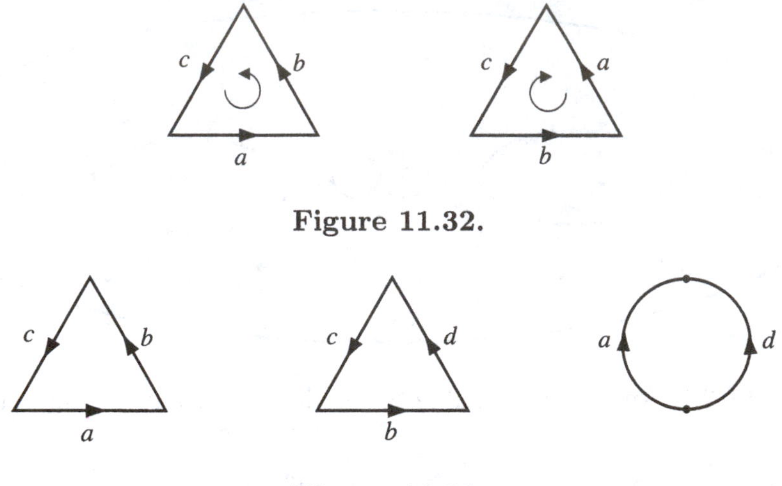

Figure 11.32.

Figure 11.33.

Example 11.2.9 For the presentation of Figure 11.33 we begin by arbitrarily choosing the traversal abc for the first triangle. The requirement of coherence (applied to both b and c) forces the choice of the traversal $d^{-1}b^{-1}c^{-1}$ for the second triangle. Once again, coherence requires the choice of $a^{-1}d$ for the third polygon. The presentation has thus been coherently oriented and is therefore orientable.

Example 11.2.10 For the presentation of Figure 11.34 we again begin by arbitrarily choosing the traversal abc for the first triangle. The requirement of coherence (applied to both b and c) forces the choice of the traversal $d^{-1}b^{-1}c^{-1}$ for the second triangle. The alternatives for the third polygon are da and $a^{-1}d^{-1}$. The first of these fails to be coherent with the first triangle along a, whereas the second of these fails to be coherent with the second triangle along d. The presentation is therefore not orientable.

We are ready to state the main theorem of the theory of surfaces.

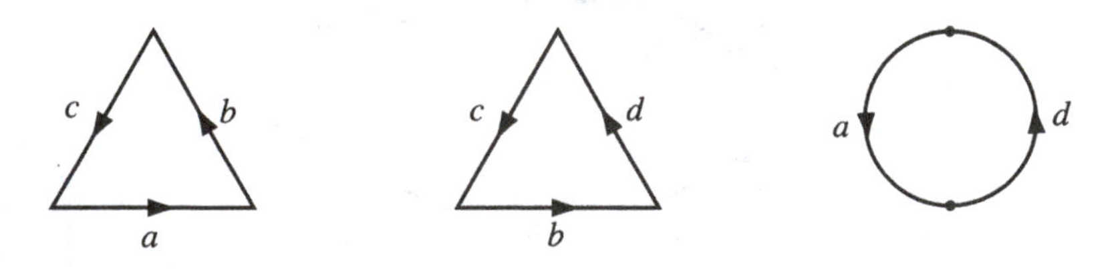

Figure 11.34.

Theorem 11.2.11 *Two polygonal presentations define homeomorphic surfaces if and only if they have the same Euler characteristic and the same orientability character.* □

The following table displays the counts, characteristics, and orientability character of the presentations in Examples 11.2.7–11.2.10. It indicates that of these four surfaces the second and the third are the only homeomorphic pair.

Example	p	q	r	Euler characteristic	Orientability character
11.2.7:	1	3	1	−1	Nonorientable
11.2.8:	1	3	2	0	Orientable
11.2.9:	1	4	3	0	Orientable
11.2.10:	1	4	3	0	Nonorientable

While Theorem 11.2.11 provides a very efficient method for deciding whether two given presentations define the same surface, it does not help visualize the different types of closed surfaces. This issue will be addressed in the next section. For this purpose it will prove useful to resort to *canonical*, or *standard*, presentations. These are the 1-polygon presentations Π^0, Π^n, $\tilde{\Pi}^n$, $n = 1, 2, 3, \ldots$ of Figures 11.35 and 11.36, which clearly have the traversals

$$\begin{aligned} \Pi_0 :& \quad aa^{-1} \\ \Pi^n :& \quad a_1 b_1 a_1^{-1} b_1^{-1} a_2 b_2 a_2^{-2} b_2^{-2} \cdots a_n b_n a_n^{-1} b_n^{-1} \\ \tilde{\Pi}^n :& \quad a_1 a_1 a_2 a_2 \cdots a_n a_n. \end{aligned}$$

The corresponding surfaces, denoted by S^0, S^n, $\tilde{S}^n$, $n = 1, 2, 3, \ldots$, respectively, are also said to be *canonical*.

Theorem 11.2.12 *Every closed surface is homeomorphic to exactly one of the canonical surfaces.* □

Some of these canonical surfaces have been encountered before. The sphere is S^0 and the torus is S^1. The projective plane is $\tilde{S}^1$ and the Klein bottle is $\tilde{S}^2$. It is clear from Figures 11.35 and 11.36 that S^0 and S^n are orientable whereas $\tilde{S}^n$ is nonorientable. The proof of the next proposition is relegated to Exercise 18.

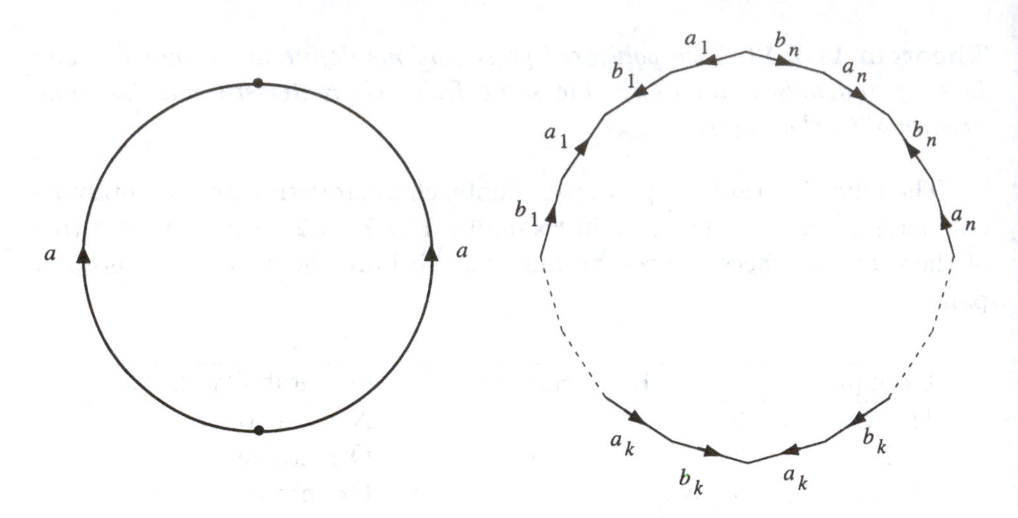

Figure 11.35. The canonical presentations of the sphere and S^n, $n \geq 1$

Proposition 11.2.13 *The Euler characteristics of the canonical surfaces are*

$$\chi(S^0) = 2$$
$$\chi(S^n) = 1 - 2n + 1 = 2 - 2n \qquad n = 0, 1, 2, \ldots$$
$$\chi(\tilde{S}^n) = 1 - n + 1 = 2 - n \qquad n = 1, 2, 3, \ldots \ . \qquad \square$$

It therefore follows that every closed surface has Euler characteristic at most 2, the sphere being the only one for which equality holds.

It will be demonstrated in the next section that all the orientable canonical surfaces can be displayed (i.e., embedded) in $\mathbb{R}^3$. As for the nonori-

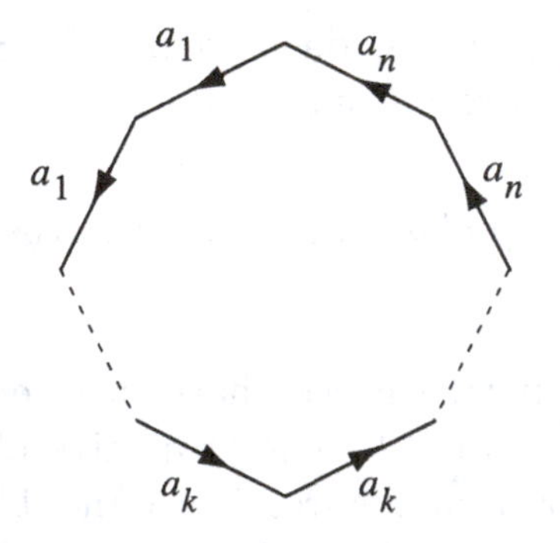

Figure 11.36. The canonical presentation of $\tilde{S}^n$

entable variety, the informal discussion in the previous chapter motivates the following observation.

Theorem 11.2.14 *The closed nonorientable surfaces cannot be embedded in* $\mathbb{R}^3$. □

As a space-saving device we note that in describing a polygonal presentation the actual polygons can be omitted. Only the traversals of their perimeters need to be displayed. Thus, the presentation of Figure 11.31 can be recorded as either $abca^{-1}b^{-1}c$ or $a^{-1}c^{-1}bac^{-1}b^{-1}$, whereas the presentation of Figure 11.34 can be recorded as abc, bdc, ad.

Exercises 11.2

1. Draw the skeletons of the following polygonal presentations.

   (a) aa (b) aa^{-1} (c) $aabb$ (d) $aa^{-1}bb$ (e) abc, $b^{-1}ed$, dc, ae
   (f) abc, $defc$, $fgi^{-1}a^{-1}h^{-1}$, $jhdg^{-1}b^{-1}$, $je^{-1}i$
   (g) abc, $defc$, $fgd^{-1}a^{-1}h^{-1}$, bgj^{-1}, $ei^{-1}j$, hai
   (h) $abcabc$ (i) $abca^{-1}b^{-1}c^{-1}$ (j) $abcdabcd$ (k) $aabbcc$
   (l) $aba^{-1}b^{-1}cdc^{-1}d^{-1}$ (m) $aba^{-1}b^{-1}cdc^{-1}d^{-1}efe^{-1}f^{-1}$

2. Identify the surfaces of the following polygonal presentations.

   (a) aa (b) aa^{-1} (c) $aabb$ (d) $aa^{-1}bb$ (e) abc, $b^{-1}ed$, dc, ae
   (f) abc, $defc$, $fgi^{-1}a^{-1}h^{-1}$, $jhdg^{-1}b^{-1}$, $je^{-1}i$
   (g) abc, $defc$, $fgd^{-1}a^{-1}h^{-1}$, bgj^{-1}, $ei^{-1}j$, hai
   (h) $abcabc$ (i) $abca^{-1}b^{-1}c^{-1}$ (j) $abcdabcd$ (k) $aabbcc$
   (l) $aba^{-1}b^{-1}cdc^{-1}d^{-1}$ (m) $aba^{-1}b^{-1}cdc^{-1}d^{-1}efe^{-1}f^{-1}$

3. For each positive integer n, identify the surface with the presentation.

   (a) $a_1a_2\cdots a_na_1a_2\cdots a_n$ (b) $a_1a_2\cdots a_na_1^{-1}a_2^{-1}\cdots a_n^{-1}$
   (c) $a_1a_2\cdots a_na_na_{n-1}a_{n-2}\cdots a_1$ (d) $a_1a_2\cdots a_na_n^{-1}a_{n-1}^{-1}a_{n-2}^{-1}\cdots a_1^{-1}$

4. For each positive integer n, identify the surface with the presentation $a_1a_1^{-1}a_2a_2^{-1}\cdots a_na_n^{-1}$.

5. A closed orientable surface has a polygonal presentation with the same number of nodes and edges. Explain why this surface must be homeomorphic to the sphere.

6. A closed nonorientable surface has a polygonal presentation with the same number of nodes and edges. Explain why this surface must be homeomorphic to the projective plane.

7. Show that for every positive integer n there is a polygonal presentation of the sphere with 2 nodes, n edges, and n polygons.
8. Show that for every positive integer n there is a polygonal presentation of the sphere with n nodes, n edges, and 2 polygons.

*9. A polygonal presentation Π is said to be *regular* if each of its polygons has the same number of edges, each node of $G(\Pi)$ has the same degree, and no arc appears twice on the perimeter of a polygon. Prove that if Π is a regular presentation of the sphere, then $(p(\Pi), q(\Pi), r(\Pi))$ is one of the triples (4, 6, 4), (6, 12, 8), (8, 12, 6), (12, 30, 20), or (20, 30, 12).

10. Show that the torus has regular presentations with an arbitrarily large number of nodes.
11. Show that the Klein bottle has regular presentations with an arbitrarily large number of nodes.

*12. Does the projective plane have any regular presentations? If so, can the number of nodes in such a presentation be arbitrarily large?

*13. Does the double torus have any regular presentations? If so, can the number of nodes in such a presentation be arbitrarily large?

14. A *triangulation* of a surface S is a presentation Π of S in which each polygon is a triangle and the skeleton is a simple graph. Prove that if Π is a triangulation with parameters p, q, r, then $3r = 2q$, $q = 3(p - \chi(S))$.
15. What are the minimum values of $p(\Pi)$, $q(\Pi)$, $r(\Pi)$ for any triangulation of the sphere?

*16. Prove that the minimum value of $p(\Pi)$ for any triangulation of the torus is 7.

*17. Prove that the minimum value of $p(\Pi)$ for any triangulation of the projective plane is 6.

18. Prove Proposition 11.2.13.

11.3 Operations on Surfaces

So far, the canonical surfaces have been displayed only as presentations. We now set out to describe them as surfaces in $\mathbb{R}^3$ whenever possible, or in $\mathbb{R}^4$ otherwise.

The *genus* $\gamma(S)$ *of the surface* S is defined as $\gamma(S^n) = \gamma(\tilde{S}^n) = n$. It follows from Theorem 11.2.12 that once the orientability character of a surface is known, its genus and Euler characteristic completely determine each other. However, the genus, as will be seen, has very natural geometric interpretations and is often the preferred parameter for describing surfaces.

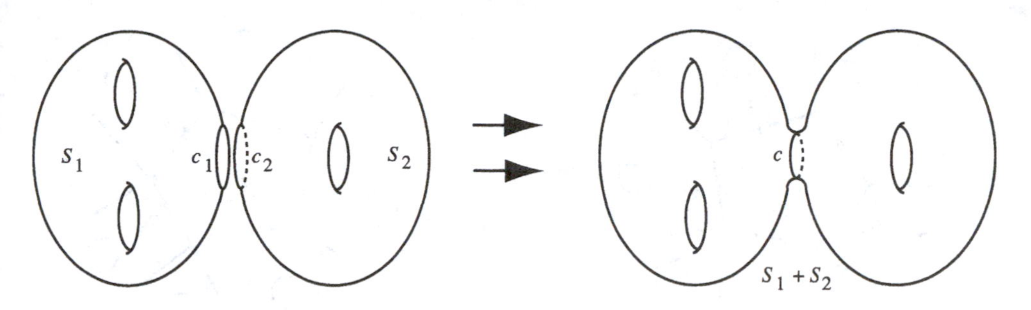

Figure 11.37. The addition of surfaces

An *excision* of a surface is a cut whose effect is permanent in the sense that it is not meant to be reversed—the borders it creates are not meant to be repasted to each other. These borders, however, may be pasted to other borders. Such will always be the case in this section, so that the end product of the operations will always be a surface without borders. For example, given two disjoint surfaces S_1 and S_2, their *connected sum* $S_1 + S_2$ is the surface obtained by excising a disk from each and pasting the resulting borders. This is illustrated in Figure 11.37, where the excised disks leave borders c_1 and c_2 which are then pasted to form the loop c on the sum $S_1 + S_2$.

It may seem at first that the sum might depend on the specific orientations that are assigned to the excisions and on their locations, but that, surprisingly, is not the case.

Proposition 11.3.1 *If S_1 and S_2 are closed surfaces, then*

$$\chi(S_1 + S_2) = \chi(S_1) + \chi(S_2) - 2$$

and $S_1 + S_2$ is orientable if and only if both S_1 and S_2 are orientable.

PROOF: Consider any presentation Π of $S_1 + S_2$ such that the excision path along which S_1 and S_2 have been pasted is a cycle c of Π with n nodes and n arcs. For each $i = 1, 2$, let Π_i be the polygonal presentation of S_i obtained by restricting Π to S_i and adding a polygon P_i that seals the hole surrounded by c_i. Then, because the nodes and arcs of c appear in both Π_1 and Π_2,

$$\begin{aligned}\chi(S_1 + S_2) = \chi(\Pi) &= p(\Pi) - q(\Pi) + r(\Pi)\\ &= (p(\Pi_1) + p(\Pi_2) - n) - (q(\Pi_1) + q(\Pi_2) - n)\\ &\quad + (r(\Pi_1) - 1 + r(\Pi_2) - 1)\end{aligned}$$

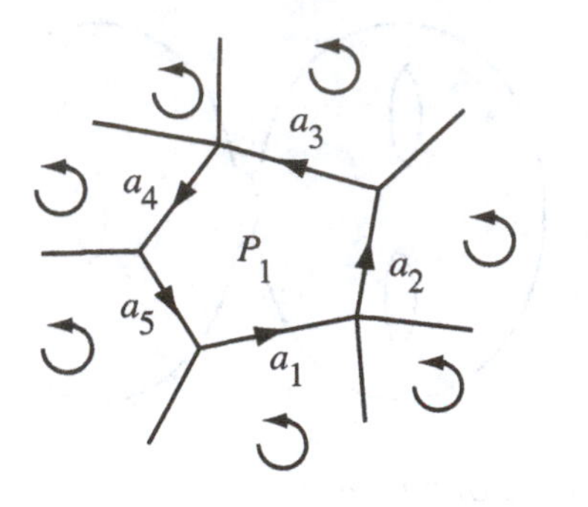

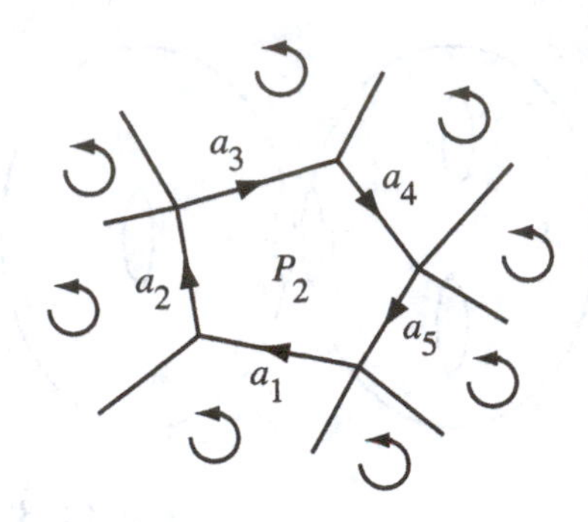

Figure 11.38. The addition of surfaces

$$= (p(\Pi_1) - q(\Pi_1) + r(\Pi_1)) + (p(\Pi_2) - q(\Pi_2) + r(\Pi_2)) - 2$$
$$= \chi(\Pi_1) + \chi(\Pi_2) - 2 = \chi(S_1) + \chi(S_2) - 2.$$

Turning to the issue of orientability, suppose c consists of the consecutive directed arcs $a_1, a_2, \ldots, a_n$ that bound the polygons P_1 and P_2 in Π_1 and Π_2, respectively (Fig. 11.38). If S_1 and S_2 are both orientable, it may be assumed that Π_1 and Π_2 have been coherently oriented as in Figure 11.38, in which case we also have here a coherent orientation of Π. Conversely, any coherent orientation of Π induces a coherent orientation of each Π_i, P_i excepted. However, as Figure 11.38 indicates, these coherent orientations can be extended to the P_i as well. Hence, the orientability of $S_1 + S_2$ entails that of both S_1 and S_2. Q.E.D.

Corollary 11.3.2 *If m, $n \geq 0$, then $S^m + S^n \approx S^{m+n}$.*

PROOF: By Proposition 11.3.1, both $S^m + S^n$ and S^{m+n} are orientable. Moreover,

$$\chi(S^m + S^n) = \chi(S^m) + \chi(S^n) - 2 = (2 - 2m) + (2 - 2n) - 2$$
$$= 2 - 2(m + n) = \chi(S^{m+n}).$$

The desired result now follows from Theorem 11.2.11. Q.E.D.

This corollary now provides a simple realization of the closed orientable surfaces in $\mathbb{R}^3$.

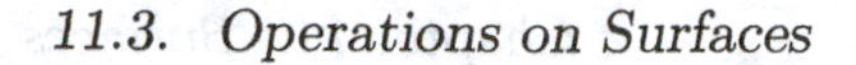

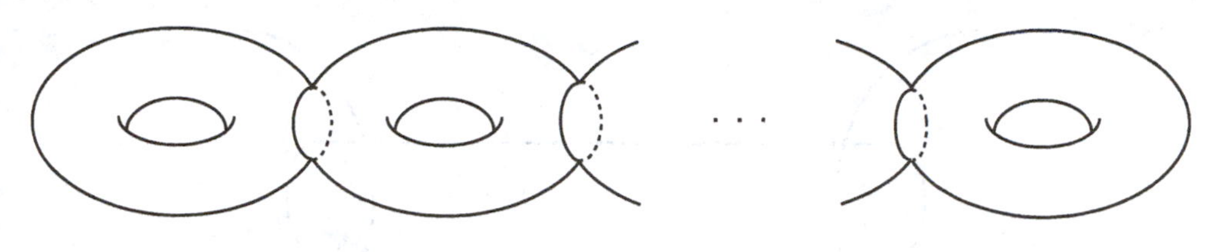

Figure 11.39. S_n as the sum of n tori

Corollary 11.3.3 *For $n = 1, 2, 3, \ldots$, the closed orientable surface S^n can be realized in $\mathbb{R}^3$ as the sum of n tori (Fig. 11.39).*

Nonorientable surfaces cannot be embedded in $\mathbb{R}^3$, and hence the preceding corollary characterizes orientability of surfaces. In other words, a closed surface is orientable if and only if it can be embedded in $\mathbb{R}^3$.

We turn next to the sums of the surfaces that are not necessarily orientable. As the proofs of the following propositions and corollaries either resemble that of Proposition 11.3.1 or else they are straightforward consequences of preceding propositions, they are mostly relegated to the Exercises.

Corollary 11.3.4 *If $m \geq 0$ and $n \geq 1$, then $S^m + \tilde{S}^n \approx \tilde{S}^{2m+n}$.*

PROOF: Exercise 18.

Corollary 11.3.5 *Every closed nonorientable surface can be realized in $\mathbb{R}^4$.*

PROOF: It was demonstrated in Section 11.1 that the projective plane $\tilde{S}^1$ and the Klein bottle $\tilde{S}^2$ can be embedded in $\mathbb{R}^4$. It follows from Corollary 11.3.4 that

$$\tilde{S}^n \approx \begin{cases} S^{(n-1)/2} + \tilde{S}^1 & \text{for } n = 1, 3, 5, \ldots \\ S^{(n-2)/2} + \tilde{S}^2 & \text{for } n = 2, 4, 6, \ldots . \end{cases}$$

In either case the four-dimensional realizations of $\tilde{S}^1$ and $\tilde{S}^2$ depicted in Figures 11.15 and 11.13 can be added to the three-dimensional realizations of $S^{(n-1)/2}$ and $S^{(n-2)/2}$ given in Figure 11.38 so as to produce a four-dimensional realization of $\tilde{S}^n$. Q.E.D.

Corollary 11.3.6 *If $m, n \geq 1$, then $\tilde{S}^m + \tilde{S}^n \approx \tilde{S}^{m+n}$.*

PROOF: Exercise 19.

Surface addition can also be described as *tube connection*, an operation wherein the excised borders are joined by a tube (Fig. 11.40) instead of

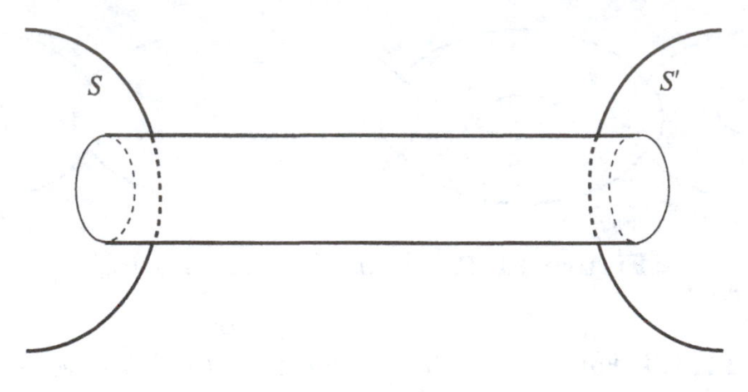

Figure 11.40. Connecting two surfaces with a tube

being pasted together. The resulting surface is homeomorphic to the sum of the two surfaces. To see this we need only imagine the length of the connecting cylinder to be shrinking to zero.

The foregoing operations created a new surface out of two distinct surfaces. Next, operations on a single surface are examined.

Suppose first that two disjoint disks are removed from a surface S and both of the resulting borders are given a direction and labeled a. Suppose further that the two borders are now pasted to each other in a manner that is consistent with their directions. Unlike the connected sum of two surfaces, in this case the nature of the resulting surface may depend on the choice of directions for the borders. If the surface S is orientable and the two borders have been assigned directions of which one is consistent and the other is inconsistent with some orientation of S, then the pasting process is illustrated in Figure 11.41, where the two borders labeled a are "pulled out" and joined. As suggested by this figure, this operation is called *handle addition* and the result of adding a handle to the surface S is denoted by $S + h$. It is clear that this process can also be visualized by simply joining the two borders a by a tube or an open-ended cylinder (Fig. 11.42).

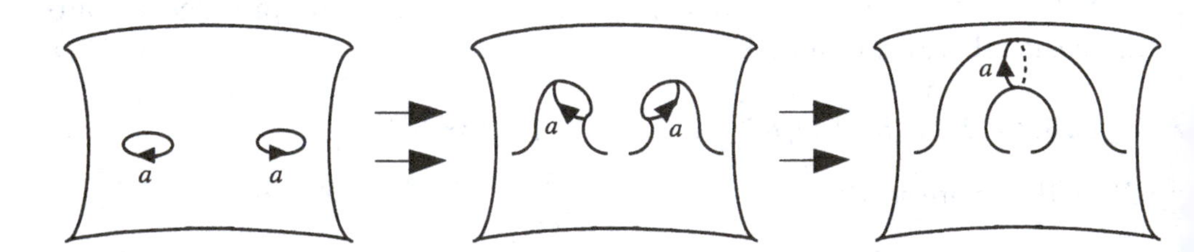

Figure 11.41. Adding a handle to a surface

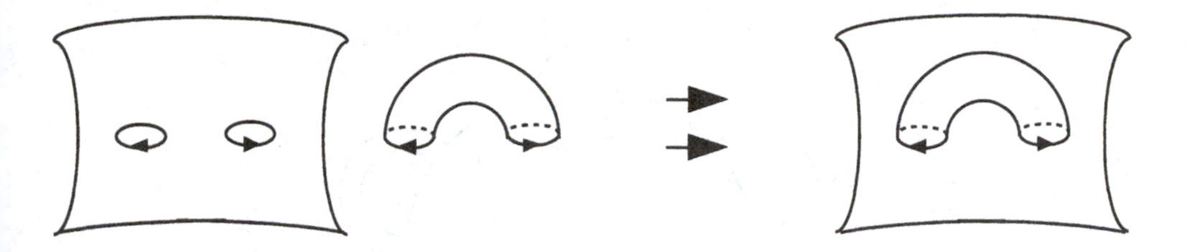

Figure 11.42. Adding a handle to a surface

Proposition 11.3.7 *If S is an orientable surface, then $S+h$ is also orientable and $\gamma(S+h)=\gamma(S)+1$.*

PROOF: Figure 11.43 indicates that any orientation of S induces an orientation of $S+h$ and hence the latter is also orientable.

Let Π^h be a presentation of $S+h$ that contains the path c in its skeleton as a cycle with n nodes and n arcs. By cutting along c and sealing the two resulting holes with two disks, we obtain a presentation Π of S such that

$$\begin{aligned}\chi(\Pi) &= p(\Pi) - q(\Pi) + r(\Pi)\\ &= \Big(p(\Pi^h)+n\Big) - \Big(q(\Pi^h)+n\Big) + \Big(r(\Pi^h)+2\Big) = \chi(\Pi^h)+2.\end{aligned}$$

It follows that

$$\gamma(S+h) = \frac{2-\chi(\Pi^h)}{2} = \frac{2-\chi(\Pi)+2}{2} = \gamma(S)+1.$$

Q.E.D.

It follows by a straightforward induction that S^n is also realizable in $\mathbb{R}^3$ as a sphere S^0 with n attached handles (see Fig. 11.44 for an example).

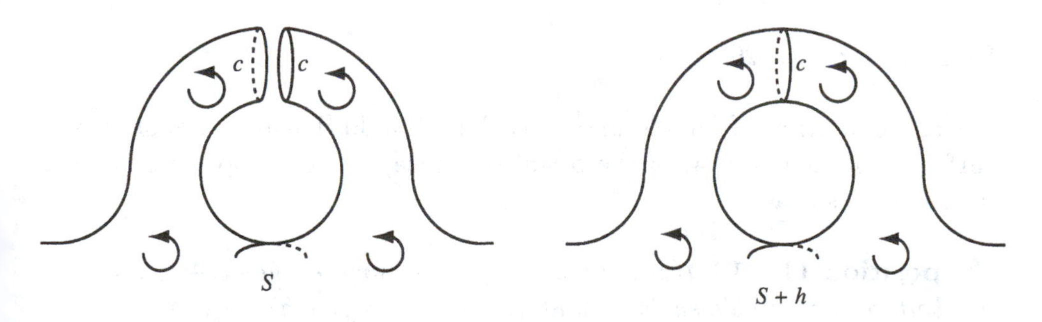

Figure 11.43.

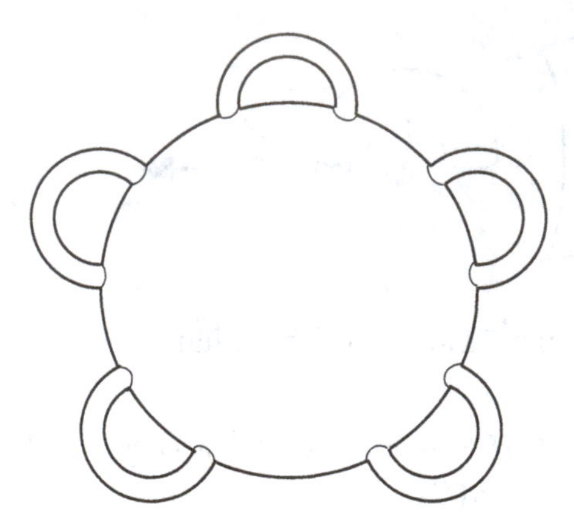

Figure 11.44. S^5 as a sphere with five handles

Example 11.3.8 The solid of Figure 11.45 is obtained from K_4 by a "fattening" process. Without the three outer tubes this is a 3-spiked sphere whose surface has genus 0. It therefore follows from Proposition 11.3.7 that the original surface has genus 3.

The handle addition process described previously was applied to an *orientable* surface S and the two borders were directed as indicated in Figure 11.41 (i.e., they had different directions relative to the orientation of S). If the two borders have the same direction relative to this orientation (Fig. 11.46), then the identification of the borders yields a different surface, denoted $S + \tilde{h}$, and the process is called *cross-handle addition.* It can be realized in four-dimensional Euclidean space by means of the same device that was used to realize the Klein bottle in Section 11.1 (Exercise 17).

Proposition 11.3.9 *If S is an orientable surface, then $S + \tilde{h}$ is a nonorientable surface and $\gamma(S + \tilde{h}) = 2\gamma(S) + 2$.*

PROOF: Exercise 20.

The definition of handle and cross-handle addition to a nonorientable surface is similar to that of the orientable case, and both operations result in the same surface.

Proposition 11.3.10 *If S is a nonorientable surface, then $S+h$ and $S+\tilde{h}$ are both nonorientable surfaces and $\gamma(S + h) = \gamma(S + \tilde{h}) = \gamma(S) + 2$.*

PROOF: Exercise 21.

Finally, suppose a single disk is excised from a surface and the remaining border is divided into two arcs, each of which is labeled b and both of which are given the same direction relative to S (Fig. 11.47). If the two b's are then pasted to each other, a new surface $S + c$ is obtained and the process is called a *cross-cap addition.*

Proposition 11.3.11 *The surface $S + c$ is nonorientable and*

$$\gamma(S+c) = \begin{cases} 2\gamma(S) + 1 & \text{if } S \text{ is orientable} \\ \gamma(S) + 1 & \text{if } S \text{ is nonorientable.} \end{cases}$$

PROOF: Exercises 22.

Corollary 11.3.12 *The nonorientable surface of genus n is homeomorphic to a sphere with n cross-caps.*

PROOF: Exercise 23.

Exercises 11.3

1. Identify the following surfaces.

   (a) $S^1 + S^1$ (b) $S^{12} + \tilde{S}^{21}$ (c) $S^{21} + \tilde{S}^{12}$
   (d) $S^1 + S^2 + S^3$ (e) $\tilde{S}^1 + S^2 + S^3$ (f) $S^1 + \tilde{S}^2 + \tilde{S}^3$
   (g) $\tilde{S}^1 + \tilde{S}^2 + \tilde{S}^3$

2. Explain why the addition of surfaces is a commutative operation.

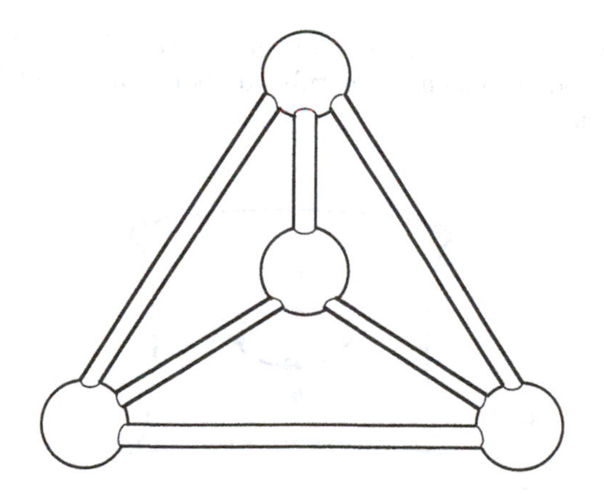

Figure 11.45. A fat K_4

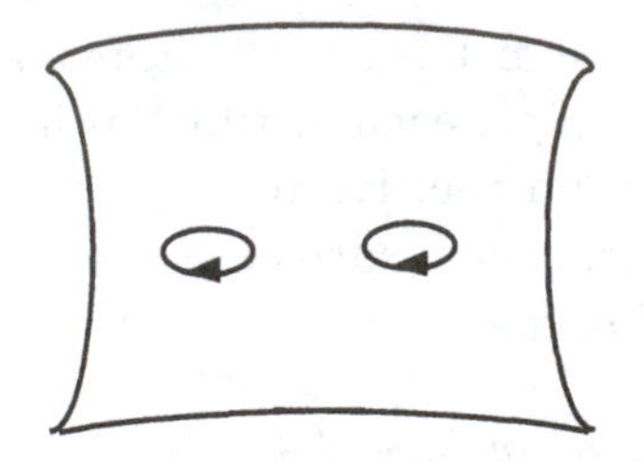

Figure 11.46. Preparing a surface for the addition of a cross-handle

3. Explain why the addition of surfaces is an associative operation.

4. Two tori are connected by means of three tubes. Identify the resulting surface.

5. The surfaces S_1, S_2, S_3 are all tori. Each is connected to the other two by means of a tube. Identify the resulting surface.

6. The surfaces S_1, S_2, S_3 are all Klein bottles. Each is connected to the other two by means of a tube. Identify the resulting surface.

7. The surfaces S_1, S_2, S_3 are all tori. Each is connected to the other two by means of two tubes. Identify the resulting surface.

8. The surfaces S_1, S_2, S_3 are all Klein bottles. Each is connected to the other two by means of two tubes. Identify the resulting surface.

9. The surfaces $S_1, S_2, \ldots, S_n$ are all tori. For each $i = 2, 3, \ldots, n-1$, S_i is connected by means of a tube to both S_{i-1} and S_{i+1}. Identify the resulting surface.

10. The surfaces $S_1, S_2, \ldots, S_n$ are all Klein bottles. For each $i = 2, 3, \ldots, n-1$, S_i is connected by means of a tube to both S_{i-1} and S_{i+1}. Identify the resulting surface.

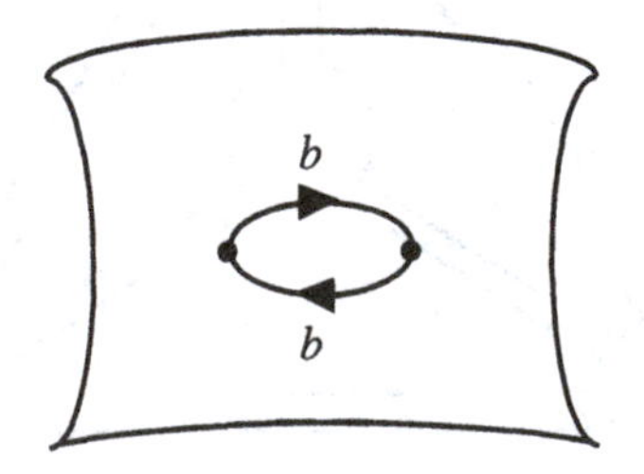

Figure 11.47. Preparing a surface for a cross-cap addition

11. The surfaces $S_1, S_2, \ldots, S_n$ are all tori. For each $i = 2, 3, \ldots, n-1$, S_i is connected by means of two tubes to S_{i-1} and two other tubes to S_{i+1}. Identify the resulting surface.
12. The surfaces $S_1, S_2, \ldots, S_n$ are all Klein bottles. For each $i = 2, 3, \ldots, n-1$, S_i is connected by means of two tubes to S_{i-1} and two other tubes to S_{i+1}. Identify the resulting surface.

See Example 11.3.8 for an illustration of a fat graph.

13. Show that surface of the fat K_5 has genus 6.
14. Show that the surface of the fat version of K_n has genus $(n-1)(n-2)/2$.
15. Show that if the connected graph G has p nodes and q arcs, then the surface of the fat version of G has genus $q - p + 1$.
16. Two distinct surfaces S and S' are connected by means of n disjoint tubes to form a new surface S''. Identify the surface S''.
17. Show by means of a diagram how the addition of a cross-handle can be realized in $\mathbb{R}^4$.
18. Prove Corollary 11.3.4.
19. Prove Corollary 11.3.6.
20. Prove Proposition 11.3.9.
21. Prove Proposition 11.3.10.
22. Prove Proposition 11.3.11.
23. Prove Corollary 11.3.12.

11.4 Bordered Surfaces

A *bordered surface* S is the topological space obtained by removing a finite number of disjoint disks from a surface. Thus, we can think of the open-ended cylinder as the bordered surface obtained from the sphere S_0 by excising (and throwing away) two disks. The rim of the "hole" left by the removal of a disk is called a *border loop* or just a *border*. Since a minute sliding of such a border along the surface results in a new surface that is still homeomorphic to the original surface, the same holds when any number of such minute slides are applied. In other words, the bordered surface obtained by the removal of one or more disks from a given surface is the same regardless of the exact location of the borders so created on the surface, so long as these borders do not intersect. Each such disk removal is also called a *perforation*. The surfaces obtained from S^n and $\tilde{S}^n$ by β perforations are

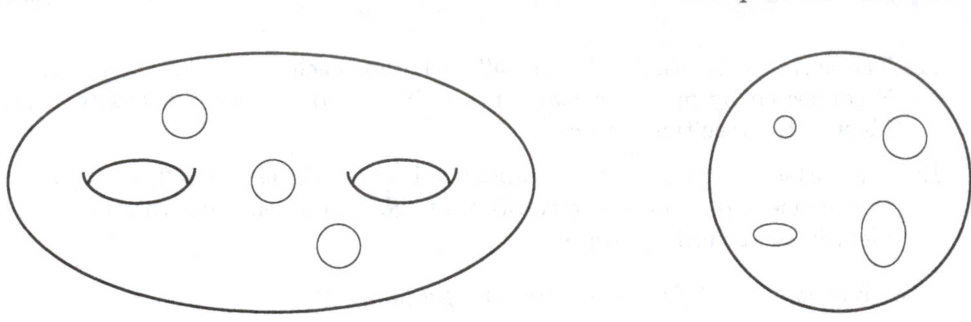

Figure 11.48. The bordered surfaces $S^{2,3}$ and $S^{0,5}$

denoted by $S^{n,\beta}$ and $\tilde{S}^{n,\beta}$, respectively (see Fig 11.48 for examples). Note that the bordered surface $S^{0,1}$ is homeomorphic to the disk and $S^{0,2}$ is homeomorphic to the open-ended cylinder of Figure 11.2. The closed surfaces S^n and $\tilde{S}^n$ are called the *closures* of $S^{n,\beta}$ and $\tilde{S}^{n,\beta}$, respectively. The *genus* of the surfaces $S^{n,\beta}$ and $\tilde{S}^{n,\beta}$ is the integer n.

The bordered surfaces with no borders are of course the closed surfaces classified in Section 11.2. Just like the closed surfaces, bordered surfaces also have polygonal presentations. These differ from the presentations of closed surfaces in that while the edges of the constituent polygons are still labeled, not all of these labels are paired; some labels appear only once, thus creating the borders of the surface. Such presentations will also be said to be *bordered*. Since these borders are arcs that are not meant to be pasted, there is no need to give them a direction. Hence border arcs are unmarked by arrowheads in the illustrations (Fig. 11.49).

The *Euler characteristic* of a bordered polygonal presentation Π is defined in the same manner as that of the closed ones, that is,

$$\chi(\Pi) = p(\Pi) - q(\Pi) + r(\Pi).$$

For example, for the four presentations of $S^{0,1}$ in Figure 11.48 we get characteristics $1 - 1 + 1$, $2 - 2 + 1$, $6 - 7 + 2$, and $4 - 5 + 2$, respectively, all

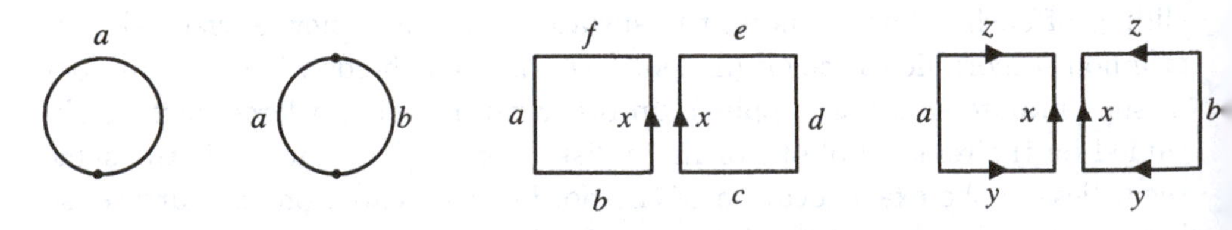

Figure 11.49. Four bordered polygonal presentations of the disk $S^{0,1}$

of which have the numerical value 1. As this example indicates, the Euler characteristic of bordered surfaces is also a topological property.

An orientation of a polygonal presentation Π of a bordered surface consists of an assignment of an orientation to each of its polygons. Such an orientation is said to be *coherent* if whenever an arc occurs twice on the polygon perimeters, one occurrence is consistent and the other is inconsistent with the orientation of the perimeters. A bordered surface is *orientable* if it has a polygonal presentation that can be coherently oriented. Otherwise it is *nonorientable.*

The number of borders of the surface S is denoted by $\beta(S)$. Accordingly, $\beta(S^n) = 0$, $\beta(\text{disk}) = 1$, and $\beta(\text{open-ended cylinder}) = 2$. It is clear that $\beta(S)$ is a topological property of S.

In view of Theorem 11.2.11, the following characterization should come as no surprise.

Theorem 11.4.1 *Two bordered polygonal presentations define homeomorphic bordered surfaces if and only if they have the same Euler characteristic, orientability character, and number of border loops.* □

It follows that the Euler characteristic of a bordered surface may be defined as the characteristic of any of its presentations. The following proposition is easily verified.

Proposition 11.4.2 *For any bordered surfaces,*

$$\chi(S^{n,\beta}) = \chi(S^n) - \beta = 2 - 2n - \beta$$
$$\chi(\tilde{S}^{n,\beta}) = \chi(\tilde{S}^n) - \beta = 2 - n - \beta.$$

PROOF: Exercise 14.

In particular, $\chi(\text{disk}) = \chi(S^0) - 1 = 1$ and $\chi(\text{open-ended cylinder}) = \chi(S^0) - 2 = 0$.

It is convenient to restate Proposition 11.4.2 in a more neutral manner.

Proposition 11.4.3 *If S^c is the closure of the bordered surface S, then*

$$\chi(S^c) = \chi(S) + \beta(S).$$

Example 11.4.4 *Identify the surface B^n obtained by introducing n similarly directed twists into a band.* (Figure 11.50 depicts B^3 and B^4.)

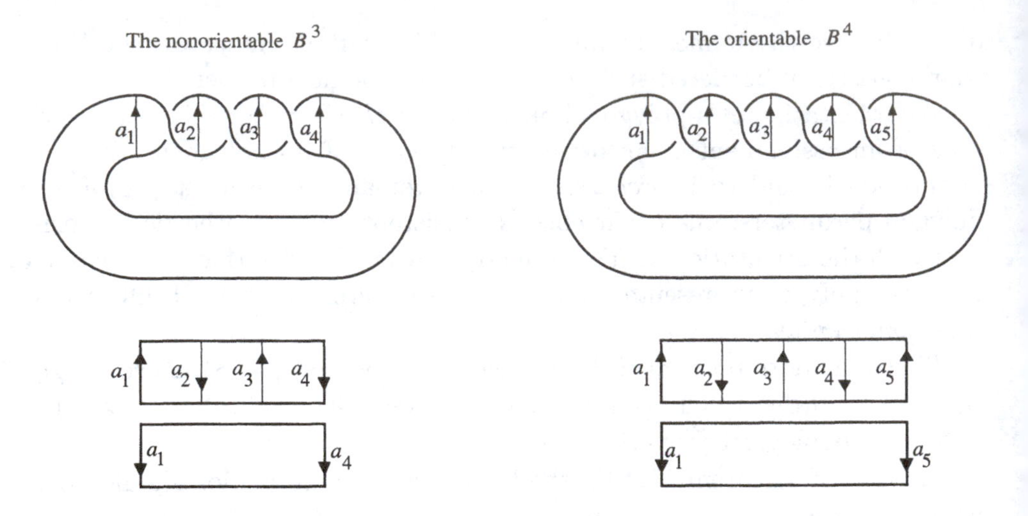

Figure 11.50. Two twisted bands

The 2-polygon presentation produced by the cuts a_1, a_{n+1} has four nodes and six arcs. It therefore has Euler characteristic $4 - 6 + 2 = 0$. When n is even, the counterclockwise traversal of both polygons yields a coherent orientation and, in addition, the band has two borders. Consequently, the closure of B^n has characteristic $0 + 2 = 2$ and genus 0 and so, for even n, $B^n \approx S^{0,2}$. When n is odd, this presentation is nonorientable and has a single border. Consequently, the closure of B^n has characteristic $(4 - 6 + 2) + 1 = 1$ and genus 1 and so, for odd n, $B^n \approx \tilde{S}^{1,1}$. These computations are, of course, consistent with the discussion in Chapter 9.

It follows from this example that bands with an odd number of twists can be used to identify nonorientable surfaces. The surface $\tilde{S}^{1,1}$ is known as the *Möbius band.*

Proposition 11.4.5 *Every bordered surface that is embedded in* $\mathbb{R}^3$ *and contains a band with an odd number of twists is nonorientable.*

PROOF: Note that two consecutive but oppositely oriented twists cancel each other out. Consequently, it may be assumed that the twists of the given band are all similarly oriented. Let Π be a presentation of such a surface in which the given band is the union of two quadrilaterals of which one contains all the twists and the other contains none. An argument similar to that employed in Example 11.4.3 leads to the conclusion that this presentation

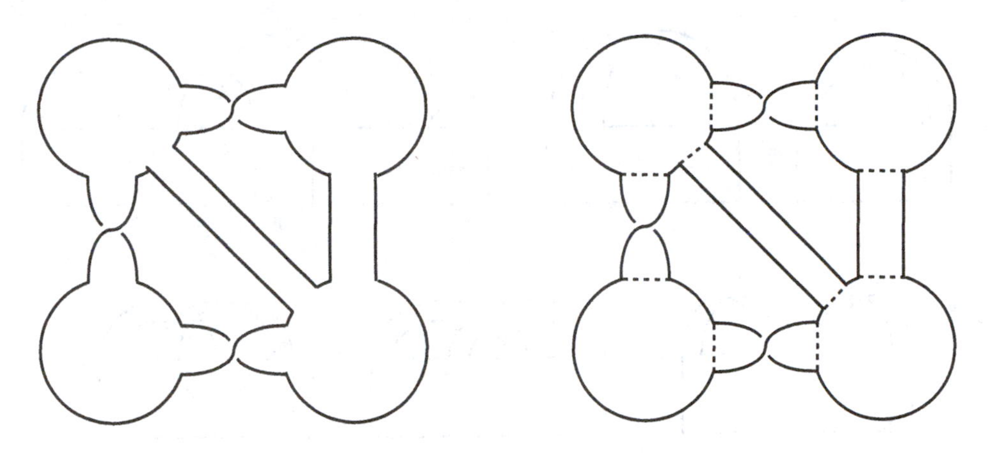

Figure 11.51. The identification of a bordered surface

cannot be coherently oriented. Hence the given surface is nonorientable. Q.E.D.

Example 11.4.6 To identify the bordered surface S on the left of Figure 11.51, we introduce the ten cuts indicated on the right. The Euler characteristic of this presentation is $20 - 30 + 9 = -1$ and there are two borders. By Proposition 11.4.3, $\chi(S^c) = \chi(S) + \beta(S) = -1 + 2 = 1$. A narrow band passing through all of the four circular portions of the surface in succession has three twists. The surface is therefore nonorientable. Hence S^c is the nonorientable surface of Euler characteristic 1, or genus $2 - 1 = 1$. It follows that S is homeomorphic to the twice perforated projective plane $\tilde{S}^{1,2}$.

Exercises 11.4

1. Prove that the Euler characteristic of every bordered surface is at most 2.
2. Describe all the bordered surfaces with the following Euler characteristics: (a) 2 (b) 1 (c) 0 (d) -1 (e) -2 (f) the arbitrary integer n
3. The bordered surfaces of Figure 11.52 consist of a rectangle with attached strips. Describe each in the form $S^{n,\beta}$. Which of these surfaces are homeomorphic to each other?
4. Describe each of the surfaces of Figure 11.53 in the form $S^{n,\beta}$.
5. Which of the surfaces in Figure 11.52 are homeomorphic to which of the surfaces in Figure 11.53?
6. Does the relation $2 - 2\gamma(S) = \chi(S)$ hold for orientable bordered surfaces?

Figure 11.52. Rectangles with strips

7. Are the two bordered surfaces in Figure 11.54 homeomorphic? Justify your answer.

8. Which of the surfaces in Figure 11.55 are homeomorphic? Justify your answer.

9. Identify the closures of the surfaces in the following exercises.

   (a) 7 (b) 8

10. Describe the topological space obtained when each of the Möbius bands in Figure 11.56 is cut along the dotted lines.

11. Suppose c is a loop of the bordered surface S that contains no border points. Prove that if the excision along c separates S into two bordered surfaces S' and S'', then $\chi(S) = \chi(S') + \chi(S'')$.

12. Suppose c is a loop of the bordered surface S that contains no border points. If the excision along c results in a single bordered surface S', find the relation between $\chi(S)$ and $\chi(S')$.

13. Is the surface of Figure 11.57 homeomorphic to the Möbius band?

14. Prove Proposition 11.4.3.

Figure 11.53. Wheel-like surfaces

Figure 11.54. Are these surfaces homeomorphic?

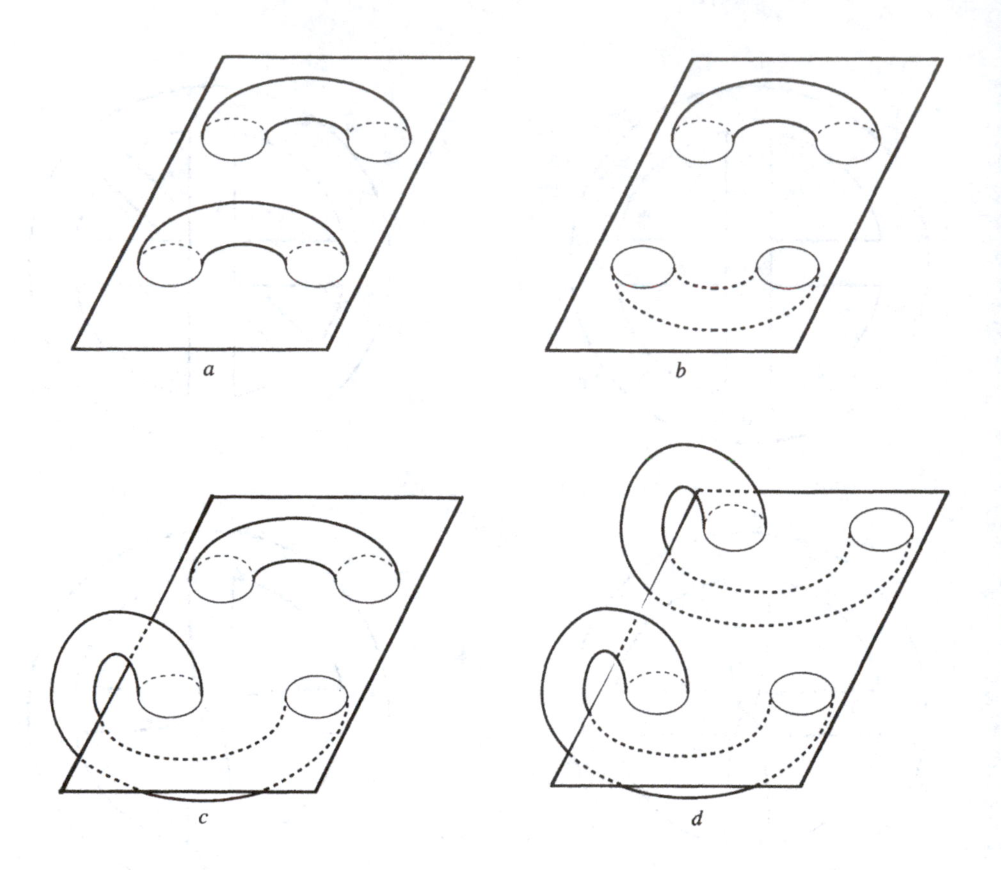

Figure 11.55. Which of these surfaces are homeomorphic?

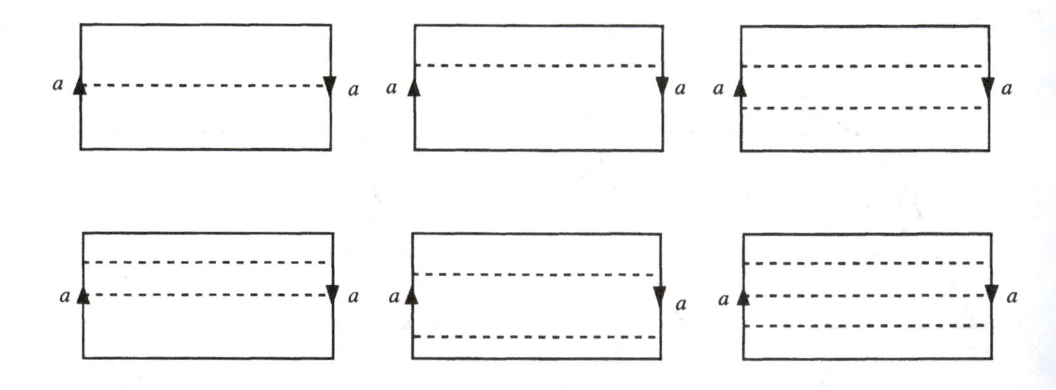

Figure 11.56. Cutting a Möbius band

Figure 11.57. Is this surface homeomorphic to the Möbius band?

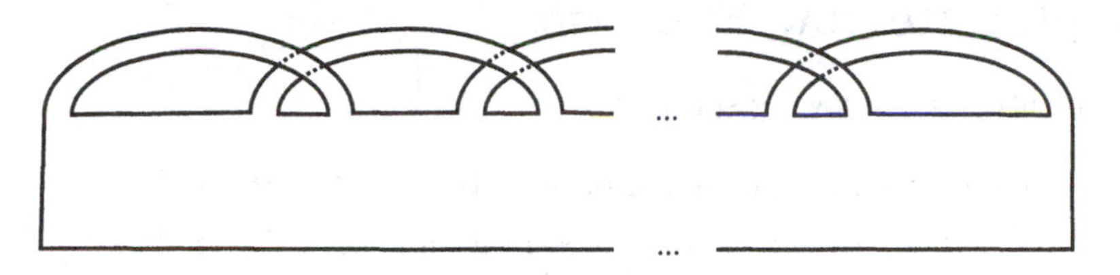

Figure 11.58.

15. The bordered surface of Figure 11.58 consists of a rectangle to which n untwisted bands have been attached. Prove that this surface is homeomorphic to $S^{n,1}$.

16. The bordered surface of Figure 11.59 consists of a rectangle to which n bands have been attached of which only the first is twisted. Prove that this surface is homeomorphic to $\tilde{S}^{n,1}$.

17. Use the previous exercise and Theorem 11.2.14 to prove that a nonorientable surface is embeddable in $\mathbb{R}^3$ if and only if it has at least one border.

18. Prove that a bordered surface is orientable if and only if it contains a subset that is homeomorphic to the Möbius band.

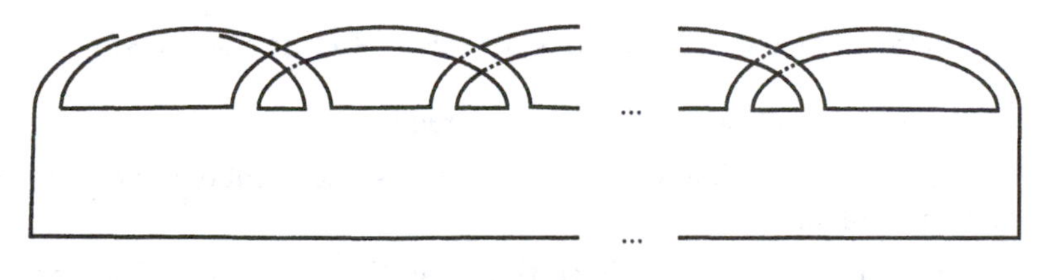

Figure 11.59.

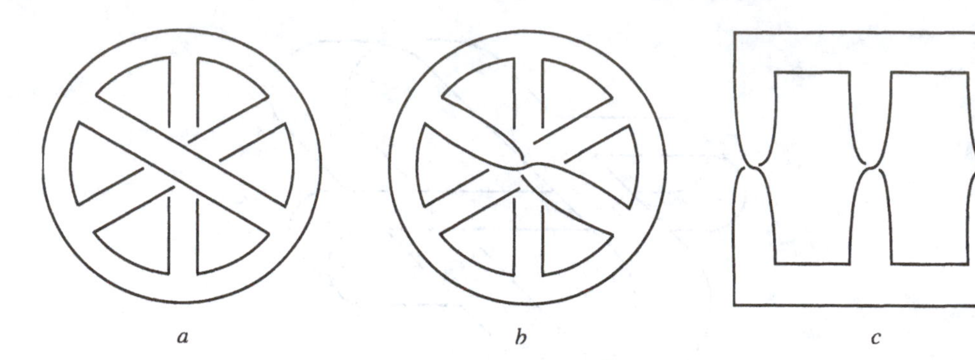

Figure 11.60.

Chapter Review Exercises

1. Identify the following surfaces.

   (a) One with the polygonal presentation $abcd, a^{-1}b^{-1}c^{-1}d^{-1}$
   (b) One with the polygonal presentation $abcd, a^{-1}b^{-1}c^{-1}d$
   (c) Surface a of Figure 11.60
   (d) Surface b of Figure 11.60
   (e) Surface c of Figure 11.60
   (f) $S^3 + S^4$
   (g) $\tilde{S}^3 + S^4$
   (h) $S^3 + \tilde{S}^4$
   (i) $\tilde{S}^3 + \tilde{S}^4$
   (j) $S^3 + h$
   (k) $S^3 + \tilde{h}$
   (l) $S^3 + c$
   (m) $\tilde{S}^3 + h$
   (n) $\tilde{S}^3 + \tilde{h}$
   (o) $\tilde{S}^3 + c$

2. Are the following statements true or false? Justify your answers.

   (a) Every two closed surfaces are homeomorphic.
   (b) Every two bordered surfaces with the same number of borders are homeomorphic.
   (c) If two closed surfaces have the same Euler characteristic, then they are homeomorphic.

(d) If two orientable closed surfaces have the same Euler characteristic, then they are homeomorphic.

(e) If two bordered surfaces have the same Euler characteristic and orientability character, then they are homeomorphic.

(f) Every closed surface has a presentation with exactly one polygon.

(g) Every closed surface has a presentation with 100 polygons.

(h) Given 200 closed surfaces each of which has characteristic greater than -100, some two of them are homeomorphic to each other.

(i) Given 300 closed surfaces each of which has characteristic greater than -100, some two of them are homeomorphic to each other.

(j) Every bordered surface has the same Euler characteristic as its closure.

(k) Every bordered surface has the same genus as its closure.

(l) Every nonorientable bordered surface is homeomorphic to some surface of the form $\tilde{S}^{n,\beta}$.

(m) $S^7 + \tilde{S}^{15} \approx \tilde{S}^{22}$

(n) $S^{15} + \tilde{S}^7 \approx \tilde{S}^{29}$

(o) $S^{15} + \tilde{S}^{15} \approx S^{45}$

Chapter 12

Knots and Links

Many of the scientists of the nineteenth century believed that what is now considered to be empty space was actually occupied by a "substance" called *ether*, which was the medium through which light waves traveled. The famous chemist Lord Kelvin (William Thomson, 1824–1907) thought that atoms might actually be vortices, or knots, in the ether. He therefore asked the mathematicians Peter G. Tait (1831–1901), Reverend Thomas. P. Kirkman (1806–1895), and C. N. Little to create a table of knots with the expectation that this would eventually become a table of atoms.

The ether hypothesis was abandoned in the early part of the twentieth century, and with it went the possibility of representing atoms as knots. Still, the questions raised in the process of organizing this knot table caught the interest of the mathematical world and gave birth to the new discipline of *knot theory*. Several of these questions have been answered, some very recently, whereas others still await resolution. Moreover, while the original idea of representing atoms as vortices in the void has gone the way of the geocentric and phlogiston theories, chemists, as well as physicists and biologists, are currently investigating a variety of other applications of the theory of knots to their respective disciplines.

12.1 Equivalence of Knots and Links

A *knot* is a loop in $\mathbb{R}^3$, and portions of the knot are referred to as its *arcs*. A *link* is a finite set of disjoint knots each of which is said to be a *component* of the link. Knots and links are generally represented by *diagrams*, which depict their projections to a plane (see Fig. 12.1). It will always be assumed that the link and its projection plane are so positioned that no three of the

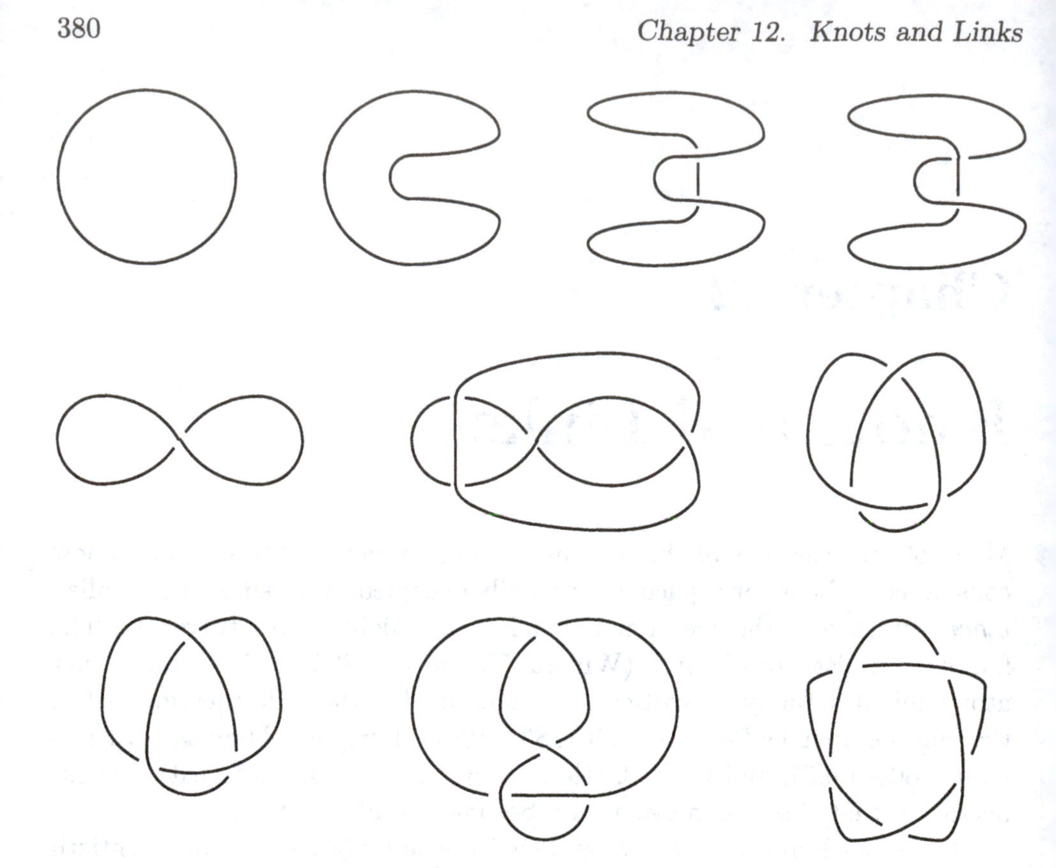

Figure 12.1. Some knot diagrams

link's arcs project to concurrent lines in the plane, and that the projections of any two arcs intersect unequivocally (Fig. 12.2).

Any knot that is isotopic to a plane loop is called an *unknot*. Any link that is isotopic to a set of disjoint plane loops is called an *unlink*. Two links are said to be *equivalent* if there is an isotopy that deforms one link into the other. Equivalent links are considered to be identical and receive the same name. Thus, all the links that are isotopic to the unlink with n

Figure 12.2. Allowed versus forbidden crossings

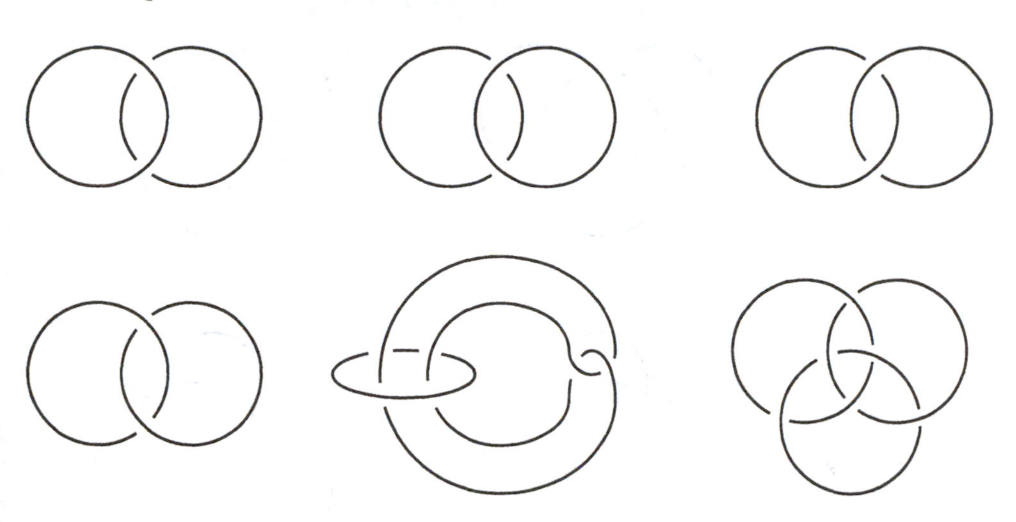

Figure 12.3. Some link diagrams

components are denoted by 0^n. The main goal of the theory of knots is to provide effective criteria for deciding which links are equivalent. A more restricted goal is to determine which links are in fact unlinks. A general procedure for resolving this last question for knots was found by Wolfgang Haken in 1961. However, this procedure is so complicated that to date it has not been implemented as a computer program.

Equivalent links may be represented by different diagrams. The first six knot diagrams of Figure 12.1 all represent the unknot 0, and the first two link diagrams in Figure 12.3 represent the unlink 0^2.

The following observation is plausible and will be taken for granted in the sequel.

Proposition 12.1.1 *If two links have the same diagram, then they are necessarily equivalent.* □

The subtlety of the issue of equivalence is brought to light by Figure 12.4, in which are exhibited two *trefoil* knots that are mirror images of each other (in a sense that will be made precise later) as well as two *figure 8* knots that are also mirror images of each other. As Figure 12.5 indicates, the mirror image figure 8 knots are equivalent to each other. The mirror image trefoil knots, however, are not equivalent—a fact that will be established in Section 12.3.

Above and beyond the fact that links with the same diagram must be equivalent, nothing more than trial and error can be offered in this text as a method for recognizing the equivalence of some two given links. The

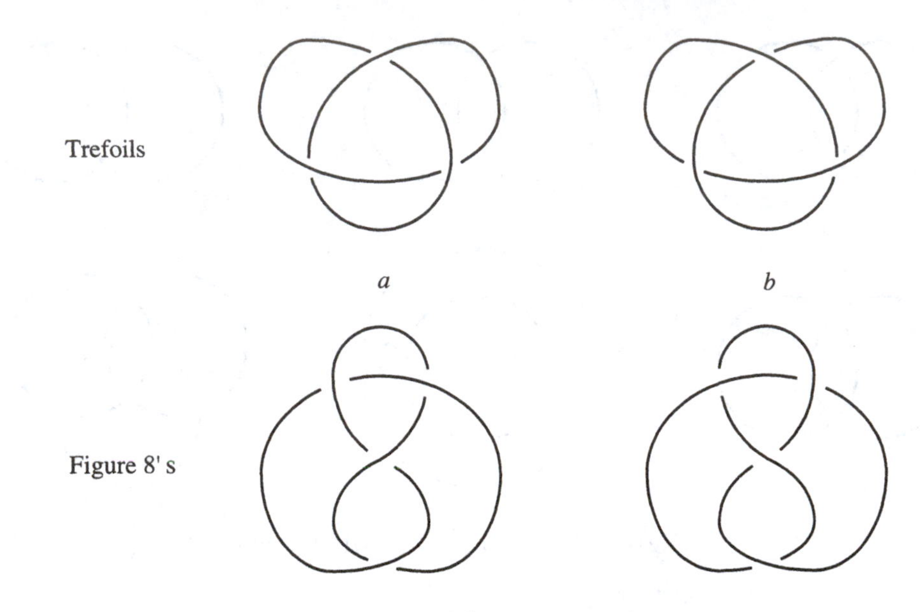

Figure 12.4. Mirror image knots

converse question, that of distinguishing between nonequivalent links, has some partial answers that are very powerful, interesting, and subject to a reasonably easy exposition. These are the subject matter of the remainder of

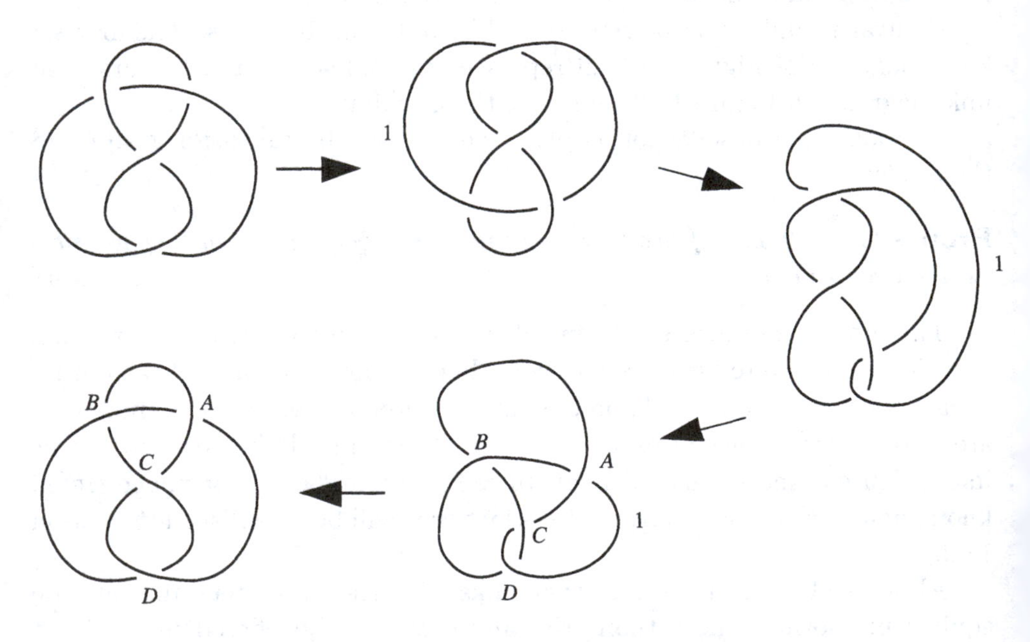

Figure 12.5. The equivalence of the figure 8 knots

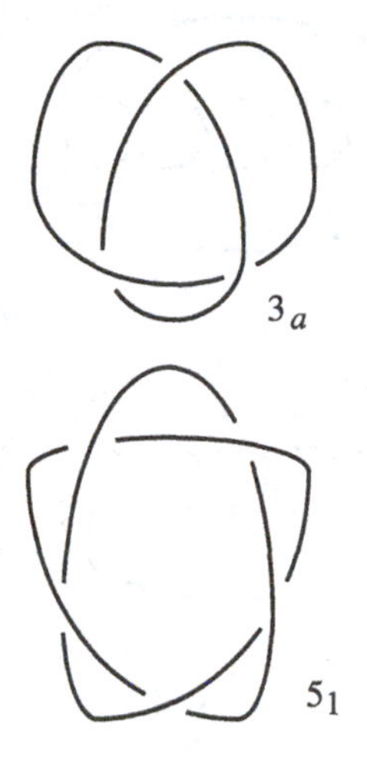

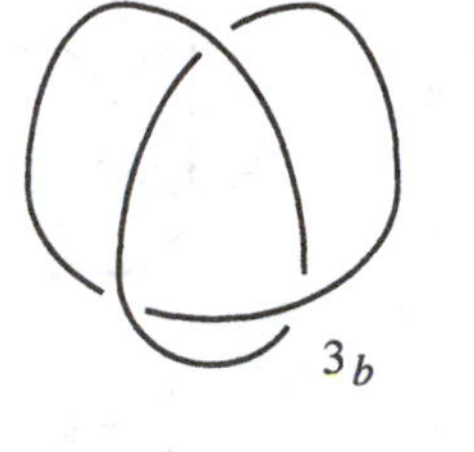

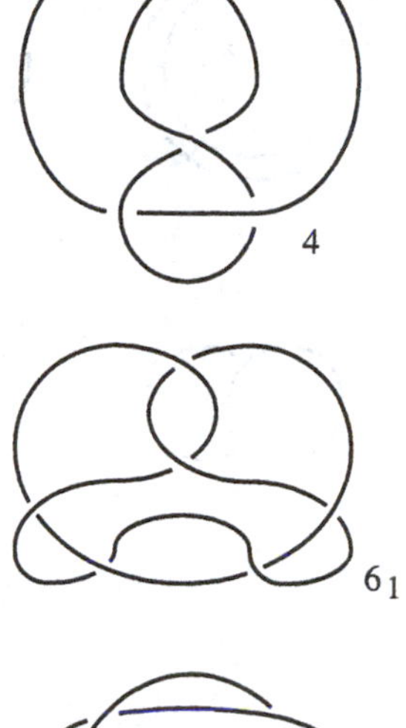

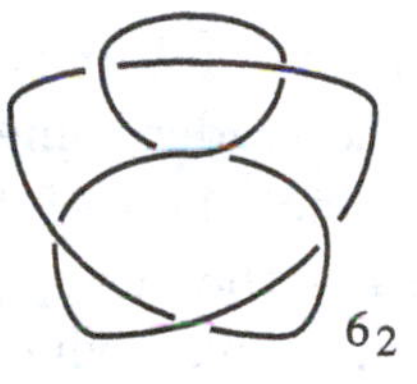

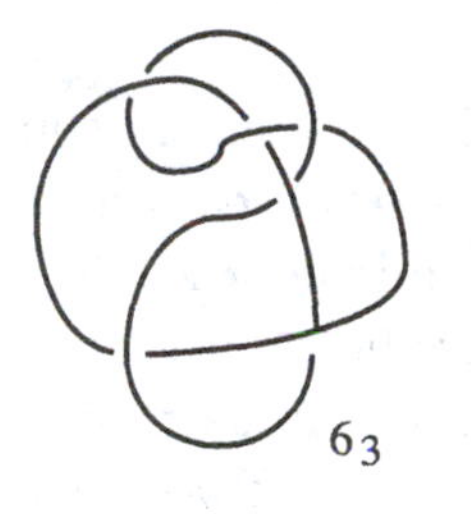

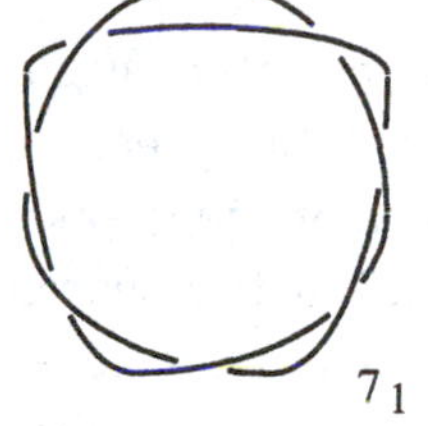

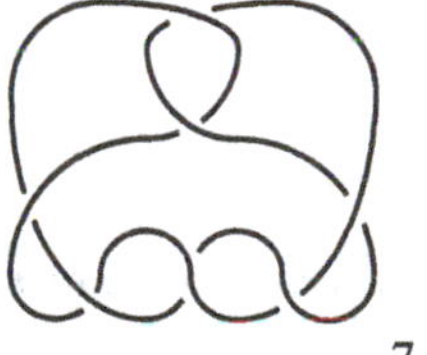

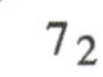

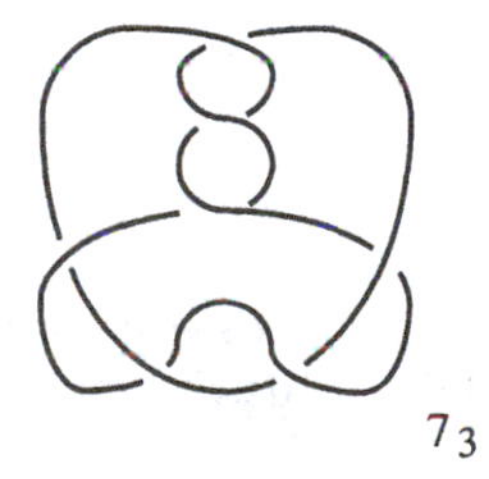

7_4

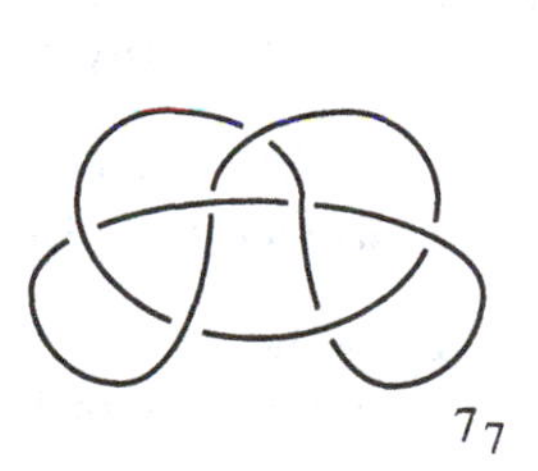

Figure 12.6. Some knot diagrams

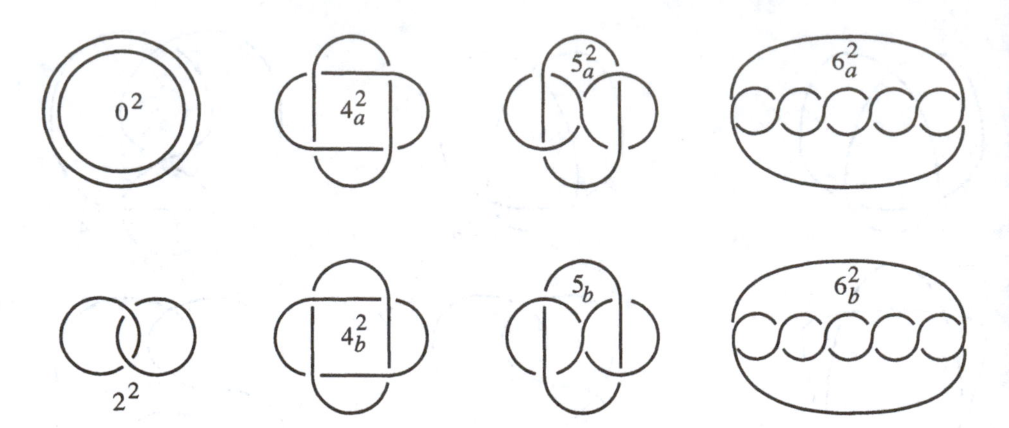

Figure 12.7. Some link diagrams

this chapter. Figures 12.6 and 12.7 contain some standard knots and links. The table preceding the exercises of Section 12.3 will give the reader some sense of the great variety of inequivalent knots and links. The most powerful tools in the search for the classification of knots and links are the *isotopy invariants.* These are properties of links that remain unchanged even when the links are subjected to isotopies. Two such invariants will be described in the next two sections.

Exercises 12.1

1. Is the statement *every knot that has a diagram with exactly one crossing is in fact an unknot* true or false? Justify your answer.

2. Is the statement *every link with exactly one crossing is an unlink* true or false? Justify your answer.

3. Is the statement *every knot that has a diagram with exactly two crossings is in fact an unknot* true or false? Justify your answer.

4. Is the statement *every 2-component link with at most two crossings is an unlink* true or false? Justify your answer.

5. How many inequivalent knots have diagrams with exactly two crossings? Justify your answer.

6. How many inequivalent links have diagrams with exactly two crossings? Justify your answer.

7. Prove that the three knots in Figure 12.8 are all equivalent.

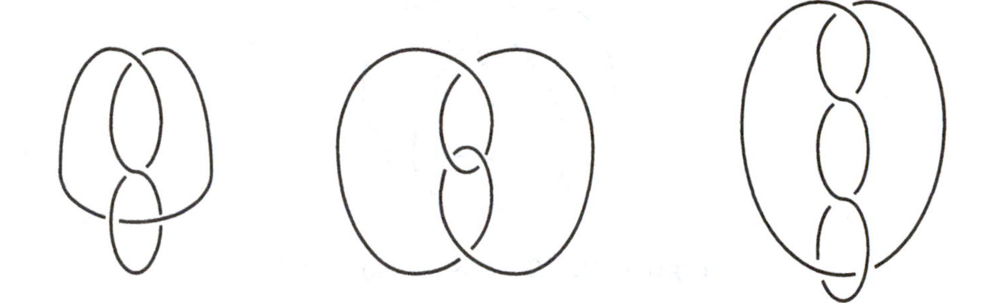

Figure 12.8. Three equivalent knots

12.2 Labelings

The most elementary procedure for distinguishing between links associates a system of linear equations to a diagram, one equation to each crossing. The nature of the solution provides information regarding the isotopy type of the link.

A *strand* of a *diagram* (not of a knot) is any of its loops and any of its arcs that begins and ends at a crossing. Accordingly, the ten knot diagrams of Figure 12.1 have 1, 1, 2, 2, 1, 4, 3, 3, 4, 5 strands respectively. In general, there are three strands, at every crossing, although it is possible for two, or all three, to be identical. If p is any odd prime number, then a *p-labeling* of a knot diagram assigns to every strand of the diagram a number from $\{0, 1, 2, \ldots, p-1\}$ so that at least two distinct numbers are used and, at each crossing, if the overstrand is labeled x and the other two strands are labeled y and z (Fig. 12.9), the *labeling equation*

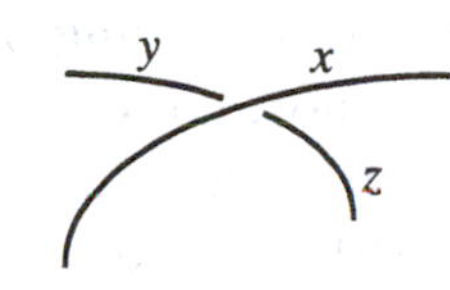

Figure 12.9. The labeling of a typical crossing

$$2x - y - z \equiv 0 \pmod{p} \tag{1}$$

holds. (The reader who is not familiar with these symbols will find them explained in Appendix G.) For example, the diagram of the trefoil knot 3_a of Figure 12.10 is 3-labeled and the three associated equations are

top crossing:	$2 \cdot 1 - 2 - 0 \equiv 0 \pmod{3}$
bottom left crossing:	$2 \cdot 2 - 1 - 0 \equiv 0 \pmod{3}$
bottom right crossing:	$2 \cdot 0 - 1 - 2 \equiv 0 \pmod{3}$.

The following theorem asserts that the sets of primes p for which p-labelings exist is an isotopy invariant.

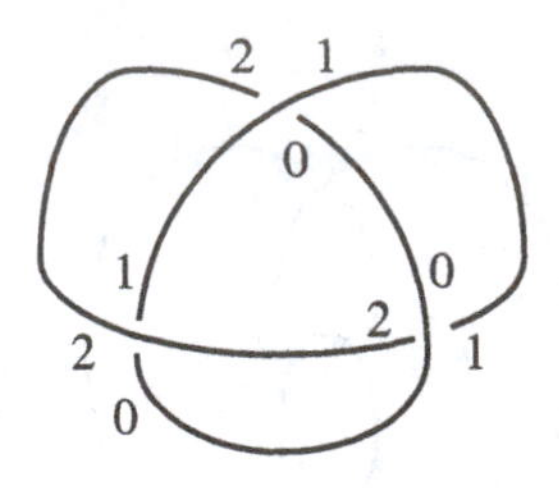

Figure 12.10. A 3-labeling of 3_a

Theorem 12.2.1 *Let p be an odd prime. If some diagram of a link can be p-labeled, then so can every diagram of that same link.* □

This theorem can be employed in different ways. First, it is used to show that two given knots are not equivalent.

Corollary 12.2.2 *The trefoil knots 3_a and 3_b of Figure 12.6 are not unknots.*

Proof: A 3-labeling of the trefoil knot 3_a is exhibited in Figure 12.10. The trivial knot, however, possesses no labelings whatsoever because its simplest diagram, a loop, has only one strand and cannot fulfill the requirement that at least two distinct labels be used in any p-labeling. It follows from Theorem 12.2.1, with $p = 3$, that the trefoil knot 3_a is not equivalent to the unknot. (See Exercise 3 regarding the trefoil knot 3_b.) Q.E.D.

In general, finding such labelings is an exercise in the solution of simultaneous equations modulo a prime number. In the interest of brevity, the expression (mod p) will be henceforth deleted from the labeling equations.

Example 12.2.3 The aforementioned 3-labeling of the trefoil knot 3_a depicted in Figure 12.10 was found by first assigning the unknowns x, y and z to the strands in the diagram (Fig. 12.11) and then writing out the labeling equations:

top crossing:	$2x - y - z \equiv 0$
bottom left crossing:	$2z - x - y \equiv 0$
bottom right crossing:	$2y - z - x \equiv 0.$

The elimination of z yields the two equations

$$3x - 3y \equiv 0 \equiv -3x + 3y, \tag{2}$$

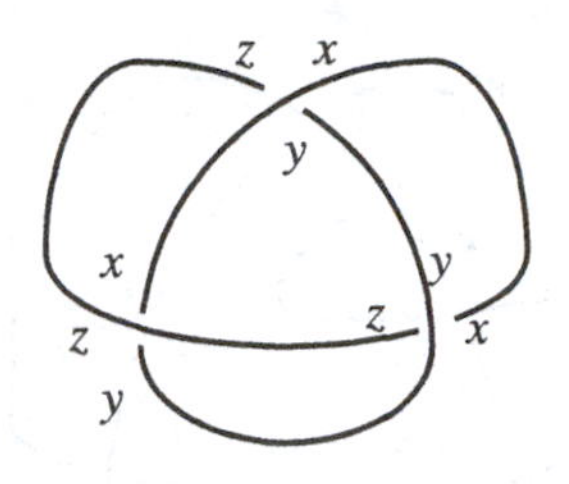

Figure 12.11. Finding a labeling of a trefoil knot

which are satisfied, modulo 3, by any values of x and y whatsoever. In particular, the 3-labeling of Figure 12.10 corresponds to the assignments $x \equiv 1$ and $y \equiv 0$ that entail $z \equiv 2$. In any other modular system the only solutions of Equation (2) have $x \equiv y$, which immediately entails $z \equiv x \equiv y$. Thus, any solution modulo $p \neq 3$ has all of its labels identical and so does not constitute a p-labeling. Consequently, the trefoil knot 3_a only has 3-labelings.

It is not at all obvious that equivalent knots should have p-labelings for identical p's. In support of this claim it is demonstrated that a certain modification of the diagram of 3_a (Fig. 12.12) is also 3-colorable, even though its labeling equations are quite different.

Example 12.2.4 The labeling equations associated with the diagram of Figure 12.12 are

$$\begin{aligned}
2v - u - v &\equiv 0 \\
2z - v - w &\equiv 0 \\
2u - z - w &\equiv 0 \\
2w - x - u &\equiv 0 \\
2w - x - y &\equiv 0 \\
2w - y - z &\equiv 0.
\end{aligned}$$

The first of these equations implies that $u \equiv v$, the fourth and the fifth imply that $u \equiv y$, and the last two that $z \equiv x$. When u, v, and z are replaced by y and x, the second, third, and fourth equations become

$$\begin{aligned}
2x - y - w &\equiv 0 \\
2y - x - w &\equiv 0 \\
2w - x - y &\equiv 0,
\end{aligned}$$

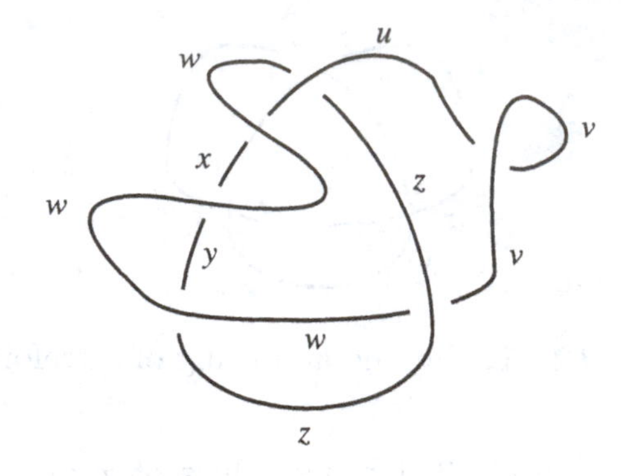

Figure 12.12. A modification of the diagram of 3_a

a system of equations that is clearly equivalent to the three equations obtained from the standard diagram of the trefoil knot 3_a. It follows that the systems of equations associated with the diagrams of Figures 12.11 and 12.12 have solutions for the same p's.

Example 12.2.5 The search for p-labelings of the knot 4 (Fig. 12.13) leads to the system of equations

$$\begin{aligned} 2x - z - w &\equiv 0 \\ 2z - x - y &\equiv 0 \\ 2y - x - w &\equiv 0 \\ 2w - y - z &\equiv 0. \end{aligned}$$

When w is eliminated from the first and third equations, we obtain

$$\begin{aligned} 2y - 3x + z &\equiv 0 \\ 2z - x - y &\equiv 0. \end{aligned}$$

The further elimination of z yields

$$5x - 5y \equiv 0.$$

This implies that $x \equiv y \equiv z \equiv w$ for any modulus $p \neq 5$, allowing for no p-labelings. For $p = 5$ there is, among others, the labeling $x \equiv 1$, $y \equiv 2$, $z \equiv 4$, $w \equiv 3$.

It may be concluded from the preceding examples that the knots 4, 3_a, and 0 are all inequivalent to each other. Unfortunately, this elegant method does not contain the whole story. Inequivalent knots that have p-labelings for the same p's are known (e.g., the two trefoil knots).

Note that the last of the equations of Example 12.2.5 was not used. It so happens that the system of labeling equations of a diagram is necessarily dependent (see Exercise 17), and hence the last equation can always be ignored.

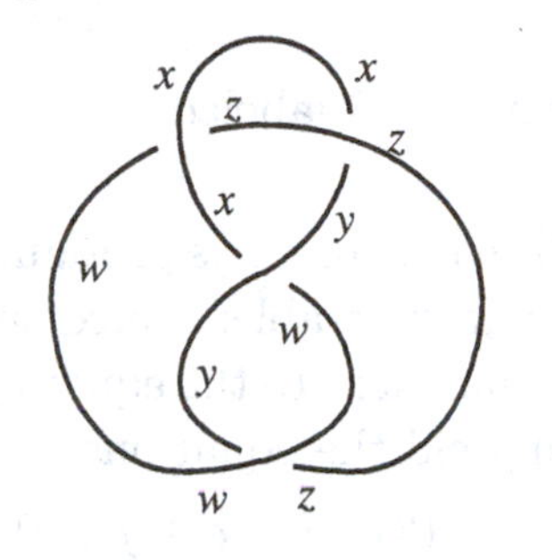

Figure 12.13. Finding a labeling for the knot 4

Links with several components are also subject to p-labelings. Surprisingly, however, the information derived from such labelings is qualitatively different from that obtained for knots. This discussion requires a definition. A link diagram is said to be *split* if there is a straight line in the diagram's plane such that portions of the diagram lie on both sides of the straight line, but the diagram does not intersect the straight line (see Figure 12.14). A link is said to be *splittable* if it is equivalent to a link that has a split diagram. It is easy to see that every split diagram has a p-labeling for every p. We need simply label all the strands on one side of the separating line 1 and all the others 2. All the p-labeling equations then have the form

$$2x - x - x \equiv 0 \qquad x = 1, 2$$

and are clearly satisfied. Hence the following conclusion can be drawn from Theorem 12.2.1.

Proposition 12.2.6 *If a link with at least two components does not have a p-labeling for even one p, then it is not splittable.*

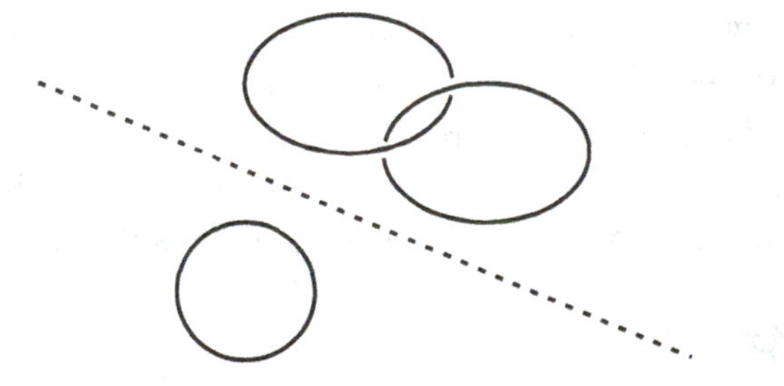

Figure 12.14. Splittings

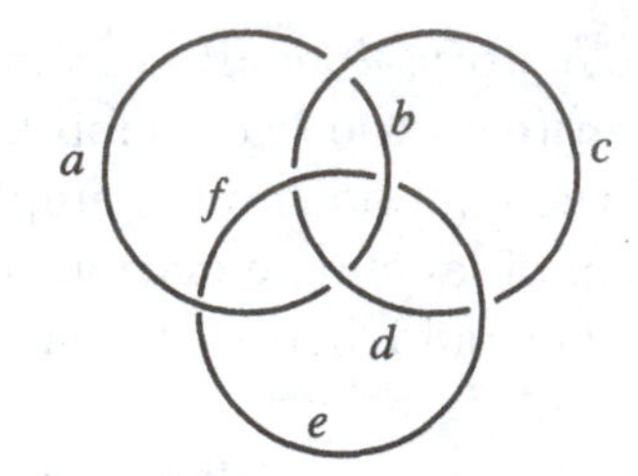

Figure 12.15. The Borromean rings cannot be 3-labeled

Example 12.2.7 It will be shown now that the *Borromean rings* of Figure 12.15 do not have a 3-labeling and are therefore not splittable. Since the modulus is 3, the labeling equation $2x-y-z \equiv 0$ is equivalent to the equation $x+y+z \equiv 0$. Hence the crossings of this diagram yield the equations

(1) $a+b+c \equiv 0$	(2) $c+d+f \equiv 0$	(3) $b+e+f \equiv 0$
(4) $a+e+f \equiv 0$	(5) $a+b+d \equiv 0$	(6) $c+d+e \equiv 0$

Equations (1) and (5) imply that $c \equiv d$, Equations (2) and (6) that $e \equiv f$, and Equations (3) and (4) that $a \equiv b$. The system therefore reduces to

$$(1') \quad 2a+c \equiv 0 \qquad (2') \quad 2c+e \equiv 0 \qquad (3') \quad 2e+a \equiv 0$$

or

$$(1'') \quad -a+c \equiv 0 \qquad (2'') \quad -c+e \equiv 0 \qquad (3'') \quad -e+a \equiv 0$$

It is now clear that $a \equiv b \equiv c \equiv d \equiv e \equiv f$ so that the requirement that at least two different labels be employed cannot be met. Hence the diagram of the Borromean rings does not have a 3-labeling and is therefore not splittable.

Regarding the Solution of Systems of Labeling Equations

The system of labeling equations of a link is dependent (see Exercise 17) and so one of the equations can be ignored. It is recommended that in solving such a system the unknowns be eliminated, one at a time, by the addition of the appropriate multiples of equations. This has the advantage of sidestepping the issue of modular inverses. Moreover, it is also suggested that the elimination process begin with the three unknowns that appear in the ignored equation. The elimination process will eventually result in an equation of the form

$$Ax - Ay \equiv 0,$$

and every odd prime divisor p of A has a corresponding p-labeling that comes from setting $x = 1$ and $y = 2$.

Exercises 12.2

1. Verify directly, without using Theorem 12.2.1, that the diagrams in Figure 12.16 do not have any p-labelings.

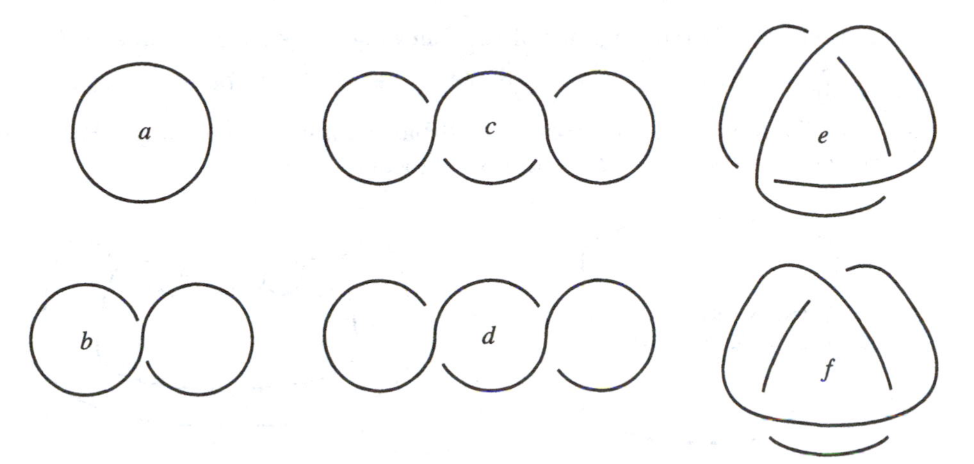

Figure 12.16.

2. For each of the diagrams in Figure 12.17 determine directly, without using Theorem 12.2.1, all the primes p for which that diagram has a p-labeling.

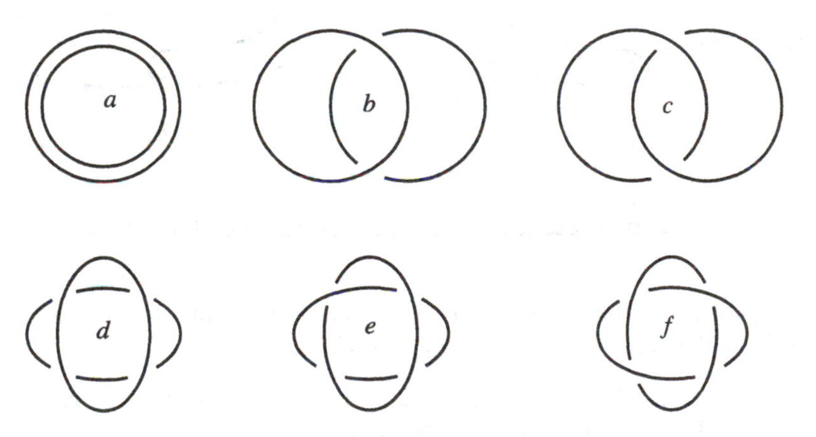

Figure 12.17.

3. Show that the knot 3_b has only 3-labelings.

4. Show that 5_1 has a p-labeling if and only if $p = 5$.

5. Show that 5_2 has a p-labeling if and only if $p = 7$.

6. Show that 6_1 has a p-labeling if and only if $p = 3$.

7. Show that 6_2 has a p-labeling if and only if $p = 11$.

8. Show that 6_3 has a p-labeling if and only if $p = 13$.

9. For which p do the diagrams of the following knots have a p-labeling?

   (a) 7_1 (b) 7_2 c) 7_3 (d) 7_4 (e) 7_5 (f) 7_6 (g) 7_7

10. The diagram of $L_{1,n}$ in Figure 12.18 has a total of n crossings. For which values of n and p does $L_{1,n}$ have a p-labeling?

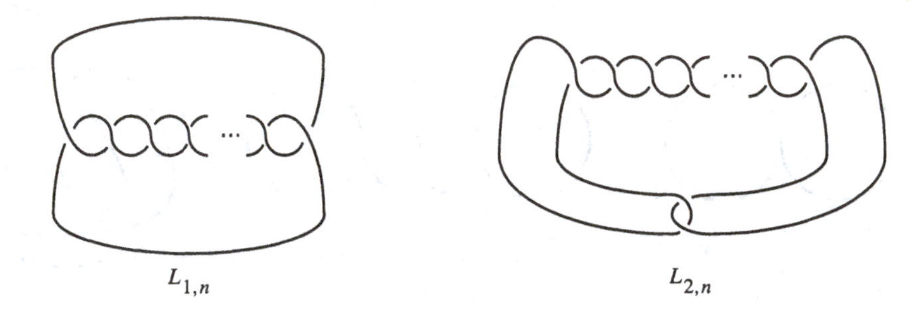

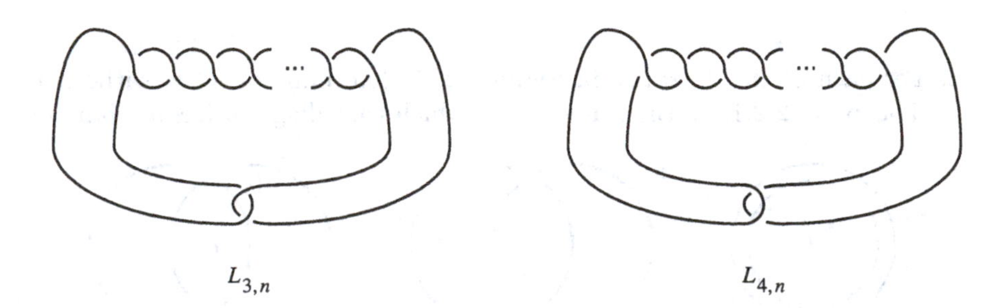

Figure 12.18. Some links in need of labelings

11. The diagram of $L_{2,n}$ in Figure 12.18 has a total of n crossings. For which values of n and p does $L_{2,n}$ have a p-labeling?

12. The diagram of $L_{3,n}$ in Figure 12.18 has a total of n crossings. For which values of n and p does $L_{2,n}$ have a p-labeling?

13. The diagram of $L_{4,n}$ in Figure 12.18 has a total of n crossings. For which values of n and p does $L_{2,n}$ have a p-labeling?

14. Show that the link of Figure 12.19 has no p-labelings.

15. Prove that no labeling of a knot uses exactly two labels. Is the same true for links?

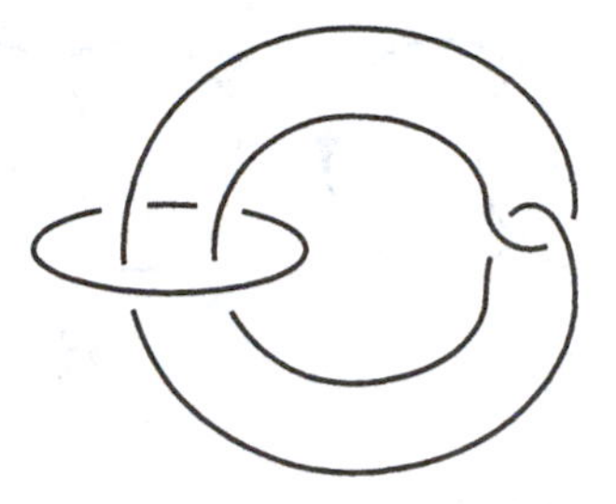

Figure 12.19. The Whitehead link

16. A link is *Brunnian* if it is not an unlink but the removal of any of its components leaves an unlink. The Borromean rings of Figure 12.15 constitute a Brunnian link with three components.

    (a) Construct a Brunnian link with four components.

    (b) Construct a Brunnian link with an arbitrary number of components.

*17. Prove that the system of labeling equations of a link diagram is dependent.

18. Suppose none of the strands of the link diagram D is a loop. Prove that the system of labeling equations associated with D has the same number of equations as unknowns.

19. Generalize Exercise 18 to arbitrary link diagrams.

12.3 The Jones Polynomial

In 1928 the topologist James W. Alexander (1888–1971) associated to each link a polynomial that was an isotopy invariant. This polynomial proved very useful in distinguishing between nonequivalent links. In 1986, as part of groundbreaking work that won him the prestigious Fields Medal, Vaughan Jones developed a new polynomial that is even more efficient at distinguishing between inequivalent links. This new isotopy invariance is the subject of this section. Its description, however, requires that the context be widened to include *oriented* links.

An *orientation* of a knot K is a choice of one of the two senses in which it can be traversed, and is denoted by $\vec{K}$. A link is said to be *oriented* if each of its components is oriented. Since each of the link's components has two possible orientations, it follows that a link with k components has 2^k possible orientations. Figure 12.20 exhibits all four orientations of the link 2^2.

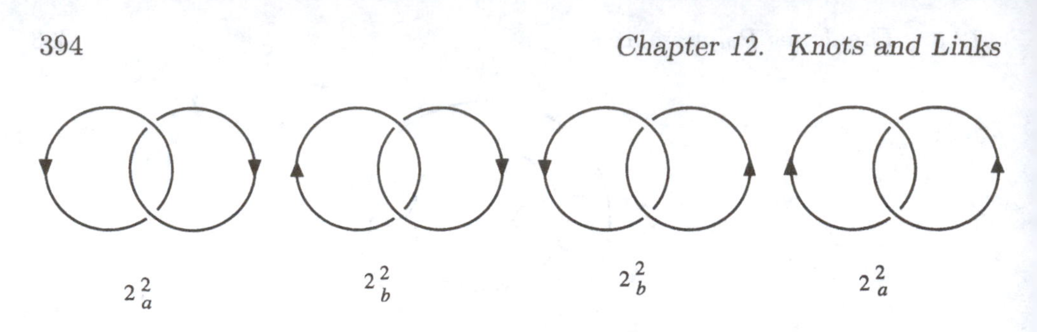

Figure 12.20. Four orientations of the link 2^2

The study of knots and links is enhanced by the widening of its scope to include oriented links as well, because the orientations lead to several new parameters that also furnish useful information about the underlying *unoriented* links. This is one of many instances in mathematics where a seemingly artificial broadening of the context results in new tools that can be brought to bear on old problems.

Two oriented links are said to be *equivalent* if there is an isotopy that transforms one to the other so that the orientations of corresponding components agree. It is clear that both orientations of 0 are equivalent to each other, as are all four orientations of 0^2. It is also easy to see that the leftmost and rightmost oriented links in Figure 12.20 are equivalent to each other, as are the two middle ones. The requisite isotopies are provided by a 180° flip around a horizontal axis. It is plausible that these two groupings are *not* equivalent to each other and a formal proof of this fact will be provided shortly.

If $\vec{L}$ is an oriented link, then the link obtained by reversing the orientations of each of the component loops of $\vec{L}$ is denoted by $\vec{L}^r$ and is called the *reverse* of $\vec{L}$. As it happens, all the oriented links displayed in this text are isotopic to their inverses, but that is not always the case.

In a diagram of an oriented link, a crossing is said to be *right* (*left*) *handed* if, when the open right (left) hand is placed between the arcs so as to separate them, with the extended thumb pointing in the direction of the overstrand's orientation and the palm facing the observer, the other four digits point in the direction of the lower arc's orientation (see Fig. 12.21).

Strictly speaking, the Jones polynomial is not a polynomial, it is a *Laurent polynomial*, as it includes powers of its variable with negative exponents. Nevertheless, this abuse of terminology is standard and will be followed here as well. The Jones polynomial of the link $\vec{L}$ is denoted by $V(\vec{L})$ and its independent variable is x. Not surprisingly, $V(\text{unknot}) = 1$. Several more polynomials are listed in Table 12.1, but this table requires a clarification as it takes advantage of hindsight to simplify the notation. Notwithstanding

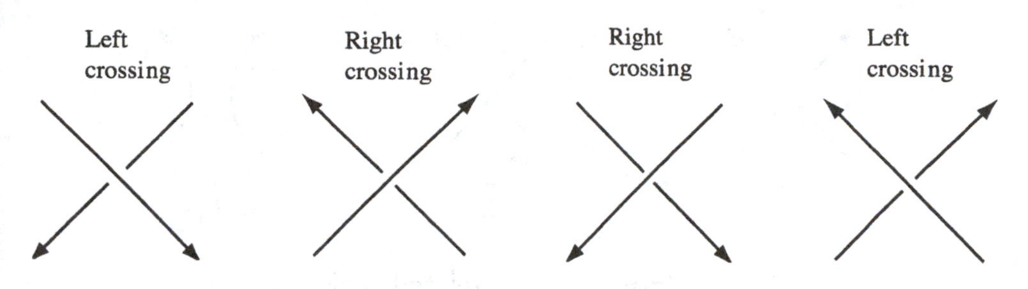

Figure 12.21. Left and right crossings

the fact that $\vec{L}$ and $\vec{L}^{\mathrm{r}}$ are not isotopic in general, it so happens (see Proposition 12.3.9) that $V(\vec{L}^{\mathrm{r}}) = V(\vec{L})$ and hence, in the important special case where $\vec{L}$ is in fact an oriented knot $\vec{K}$, we may speak of the Jones polynomial

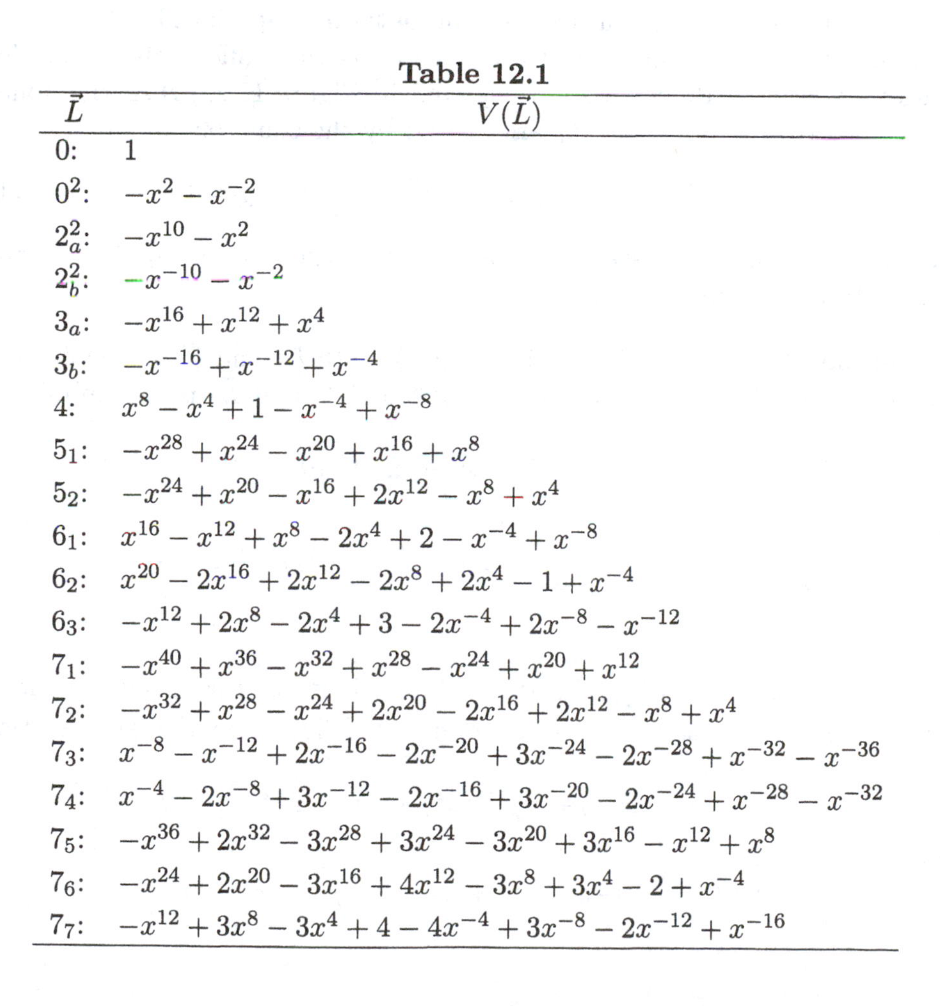

Table 12.1

$\vec{L}$	$V(\vec{L})$
0:	1
0^2:	$-x^2 - x^{-2}$
2_a^2:	$-x^{10} - x^2$
2_b^2:	$-x^{-10} - x^{-2}$
3_a:	$-x^{16} + x^{12} + x^4$
3_b:	$-x^{-16} + x^{-12} + x^{-4}$
4:	$x^8 - x^4 + 1 - x^{-4} + x^{-8}$
5_1:	$-x^{28} + x^{24} - x^{20} + x^{16} + x^8$
5_2:	$-x^{24} + x^{20} - x^{16} + 2x^{12} - x^8 + x^4$
6_1:	$x^{16} - x^{12} + x^8 - 2x^4 + 2 - x^{-4} + x^{-8}$
6_2:	$x^{20} - 2x^{16} + 2x^{12} - 2x^8 + 2x^4 - 1 + x^{-4}$
6_3:	$-x^{12} + 2x^8 - 2x^4 + 3 - 2x^{-4} + 2x^{-8} - x^{-12}$
7_1:	$-x^{40} + x^{36} - x^{32} + x^{28} - x^{24} + x^{20} + x^{12}$
7_2:	$-x^{32} + x^{28} - x^{24} + 2x^{20} - 2x^{16} + 2x^{12} - x^8 + x^4$
7_3:	$x^{-8} - x^{-12} + 2x^{-16} - 2x^{-20} + 3x^{-24} - 2x^{-28} + x^{-32} - x^{-36}$
7_4:	$x^{-4} - 2x^{-8} + 3x^{-12} - 2x^{-16} + 3x^{-20} - 2x^{-24} + x^{-28} - x^{-32}$
7_5:	$-x^{36} + 2x^{32} - 3x^{28} + 3x^{24} - 3x^{20} + 3x^{16} - x^{12} + x^8$
7_6:	$-x^{24} + 2x^{20} - 3x^{16} + 4x^{12} - 3x^8 + 3x^4 - 2 + x^{-4}$
7_7:	$-x^{12} + 3x^8 - 3x^4 + 4 - 4x^{-4} + 3x^{-8} - 2x^{-12} + x^{-16}$

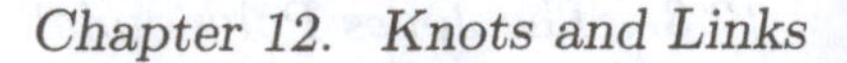

Figure 12.22. Related Links

of the unoriented knot K and define it as

$$V(K) = V(\vec{K}) = V(\vec{K}^{\mathrm{r}}).$$

The Jones polynomial satisfies a relation that is more useful than the actual definition for the purposes of computation. Specifically, if $\vec{L}_+$, $\vec{L}_-$, and $\vec{L}_0$ are three oriented links with diagrams that differ only at one location, where the differences are depicted in Figure 12.22, then the Jones polynomials of $\vec{L}_+$, $\vec{L}_-$, and $\vec{L}_0$ are related by the equation

$$x^4 V(\vec{L}_+) - x^{-4} V(\vec{L}_-) + (x^2 - x^{-2}) V(\vec{L}_0) = 0, \tag{1}$$

which is known as the *skein relation.* This relation is now applied to the computation of some specific polynomials.

Example 12.3.1 ($V(0^2)$) In Figure 12.23, both $\vec{L}_+$ and $\vec{L}_-$ are equivalent to the oriented unknot 0 that has polynomial 1, and $\vec{L}_0$ is the unlink 0^2. Hence,

$$x^4 \cdot 1 - x^{-4} \cdot 1 + (x^2 - x^{-2}) V(0^2) = 0$$

or

$$V(0^2) = -\frac{x^4 - x^{-4}}{x^2 - x^{-2}} = -(x^2 + x^{-2}) = -x^2 - x^{-2}.$$

Example 12.3.2 ($V(2^2_a)$) In Figure 12.24, it is $\vec{L}_+$ that is equivalent to the unlink 0^2 whereas $\vec{L}_0$ is now the unknot. It follows from the skein relation that

$$x^4(-x^2 - x^{-2}) - x^{-4} V(2^2_a) + (x^2 - x^{-2}) \cdot 1 = 0$$

or

$$V(2^2_a) = x^4(-x^6 - x^2 + x^2 - x^{-2}) = -x^{10} - x^2.$$

Example 12.3.3 ($V(2_b^2)$) In Figure 12.25, $\vec{L}_-$ is equivalent to the unlink 0^2 (and so, by Example 12.3.1, has polynomial $-x^2 - x^{-2}$), $\vec{L}_0$ is the unknot with polynomial 1, and $\vec{L}_+$ is the link 2_b^2. It follows from the skein relation that

$$x^4 V(2_b^2) - x^{-4}(-x^2 - x^{-2}) + (x^2 - x^{-2}) \cdot 1 = 0$$

or

$$V(2_b^2) = x^{-4}(-x^{-2} - x^{-6} - x^2 + x^{-2}) = -x^{-10} - x^{-2}.$$

Since $V(2_a^2) \neq V(2_b^2)$, it follows that 2_a^2 and 2_b^2 are inequivalent links, an intuitively plausible observation. We now go on to show that the trefoil knots 3_a and 3_b are also inequivalent.

Example 12.3.4 ($V(3_a)$) The polynomials $V(0)$ and $V(2_a^2)$ can be substituted into the skein relation of Figure 12.26 to obtain the Jones polynomial of the trefoil knot 3_a:

$$x^4 \cdot 1 - x^{-4} V(3_a) + (x^2 - x^{-2}) V(2_a^2) = 0$$

or

$$\begin{aligned} V(3_a) &= x^4 \left[x^4 + (x^2 - x^{-2})(-x^{10} - x^2)\right] \\ &= x^4(x^4 - x^{12} - x^4 + x^8 + 1) \\ &= x^4 + x^{12} - x^{16}. \end{aligned}$$

Example 12.3.5 ($V(3_b)$) The polynomials $V(0)$ and $V(2_b^2)$ can be substituted into the skein relation of Figure 12.27 to obtain the Jones polynomial of the trefoil knot 3_b:

$$x^4 V(3_b) - x^{-4} \cdot 1 + (x^2 - x^{-2}) V(2_b^2) = 0$$

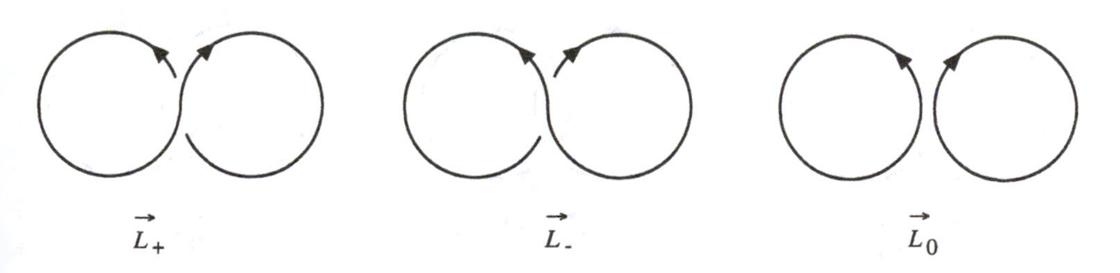

Figure 12.23. A derivation of $V(0^2)$

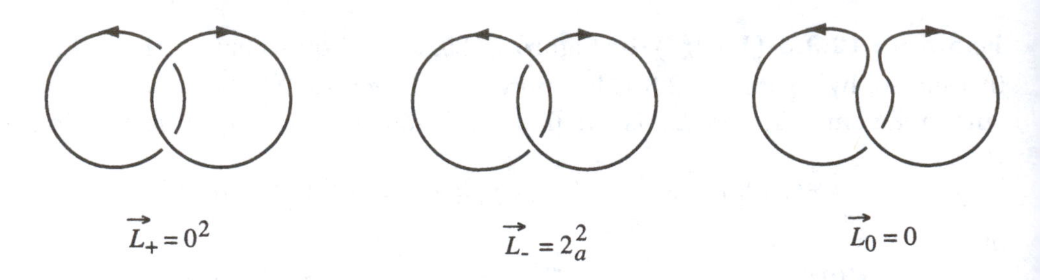

Figure 12.24. Deriving the Jones polynomial of 2^2_a

or

$$\begin{aligned} V(3_b) &= x^{-4}\left[x^{-4} + (x^2 - x^{-2})(-x^{-10} - x^{-2})\right] \\ &= x^{-4}(x^{-4} + x^{-8} + 1 - x^{-12} - x^{-4}) \\ &= x^{-4} + x^{-12} - x^{-16}. \end{aligned}$$

Since $V(3_a) \neq V(3_b)$, it follows that the trefoil knots 3_a and 3_b are not isotopic.

The preceding examples may be misleading. The skein relation does not necessarily provide an algorithm for computing the Jones polynomial. The reason this relation fails to yield an algorithm is that $\vec{L}_+$ and $\vec{L}_-$ have the same number of crossings and so there is no way of predicting which of the links $\vec{L}_+$ or $\vec{L}_-$ is "simpler." Nevertheless, the skein relation is sufficient for our purposes here, and, with the judicious choice of crossings, can be made to yield an algorithm. Another question left unanswered by the skein relation is that of invariance. Even in those cases where the skein relation can be used to compute the Jones polynomial, it is not at all obvious that the derived polynomial is independent of the crossings at which the relation is applied. Nevertheless, such is indeed the case (see Exercise 2).

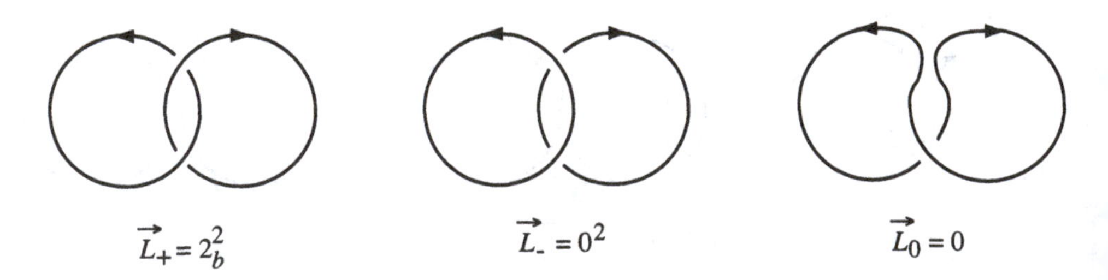

Figure 12.25. Deriving the Jones polynomial of 2^2_b

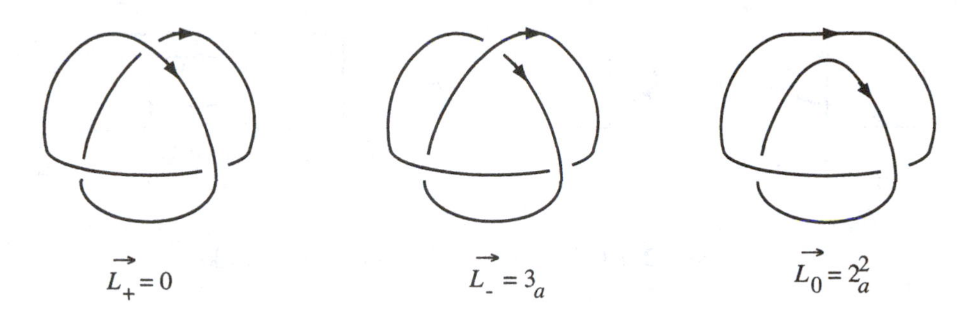

Figure 12.26. Deriving the Jones polynomial of the trefoil knot 3_a

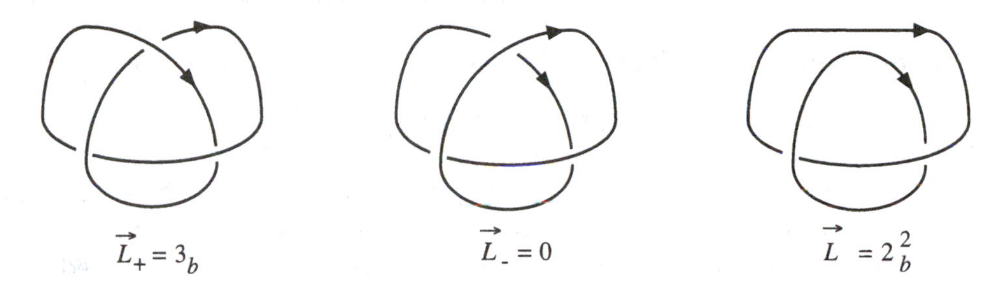

Figure 12.27. Deriving the Jones polynomial of the trefoil knot 3_b

The following three propositions are very helpful in the computation of Jones polynomials. Suppose $\vec{L}_1$ and $\vec{L}_2$ are two oriented links whose diagrams D_1 and D_2 are drawn in the same plane so that they can be separated by a single straight line. The *disjoint union* $\vec{L}_1 \cup \vec{L}_2$ is the link whose diagram is $D_1 \cup D_2$.

Proposition 12.3.6 $V(\vec{L}_1 \cup \vec{L}_2) = (-x^2 - x^{-2})V(\vec{L}_1)V(\vec{L}_2)$. □

The *sum* $\vec{L}_1 + \vec{L}_2$ of two oriented links is defined by means of Figure 12.28. The connected sum of two oriented *knots* is well defined in the sense that when the exact locations on the diagrams where the two summand knots are connected are allowed to vary, the results remain isotopic (see Exercise 7).

Proposition 12.3.7 $V(\vec{L}_1 + \vec{L}_2) = V(\vec{L}_1)V(\vec{L}_2)$. □

The *obverse* or *mirror image* of L (or $\vec{L}$) is the link L^m (or $\vec{L}^m$) obtained by looking at a diagram from the other side of the page. For example, the

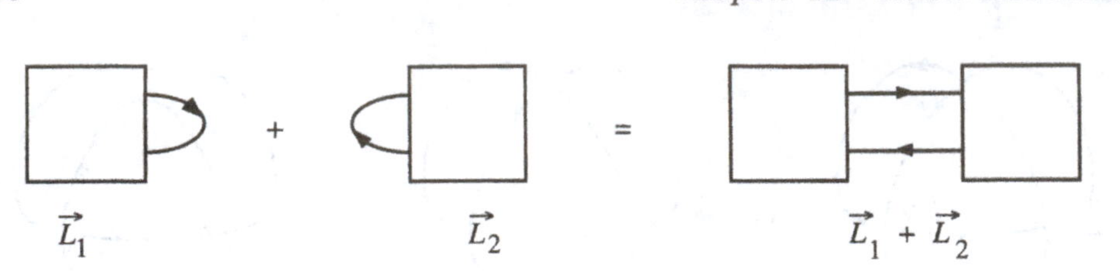

Figure 12.28. The sum of two oriented links

links 2_a^2 and 2_b^2 are obverses of each other as are the knots 3_a and 3_b. On the other hand, Figure 12.5 demonstrates that the knot 4 is its own obverse.

Proposition 12.3.8 *If $\vec{L}$ and $\vec{L}^m$ are obverse oriented links, then*

$$V(\vec{L})(x) = V(\vec{L}^m)(x^{-1}).$$ □

Note that this claim is supported by the polynomials of the knots 3_a, 3_b and the self-obverse 4, as well as by those of the links 2_a^2 and 2_b^2.

A knot whose orientations cannot be expressed as the connected sums of nontrivial knots is said to be *prime*. The *crossing number* of a knot is the smallest number of crossings in any of its diagrams. For each integer $n = 3, 4, \ldots, 13$, the following table lists the number of inequivalent prime knots that have crossing number n, with the stipulation that only one of each obverse pair of knots is counted.

Crossing number	3	4	5	6	7	8	9	10
Number of prime knots	1	1	2	3	7	21	49	165

Crossing number	11	12	13	14	15
Number of prime knots	552	2176	9988	46972	253293

Exercises 12.3

1. Show that the Jones polynomials of the links in Figure 12.29 are
   (a, d) $-x^{22} + x^{18} - x^{14} - x^6$ (b, c) $-x^{-2} + x^{-6} - x^{-10} - x^{-18}$
   (e, h) $-x^{-6} - x^{-14} + x^{-18} - x^{-22}$ (f, g) $-x^{18} - x^{10} + x^6 - x^2$
   Explain why these results prove that the unoriented links 4_a^2 and 4_b^2 of Figure 12.7 are not equivalent.

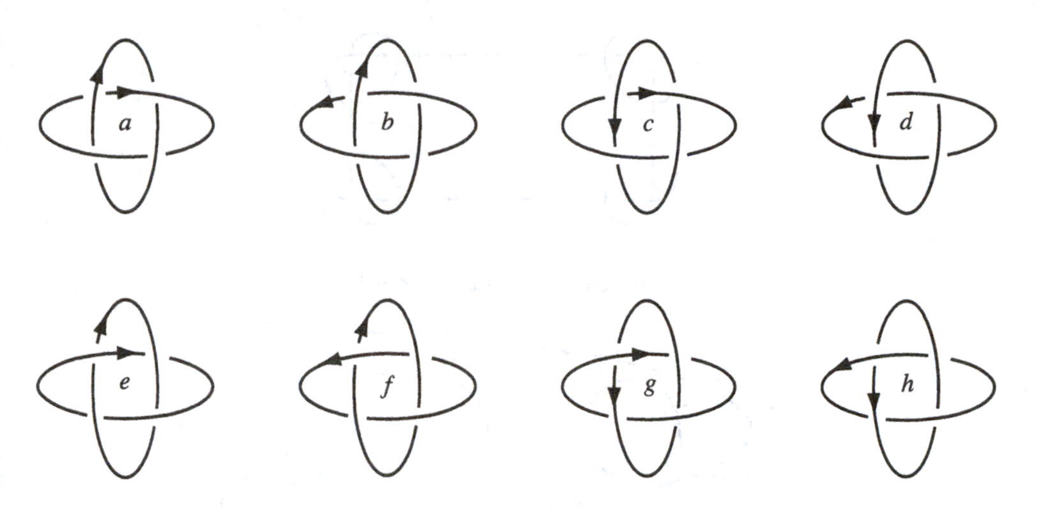

Figure 12.29. Some links

2. Compute the Jones polynomial of the knot 4 of Figure 12.30 by applying the skein relation to the crossing with the following label:

   (a) A (b) B (c) C (d) D.

3. Compute the Jones polynomials of the following knots (you may use Exercise 1):

   (a) 5_1 (b) 5_2

4. Compute the Jones polynomials of the following knots (you may use Exercise 1):

   (a) 6_1 (b) 6_2 (c) 6_3

5. Compute the Jones polynomials of the following knots:

   (a) 7_1 (b) 7_2 (c) 7_3 (d) 7_4 (e) 7_5 (f) 7_6 (g) 7_7

6. Use Proposition 12.3.7 to find the Jones polynomial of the knot of Figure 12.31.

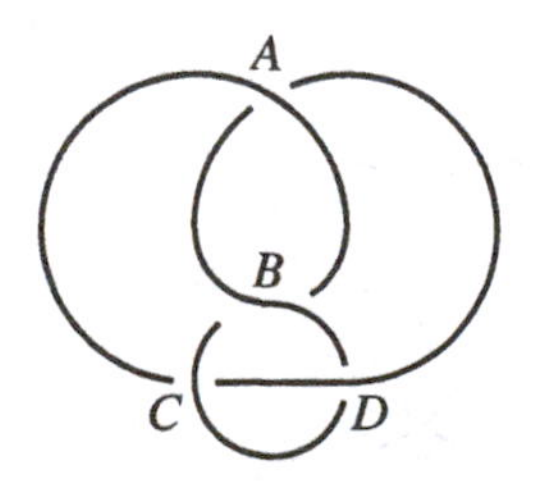

Figure 12.30. Some links

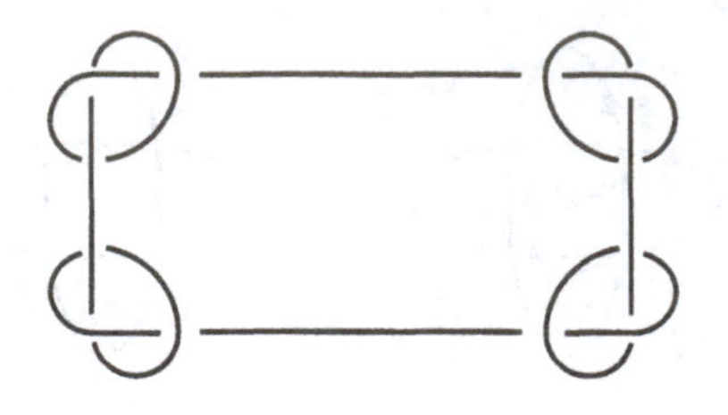

Figure 12.31. A knot

Figure 12.32. A generalization of the knots 3_b and 4

7. Prove that the sum of two oriented knots $\vec{K}_1$ and $\vec{K}_2$ is a well-defined oriented knot.
8. Let $\vec{K}_n$ be the oriented knot given by the n-crossing diagram of Figure 12.32. Show that $\vec{K}_3$ and $\vec{K}_4$ are equivalent to the knots 3_b and 4 and compute $V(\vec{K}_n)$ for all values of n.
9. Use Proposition 12.3.7 to find the Jones polynomial of the n-component oriented link of Figure 12.33.
10. Use Proposition 12.3.6 to prove that $V(0^n) = (-x^2 - x^{-2})^{n-1}$.
11. Show by means of an example that the sum of two unoriented knots need not be well defined.
12. Show by means of an example that the sum of two oriented links need not be well defined.
13. Construct two inequivalent oriented links that have the same Jones polynomials.

Figure 12.33. An oriented links with n components

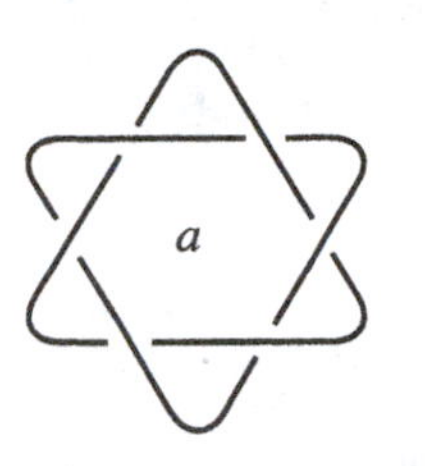

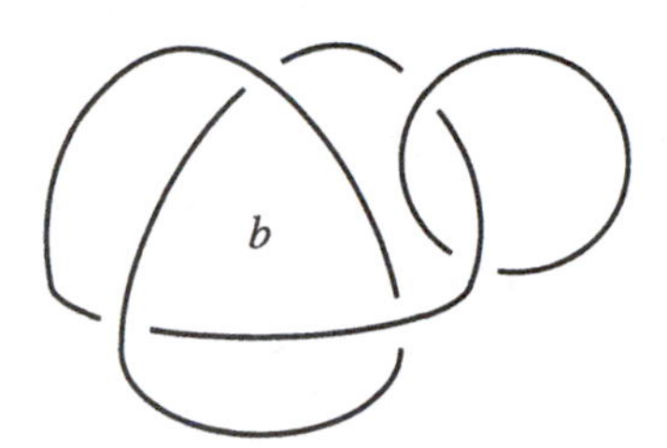

Figure 12.34.

Chapter Review Exercises

1. Show that 6_3 is its own obverse.
2. Show that link a of Figure 12.34 has no p-labelings.
3. Show that link b of Figure 12.34 has a 3-labeling.
4. Find the Jones polynomials of all the orientations of link a of Figure 12.34.
5. Find the Jones polynomials of all the orientations of link b of Figure 12.34.
6. Are the following statements true or false? Justify your answers.

   (a) Every two knots are homeomorphic.
   (b) Every two knots are isotopic.
   (c) Every two links are homeomorphic.
   (d) Every two 2-component links are homeomorphic.
   (e) Every knot diagram has a p-labeling for some odd prime p.
   (f) Every split diagram has a p-labeling for some odd prime p.
   (g) Every split diagram has a p-labeling for every odd prime p.
   (h) Every knot has a Jones polynomial.
   (i) Every oriented knot has a Jones polynomial.
   (j) Every knot diagram has a Jones polynomial.
   (k) Every link has a Jones polynomial.
   (l) Every oriented link has a Jones polynomial.
   (m) Isotopic oriented knots have the same Jones polynomials.
   (n) Isotopic oriented links have the same Jones polynomials.
   (o) Homeomorphic knots have the same Jones polynomials.
   (p) Isotopic knots have the same labeling equations.
   (q) The unknot is not a knot.

Appendix A

A Brief Introduction to The Geometer's Sketchpad®

The Geometer's Sketchpad® is a computer application for sketching and manipulating geometrical figures that is distributed by the Key Curriculum Press. When it is opened, a window appears that is called a *sketch*. The sketch has a toolbox with several icons that are the tools for the basic constructions of geometry.

Arrow tool: In order to select a figure, click first on the arrow tool and then on the figure. To select several objects, hold down the shift key while selecting them.

Point tool: To draw a point, click the point tool and then the location where the point is to fall.

Circle tool: To draw a circle, click the circle tool, place the cursor at the location where the center is to fall, and then drag to create the desired radius.

Line tool: Holding the mouse down on this tool reveals a pop-up menu with three options: segment, ray, and (infinite) line. The appropriate tool is selected by dragging to it. Dragging the mouse in the window then creates the desired line.

Text tool: This is the tool for manipulating labels. When this tool is chosen, selecting any unlabeled object will create a label, and selecting a labeled object will hide the label.

Erasing: To erase an object, select it and then drag down to the appropriate **Hide** command in the **Display** menu.

Measurement: To measure any aspect of a figure, select that figure and then drag down to the appropriate function in the **Measure** menu (an angle is measured by selecting three points on its sides; to measure the area of a selected polygon first fill it with **Polygon Interior** from the **Construct** menu). The value of the function will appear in the upper left-hand corner of the window. In order to algebraically manipulate such measurements, display the calculator in the Measure and select the desired measurement.

More constructions:

1. A variety of standard constructions can be performed by the **Construct** menu. Select the objects to which the construction is to be applied and then drag down in the menu to the appropriate command.

2. Users can encode and keep their own constructions as **scripts**. To create a script, use the **File** menu to open both a new sketch and a new script. Draw and select some points in the sketch that are to serve as the *input* for the construction, and click the **REC** button in the script window. Apply the construction to these points in the sketch (note that *Sketchpad* automatically records the steps of the construction in the script). Signify the completion of the construction by clicking the **STOP** button in the script. Save the script. In order to apply this construction script to any set of points, select these points, open the script, and click the **PLAY** button in the script. If the script concludes with the selection of a set of points, then these are termed the *output*. If the number of points in the output equals that of the input, the script can be reapplied to the output. Such scripts are said to be **reapplicable**.

Example A.1. The following is the script that takes two (selected) points at its input and constructs an equilateral triangle on them.

Given:

1. Point A

2. Point B

Steps:

1. Let [j] = Segment between Point B and Point A.

2. Let [c1] = Circle with center at Point B passing through point A (hidden).
3. Let [c2] = Circle with center at Point A passing through point B (hidden).
4. Let [C] = Intersection of Circle [c2] and Circle [c1].
5. Let [k] = Segment between Point [C] and Point B.
6. Let [l] = Segment between Point [C] and Point A.

Exercises

Use Sketchpad to verify the following propositions.

1. The three perpendicular bisectors of the sides of a triangle are concurrent.
2. The three bisectors of the interior angles of a triangle are concurrent.
3. The three medians of a triangle are concurrent.
4. The three altitudes of a triangle are concurrent.
5. The line joining the midpoints of two sides of a triangle is parallel to the third side and has half its length.
6. The triangle formed by joining the midpoints of the sides of a given triangle has area one fourth that of the original triangle.
7. The quadrilateral formed by joining the midpoints of the successive sides of an arbitrary quadrilateral is a parallelogram.

Exploration Exercises

Use Sketchpad to locate and/or investigate the special properties of the points or figures specified.

1. That point the sum of whose distances from the vertices of a triangle is minimum
2. That point the sum of whose distances from the sides of a triangle is minimum
3. A triangle of minimum perimeter with its vertices on the sides of a given triangle
4. A triangle of minimum area with its vertices on the sides of a given triangle
5. Triangle $A'B'C'$, where ABC is a counterclockwise triangle and $\triangle A'BC$, $\triangle AB'C$, $\triangle ABC'$ are clockwise equilateral triangles.

6. Triangle $A''B''C''$, where A'', B'', C'' are the respective centers of the equilateral triangles of Exercise 5
7. The centers of the squares constructed on the sides of (and outside of) a convex quadrilateral
8. Replace the equilateral triangles of Exercise 6 with similar triangles.
9. Replace the squares of Exercise 7 with similar rectangles.
10. The quadrilateral formed by the perpendicular bisectors of a given quadrilateral
11. The quadrilateral formed by the angle bisectors of a given quadrilateral
12. That point on a given line the sum of whose distances from two points not on the line is minimum

Appendix B

Summary of Propositions

1.1.1	Spherical geodesics
1.1.2	Spherical trigonometry
1.1.5	Angles of a spherical triangle
1.1.6	Area of a spherical triangle
1.2.1	Hyperbolic geodesics
1.2.2	Hyperbolic trigonometry
1.2.5	Angles of a hyperbolic triangle
1.2.7	Area of a hyperbolic triangle
1.3.1	Betweenness in taxicab geometry
2.3.1	Equilateral triangles
2.3.2	Copy a segment
2.3.3	Lay off a segment
2.3.4	SAS
2.3.5	Equal sides $\Rightarrow$ equal angles
2.3.6	Equal angles $\Rightarrow$ equal sides
2.3.7	Lemma for 2.3.8
2.3.8	SSS
2.3.9	Bisection of angle
2.3.10	Bisection of segment
2.3.11	Raise perpendicular
2.3.12–14	Perpendicular bisector
2.3.15	Drop perpendicular
2.3.17	Supplementary angles equal 180°
2.3.18	180° $\Rightarrow$ supplementary angles
2.3.19	Vertically opposite angles

2.3.20	Exterior angle $>$ opposite interior angle
2.3.21	Sum of two angles in triangle $<180^\circ$
2.3.22	$a > b \Rightarrow \alpha > \beta$
2.3.23	$\alpha > \beta \Rightarrow a > b$
2.3.24	The perpendicular is the shortest line
2.3.25	$a + b > c$
2.3.26	Inside line is shorter
2.3.27	$a + b > c \Rightarrow$ triangle exists
2.3.28	Copy an angle
2.3.29	ASA
2.3.30	AAS
2.3.31	SSA for right triangles
2.3.32–34	Angle bisector
2.3.35	Equal alternate angles $\Rightarrow$ parallel lies
2.3.36	Supplementary interior, equal corresponding angles $\Rightarrow$ parallel
2.3.37	Existence of parallels
3.1.1	Parallel lines $\Rightarrow$ equal or supplementary angles
3.1.2	Equidistant straight lines
3.1.3	Parallelism is transitive
3.1.4	Construction of parallel lines
3.1.5	Uniqueness of parallel lines
3.1.6	Exterior angle equals sum of interior angles
3.1.6	Sum of three angles of triangle
3.1.7	Opposite and equal sides $\Rightarrow$ parallelogram
3.1.8	Parallelograms have equal sides & angles
3.2.1	Area of rectangle $= ab$
3.2.2	Area of parallelogram $= bh$
3.2.3–4	Equal parallelograms
3.2.5	Area of triangle $= bh/2$
3.2.6–7	Equal triangles
3.2.8–9	Equal triangles and bases $\Rightarrow$ equal altitudes
3.2.10	Parallelogram $=$ double of triangle
3.2.11	Area of spherical lune
3.2.12	Area of spherical triangle
3.2.13	Convert triangle to parallelogram
3.2.14	Convert parallelogram to parallelogram
3.2.15	Convert triangle to parallelogram
3.2.16	Convert polygon to parallelogram

3.3.1	Construct a square
3.3.2	Pythagoras
3.3.3	Converse of Pythagoras
3.3.4	Spherical Pythagoras
3.3.5	Hyperbolic Pythagoras
3.4.1	Golden ratio
3.4.2–3	Law of cosines
3.4.4	Convert rectangle to square
3.5.1	Distributive law
3.5.2–5	Laws of proportions
3.5.6	Parallel lines $\Leftrightarrow$ proportional cuts
3.5.7	AAA similarity
3.5.8	SSS similarity
3.5.9	SAS similarity
4.1.1	Equal chords $\Leftrightarrow$ equal arcs $\Leftrightarrow$ equal angles
4.1.2	Semicircles are equal
4.1.3	Radius bisects chord
4.1.4	Tangent to circle
4.1.5	Proportional arcs and angles
4.2.1	Central angle equals twice the angle at the circumference
4.2.2	Angles at the circumference are equal
4.2.3	Angle on diameter equals right angle
4.2.4	Tangent and chord
4.2.5	Tangent and secant
4.2.6	Angles in cyclic quadrilateral
4.2.7	Perpendicular bisectors of Δ are concurrent
4.2.8	Circumscribed circle
4.2.9	Angle bisectors of Δ are concurrent
4.2.10	Inscribed circle
4.3.1	Regular hexagon
4.3.2–3	$36°$, $72°$,$72°$ triangle
4.3.4	Regular pentagon
4.3.5	Regular polygons
4.4.1–6	Circumference and area of circle
4.4.7–8	Estimating π
4.5.1	Constructible configurations
4.5.2	Hippasian solutions of cubics
4.5.3	The rational zero theorem

5.1.1–4	Construction and uniqueness of division
5.2.1	Theorem of Menelaus
5.2.2	Theorem of Ceva
5.2.3	Theorem of Pappus
5.2.4	Theorem of Desargues
5.2.5	Theorem of Pascal
5.3.1–2	Projective points and lines
5.3.4	Theorem of Menelaus in the projective plane
6.1.1	The composition of translations is a translation
6.1.2	Rigid motions preserve straight lines
6.1.3	Agreement at two points $\Rightarrow$ agreement on their line
6.1.4	Agreement at three noncollinear points $\Rightarrow$ agreement on plane
6.1.5	Fixing three noncollinear points $\Rightarrow$ identity
6.2.1–2	Composition of reflections
6.2.3	Composition of rotations
6.2.5	Composition of translations with rotations
6.3.1,3,5	Composition of glide reflections
6.4.1–2	Every rigid motion is the composition of at most three reflections
6.4.3	The rigid motions are translations, rotations, or glide reflections
6.6.1	The classification of the frieze patterns
6.7.1	There are exactly 17 wallpaper symmetry groups
7.1.1	The effect of inversions on straight lines and circles
7.1.5	Inversions preserve angles
7.2.3	Ptolemy's theorem
7.3.1	Hyperbolic rigid motions
8.1.1	Euler's equation for polyhedra: $v - e + f = 2$
8.2.1	The composition of rotations with intersecting axes is a rotation
8.2.2	The rotation group of the cube
8.2.3	The rotation group of the tetrahedron
8.2.4	The rotation group of the octahedron
10.1.1	The sum of the degrees in a graph equals twice the number of arcs

10.2.2	The characterization of Eulerian graphs
10.2.3	A sufficient condition for a simple graph to be Hamiltonian
10.3.1	A graph is 2-colorable $\Leftrightarrow$ it has no odd cycles
10.3.2	Every loopless k-degenerate graph is $(k+1)$-colorable
10.3.3	If a graph contains the complete graph K_{k+1} as a subgraph, then it is not k-colorable
10.4.1	The Jordan curve theorem
10.4.2	Euler's equation for plane graphs: $p - q + r = 2$
10.4.5	The graphs K_5 and $K_{3,3}$ are not planar
10.4.6	The containment of K_5 and $K_{3,3}$ characterizes nonplanar graphs
10.4.9	Every plane map is 6-colorable
10.5.2	Homeomorphic graphs have identical reduced degree sequences
11.2.11	Surfaces are classified by the Euler and orientability characteristics
11.2.12	Every closed surface is homeomorphic to exactly one of the canonical surfaces
11.2.13	The Euler characteristics of the canonical surfaces
11.2.14	The closed nonorientable surfaces cannot be embedded in $\mathbb{R}^3$
11.3.1–4	Sums of closed surfaces
11.3.5	Every closed nonorientable surface can be realized in $\mathbb{R}^4$
11.3.6	The sum of nonorientable surfaces
11.3.7	Handle addition
11.3.9,10	Cross-handle addition
11.3.11,12	Cross-cap addition
11.4.1	The classification of bordered surfaces
11.4.5	Twisted bands and nonorientability
12.1.1	Links with identical diagrams are equivalent
12.2.1	p-Colorability is an isotopy invariant
12.2.2	Trefoils are not unknots
12.2.6,7	The Borromean rings are not splittable
12.3.1–5	The obverse trefoil knots have different Jones polynomials and are not equivalent
12.3.6	The Jones polynomial of the disjoint union of links
12.3.7	The Jones polynomial of the sum of links
12.3.8	The Jones polynomial of obverse links

Appendix C

George D. Birkhoff's Axiomatization of Euclidean Geometry

I. The points on any straight line can be numbered so that number differences measure distances.

II. There is one and only one straight line through two distinct given points.

III. All rays emanating from the same end point can be numbered so that number differences (mod 2π) measure angles.

IV. All straight angles have the same measure.

V. Two triangles are similar if an angle of one equals an angle of the other and the sides including these angles are proportional.

Appendix D

The University of Chicago School Mathematics Project's Geometrical Axioms

I. Point-Line-Plane Postulate:

a. Unique Line Assumption: Through any two points, there is exactly one line.

b. Dimension Assumption: Given a line in a plane, there exists a point in the plane not on the line. Given a plane in space, there exists a point in space not on the plane.

c. Number Line Assumption: Every line is a set of points that can be put in a one-to-one correspondence with the real numbers, with any point on it corresponding to 0 and any other point corresponding to 1.

d. Distance Assumption: On a number line there is a unique distance between two points.

II. Triangle Inequality Postulate:

The sum of the lengths of two sides of any triangle is greater than the length of the third side.

III. Angle Measure Postulate:

a. Unique Measure Assumption: Every angle has a unique measure form 0° to 180°.

b. Two Sides of Line Assumption: Given any ray $\overrightarrow{VA}$ and any number x between 0 and 180, there are unique rays $\overrightarrow{VB}$ and $\overrightarrow{VC}$ such that BC intersects line $\overleftrightarrow{VA}$ and $m\angle BVA = m\angle CVA = x°$.

c. Zero Angle Assumption: If $\overrightarrow{VA}$ and $\overrightarrow{VB}$ are the same ray, then $m\angle AVB = 0°$.

d. Straight Angle Assumption: If $\overrightarrow{VA}$ and $\overrightarrow{VB}$ are opposite rays, then $m\angle AVB = 180°$.

e. Angle Addition Property: If $\overrightarrow{VC}$ (except for point V) is in the interior of $\angle AVB$, then $m\angle AVC + m\angle CVB = m\angle AVB$.

IV. Corresponding Angles Postulate:

If two coplanar lines are cut by a transversal so that two corresponding angles have the same measure, then the lines are parallel.

V. Parallel Lines Postulate:

If two lines are parallel and are cut by a transversal, corresponding angles have the same measure.

VI. Reflection Postulate:

Under a reflection,

a. There is a one-to-one correspondence between points and their images.

b. If three points are collinear, then their images are collinear.

c. If B is between A and C, then the image of B is between the images of A and C.

d. The distance between two preimages equals the distance between their images.

e. The image of an angle is an angle of the same measure.

VII. Area Postulate:

a. Uniqueness Property: Given a unit region, every polygonal region has a unique area.

b. Rectangle Formula: The area of a rectangle with dimensions l and w is lw.

c. Congruence Property: Congruent figures have the same area.

d. Additive Property: The area of the union of two nonoverlapping regions is the sum of the areas of the regions.

VIII. Volume Postulates:

a. Uniqueness Property: Given a unit cube, every polyhedral solid has a unique volume.

b. Box Volume Formula: The volume of a box with dimensions l, w and h is lwh.

c. Congruence Property: Congruent figures have the same volume.

d. Additive Property: The volume of the union of two nonoverlapping solids is the sum of the volumes of the solids.

e. Cavalieri's Principle: Let I and II be two solids included between two parallel planes. If every plane P parallel to the given planes intersects I and II in sections with the same area, then volume(I) = volume(II).

Appendix E

David Hilbert's Axiomatization of Euclidean Geometry

0. Undefined Elements and Relations

A class of undefined objects called points, denoted by $A, B, C, \ldots$.

A class of undefined objects called lines, denoted by $a, b, c, \ldots$.

A class of undefined objects called planes, denoted by $\alpha, \beta, \gamma, \ldots$.

Relations: *Incidence* (being *incident, lying on, containing*); being *between* (for points on a line), *congruence*.

I. Axioms of Incidence

I.1. For every two points A and B there exists a line that contains each of the points A, B.

I.2. For every two points A and B there exists no more than one line that contains each of the points A, B.

I.3. There exist at least two points on a line. There exist at least three points that do not lie on a line.

I.4. For any three points A, B, C that do not lie on the same line there exists a plane α that contains each of the points A, B, C. For every plane there exists a point which it contains.

I.5. For any three points A, B, C that do not lie on the same line there exists no more than one plane that contains each of the points A, B, C.

I.6. If two points A and B of a line a lie on a plane α, then every point of a lies in the same plane α.

I.7. If two planes α, β have a point A in common, then they have at least one more point B in common.

I.8. There exist at least four points which do not lie in a plane.

II. Axioms of Order

II.1. If a point B lies between a point A and a point C then the points A, B, C are three distinct points of a line, and B also lies between C and A.

II.2. For two points A and C, there always exists at least one point B on the line AC such that C lies between A and B.

II.3. Of any three points on a line there exists no more than one that lies between the other two.

DEFINITIONS: Segment, point of segment, inside of segment, outside of segment, side of line in a plane, ray, polygon, side of polygon, vertex of polygon, triangle, quadrilateral, n-gon.

II.4. Let A, B, C be three points that do not lie on a line and let a be a line in the plane ABC which does not meet any of the points A, B, C. If the line a passes through a point of the segment AB, it also passes through a point of the segment AC or through a point of the segment BC.

III. Axioms of Congruence

III.1. If A, B are two points on a line a, and A' is a point on the same or on another line a', then it is always possible to find a point B' on a given side of the line a' through A' such that the segment AB is congruent to the segment $A'B'$. In symbols $AB \equiv A'B'$.

III.2. If a segment $A'B'$ and a segment $A''B''$ are congruent to the same segment AB, then the segment $A'B'$ is also congruent to the segment $A''B''$.

III.3. On a line let AB and BC be two segments which except for B have no point in common. Furthermore, on the same or another line a' let $A'B'$ and $B'C'$ be two segments which except for B' also have no point in common. In that case, if

$$AB \equiv A'B' \quad \text{and} \quad BC \equiv B'C'$$

then

$$AC \equiv A'C'.$$

DEFINITION: Angle.

III.4. Let $\angle(h,k)$ be an angle in a plane α and a' a line in a plane α' and let a definite side of a' in α' be given. Let h' be a ray on the line a' that emanates from the point O'. Then there exists in the plane α' one and only one ray k' such that the angle $\angle(h,k)$ is congruent to the angle $\angle(h',k')$ and at the same time all interior points of the angle $\angle(h',k')$ lie on the given side of a'.

III.5. If for two triangles ABC and $A'B'C'$ the congruences

$$AB \equiv A'B', \quad AC \equiv A'C', \quad \angle BAC \equiv \angle B'A'C'$$

hold, then the congruence

$$\angle ABC \equiv \angle A'B'C'$$

is also satisfied.

IV. Axiom of Parallels

IV.1. Let a be any line and A a point not on it. Then there exists at most one line in the plane determined by a and A that passes through A and does not intersect a.

V. Axioms of Continuity

V.1. **(Axiom of Archimedes)** If AB and CD are any segments then there exists a number n such that n segments [congruent to] CD constructed contiguously from A, along a ray from A through B, will pass beyond B.

V.2. **(Axiom of completeness)** An extension of a set of points on a line with its order and congruence relations existing among the original elements as well as the fundamental properties of line order and congruence that follows from Axioms I–III and from V.1 is impossible.

Appendix F

Permutations

If S is a finite set, then a *permutation* of S is a function $f : S \to S$ that has the following two properties:

1. If a and b are distinct elements of S, then $f(a)$ and $f(b)$ are also distinct elements of S;

2. For every element y of S there is an element x of S such that $y = f(x)$.

It is customary to display permutations as a collection of cycles. A *cycle* of a permutation f is a cyclic sequence

$$(a_1 \quad a_2 \quad \dots \quad a_k)$$

where

$$a_{i+1} = f(a_i) \qquad \text{for } i = 1, 2, \dots, k-1$$

and

$$a_1 = f(a_k).$$

Example F.1 If $S = \{1, 2, 3, 4, 5, 6, 7\}$ and $f(1) = 6$, $f(2) = 5$, $f(3) = 7$, $f(4) = 4$, $f(5) = 3$, $f(6) = 1$, $f(7) = 2$, then $(1 \quad 6)$, $(5 \quad 3 \quad 7 \quad 2)$, and (4) are cycles of f, as are $(6 \quad 1)$ and $(3 \quad 7 \quad 2 \quad 5)$. However, since cycles are, by their definition, cyclically ordered, it follows that

$$(1 \quad 6) = (6 \quad 1)$$

and

$$(5 \quad 3 \quad 7 \quad 2) = (3 \quad 7 \quad 2 \quad 5) = (7 \quad 2 \quad 5 \quad 3) = (2 \quad 5 \quad 3 \quad 7).$$

Hence $(1\quad 6)$, $(5\quad 3\quad 7\quad 2)$, and (4) are the complete set of cycles of f and we write

$$f = (1\quad 6)(5\quad 3\quad 7\quad 2)(4).$$

The order of the cycles is immaterial. Thus,

$$\begin{aligned} f &= (1\quad 6)(5\quad 3\quad 7\quad 2)(4) = (5\quad 3\quad 7\quad 2)(4)(1\quad 6) \\ &= (3\quad 7\quad 2\quad 5)(4)(6\quad 1) = \cdots . \end{aligned}$$

Example F.2 If $S = \{1, 2, 3, 4, 5, 6, 7, 8, 9, a, b, c\}$ and $f(1) = a$, $f(2) = 1$, $f(3) = c$, $f(4) = 8$, $f(5) = 9$, $f(6) = 7$, $f(7) = 3$, $f(8) = 4$, $f(9) = 6$, $f(a) = 2$, $f(b) = b$, $f(c) = 5$, then

$$f = (2\quad 1\quad a)(7\quad 3\quad c\quad 5\quad 9\quad 6)(4\quad 8)(b).$$

If f and g are permutations of the same set S, then the *composition* $f \circ g$ is also a permutation of S such that

$$(f \circ g)(x) = f(g(x)) \qquad \text{for all } x \text{ in } S.$$

Example F.3 If $f = (1\quad 6)(5\quad 3\quad 7\quad 2)(4)$ and $g = (1\quad 7\quad 2\quad 6\quad 3\quad 5\quad 4)$, then

$$\begin{aligned} (f \circ g)(1) &= f(g(1)) = f(7) = 2 \\ (f \circ g)(2) &= f(g(2)) = f(6) = 1 \\ (f \circ g)(3) &= f(g(3)) = f(5) = 3 \\ (f \circ g)(4) &= f(g(4)) = f(1) = 6 \\ (f \circ g)(5) &= f(g(5)) = f(4) = 4 \\ (f \circ g)(6) &= f(g(6)) = f(3) = 7 \\ (f \circ g)(7) &= f(g(7)) = f(2) = 5. \end{aligned}$$

Consequently, $f \circ g = (1\quad 2)(3)(4\quad 6\quad 7\quad 5)$. Similarly,

$$\begin{aligned} (g \circ f)(1) &= g(f(1) = g(6) = 3 \\ (g \circ f)(2) &= g(f(2) = g(5) = 4 \\ (g \circ f)(3) &= g(f(3) = g(7) = 2 \\ (g \circ f)(4) &= g(f(4) = g(4) = 1 \\ (g \circ f)(5) &= g(f(5) = g(3) = 5 \\ (g \circ f)(6) &= g(f(6) = g(1) = 7 \\ (g \circ f)(7) &= g(f(7) = g(2) = 6. \end{aligned}$$

Consequently, $g \circ f = (1\quad 3\quad 2\quad 4)(5)(6\quad 7)$.

Exercises F

Rewrite the functions of Exercises 1–5 in terms of their cycles.

1. $f(1) = 6,\ f(2) = 5,\ f(3) = 7,\ f(4) = 2,\ f(5) = 3,\ f(6) = 1,\ f(7) = 4$
2. $f(1) = 6,\ f(2) = 5,\ f(3) = 7,\ f(4) = 8,\ f(5) = 3,\ f(6) = 1,\ f(7) = 2,\ f(8) = 4$
3. $f(1) = 9,\ f(2) = 5,\ f(3) = 7,\ f(4) = 8,\ f(5) = 3,\ f(6) = 1,\ f(7) = 2,\ f(8) = 4,\ f(9) = 6$
4. $f(1) = 9,\ f(2) = 5,\ f(3) = 7,\ f(4) = 8,\ f(5) = 3,\ f(6) = 1,\ f(7) = a,\ f(8) = 4,\ f(9) = 6,\ f(a) = 2$
5. $f(1) = 9,\ f(2) = 5,\ f(3) = b,\ f(4) = 8,\ f(5) = 3,\ f(6) = 1,\ f(7) = a,\ f(8) = 4,\ f(9) = 6,\ f(a) = 2,\ f(b) = 7$
6. Suppose $f = (1\ \ 2\ \ 3\ \ 4\ \ 5\ \ 6\ \ 7\ \ 8\ \ 9)$, $g = (4\ \ 3\ \ 2\ \ 1)(5)(9\ \ 8\ \ 7)\cdot(6)$, $h = (1\ \ 2)(3\ \ 4)(5\ \ 6)(7\ \ 8)(9)$. Display the following compositions in terms of their cycles.

   (a) $f \circ g$ (b) $g \circ f$ (c) $f \circ h$ (d) $h \circ f$ (e) $g \circ h$ (f) $h \circ g$
   (g) $f \circ f$ (h) $g \circ g$ (i) $h \circ h$

Appendix G

Modular Arithmetic

Let n, a, b be any integers. We define

$$a \equiv b (\mathrm{mod}\, n)$$

(pronounced "a is *congruent* to b *modulo* n") provided that $a - b$ is divisible by n [i.e., $(a - b)/n$ is an integer]. For instance,

$$\begin{array}{llll} 7 \equiv 3(\mathrm{mod}\, 4) & 7 \equiv 2(\mathrm{mod}\, 5) & 7 \equiv 1(\mathrm{mod}\, 3) & 7 \equiv 1(\mathrm{mod}\, 2) \\ 12 \equiv 0(\mathrm{mod}\, 4) & 17 \equiv -3(\mathrm{mod}\, 5) & -7 \equiv 1(\mathrm{mod}\, 4) & -7 \equiv -4(\mathrm{mod}\, 3) \\ 7x \equiv 3x(\mathrm{mod}\, 4) & 7x \equiv 2x(\mathrm{mod}\, 5) & 7x \equiv x(\mathrm{mod}\, 3) & -17x \equiv -x(\mathrm{mod}\, 16), \end{array}$$

where x is also any integer.

If p is a prime integer, then *arithmetic modulo* p consists of the integers $\{0, 1, 2, \ldots, p-1\}$, subject to the usual four arithmetic operations, with the added stipulation that equality is to be replaced by congruence. Thus, in arithmetic *modulo 5*,

$$3 + 4 \equiv 2 \quad 3 \cdot 4 \equiv 2 \quad 3 - 4 \equiv 4 \quad 4x + 4x \equiv 3x$$

and

$$2 \div 4 \equiv 3 \qquad \text{(because } 3 \cdot 4 \equiv 2\text{)},$$

whereas in arithmetic *modulo 7*,

$$3 + 4 \equiv 0 \quad 3 \cdot 4 \equiv 5 \quad 3 - 4 \equiv 6 \quad 4x + 4x \equiv x$$

and

$$5 \div 4 \equiv 3 \qquad \text{(because } 3 \cdot 4 \equiv 5\text{)}.$$

Arithmetic modulo a prime p is very similar to the standard arithmetic. The operations of addition and multiplication are commutative, associative, and distributive. Additive and multiplicative identities and inverses exist and are unique. The additive inverse of a is, of course, $p - a$. The multiplicative inverse, on the other hand, is harder to find. These inverses are tabulated here for $p = 3, 5, 7, 11, 13, 17, 19$:

$$
\begin{aligned}
&1 \cdot 1 \equiv 2 \cdot 2 \equiv 1 (\bmod 3) \\
&1 \cdot 1 \equiv 2 \cdot 3 \equiv 4 \cdot 4 \equiv 1 (\bmod 5) \\
&1 \cdot 1 \equiv 2 \cdot 4 \equiv 3 \cdot 5 \equiv 6 \cdot 6 \equiv 1 (\bmod 7) \\
&1 \cdot 1 \equiv 2 \cdot 6 \equiv 3 \cdot 4 \equiv 5 \cdot 9 \equiv 7 \cdot 8 \equiv 1 (\bmod 11) \\
&1 \cdot 1 \equiv 2 \cdot 7 \equiv 3 \cdot 9 \equiv 4 \cdot 10 \equiv 5 \cdot 8 \equiv 6 \cdot 11 \equiv 12 \cdot 12 \equiv 1 (\bmod 13) \\
&1 \cdot 1 \equiv 2 \cdot 9 \equiv 3 \cdot 6 \equiv 4 \cdot 13 \equiv 5 \cdot 7 \equiv 8 \cdot 15 \equiv 10 \cdot 12 \equiv 11 \cdot 14 \\
&\qquad \equiv 16 \cdot 16 \equiv 1 (\bmod 17) \\
&1 \cdot 1 \equiv 2 \cdot 10 \equiv 3 \cdot 13 \equiv 4 \cdot 5 \equiv 6 \cdot 16 \equiv 7 \cdot 11 \equiv 8 \cdot 12 \equiv 9 \cdot 17 \equiv 14 \cdot 15 \\
&\qquad \equiv 18 \cdot 18 \equiv 1 (\bmod 19).
\end{aligned}
$$

Example G.1 Solve the modular equation $5x \equiv 4 (\bmod 17)$. Here

$$x \equiv 4 \cdot 5^{-1} \equiv 4 \cdot 7 = 28 \equiv 11 (\bmod 17).$$

Exercises G

1. Evaluate the following in modulo 5 arithmetic:

   (a) $2 + 4$ (b) $2 \cdot 4$ (c) $2 - 4$ (d) $3 + 4$
   (e) $3 - 4$ (f) $3 \cdot 4$ (g) $3 \cdot 3$ (h) $4 \cdot 4$

2. Evaluate the following in modulo 7 arithmetic:

   (a) $4 + 5$ (b) $4 - 5$ (c) $4 \cdot 5$ (d) $3 + 5$
   (e) $3 - 5$ (f) $3 \cdot 5$ (g) $4 \cdot 4$ (h) $5 \cdot 5$

3. Evaluate the following in modulo 11 arithmetic:

   (a) $4 + 8$ (b) $4 - 8$ (c) $4 \cdot 8$ (d) $3 + 10$
   (e) $3 - 10$ (f) $3 \cdot 10$ (g) $4 \cdot 4$ (h) $5 \cdot 5$

4. Simplify the following expressions in modulo 5 arithmetic:

   (a) $2x + 4x$
   (b) $2x + 3y - 3x - y$
   (c) $2x - 3y - (4x - y)$

(d) $3x - y + 4z + 2(x - 2y + 3z)$

(e) $3(2x - 3y + 4z) - 2(4x - y + 3z)$

5. Simplify the following expressions in modulo 7 arithmetic:

(a) $2x + 6x$

(b) $2x + 3y - 3x - y$

(c) $2x - 3y - (4x - y)$

(d) $3x - y + 4z + 3(x - 2y + 3z)$

(e) $4(2x - 3y + 4z) - 3(4x - y + 3z)$

6. Simplify the following expressions in modulo 3 arithmetic:

(a) $2x + 2x$

(b) $2x + y - x - 2y$

(c) $2x - y - (2x - y)$

(d) $2x - y + 2z + 2(x - 2y + z)$

(e) $2(2x - y + 2z) - (2x - y + 2z)$

Solutions and Hints to Selected Problems

EXERCISES 1.1

2a. $.92169R$, $1.04393R$, $1.13347R$

3b. $51.1°$, $67.3°$,$122.8°$

4a. $\alpha = 48.6°$, $b = 0.58683R$, $c = 0.688113R$

5a. $c = 1.93661R$; $\alpha = 106.5°$, $\beta = 62.5°$

6a. $77.4°$, $91.7°$, $106.1°$

7a. $\pi/3$

EXERCISES 1.2

2a. 1.32582, 1.221, 1.0771

3b. $7.3°$, $20.6°$, $73.1°$

4a. $\alpha = 75.8°$, $b = 0.45008$, $c = 0.33893$

6a. $c = 1.30847$; $\alpha = 103.2°$, $\beta = 18.4°$

EXERCISES 1.3

1a. $30° = \pi/6$ radians

5b. Perimeter = 22, curvature = 0

10b. Perimeter = 16, curvature = 0

EXERCISES 1. REVIEW

1. (a) 1.5 (b) 1.45242 (c) 1.54612

5. (a) Euclidean triangles: 0.433 and 0.108; spherical triangles: 0.496 and 0.105, hyperbolic triangles: 0.385, 0.111

7. (e) True (j) False (n) False

EXERCISES 2.2A

1. If O is the center of the sphere, rotate the sphere around an axis through O perpendicular to both OA and OB.

EXERCISES 2.2B

2c. The maxi circle looks like a Euclidean square with sides parallel to the coordinate axes.

EXERCISES 2.3A

1. Using $CD = AB$, modify the construction of PN 2.3.3 so as to add CD to AB rather than subtract.

7. (a) Valid (b) Valid (c) Valid (d) Valid

EXERCISES 2.3B

2. Apply $\triangle ABC$ to $\triangle DEF$ so that BC falls on DE. Conclude that AB and AC fall along DE and DF respectively.

6. Rotate the triangle slightly around one of its vertices.

EXERCISES 2.3C

3. Show that $\triangle ABD$ and $\triangle ACE$ are congruent.

EXERCISES 2.3E

2. (a) Valid (b) Valid (c) Not valid (d) Not valid

EXERCISES 2.3G

2. Suppose in $\triangle ABC$, $AB = AC$ and BE is the bisector of $\angle ABC$. Let D be a point on AB such that $AD = AE$. Prove that CD is the bisector of $\angle ACB$.

3. Show that the triangles so formed are congruent.

4. (a) Valid (b) Valid (c) Not valid (d) Not valid

EXERCISES 2.3H

1. Show that the two triangles standing on the halves of the base of the given isosceles triangle are congruent.

EXERCISES 2.3I

2. Join the given point to the midpoint of the proposed line segment and prove that this new line is perpendicular to it.

8. Start with a right angle with vertex D and on one side find B such that $DB = h_b$. Let C be the intersection of $(B; a)$ with the right angle's second side. If E is the midpoint of BC, let A be the intersection of $(E; m_a)$ with the right angle's second side.

12. First construct an auxiliary $\triangle BCD$ such that $BC = a$, $\angle CBD = \beta$, and $BD = b+c$. Then let A be the intersection of BD with the perpendicular bisector to CD.

13. First construct an auxiliary $\triangle BCD$ such that $BC = a$, $\angle BCD = \gamma$, and $CD = b-c$. Then let A be the intersection of the extension of CD with the perpendicular bisector to BD.

EXERCISES 2.3K

3. Extend the median by its own length.

EXERCISES 2.3L

3. If there were two, they would form a triangle in which both an exterior angle and an opposite interior angle are right.

4. (a) Not valid (b) Valid (c) Valid (d) Valid

EXERCISES 2.3M

4. Observe that if the altitude from a vertex falls outside the triangle then one of the interior angles not at that vertex is obtuse. Now use Exercise 1.

EXERCISES 2.3N

1. $\angle BDA > \angle CAD = \angle BAD$.

2. Let D be a point on AC such that $AD = A'C'$. Then $\triangle ABD \cong \triangle A'B'C$ so that if $D \neq C$, then $\triangle BCD$ is isosceles. PNs 2.3.20 and 2.3.23 can then be used to obtain a contradiction of $AB < BC$.

7. Extend BE by its own length.

8. Use Exercise 7.

9. Use Exercise 2.3K.4

EXERCISES 2.3O

4. Extend the median by its own length.

9. Extend BA to a point F such that $AF = AC$.

10. Use the previous exercise.

EXERCISES 2.3P

4. If $a \geq b$, then x can be any number between $a + b$ and $a - b$.

11. (a) Valid (b) Valid (c) Valid (d) Valid

EXERCISES 2.3Q

5. Construct $\triangle DEG \cong \triangle ABC$.

6. Use a proof by contradiction.

EXERCISES 2.3R

1. Assume the triangle sides joining the given angles are not equal and obtain a contradiction using PN 2.3.20.

9. (a) Valid (b) Valid (c) Not valid (d) Not valid

EXERCISES 2.3S

1. Draw a diagonal.

EXERCISES 2. REVIEW

1. Observe that BE is the hypotenuse of a right triangle.

7. The diagonals divide the quadrilateral into four triangles. Apply PN 2.3.25 to two of these triangles.

14. Proceed by contradiction.

17. (d) True (h) True (l) False

EXERCISES 3.1A

5. Join the vertices of the two angles.

12. Since the sum of two angles of a triangle is less than two right angles, so is the sum of their halves. Now apply Postulate 5.

14. (a) Not valid (b) Not valid (c) Valid (d) Valid

EXERCISES 3.1C

1. Use PNs 3.1.3–4.

7. "Reflect" $\triangle ABC$ in the side AC.

15. Begin by using SSS to prove some triangles are congruent.

16. Go ahead and assume that every quadrilateral has an interior diagonal.

18. Go ahead and assume that every polygon has an interior diagonal.

23. First construct an auxiliary ΔBCD such that $BC = a$, $CD = b + c$, $\angle BDC = \alpha/2$.

30. Yes.

EXERCISES 3.1D

7. Extend this line segment by its own length and prove that it forms a parallelogram with the third side.

10. Apply Exercise 7 to the triangle formed by the point of intersection and the third side.

11. Use Exercise 10.

14. Use Exercise 7.

15. Through the middle intersection point on one straight line draw a line parallel to the other straight line.

16. Join a vertex to the midpoint of the opposite non-parallel side and extend to intersect the other parallel side. Then use Exercise 7.

19. Use Exercise 7.

21. Use Exercise 12.

EXERCISES 3.2A

6. See Exercise 3.1D.7.

8. Recall that according to Exercise 3.1D.11 these three medians are concurrent.

13. None of these propositions is valid in this context.

EXERCISES 3.2B

2. Compare the triangles on one side of the diagonal AC with their counterparts on the diagonal's other side.

6. First convert the parallelogram into a $\square ABCD$ with the same base, say AB. Next, intersect the side CD with the circle $(A; AB)$.

13. This point is the intersection of the triangle's medians.

EXERCISES 3.3A

3. $c/\sqrt{2}$, $c^2/4$

7. Since $(30/2)^2 + (16/2)^2 = 17^2$, it follows that the diagonals are perpendicular to each other.

12. Show first how to construct a square whose area equals that of two given squares and then use mathematical induction.

EXERCISES 3.3B

2. 1.27456, 0.14130, 0.0141420 compared to 1.41421, 0.141421, 0.0141421

5. No

EXERCISES 3.4B

1. (a) Use the method of Exercise 3.3A.12.

EXERCISES 3.5A

1. Use the method of PN 3.5.2.

EXERCISES 3.5B

6. Extend BA to C' so that $AC' = AC$.

8. Through the intersections on the second side draw lines parallel to the third side, thus forming congruent triangles.

14. See PN 3.4.4.

23. See Exercise 3.1D.10.

EXERCISES 3.5C

4. On one of the sides of the second triangle construct a triangle that is similar to the first one. Then show that the new triangle is in fact congruent to the second one.

5. See hint to Exercise 4.

11. Through B draw a straight line parallel to n.

17. Use Exercise 10 above.

23. Use the previous exercise.

EXERCISE 3. REVIEW

1. Draw a straight line $m \parallel BC$ through P and show that the sum in question is constant as long as P remains on m.

8. Extend CA to D so that $AD = AB$.

11. Show that this perimeter actually equals the sum of the two equal sides of the given triangle.

15. (e) True (j) True (o) True

EXERCISES 4.1A

2. You may find it necessary to use PN 2.3.30 to produce a neutral proof.

3. Apply PN 2.3.25 to the triangle formed by a chord and two radii.

7. Use Exercise 4.

11. The center of this circle is on the perpendicular bisectors of AB and CD.

EXERCISES 4.1B

1. For a neutral proof use PN 2.3.31.

4. Use Exercise 3.

5. Proceed by contradiction and examine the triangle formed by the centers and the contact point.

17. Use Exercise 11.

EXERCISES 4.1C

1. (a) 3.33 inches

6. 18° N

EXERCISES 4.2A

7. Prove that $\triangle PAC$ and $\triangle PDB$ are similar.

15. Use Exercises 3.5B6–7.

18. If O is the center of this arc, then $\triangle AOB$ has angles 30°, 30°, and 120°.

21. (d) First construct an auxiliary $\triangle ADE$ such that $DE = a + b + c$, $\angle DAE = \alpha$, and the altitude from A to DE has length h_a. Then intersect DE with the perpendicular bisectors to AD and AE.

25. (a) Not valid (b) Not valid (c) Not valid (d) Not valid

32. Use Exercise 3.5C.16 and PN 4.2.3.

EXERCISES 4.2B

4. First show that the angle bisectors of similar triangles divide them into respectively similar triangles. Then use Exercise 3.5C.6.

9. Examine the three triangles formed by the center of the circle with the triangle's three sides.

11. Through each vertex of the triangle draw a line parallel to the opposite side to form a new triangle. Then show that the altitudes in question are the perpendicular bisectors of the new triangle.

EXERCISES 4.3

3. Note that $24° = 2 \cdot 12° = 2(72° - 60°)$ is a constructible angle.

6. It is known from the theory of numbers that there exist integers a, b, such that $ag + bh = 1$. It follows that $360/gh = a(360/h) + b(360/g)$ is a constructible angle.

EXERCISES 4.4

3. (a) 10.47 inches, 52.33 square inches

9. Use Exercise 1.

EXERCISES 4.5

5. Make use of Exercise 4.

7. Substitute p/q into the equation $P(x) = 0$ and clear the denominators to obtain $a_n p^n + a_{n-1}p^{n-1}q + \cdots + a_1 pq^{n-1} + a_0 q^n = 0$. Since p divides the first $n - 1$ terms and is relatively prime to q, it must divide a_0. Ditto q.

11. Use the facts: $72° = 90° - 18°$, $3(18°) = 90° - 2(18°)$, as well as Exercise 8.

13. Note that $360°/9 = 40° = 2(20°)$.

17. No

EXERCISES 4. REVIEW

6. If the polygon has n sides, then there are $2n$ triangles each of which has as vertices the center of the circle, a point of contact, and a vertex of the polygon. Prove that these triangles are all congruent.

9. Let O_1, O_2, O_3, be the circumcenters of triangles OBC, OCA, OAB, respectively. Prove that $\triangle ABC \cong \triangle O_1O_2O_3$.

14. Use the circle of Apollonius of Exercise 4.2A.15.

19. (e) True (j) False (o) True

EXERCISES 5.1

9. Let AC be a line segment perpendicular to AB and consisting of n copies of some unit segment. Draw BC and through the $n-1$ division points on AC draw lines parallel to BC.

10. Draw a line segment AQ' perpendicular to AB and equal to PQ. Let X' be a point on AQ' such that $AX' = PX$. Through X' draw a line parallel to $Q'B$.

EXERCISES 5.2A

3. Use Exercise 3.5B.6–7.

4. Use Exercise 3.5B.6–7.

5. Apply the theorem of Menelaus to two different triangles.

10. Apply the theorem of Menelaus to $\triangle ABC$ with each of the transversals $\{P, Q, R\}$, $\{P', Q', R'\}$, $\{Z, Q, R'\}$, $\{Y, Q', P\}$, $\{R, X, P'\}$.

11. Let $X = \overleftrightarrow{BC} \cap \overleftrightarrow{EF}$. Apply the theorem of Menelaus to both $\triangle BEX$ and $\triangle CFX$ with the transversal $\{A, G, M\}$.

12. If $X = m \cap n$, apply the theorem of Menelaus to each of $\triangle AA'X$, $\triangle CC'X$ with each of the lines $\overleftrightarrow{BB'}$ and $\overleftrightarrow{DD'}$.

18. Use the spherical law of sines.

EXERCISES 5.2B

1. A Cevian forms two triangles with the sides of the given triangle. Apply the theorem of Menelaus to these two. Prove the converse in the same indirect manner used to prove the converse part of the theorem of Menelaus.

3. Use Exercise 3.5B.6.

4. Use Exercise 3.5B.7.

6. Each of the three altitudes "divides" the triangle into two triangles and some of these six triangles are similar.

14. Use Exercise 5.2A.18.

17. Use Exercise 5.2A.19.

EXERCISES 5.2C

1. Let $U = \overleftrightarrow{AB'} \cap \overleftrightarrow{BC'}$, $V = \overleftrightarrow{AB'} \cap \overleftrightarrow{A'C}$, $W = \overleftrightarrow{BC'} \cap \overleftrightarrow{A'C}$ and apply the Theorem of Menelaus to this triangle with each of the straight lines $\overleftrightarrow{AC'Q}$, $\overleftrightarrow{B'CR}$, $\overleftrightarrow{A'BP}$, $\overleftrightarrow{ABC}$, $\overleftrightarrow{A'B'C'}$.

2. Apply the theorem of Menelaus to $\triangle BCO$ with line $\overleftrightarrow{B'C'Q}$, $\triangle ACO$ with line $\overleftrightarrow{A'B'R}$, $\triangle ABO$ with line $\overleftrightarrow{A'B'P}$, where O is the common intersection of $\overleftrightarrow{AA'}$, $\overleftrightarrow{BB'}$, $\overleftrightarrow{CC'}$.

3. Apply the first half to $\triangle BB'P$ and $\triangle CC'R$.

4. Set $P = \overleftrightarrow{AB} \cap \overleftrightarrow{DE}$, $Q = \overleftrightarrow{CD} \cap \overleftrightarrow{AF}$, $R = \overleftrightarrow{BC} \cap \overleftrightarrow{EF}$, $U = \overleftrightarrow{CD} \cap \overleftrightarrow{EF}$, $V = \overleftrightarrow{AB} \cap \overleftrightarrow{EF}$, $W = \overleftrightarrow{AB} \cap \overleftrightarrow{CD}$, and apply the theorem of Menelaus to $\triangle UVW$ and each of the lines $\overleftrightarrow{PDE}$, $\overleftrightarrow{AQF}$, $\overleftrightarrow{BCR}$. You will also want to make use of Exercises 4.2A.7–8.

EXERCISES 5.3

4. If $\{A, B, C\}$ and $\{A', B', C'\}$ are two sets of collinear points such that $\overleftrightarrow{AB'} \parallel \overleftrightarrow{A'B}$, then also $QR \parallel \overleftrightarrow{AB'} \parallel \overleftrightarrow{A'B}$, where $Q = \overleftrightarrow{AC'} \cap \overleftrightarrow{A'C}$, and $R = \overleftrightarrow{BC'} \cap \overleftrightarrow{B'C}$.

5. If $\{A, B, C\}$ and $\{A', B', C'\}$ are two sets of collinear points such that $\overleftrightarrow{AB'} \parallel \overleftrightarrow{A'B}$, and $\overleftrightarrow{AC'} \parallel \overleftrightarrow{A'C}$, then also $\overleftrightarrow{BC'} \parallel \overleftrightarrow{B'C}$.

EXERCISES 5. REVIEW

1. Use the theorem of Menelaus twice.

4. See Coxeter and Greitzer's book.

8. (e) False (j) True (l) False

EXERCISES 6.1

4. Let $B' = R_{A,-2\alpha}(B)$ and show that $R_{C,2\gamma} \circ R_{B,2\beta} \circ R_{A,2\alpha}$ fixes the three noncollinear points A, B', C.

8. Construct a triangle $\triangle$ two of whose sides are on m and n and use the fact that $\triangle$ and $f(\triangle)$ are congruent.

EXERCISES 6.2

1. This is a 180° rotation.
6. This is a 90° rotation.
11. This is a translation.
16. This is a 120° rotation.
20. Identity
24. Let n be the straight line through P such that the angle from n to m is $\theta/2$. Then use the fact that $R_{P,\theta} = \rho_m \circ \rho_n$.

EXERCISES 6.3

1. This is a glide reflection.
6. This is a glide reflection.
11. This is a 180° rotation.
16. This is a glide reflection.
21. Show that this composition fixes the triangle's incenter.
26. Assume the lines intersect at P and let the angle from k to m be α. Then it is necessary to prove that $R_{P,-2\alpha} \circ \rho_n = \rho_n \circ R_{P,2\alpha}$. Now use Exercise 6.2.24.

EXERCISES 6.4

1. Use Exercises 6.3.27–28.
2. Let P be an arbitrary point. Show that g fixes P if and only if $f \circ g \circ f^{-1}$ fixes $f(P)$.
7. Yes
10. No

EXERCISES 6.5

1. (c) There are twelve of these.
2. (e) ρ_n
3. (e) ρ_{36}

EXERCISES 6.6

5. Γ_7
10. Γ_5
14. Γ_4
19. Γ_1
26. Γ_3

EXERCISES 6.7

5. pgg
10. $p2$
14. $p4g$
23. $p3m1$
26. $p6m$
27. $p4m$
28. $p3$

EXERCISES 6. REVIEW

2. $y = -x - 1$
5. (f) ρ_{AC}, where $C = (2, 2)$
12. (e) True (j) False (o) True

EXERCISES 7.1

1. (e) The line $y = -2x$ (j) The circle $((0, -1); 1)$ (o) The circle $(O; 16/3)$ (t) The circle $((0, 8/3); 4/3)$ (y) The circle $((3.2, 3.2); 3.2)$

2. (a) No (d) $I_{((0,-5);\sqrt{150})}$

5. The center of the inversion lies on the intersection of the the external common tangents of the given circles.

EXERCISES 7.2

3. Let the circles circumscribing $ABCD$, ABT, CDT, be p, q, r, respectively. Let I be any inversion centered at A. Then $I(p)$ and $I(q)$ are straight lines through $I(B)$ and $I(r)$ is a circle. Let $B' = I(B)$, $T' = I(T)$, and so on. Since $B'T'^2 = B'D' \cdot D'C'$ and B', C', D' are fixed points, it follows that the locus of $T' = I(T)$ is a circle centered at B'. Consequently, the locus of T is also a circle.

5. $P' = I_p(P)$

EXERCISES 7.3

1. $I_{((2,0);\sqrt{10})}$
8. $I_{((15,0);4\sqrt{5})}$

EXERCISE 7. REVIEW

2. No. If I is any inversion and τ any translation, then $I \circ f \circ I^{-1}$ is not an Euclidean rigid motion.

5. (e) False (j) True

EXERCISES 8.1

1. Method I: (a) 12 (b) 24 (c) 8 triangles and 6 squares (d) cuboctahedron (e) $12 - 24 + 14 = 2$. Method II: (a) 24 (b) 36 (c) 8 hexagons and 6 squares (d) truncated octahedron (e) $24 - 36 + 14 = 2$

5. (a) The truncation of cuboctahedron has 24 vertices, 48 edges, 8 triangles, and 18 squares. The truncation of the truncated cube has 36 vertices, 72 edges, 32 triangles and 6 octagons.

10. 16 vertices, 32 edges, 16 quadrilaterals. $16-32+16=0$. The tunnel constitutes a hole.

15. Since the angle of the square is a right angle, each vertex of such a polyhedron must be surrounded by 3 squares.

EXERCISES 8.2

4. (b) (i) $R_{1,-120°}$ (ii) $R_{34,180°}$ (iii) $R_{1265,90°}$ (iv) $R_{2,-120°}$

10. (b) (i) $R_{5,-120°}$ (ii) $R_{5d9g7,-72°}$ (iii) $R_{57,180°}$ (iv) $R_{f,-120°}$

15. 6

EXERCISES 8. REVIEW

1. Method I: There are $2n$ vertices, $4n$ edges, $2n$ triangles and two n-gons. Method II: There are $4n$ vertices, $6n$ edges, n triangles, n hexagons, and two n-gons.

4. These are all rotations whose axes join the apex to the center of the base.

9. (e) False (j) True (o) True

EXERCISES 9

2. a, b, c, e are homeomorphic; d, f are homeomorphic; h, k are homeomorphic, i, j are homeomorphic.

5. (e) False (f) True

EXERCISES 10.1

1. (g) (four 210's, five 168's, six 140's, seven 120's)

2. (h) Yes (i) No (j) Yes (k) Yes

3. (h) Yes (i) No (j) Yes (k) No

4. (a) Yes for $n = 3, 4, 7, 8, 11, 12, 15, 16, \ldots$ and no otherwise

8. The necessity of the sum being even follows from PN 10.1.1. The sufficiency can be proved by induction on n, but different arguments may be needed depending on the parity of n.

10. If $N(G) = \{v_1, v_2, \ldots, v_p\}$, place v_k at the point (k, k^2, k^3).

EXERCISES 10.2

3. (k) Yes (l) No

6. The graph is Eulerian if and only if n_1, n_2, and n_3, all have the same parity.

15. Assuming that $n_1 \geq n_2 \geq \cdots \geq n_m$, then $n_2 + n_3 + \cdots + n_m \geq n_1$ is a necessary and sufficient condition for the graph to be Hamiltonian.

EXERCISES 10.3

2. Show that the endnodes of a path of maximum length must have degree 1.

5. Use PN 10.1.1 to show that there are at most k nodes of degree at least k. Then apply PN 10.3.2.

EXERCISES 10.4

1. e is the only nonplanar one.

7. The graph is planar if and only if minimum$\{n_1, n_2\} \leq 2$.

15. Color each region by the parity of the number of circles that contain that region in their interior.

16. Puncture one of faces of the solid and then stretch out and flatten the surface of the solid until its edges form a plane graph.

EXERCISES 10.5

2. c, e are homeomorphic, as are d, f.

EXERCISES 10. REVIEW

2. (a) Yes (b) Yes (c) For $k \geq 3$ (d) Yes

4. (e) False (j) False (m) True

EXERCISES 11.1

2. Left sphere: one rectangle and four triangles; Right sphere: three rectangles and two triangles

3. The presentation will consist of one octagon.

EXERCISES 11.2

2. (e) abc, $b^{-1}ed$, $c^{-1}d^{-1}$, $a^{-1}e^{-1}$ is a coherent orientation. Since $p = 3$, $q = 5$, and $r = 4$, the Euler characteristic is 2 and so this is a sphere.

2. (g) This is not orientable with Euler characteristic $4 - 10 + 5 = -1$.

3. (a) This is not orientable with Euler characteristic $n - n + 1 = 1$.

6. The Euler characteristic is 1.

10. Impose a rectangular grid on the 1-rectangle presentation of the torus.

16. For any such triangulation Exercise 14 implies $q = 3p$ and the simplicity of the skeleton implies that $p(p-1)/2 \geq q$. It follows that $p \geq 7$. To produce such a triangulation, start with the standard 1-rectangle presentation of the torus. Place the same node at each of the four vertices of the rectangle, two nodes on each of the two arcs of the presentation, and two more nodes inside the rectangle.

EXERCISES 11.3

1. (e) $\tilde{S}^{11}$ (g) $\tilde{S}^{6}$

6. $\tilde{S}^8$
11. S^{2n-1}

EXERCISES 11.4

2. (c) $S^{0,2}$, $\tilde{S}^{1,1}$, S^1, $\tilde{S}^2$
4. (a) $S^{1,2}$ (c) $\tilde{S}^{2,2}$
12. This could be the reversal of either a handle or a cross cap addition. The answer depends on the orientability characters of S and S'
13. No

EXERCISES 11. REVIEW

2. (d) True (e) False (i) True (j) False (o) False

EXERCISES 12.1

1. Start with a crossing and list all the possible ways in which its ends can be connected to form a diagram.
2. True
5. One
6. Infinitely many

EXERCISES 12.2

10. A p-labeling exists if and only if p is an odd prime factor of n.

EXERCISES 12.3

7. Imagine that $\vec{K}_1$ is small compared to $\vec{K}_2$ and attach it to the latter at any location. Now perform the isotopy of "sliding" $\vec{K}_1$ along $\vec{K}_2$.

EXERCISES 12. REVIEW

1. Take a cue from Figure 12.5.
6. (d) True (e) False (j) True (k) False (o) False

Bibliography

Adams, Colin C., *The Knot Book: An Elementary Introduction to the Mathematical Theory of Knots*, 2nd edition. New York: W. H. Freeman and Co., 2001.

Arnold, Bradford H., *Intuitive Concepts in Elementary Topology.* Upper Saddle River, NJ: Prentice Hall, 1962.

Ausdley, W. and G., *Designs and Patterns from Historic Ornament.* New York: Dover Publications, Inc.,1968.

Barr, Stephen, *Experiments in Topology.* New York: Dover Publications, Inc., 1989.

Bennett, Dan, Dynamic Geometry Renews Interest in an Old Problem, *Geometry Turned On!* James R. King and Doris Schattschneider, editors, MAA Notes 41. Washington, DC: Mathematical Association of America, 1974.

Birkhoff, George D. and Beatley, Ralph, *Basic Geometry*, 3rd edition. New York: Chelsea Pub. Co., 1959.

Boehm, Kathryn W., Experiences with *The Geometer's Sketchpad* in the Classroom, *Geometry Turned On!* James R. King and Doris Schattschneider, editors, MAA Notes 41. Washington, DC: Mathematical Association of America, 1974.

Bonola, Roberto, *Non-Euclidean Geometry, a Critical and Historical Study of its Developments.* New York: Dover Publications, Inc., 1955.

Borofsky, Samuel, *Elementary Theory of Equations.* New York: The MacMillan Company, 1950.

Chinn, W. G., and Steenrod, N. E., *First Concepts in Topology*, New Mathematical Library, Vol. 18. New York: Random House, Inc., 1966.

Conway, John H., Monsters and moonshine. *Math. Intelligencer* **2**(1979/1980), no. 4, 165–171.

Coxeter, H. S. M., *Introduction to Geometry*, 2nd edition. New York: John Wiley and Sons, 1989.

Coxeter, H.S.M., *Non-Euclidean Geometry*, 5th edition. Toronto: The University of Toronto Press, 1968.

Coxeter, H. S. M., *The Real Projective Plane*, 3rd edition. New York: Springer-Verlag, 1992.

Coxeter, H. S. M. and Greitzer, S. L., *Geometry Revisited.* Washington, DC: The Mathematical Association of America, 1967.

Dalle, Antoine and De Waele, C., *Géométrie Plane et Eléments de Topographie*, 18th edition. Namur: Maison d'Edition Ad. Wesmael-Charlier (S.A.), 1946.

De Villiers, Michael D., The Role of Proof in Investigative, Computer-Based Geometry: Some Personal Reflections, *Geometry Turned On!* James R. King and Doris Schattschneider, editors, MAA Notes 41. Washington, DC: Mathematical Association of America, 1974.

Euclid, *The Elements*, Sir Thomas L. Heath editor. New York: Dover Publications, Inc., 1956.

Eves, Howard, *An Introduction to the History of Mathematics.* Philadelphia: Saunders College Publishing, 1990.

Eves, Howard, *A Survey of Geometry*, 2 vols. Boston: Allyn and Bacon, Inc., 1966.

Francis, George K. and Weeks, Jeffrey, Conway's ZIP proof. *Amer. Math. Monthly* **106**(1999), no. 5, 393–399.

Greenberg, Marvin, J., *Euclidean and non-Euclidean Geometries: Development and History*, 3rd edition. San Francisco: W. H. Freeman and Co., 1995.

Gerber, David and Tsaban, Boaz, A Circular Booth II, *Magal* **11**(1995), 127–134. [In Hebrew]

Grünbaum, Branko and Shepard, G. C., *Tilings and Patterns.* New York: W. H. Freeman and Co., 1987.

Haeckel, Ernst H. P. A, *Report on the Scientific Expedition of the Voyage of H. M. S. Challenger*, Vol. XVIII. Edinburgh: H. M. Stationary Office, 1880.

Henderson, David W., *Experiencing Geometry in Euclidean, Spherical, and Hyperbolic Spaces*, 2nd edition. Upper Saddle River, NJ: Prentice Hall, 2001.

Hilbert, David, *The Foundations of Geometry.* Chicago: The Open Court Publishing Co., 1910.

Hilbet, David and Cohn-Vossen, S., *Geometry and the Imagination*, P. Nemenyi trans. New York: Chelsea Publishing Co., 1983.

Hilton, Peter and Pedersen, Jean, *Build Your Own Polyhedra.* Reading, MA: Addison-Wesley, 1994.

Jiang, Zhonghong and McClintock, Edwin, Using The Geometer's Sketchpad with Preservice Teachers, *Geometry Turned On!* James R. King and Doris Schattschneider, editors, MAA Notes 41. Washington, DC: Mathematical Association of America, 1974.

Jones, Owen, *The Grammar of Chinese Ornament.* London: Studio Editions, 1987.

Katz, Victor J., *A History of Mathematics: An Introduction*, 2nd edition, Reading, MA: Addison-Wesley Longman, 1998.

Kells, Lyman M., *Plane and Spherical Trigonometry*, 2nd edition. New York: McGraw-Hill, 1940.

Keyton, Michael, Students Discovering Geometry Using Dynamic Geometry Software, *Geometry Turned On!* James R. King and Doris Schattschneider, editors, MAA Notes 41. Washington, DC: Mathematical Association of America, 1974.

King, James, Quadrilaterals Formed by Perpendicular Bisectors, *Geometry Turned On!* James R. King and Doris Schattschneider, editors, MAA Notes 41. Washington, DC: Mathematical Association of America, 1974.

Krause, Eugene F., *Taxicab Geometry: An Adventure in Non-Euclidean Geometry.* New York: Dover Publications, Inc., 1986.

Ladizhenski, Y., *Geometriat Hamishor*, 12th edition. Jerusalem: Gesher, 1954. [In Hebrew]

Loomis, Elisha, S., *The Pythagorean Proposition; its Demonstrations Analyzed and Classified and Bibliography of Sources for Data of the Four Kinds of "Proofs."* Ann Arbor, MI: Edwards Brothers, Inc., lithoprinters, 1940.

Martin, George E., *The Foundations of Geometry and the Non-Euclidean Plane.* New York: Springer-Verlag, 1975.

Martin, George E., *Transformation Geometry: An Introduction to Symmetry.* New York: Springer-Verlag, 1982.

Massey, William S., *Algebraic Topology: An Introduction.* New York: Harcourt, Brace & World, Inc., 1967.

Morley, Frank V., *Inversive Geometry.* Boston: Ginn & Company, 1933.

Nanyes, Ollie, An Elementary Proof that the Borromean Rings are Non-Splittable, *Amer. Math. Monthly*, **100**(1993), no. 8, 786–789.

Pedoe, Daniel, *Geometry: A Comprehensive Course.* New York: Dover Publications, Inc., 1970.

Pedoe, Daniel, *Circles.* New York: Pergamon Press, 1957.

Plato, *The Republic of Plato*, Allan Bloom trans. New York: Basic Books Inc., 1968.

Proclus, *Commentary on the First Book of Euclid*, Glenn R. Morrow, trans. Princeton, NJ: Princeton University Press, 1970.

Robins, Gay and Shute, Charles, *The Rhind Mathematical Papyrus.* New York: Dover Publications, Inc., 1987.

Row, T. Sundra, *Geometric Exercises in Paper Folding.* New York: Dover Publications, Inc., 1966.

Schattschneider, Doris, The Plane Symmetry Groups: Their Recignition and Notation, *Amer. Math. Monthly*, **85**(1978), no. 6, 439–450.

Seifert, H. and Threllfall, W., *A Textbook of Topology*, Michael A. Goldman trans. New York: Academic Press, 1980.

Senechal, Marjorie and Fleck, George, editors, *Shaping Space: A Polyhedral Approach.* Boston: Birkhäuser, 1988.

Stahl, Saul, *The Poincaré Half-Plane: A Gateway to Modern Geometry.* Boston: Jones & Bartlett Publishers, 1993.

Todhunter, Isaac, *The Elements of Euclid for the Use of Schools and Colleges.* London: Macmillan & Co., 1869.

Wallace, Edward C. and West, Stephen, *Roads to Geometry*, 2nd edition. Upper Saddle River, NJ: Prentice Hall, 1997.

Weeks, Jeffrey, *The Shape of Space: How to Visualize Surfaces and Three-Dimensional Manifolds.* New York: Marcel Dekker, 1985.

Wells, David, *The Penguin Dictionary of Curious and Interesting Geometry.* London: Penguin Books Ltd., London, 1991.

West, Douglas B., *Introduction to Graph Theory*, 2nd edition. Upper Saddle River, NJ: Prentice Hall, 2000.

Weyl, Hermann, *Symmetry.* Princeton, NJ: Princeton University Press, 1952.

Yale, Paul B. *Geometry and Symmetry.* New York, Dover Publications, Inc., 1988.

Index